Insect Molecular
Biology and Ecology

Insect Molecular Biology and Ecology

Editor

Klaus H. Hoffmann
Animal Ecology I
Bayreuth Center of Ecology and Environmental
Research (BayCEER)
University of Bayreuth
Bayreuth
Germany

CRC Press
Taylor & Francis Group
Boca Raton London New York

CRC Press is an imprint of the
Taylor & Francis Group, an **informa** business

A SCIENCE PUBLISHERS BOOK

Cover illustrations reproduced by kind courtesy of Dr. Yuichi Oba, Nagoya University, Nagoya 464-8601, Japan

CRC Press
Taylor & Francis Group
6000 Broken Sound Parkway NW, Suite 300
Boca Raton, FL 33487-2742

First issued in paperback 2020

© 2015 by Taylor & Francis Group, LLC
CRC Press is an imprint of Taylor & Francis Group, an Informa business

No claim to original U.S. Government works

ISBN-13: 978-1-4822-3188-5 (hbk)
ISBN-13: 978-0-367-73861-7 (pbk)

Library of Congress Cataloging-in-Publication Data

Insect molecular biology and ecology / editor: Klaus H. Hoffmann.
    pages cm
  Includes bibliographical references and index.
  ISBN 978-1-4822-3188-5 (hardcover : alk. paper) 1. Insects--Molecular aspects. 2. Insects--Ecology. I. Hoffmann, Klaus H. (Klaus Hubert), 1946-

QL493.5.I574 2015
595.7--dc23   2014026003

Visit the Taylor & Francis Web site at
http://www.taylorandfrancis.com

and the CRC Press Web site at
http://www.crcpress.com

Dedication

Dedicated to my grand daughters

Preface

A major challenge in current entomology is to integrate different levels of organization, from cellular mechanism to function in ecosystem. In the postgenomic era of the 21st century various fields of study have become possible, which use the information of fully sequenced insect genomes (https://www.hgsc.bcm.edu/arthropods/i5k-pilot-project-summary). However, the rapid development of molecular techniques for studying gene functions will revolutionize entomology not only for the insect model organisms, but in general. The majority of these techniques can also be applied if only partial sequence information is available. With these tools, entomologists are able to answer questions in insect biochemistry, physiology, and endocrinology, but also illuminate very complex behavioral and ecological aspects.

When I edited a book on "Environmental Physiology and Biochemistry of Insects" in 1985 for the Springer-Verlag, Berlin, mechanisms of environmental adaptation in growth and development, energy metabolism, or respiration to temperature, oxygen tension, food supply or salt concentrations were in the focus of interest. It was at the time of "Physiological Ecology". About 30 years later, the omics era gives us the opportunity to gain deeper insight into different aspects of insect physiology and environmental adaptation, for example, by overexpression or silencing of candidate genes of interest. When we understand, how physiological processes are regulated and at what time, we will be able to manipulate them, hereby providing attractive potential for practical application, for example, in an ecologically friendly insect pest control.

In 2008, we started with a Master program in "Molecular Ecology" at our University of Bayreuth, which has become very successful during the last 6 years. The Master's program was designed to play a special role in the synergistic cross-linking of the two focal points at our University, "Ecology and Environmental Sciences" and "Molecular Biosciences". The focus of interest is the functions of organisms—and especially of insects—in their environment and the analysis of (bio)chemical interactions in complex ecosystems. "Molecular" should mean not only to study the function of macromolecular compounds such as proteins and nucleic acids, but to analyze also the structure and capacity of low molecular weight substances like signal molecules, toxins or drugs. This Master program inspired me to edit the present book on Insect Molecular Biology and Ecology.

The book provides a mix of topical review articles and current research work. In several chapters previously unpublished data are presented showing novel applications for the use of omics technologies in the postgenomic era. The book should prove useful not only to researchers of the Insecta, but also to teachers and graduate students who are interested in understanding the molecular basis of insect functioning in their natural environment.

I acknowledge the support received from the authors who accepted the invitation to write an article on their area of expertise and for delivering the manuscripts in due time. Any success this book may achieve has to be attributed to their efforts.

Bayreuth, July 17, 2014

Contents

1

Mechanisms of Polyphenism in Insects

Stephen M. Rogers

Introduction: What is Polyphenism?

Polyphenism, broadly defined, is where two or more distinct phenotypes can be produced by the same genotype. Woltereck (1909) coined the term 'reaction norm' (Reaktionsnorm) to describe how the phenotype of an individual depends on the interaction between its particular genotype and environmental cues. In practice many or most genes with pleiotropic effects have context-dependent expression or action during development, i.e., the internal environment in which they are expressed. Phenotypic plasticity is therefore usually defined as a change in phenotype driven by cues in the external environment, which may be abiotic, such as temperature or photoperiod, or biological in origin, deriving from other species or even members of the same species. Phenotypic plasticity can result from variation in developmental, physiological, biochemical and behavioral processes that are sensitive to these environmental variables (Nijhout and Davidowitz 2009).

Polyphenism, in one sense, occurs very widely in insects. In holometabolous insects, larvae and pupal stages often differ radically in appearance and function from the adults; this is a clear example of very different phenotypes being generated from the same genotype during the normal course of post-embryonic development. Even in the

School of Biological Sciences, The University of Sydney, Sydney, NSW 2006, Australia.
Email: Stephen.rogers@sydney.edu.au

hemimetabolous insects where the distinction between larval and adult forms is usually less dramatic, there is typically a clear distinction between specializations for feeding in larvae/nymphs and for dispersal and reproduction in the winged adults, with clear differences in phenotype to accomplish these tasks. Polyphenism in insects, however, is usually taken to designate the occurrence of alternate developmental pathways leading to an endpoint of two or more distinct phenotypes, rather than successive stages of post-embryonic growth (Fig. 1A). These may be discrete mutually exclusive alternatives, or range across a continuum between two extreme phenotypic outcomes. In practice, environmental variables may tend to canalize the development towards discrete and distinctive final phenotypes, even though the underlying developmental and physiological processes may allow for more variety in outcome.

Polyphenisms are a major reason for the success of the insects (Simpson et al. 2011), allowing them to exploit the advantages of being both specialists and generalists without the disadvantages of either strategy. The advantages of specialism is that morphology, physiology, biochemistry and behavior can be tailored to the demands of a particular environment, allowing an insect to efficiently exploit resources whilst being protected from predators (for example, camouflage coloration or exploitation of secondary plant defenses); but the disadvantage is that particular environments may change or only be periodically available (for example, seasonal, or host plant abundance). Generalists are more likely to find an environment that will support them, but they may be inefficient or conspicuous in any one environment and find themselves outcompeted by specialists.

Polyphenism may therefore allow insects to adopt phenotypes that best suit predictable and regular environmental changes, or respond to 'predictably unpredictable' transformations in the environments such as exploiting the increase in plant abundance in deserts after rains, or conversely escaping the degradation of an environment following overcrowding (Simpson et al. 2011). Eusocial insects, bees, ants, wasps and termites are some of the most abundant and successful insect species, and polyphenism has been central to this success allowing them to partition labor between individuals within social groups.

The expression of alternative phenotypes suited to different environments requires exposure to salient trigger stimuli related to or anticipatory of those environments during a critical window, which then sets the insect onto an alternate developmental pathway (Brakefield and Frankino 2009; Fig. 1A). This may be during an early period of development, or may even derive from the experience of a parent who then transmits an epigenetic change to their offspring. These critical windows may occur during a fixed period in ontogeny where an animal is only

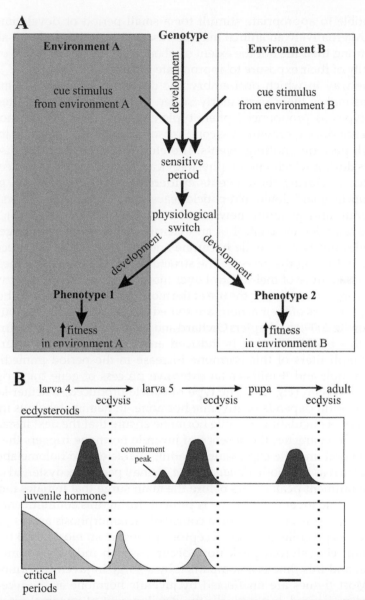

Figure 1. A. Schematic showing how different environments can redirect development towards two or more phenotypic outcomes depending upon suitable stimuli being received at a critical point in development. The environmental cue stimuli must reliably anticipate the environment to be faced by the organism after the induction of plasticity in order to increase fitness within that environment. B. A simplified diagram showing variation in the titers of ecdysteroids and juvenile hormone during molting events (ecdysis) across normal postembryonic development in a holometabolous insect, such as *Manduca sexta*. See text for more details. B. Derived from Reynolds (2013).

susceptible to appropriate stimuli for a small period or developmental stage. Alternatively an animal may be susceptible to these stimuli at any time during their life but the extent of phenotypic change depends on the longevity of their exposure to appropriate stimuli.

The way in which post-embryonic development occurs in insects perhaps makes them particularly suited to the expression of alternate phenotypes as pronounced phenotypic change is part of their normal experience during growth. A second reason is that insects change in size through periodic molting events, therefore the occurrence of discrete periods during which intense physiological activity occurs are a frequent occurrence. During these events epidermal tissues divide and grow, new cuticle is laid down, old cuticle reabsorbed and imaginal discs may proliferate and generate new adult structures (Reynolds 2013). This process is under the control of an extensive network of morphogenetic hormones but two, juvenile hormone and ecdysone, have an executive role in controlling the process. The structures of these two hormones and the precise nature of their control over moulting and development varies between insect taxa (hence the use of the more generic 'ecdysteroid') but the broad principles of their actions are shared across the insects (Nijhout and Davidowitz 2009; see chapters Orchard and Lange, this book and Hoffmann et al., this book). Molting is induced and regulated by ecdysteroids: hemolymph titers of this hormone increase in the period immediately before a molt and it initiates an extensive process of gene transcription and cell division (Fig. 1B). The type of cuticle produced, whether larval, pupal or adult depends on juvenile hormone. In hemimetabolous insects high titers of circulating juvenile hormone ensure that the next instar will be larval in character; the absence of juvenile hormone triggers the molt to adulthood and the expression of adult characters. In holometabolous insects, such as the moth *Manduca sexta*, a small peak of ecdysteroid called the commitment peak occurs before the main surge that initiates molting (Fig. 1B). If no juvenile hormone is present during this commitment peak the insect is set irretrievably on a course of metamorphosis and as part of this process juvenile hormone receptors are lost from most larval tissues. The main ecdysteroid peak now occurs, which induces the molting process, and this time is accompanied by a surge in juvenile hormone as well. Most tissues are unaffected by juvenile hormone as the receptors have been silenced, but it blocks the development of imaginal discs and causes the larva to molt into a pupa. The next molt to adulthood then follows the same pattern as in hemimetabolous insects; ecdysteroids in the absence of juvenile hormone. The key point about these hormones is that they are physiologically potent and have executive roles in the control of morphogenesis, and it therefore unsurprising that these two hormones have often been the first target of analysis for those looking to understand the physiological basis of polyphenism.

Behavior is a manifestation of the phenotype, and learning, defined as an experience-dependent change in behavior, is one of the most universal and basic expressions of phenotypic plasticity. Some form of learning has been found in all insect orders where it has been looked for (Jermy 1987). Learning and memory operate over many different timescales from seconds to lifetimes, and may or may not be reversible. Mechanistically, it also operates at many levels, from phosphorylation or other short-term modification of existing neuronal structures through to new gene expression and extensive remodeling and growth of neuronal tissues. The mechanisms underlying learning and memory have been extensively studied in insects and other organisms, and there is a good understanding of some of the molecular pathways involved linking initial stimulus through to altered gene expression (Kandel 2001). Some forms of phenotypic plasticity leading to polyphenism resemble learning. This understanding has also been a useful starting point for those wishing to analyze the mechanistic basis of polyphenism. In this chapter I review progress in understanding the mechanistic basis of polyphenism in several classic insect systems: seasonal polyphenism in butterflies and moths; phase change in locusts; flight/flightless polyphenism in aphids and crickets, and caste determination in two very different groups of social insect—the termites and bees.

Why study mechanisms underlying polyphenism? The case was succinctly put by West-Eberhard (2003): 'For evolutionary biology, proximate mechanisms represent more than just different levels of analysis or research styles. They are *the* causes of variation upon which selection acts'. She goes on to say that 'Among the consequences of the neglect of mechanisms in modern evolutionary biology are the problems that arise when the black box of mechanism is filled with imaginary devices'. This is a survey of the state of current attempts to open that black box.

Seasonal Morphs in Lepidoptera

Seasonal morphs are a frequent occurrence in Lepidoptera with two or more generations per year, in larvae (e.g., Greene 1989, Hazel 2002), pupae (e.g., Angersbach 1975, Hazel and West 1979) and as adults (e.g., Brakefield and Reitsma 1991, Kingsolver 1995, 1996). One of the best characterized examples is that shown by the African Satyrine butterfly, *Bicyclus anynana* (Fig. 2A). This tropical woodland species experiences two distinct seasons: a warm wet season when food is abundant and during which egg laying and larval growth can occur, and a cooler dry season when vegetation is withered and through which the butterflies

Figure 2. A. The wet season (upper panel) and dry season morphs (lower panel) of the Satyrine butterfly *Bicyclus anynana*. B. Developmental control of eyespot size and features on the wing of a developing butterfly. Chemical signals diffuse out down a concentration gradient (shown as the curves) from an organizing focus (F) on the developing wing. Large quantities diffuse further and induce larger eyespots. Ring structures are produced according to the threshold (T) of induction in the surrounding tissue. If the threshold is high (T1) then large amounts a chemical signal near the point of origin are necessary in order to induce a pattern. Chemical signals to which the tissue has a low threshold (T2) will still induce a phenotypic effect when they have diffused far from the focus and are at a low concentration. Photographs are courtesy of Oskar Brattström and Andre Coetzer. B. Redrawn from Beldade and Brakefield (2002).

must survive until the onset of the next rains several months later (Shapiro 1976, Brakefield and Reitsma 1991, Brakefield and Mazzotta 1995). Each season is associated with a distinct morph. The wet season morph has a distinctive pale medial stripe and rows of conspicuous 'eyespot' patterns on the ventral surfaces of its wings. The dry season morph is much less conspicuous with highly reduced markings and is mostly a drab brown color. There are also differences in behavior and physiology: the wet season form is highly active, reproduces quickly and has a short lifespan; the dry season is relatively long-lived and fairly inactive, sitting out the dry season and only reproducing once the wet season recommences (Brakefield et al. 2007). Typically, two rapid generations of the wet season morph are followed by a single generation of the dry season morph that survives the rest of the year.

The dry season form has been suggested to be a stress-tolerant form (Brakefield et al. 2007), and various field experiments have attempted to establish the adaptive value of each morph. Although in nature the two morphs are distinctive, in the laboratory a continuum of forms can be generated between the two extreme morphs by adjusting rearing conditions (Brakefield and Reitsma 1991). Releasing these lab-reared forms into the field in different seasons allow the relative success of each morph to be assessed. From these, and other experiments with direct experimental alteration of the wing patterns (Brakefield and Frankino 2009), it is clear that the conspicuous wet season morphs have much lower survivorship in the dry season than the cryptic dry season morph, which benefits from its enhanced camouflage. The converse fitness effect, increased survivorship of the dry season morph in the wet season, is much weaker. There is some evidence that the wing eyespots confuse naïve predators and enhance the chance of escape (Lyytinen et al. 2003), but it may be that the conspicuous patterns serve another purpose entirely, for example making the butterflies more obvious to potential mates in a form where rapid reproduction is prioritized over prolonged survival (Brakefield and Frankino 2009).

Temperature is the major environmental determinant of adult morphology (Brakefield and Reitsma 1991). The second generation of wet season morphs develops through a period of declining ambient temperature as the wet season gives way to the dry season, which induces the larvae to metamorphose into the dry season form. Conversely, the larvae of dry season morph butterflies develop in the uniformly warm conditions of the wet season and develop into the wet season morph. Adults of either morph can be generated in the laboratory by altering the developmental temperature (Brakefield and Reitsma 1991). Other factors that retard or accelerate development can also influence the final form of the adult. For example, caterpillars reared on a low-quality artificial diet that slowed the rate of development developed into adults similar to the dry season morph, despite being kept at temperatures that would normally induce the wet season morph (Holloway et al. 1992). Artificial selection experiments with the aim of producing separate rapidly and slowly developing lineages induced corresponding changes in adult morphology; those of the slowly developing lineage had smaller eyespots than wild type butterflies reared at the same temperature whereas the rapidly developing lineage had a hypertrophied wet season phenotype (Zijlstra et al. 2003, 2004). So, even though ambient temperature is the normal driver of plasticity, the underlying arbiter is the rate of development.

The wing pattern is determined during late larval and pupal development and is controlled by morphogenetic hormones, especially ecdysteroids and these same hormones also appear to control seasonal polyphenism (Koch 1992, Koch et al. 1996). There is a surge of circulating ecdysone in the hemolymph after molting to the pupal stage which cues the development of adult structures. The timing of this surge relative to the start of pupation appears to be critical for cueing which form the adult will assume in a number of different butterfly species (Koch and Bückmann 1987, Rountree and Nijhout 1995, Koch et al. 1996). Thus, seasonal differences are represented internally by circulating levels of hormones (Koch 1992).

Ecdysteroid titer appears to have a key role in determining whether a *B. anynana* larva develops to have the wet or dry season morph as an adult (Koch et al. 1996), with lower rearing temperatures prolonging pupal development so that an increase in circulating ecdysone titer occurs later in development, which in turn induces the dry season morph. Microinjecting ecdysone into early stage pupae reared at low temperatures led to the emerging adults having larger eyespots and markings akin to that of the wet season form. In an artificial selection experiment designed to canalize larvae into consistently producing wing patterns with either a wet or dry season morph at a constant rearing temperature of 20°C, larvae selected for the wet season form also had faster pupal development times and higher ecdysteroid titers shortly after pupation (Koch et al. 1996). In other butterfly species other signaling mechanisms may act in conjunction with ecdysone. For example, neuropeptides may also alter the sensitivity of different tissues to ecdysteroids in some species and therefore have a role in determining the final adult morph (Endo and Kamata 1985).

How do ecdysteroids and other hormones determine wing pattern in butterflies? Butterfly wings are covered in minute scales, each of which has a single color (Nijhout 1985). The regulation of color pattern was investigated by Sawada et al. (2002) who examined the expression of guanosine triphosphate cyclohydrolase I (GTP-CH I), an enzyme with a key role in the biosynthesis of white pteridine pigments during wing pattern formation in the butterfly *Precis coenia*. In this butterfly different types of pigments are synthesized sequentially during late pupal development, beginning with white (pterins) followed by red (ommatins) and then black (melanin). Expression of this gene was strongest at the time of pigment formation, about one day before emergence of the adult. They also investigated the effect of an ecdysteroid on the expression of GTP-CH I mRNA and on the enzyme's activity during wing pattern formation. The results suggest that the onset and duration of the expression of a GTP-CH I mRNA is triggered by a declining ecdysteroid hormone titer during late pupal development. Similar sequences of activation of pigment synthesis pathways are also reported in other butterflies (Koch et al. 2000).

Eyespot development is the butterfly wing-pattern element about which most is known. This derives in most part from experimental manipulation of pupal wings and from the study of gene-expression patterns in larval and pupal wing primordia (French 1997, McMillan et al. 2002). Eyespots develop around a central focus, or 'organizer'. It has been suggested that the cells within these foci produce diffusible signals that form a concentration gradient in the surrounding tissue of the pupal epidermis (Fig. 2B). These gradient contours of morphogenic hormones then define the rings of color that make up an eyespot by organizing the synthesis of particular pigments by the neighboring cells shortly before adult emergence (Beldade and Brakefield 2002). Eyespot foci were transplanted between pupae of phenotypically divergent lines of *B. anynana* to examine eyespot development (Monteiro et al. 1994, 1997). These transplant experiments showed that eyespot size was primarily dependent on properties of the transplanted focal tissue, suggesting that higher production of the focal signal diffused further and extended the range over which it could affect development (Fig. 2B). The eyespot color composition, however, was entirely determined by the host tissue surrounding the focal graft, suggesting that altering the color of eyespot must involve changes in the response threshold to the focal signal. The identity of these focal signals and how they interact with circulating levels of ecdysteroids to determine specific patterns remain largely unknown.

Studies of gene expression in larval and pupal wing primordia have identified a series of wing-patterning genes implicated in the developmental pathways involved in eyespot formation. Amongst these are genes that encode for transcription factors that show characteristic expression patterns during whole-wing development in insects: *apterous*, expressed on the dorsal surface; *wingless* and *distal-less*, along the wing margin; *engrailed* in the posterior of the wing; and *cubitus interruptus*, in the anterior (Carrol et al. 1994, Brakefield et al. 1996, Keys et al. 1999). These same genes also have a role in eyespot formation, presumably after the major axes of the wing have been established. *Distal-less*, *engrailed* and *spalt* are all expressed together within the organizing foci of eyespots, but are separately expressed in each one of the colored rings that make the final eyespot pattern (Brunetti et al. 2001). Other genes, such as *hedgehog, patched* and *cubitus interruptus* have specific windows of expression in or around the eyespot foci (Carrol et al. 1994, Keys et al. 1999).

Variation in the expression of some of these genes may underlie differences in wing pattern arising from phenotypic plasticity. *Distal-less* expression in the eyespot-organizing centers of wing primordia in *B. anynana* changes in parallel with adult changes in eyespot morphology (Brakefield et al. 1996, Beldade et al. 2002). This applies both to the environmentally induced differences between wet and dry season morphs and lineages selected to have large or small eyespots. Polymorphisms of this gene also correspond with eyespot variations, suggesting underlying genetic as well as environmental determinants for variation in eyespot developments.

Phase Change in Locusts

Swarms of locusts (Orthopteroidea/Orthoptera: Acrididae) are some of the best-known and iconic pests of worldwide agriculture. Locusts are defined as grasshoppers that express a form of phenotypic plasticity known as 'phase polyphenism' which entails progressive changes in behavior, physiology and morphology according to local population density and is mediated by a socially driven mechanism (Uvarov 1966, 1977, Simpson et al. 1999, Simpson and Sword 2009). Swarming is not an inevitable part of their life history and locusts can spend many generations in a form that not only does not swarm, but actively avoids other locusts. All locust species, of which there are only approximately 20 out of >10,000 Acridid species, however, are able to transform reversibly from a completely asocial solitarious phase to a highly social gregarious phase (Fig. 3A) depending upon environmental cues. In several species the two phases are so different that until the 1920s they were thought to be separate species (Uvarov 1921). The different species of locust are not closely related and phase change appears to have evolved independently in several different subfamilies (Song 2011); the best studied species are the Desert Locust (*Schistocerca gregaria*) and the Migratory Locust (*Locusta migratoria*).

Solitarious locusts are typical grasshoppers in that they are cryptic in both coloration and behavior, relying on inconspicuousness to avoid predators. They walk with a slow, creeping gait, fly predominately at night, have narrow dietary preferences and actively avoid other locusts (unless seeking a mate), thus spreading themselves out thinly in the environment and maintaining their low population density. By contrast,

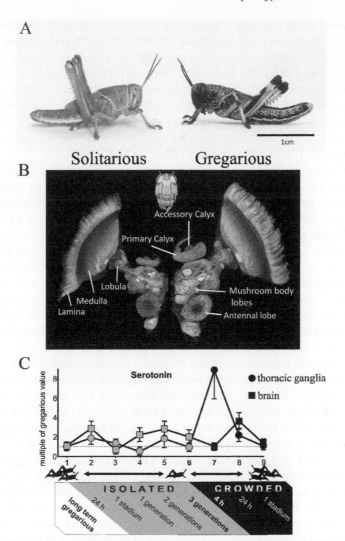

Figure 3. A. Final instar nymphs of the Desert Locust (*Schist cerca gregaria*) in the solitarious (left) and gregarious (right) phases. The solitarious phase is predominately green; the gregarious has aposematic black and yellow coloration. B. Differences in brain structure between solitarious (left half) and gregarious (right half) Desert Locusts. These two insects had the same body size. Higher association centers associated with vision (medulla, lobula) and olfaction (primary calyces and mushroom body lobes) are substantially larger in gregarious locusts. C. Changes in the amount of serotonin in the brain (squares) and thoracic ganglia (circles) during the entire process of phase change from isolation of gregarious individuals through to the crowding of 3rd generation solitarious locusts. The substantial peak in the thoracic ganglia after four hours crowding coincides with the critical period during which behavioral gregarization is established. B. Derived from Ott and Rogers (2010), C. Redrawn from Rogers et al. (2004).

gregarious Desert Locusts walk with a rapid, upright gait, fly during the day, have a broad diet and, most importantly, are attracted to other locusts such that they may eventually aggregate into vast migratory swarms (Simpson et al. 1999, 2002, Pener and Simpson 2009, Blackburn et al. 2010). With inconspicuousness no longer an option, they have bright warning colors, which advertise that they eat small quantities of poisonous plants to make themselves distasteful to predators (Sword 1999). They also have shorter wings and hind legs, smaller eyes and antennae, but more close-proximity touch and taste receptors (Rogers et al. 2003).

The brains of gregarious Desert Locusts are 30% larger than those of solitarious locusts (Ott and Rogers 2010; Fig. 2B) driven by a combination of allometric growth across the brain and discrete changes in specific brain regions (grade shifts). The larger brains of gregarious locusts prioritize higher levels of integration in both the visual and olfactory system as well as other important integration centers such as the central complex. Solitarious locusts have relatively more developed low-level sensory processing, having disproportionally larger primary visual and olfactory neuropiles. The reason for this may lie in the need to increase sensitivity since the solitarious lifestyle entails greater activity at lower light levels and the need to detect stimuli over greater distances since they are subject to increased individual predation risk (Rogers 2014).

Locusts are typically insects of arid environments punctuated by brief periods of superabundant vegetation growth induced by infrequent heavy rain falls, leading to a concomitant rapid rise in the local locust population. Sheer population size and local shifts in nutritional resource distribution cause solitarious locusts to converge upon limiting patches of vegetation despite their natural aversion to one another (Collett et al. 1998, Despland and Simpson 2000a,b). This forced contact triggers a far-reaching transformation that results in the gregarious phase.

Stimuli inducing phase change have been most extensively studied in *S. gregaria*. Behavior emerges as the pivotal aspect of the process because it is the first characteristic to change: after just two to four hours of forced crowding previously solitarious locusts behave in many respects like locusts that have been in the gregarious phase their entire lives (Roessingh et al. 1993, Simpson et al. 1999, Rogers et al. 2014). This sets up conditions for a positive feed-back loop in which the continual presence of other locusts provides the stimuli necessary to drive on the process with successive changes in behavior, physiology and coloration. Some of these changes will occur within the lifespan of an individual, while others develop over the course of several generations through epigenetic mechanisms (Uvarov 1966, 1977, Pener and Simpson 2009). This process may be arrested and reversed at any stage, but the longer a locust has been canalized in a particular direction, the longer it takes to change towards the alternative (Roessingh and Simpson 1994, Bouaichi et al. 1995, Simpson et al. 1999, Gray et al. 2009).

Stimuli provided by other locusts are entirely responsible for starting this transformation and not any other indicator of environmental stress. In *S. gregaria*, there are two distinct sensory pathways through which behavioral gregarization can be induced: first, mechanosensory stimulation of the hind femora brought about by locusts jostling each other (Simpson et al. 2001, Rogers et al. 2003); second, the combined sight and smell of other locusts with neither stimulus on its own being effective (Roessingh et al. 1998, Anstey et al. 2009). In the distantly related Australian Plague Locust (*Chortoicetes terminifera*) behavioral gregarization is induced by mechanosensory stimulation of the antennae, not the hind legs and there is no visual or olfactory component (Cullen et al. 2010). The stimuli responsible for inducing behavioral gregarization in *L. migratoria* are still not fully characterized. Other aspects of phase transformation, notably the longer term changes in coloration and morphology, appear to be mediated by separate, parallel pathways with independent induction mechanisms and hormonal control (Lester et al. 2005). For example, the neuropeptide [*His*[7]]-corazonin mediates the melanization of gregarious locusts and some morphometric changes in body proportions, but has little effect on behavior (Tawfik et al. 1999a, Hoste et al. 2002, Maeno and Tanaka 2004).

Surprisingly, there is little evidence for a strong role for juvenile hormone and ecdysteroids in phase change, and experimental studies have produced conflicting and contradictory results (Pener and Simpson 2009). This is despite an early and somewhat persistent idea that the solitarious phase was in some respects neotenous, deriving from the observation that solitarious locusts occasionally undergo an extra larval molt and may have persistent prothoracic (ventral) glands as adults whereas most insects lose them (Kennedy 1956). Joly (1954) implanted corpora allata, a pair of endocrine glands that secrete the juvenile hormone, into gregarious locusts and showed that they induce solitarious green coloration in *L. migratoria*. It is now clear that juvenile hormone is responsible for induction of green coloration in grasshoppers, even in non-locust species that show green/brown polyphenism in response to ambient humidity (Pener 1991). Wiesel et al. (1996) investigated the effect of juvenile hormone and juvenile hormone analogs on phase-related behavior in *S. gregaria* and *L. migratoria*, including aggregation and marching. They found juvenile hormone reduced aggregation of crowded hoppers in both species, suggestive of a solitarizing effect. Marching behavior however, which is a gregarious behavior, was stimulated; even surpassing that of the crowded controls. Applebaum et al. (1997) also studied the effect of a juvenile hormone analog (methoprene) on both activity and aggregation behavior in nymphs of both *S. gregaria* and *L. migratoria*. Five hours after treatment, the behavior was shifted in a solitarious direction in both species, but the effect was temporary, by 72 hours they exhibited gregarious behavior, and indeed were more active than controls.

Several studies have demonstrated that ecdysteroid titers do not differ much in crowded and isolated nymphs of *L. migratoria* (Joly et al. 1977, Roussel 1993) and *S. gregaria* (Wilson and Morgan 1978, Tawfik et al. 1996), although the last study found that the molt-inducing peak of ecdysteroid lasted longer in crowded than in isolated hoppers. Total ecdysteroid content of the ovaries with developing oocytes, however is about four times higher in crowded than in isolated females (Tawfik et al. 1999b) and total ecdysteroid content in freshly laid eggs from crowded females was over six times higher than in eggs from isolated females, and the resulting hatchlings had ecdysteroid titers five times higher than in those from eggs laid by isolated mothers. Are these differences in ecdysteroids a causal factor in inducing phase characteristics in hatchling locusts? Locusts can affect the phase of their offspring through a gregarizing factor, added by a gregarious mother to the foam surrounding her egg pod (Islam et al. 1994a,b, McCaffery et al. 1998), which transmits certain gregarious characteristics from parent to progeny. Hägele et al. (2004) studied the effects of crowding and isolation on ecdysteroid content of eggs of *S. gregaria* in the presence or absence of the gregarizing factor. Crowding isolation-reared females for a brief period shortly before egg laying induces gregarious behavior and gregarious hatchling color in her progeny (Islam et al. 1994b). Ecdysteroid content in eggs from briefly crowded females was the same as that in eggs from long-term solitarious females, despite the clear differences in hatchling phenotype (Hägele et al. 2004), excluding the possibility that the induction of gregarious characteristics is mediated by ecdysteroids.

In *S. gregaria* the entire process of phase change is accompanied by extensive changes in neurochemistry. In an analysis of the amounts of 13 different neurotransmitters and/or neuromodulators in different parts of the CNS using high-pressure liquid chromatography (HPLC) eight were more abundant in the brain and the thoracic ganglia of long-term solitarious locusts than in long-term gregarious ones, and three were less abundant, with only two showing no significant difference (Rogers et al. 2004). Isolating larval gregarious locusts led to rapid changes in seven neurochemicals within 24 hours, equal to or in most cases exceeding the differences seen between long-term solitarious and gregarious locusts. During the process of gregarization, crowding larval 3rd generation solitarious locusts led to rapid changes in six neurochemicals within the first 24 hours, but only serotonin (5-hydroxytryptamine, 5HT) showed a dramatic increase during the critical four hour window following crowding during which behavioral gregarization occurs (Fig. 3C). This increase in serotonin was confined to the thoracic ganglia and could be induced by either the mechanosensory or the sight + smell gregarizing pathway, and the amount was correlated with the extent of behavioral change (Anstey et al. 2009). Locusts that behaved the most gregariously had approximately three times more serotonin than the most solitariously behaving locusts.

Blocking the action of serotonin, either through the use of receptor antagonists or by using α-methyltryptophan (AMTP) to inhibit serotonin synthesis, prevented gregarization. Application of serotonin or serotonin receptor agonists however, induced a shift towards gregarious behavior in the complete absence of stimuli associated with other locusts. Finally treatment with a serotonin precursor, 5-hydroxytryptophan (5-HTP), accelerated the acquisition of gregarious behavior (Anstey et al. 2009, Rogers et al. 2014).

L. migratoria represents an independent evolution of the full phase-change phenotype. The time course and mechanism of phase change show similarities and differences with that of *S. gregaria*. In *Locusta* behavioral gregarization is much slower than solitarization and is not fully achieved even after 64 h of crowding (Guo et al. 2011). Several olfaction-related genes are strongly regulated and display opposite expression trends in response to changes in population density during both gregarization and solitarization. These include several *Chemosensory Protein* (*CSP*) genes that code for proteins secreted into the lymph of insect chemosensory organs where they are thought that they assist in ligand binding (Jacquin-Joly et al. 2001, Ban et al. 2003), but they are also more widely expressed in insect tissue and may have a role in development (Kitabayashi et al. 1998). Another gene of interest is a *takeout* gene, which is also expressed in insect chemosensory organs and has a role in modulating taste sensitivity according to nutritional status (Meunier et al. 2007) and in a circadian manner (Sarov-Blat et al. 2000). In *Locusta* the *CSP* gene *LmigCSP3* and the *takeout* gene *LmigTO1* are implicated in the behavioral change between avoidance of and attraction to conspecifics by regulating peripheral odor sensitivity (Guo et al. 2011). Olfactory repulsion by solitarious nymphs was reported for *S. gregaria* by Despland (2001), with visual repulsion being clearly evident as well.

Ma et al. (2011) employed RNA interference (RNAi; see Orchard and Lange, this book and Hoffmann et al., this book) to implicate dopamine as a modulator in the gregarization process, with evidence for an influence of serotonin as well in pharmacological manipulation studies. They identified *pale*, *henna* and *vat1*, genes involved in dopamine biosynthesis, synaptic release and cuticular melanization, as critical targets related to behavioral phase changes. These results might suggest a difference in the mechanism of gregarization between the two species, though the experiments were not designed to track the time course of behavioral phase change in detail. Furthermore, the behavioral phenotyping analyses employed in these studies suggest that RNAi intervention did not completely arrest the process of phase transition. A recent study by the same group (Guo et al. 2013) presented evidence that serotonin may mediate solitarization in *L. migratoria*. The role of serotonin may be as a facilitator of neuronal plasticity during phase change rather than being involved in the expression of gregarious behavior itself. The initial increase in serotonin that drives gregarization in *S. gregaria* decays to near baseline levels within 24 hours. Furthermore, *S. gregaria* that have spent their entire lives in the gregarious phase have half the amount of serotonin in their CNS compared with long-term solitarious locusts. It is therefore possible that serotonin could also have a role in the reverse process of solitarization (Rogers et al. 2014). The establishment of the full gregarious phase is accompanied by extensive changes in gene expression in both *L. migratoria* and *S. gregaria* (Kang et al. 2004, Badisco et al. 2011, Guo et al. 2011) so that signaling systems are acting on a different underlying physiology in the two phases. Furthermore, the same transmitter could act through different receptor subtypes and/or target different regions of the CNS in the two processes.

Gregarious behavior is reinforced by the continuing presence of stimuli from other locusts, but with time gregariousness in *S. gregaria* becomes more robust and is retained even after periods of isolation from other locusts. Newly gregarious locusts that have been crowded for just 24 h lose their gregarious behavior after only 4 h of isolation (Roessingh and Simpson 1994), but locusts that have been gregarious for many generations only start to become behaviorally solitarious after days of isolation. Evidence suggests that the cAMP-dependent protein kinase A (PKA), an effector protein with a pre-eminent role in different forms of learning is involved in the consolidation of the gregarious phase in *S. gregaria* (Ott et al. 2012). Injecting the PKA inhibitor KT5720 prior to crowding prevented locusts becoming gregarious but had no effect on the behavior of long-term gregarious locusts. Solitarious locusts injected with an RNAi construct against the PKA catalytic subunit C1 behaved less gregariously after crowding. Furthermore, RNAi against the PKA regulatory R1 subunit, which inhibits catalytic activity by PKA, promoted more extensive gregarization following a brief period of crowding. PKA and serotonin have a highly evolutionarily-conserved role in mediating neuronal plasticity in many species (Kandel 2001), and it appears it has been co-opted into meditating the transformation from solitarious to gregarious behavior in *S. gregaria*.

The PKA effector pathway causes rapid but short-term changes in neuronal function through phosphorylation of existing proteins and initiates long-term changes in gene expression (Müller and Carew 1998). Several recent studies have highlighted the extensive differences in gene expression between the solitarious and gregarious phases of both *S. gregaria* (Badisco et al. 2011, 2012) and *L. migratoria* (Kang et al. 2004, Ma et al. 2006, Guo et al. 2011). A study using microarrays developed from an Expressed Sequence Tag (EST) database of *S. gregaria* central nervous tissue, detected 214 differentially expressed genes between long-term solitarious and gregarious locusts, of which only 40% have been annotated to date (Badisco et al. 2012). These included genes encoding proteins that are associated with CNS development and modelling, sensory perception, stress response and resistance, and fundamental cellular processes. Genes for heat shock proteins and proteins which confer protection from infection were up-regulated in gregarious locusts, which may allow them to respond to acute physiological challenges. By contrast the longer-lived solitarious locusts appear to be more strongly protected from the slowly accumulating effects of ageing by an up-regulation of genes related to anti-oxidant systems, detoxification and anabolic renewal. Gregarious locusts also had a greater abundance of transcripts for proteins involved in sensory processing and in nervous system development and plasticity. An earlier study in *L. migratoria* identified 532 differentially expressed genes between phases (Kang et al. 2004). Chen et al. (2010) used next-generation deep sequencing to identify *de novo* sequences involved in phase transition in *L. migratoria*. These studies probably only provide a flavour of the true scale of differential gene expression that accompanies phase change. A major breakthrough has been the sequencing of the genome of *L. migratoria* (Wang et al. 2014). This is a considerable achievement since locusts have the largest genome of any animal sequenced so far. Berthier et al. (2011) suggested that these inflated genomes could result from a disproportionately high number of ancestral insertions of mitochondrial pseudogenes.

The presence of DNA methylation and DNA methyltransferase encoding genes have been found now in both *S. gregaria* (Boerjan et al. 2011) and *L. migratoria* (Robinson et al. 2011), with both species having genes that putatively encode methylation machinery (DNA methyltransferases and a methyl-binding domain protein). Robinson et al. (2011) showed the existence of genomic methylation, some of which appears to be localized to repetitive regions of the genome and identified a distinct group of genes within the *L. migratoria* genome that appear to have been historically methylated and show some possible functional differentiation. The genetic basis of locust phase transition has been described as the 'final frontier' in locust research (Pener and Simpson 2009) and the sequencing of a locust genome should greatly accelerate future research.

Dispersal Polyphenism

The ability to produce winged or wingless morphs depending upon local environmental conditions is found across many groups of insects including aphids, planthoppers (Fulgoroidea) and pond skaters (Gerridae) within the Hemiptera, various families of beetle (Coleoptera), and grasshoppers and crickets within the Ensifera (Harrison 1980, Roff 1986). In most cases it appears that this polyphenism enables insect to alter the trade-off between the ability to disperse to new environments against greater reproductive capacity *in situ* and like most kinds of phenotypic plasticity, environmental heterogeneity is the driver. Stable environments with plentiful resources support reproduction and the greatest fitness benefit accrues to those individuals that can produce the most offspring to exploit the available food supply. The deterioration of the local environment (which may in part be due to population pressure diminishing resources), elicits the production of alternative phenotypes capable of dispersing to new habitats.

Aphids have complex life cycles that show considerable variation between species (Moran 1992), and many exhibit alternating generations of parthenogenesis and sexual reproduction. Wing polyphenism in this group has been intensively studied. Most reproduction during the summer occurs through parthenogenesis, with the rapid production of multiple generations of females able to give birth to fully formed first-instar larvae. Falling temperatures or the decreasing photoperiod at the end of the season stimulate the production of a sexual generation (often winged) that mate and produce overwintering eggs.

Environmentally induced wing polyphenism may occur in females during the parthenogenetic reproduction stage. The fully winged morphs are generally more heavily sclerotized, have larger compound eyes, functional ocelli and longer antennae with more olfactory receptors than their flightless congeners (Kalmus 1945, Kring 1977, Kawada 1987, Miyazaki 1987, Braendle et al. 2006). These changes equip the winged morphs to explore the wider environment in order to find new host plants, to which end the winged forms are also more resistant to starvation (Tsuji and Kawada 1987, Hazell et al. 2005).

It appears however that in males the winged/wingless phenotype is entirely genetically determined (Lambers 1966), thus aphids may combine both genetically and environmentally determined mechanisms for the expression of different phenotypes within the same species. Various environmental triggers have been implicated in inducing the production of wings in parthenogenetic females depending on species or even clonal line (Braendle et al. 2006). These include: temperature, photoperiod, host plant quality, population density, and interactions with predators/

parasites or pathogens perhaps mediated via alarm pheromones (Müller et al. 2001, Braendle et al. 2006). Although the environmental triggers have been well characterized in different species, little is known of how these stimuli are detected and then how they in turn induce the development of alternative phenotypes. The disparate range of environmental stimuli suggests considerable evolutionary innovation, but it is possible that this diversity masks underlying mechanistic similarities. For example, greater tactile stimulation of aphids by each other as a result of increased activity may be the immediate proximate stimulus that induces wing formation, but this in turn may arise as consequence of reduced host plant quality, responses to predators or alarm pheromones (Müller et al. 2001, Kunert and Weisser 2005, Simpson et al. 2011). A similar mechanism is apparent in locusts where tactile stimuli are the immediate trigger for gregarization, but environmental structure and motivational state affect the likelihood of such contacts (Collett et al. 1998, Simpson et al. 2001, Cullen et al. 2010).

Asexual females can give birth to offspring that go on to have the winged phenotype despite being genetically identical to their parent, showing that this polyphenism is under epigenetic control. This winged generation is however resistant to producing winged progeny itself, even in the face of the same environmental stimuli that induced the flight phenotype in themselves (Sutherland 1969).

The small size of aphids has hampered analysis of the downstream mechanisms driving the development of the alternative phenotypes, despite its considerable ecological importance and many years of research. Winglessness is thought to be the derived condition in aphids (Johnson and Birks 1960) and this suggests that it is achieved in part through an interruption of the developmental processes leading to wing formation, but as detailed above there are other differences between winged and wingless morphs.

The apparent similarity in form between wingless adults and nymphs, including not only their lack of wings but also the underdeveloped nature of other morphological characters compared with winged adults, has led investigators to analyze the potential role of juvenile hormone and ecdysteroids. The reasoning being that the 'nymphal' character of wingless adults was the result of unusually high juvenile hormone titers in late larval development that suppress wing development in the wingless morphs, but nymphs destined to become winged adults experience the usual collapse in juvenile hormone titer which cues the development of typical adult structures (Reynolds 2013). In adults, juvenile hormone is a gonadotropin and it is hypothesized that not only adult winglessness but the earlier and more developed ovarian growth seen in apterous insects could also be explained through higher and persistent juvenile hormone titers (Nijhout 1994). Experimental support for this model has been ambiguous (Hardie and Lees 1985). This has arisen partly due to mistaking juvenilization induced by juvenile hormone for the genuine wingless phenotype in some cases (Lees 1977), but also because of differences between species and experimental designs. The small size of aphids poses great technical difficulties in making direct measurements of juvenile hormone titers, and administration of juvenile hormone or its analogs can induce a cascade of knock-on effects in other hormones and signaling systems, making it difficult to know if phenotypic effects are direct or secondary consequences of the experimental manipulation (Stay et al. 1994, Zera 2009).

More success in measuring the effects of morphogenic hormones and their potential role in generating flightless phenotypes has been achieved in crickets (*Gryllus* spp.: Orthoptera, Ensifera), largely through the work of Zera and colleagues. The larger size of crickets makes them more amenable to experimental manipulation and measurement of hormone titers. The occurrence of flightless morphs occurs in many *Gryllus* species (Zera and Denno 1997), in some species the two morphs occur as genetically determined polymorphism, in others it is an environmentally determined polyphenism. Flightless morphs in crickets typically have reduced rather than entirely absent wings. The environmental trigger that induces the short-winged morph in species such as *Gryllus rubens* or *G. firmus* is high population density, contrary to the situation seen in aphids, and it is possible that the short-winged phenotype is a stress resistant form rather than one specialized specifically for increased reproduction (Zera 2009). Short-winged morphs may be more energy efficient since they do not have to divert resources into the production and maintenance of the massive flight muscles (Mole and Zera 1994).

Measurements of morphogenic hormones in crickets have revealed that the endocrine control of wing polyphenism is more complex than previously suspected. Injecting juvenile hormone into nymphs destined to become fully-winged adults redirected development towards the flightless morph (Zera and Tiebel 1988), but *in vivo* differences in juvenile hormone titer in late-nymphal crickets between the two incipient morphs are small (Zera et al. 1989, Zera 2009). This may be because juvenile hormone titers are typically very low in late-stage larvae and such is its potency as a developmental effector that small changes in the circulating hormone could still have profound effects on adult morphology (Zera 2009). A much more striking difference was seen in the amount and activity of juvenile hormone esterase, an enzyme which degrades juvenile hormone (Hammock 1985), which corresponded almost exactly with the final phenotype. Activity of this enzyme in the hemolymph was 3 to 6-fold higher in incipient fully-winged nymphs (Zera and Tiebel 1989, Zera and Huang 1999) a difference driven by increased transcription of the gene (Zera et al. 2007). Similar differences in juvenile hormone esterase activity between flying and flightless morphs of cricket were found in studies comparing populations where the difference arose through an environmentally driven polyphenism and populations where the difference was caused by a genetic polymorphism (Zera et al. 1989, Evans and Wheeler 2001). This suggests that there may be an underlying similarity in physiological mechanism despite having such different causal agencies.

A much greater difference was also detected in ecdysteroid titer in the penultimate and last larval instar of crickets destined to become the two morphs (Zera et al. 1989). During the penultimate instar ecdysteroid titer peaked later and remained elevated for a shorter period of time in flightless morphs than in the flying morph. In the last nymphal instar the timing of the ecdysteroid peak was similar in both morphs, but approximately twice the amount was in circulation in the hemolymph of the flying morph. It is therefore possible that ecdysteroids have a role in specifying phenotype in this system.

In common with the flightless morphs of other insect species, ovarian development is much faster (100–400%) in flightless crickets than in flying ones and after one week constitute approximately 20% of total body mass of flightless crickets as opposed to 10% of flying morphs (Zera and Cisper 2001, Zhao and Zera 2004). Injecting juvenile hormone into young adult female crickets of the flying morph accelerates ovarian development to a level typical of that of the flightless morph (Tanaka 1994, Zera 2009). Again, measurement of hormone levels within the insects reveals a more complex picture: surprisingly, juvenile hormone titers in flying morph crickets undergo a 10 to 100-fold circadian cycle of increase and decrease (with the peak occurring at dusk), whereas flightless morph crickets have an approximately constant (and fairly low) level of circulating hormone (Zera and Cisper 2001, Zhao and Zera 2004, Zera et al. 2007). Ecdysteroid titers remain strongly different between flying and flightless adults, with flightless adults having consistently 30–200% more circulating ecdysteroids with no circadian variation (Zhao and Zera 2004). These data suggests that neither winglessness as a juvenile character nor hypertrophied gonad development can be thought to arise through a simple manipulation of juvenile hormone signaling in crickets and by implication in other insects as well.

Genomics approaches have recently been used to analyze the mechanistic basis of wing polyphenism as the complete genome sequence is now known for the pea aphid, *Acyrthosiphon pisum* (The International Aphid Genomics Consortium 2010). Genes have been identified in the pea aphid that are orthologous to those responsible for wing development in the *Drosophila* and *Tribolium* genomes (Brisson et al. 2010). The expression profile of eleven genes was compared between morphs and two paralogs of the gene *apterous* were shown to be differentially expressed during the first and second stadia. Transcript profiles of species in which flying and flightless morphs may be environmentally or genetically determined (Ghanim et al. 2006, Brisson et al. 2007) have revealed that both share similar patterns of large-scale of gene expression (Brisson 2010), which is reminiscent of the shared differences in juvenile hormone and ecdysteroids signaling underlying flightlessness in crickets regardless of its underlying genetic or environmental induction mechanism.

The possible role of DNA methylation in regulating gene expression and the epigenetic control of wing morph has been analyzed in the pea aphid (Walsh et al. 2010), which possesses the full set of DNA methyltransferase genes and its genome has been shown to be methylated. Walsh et al. (2010) looked for possible differences in methylation status in genes involved in juvenile hormone synthesis and transport using bisulfite sequencing but failed to find any. This may however, have been because the analysis was carried out on whole aphids, which may have masked tissue-specific methylation patterns.

Caste Determination

Cooperative brood care, a reproductive division of labor and overlapping generations are the defining features of eusociality. In the majority of eusocial insects this is also accompanied by the occurrence of morphologically as well as behaviorally distinct castes, which are generally environmentally determined. A fundamental distinction is between reproductive and non-reproductive members of the colony, but other distinct functional and morphological phenotypes that are specialized to different tasks may also exist. Here I provide a brief overview of some key aspects of caste determination in two very different and unrelated social insects, the hemimetabolous termites (Isoptera) and the holometabolous honey bee (Hymenoptera: *Apis mellifera*). In such a large group of animals there is considerable variation in caste-determination mechanisms, and in ants there is recent evidence for a genetic role in caste determination in some species, and it seems that not all social insect castes are clear-cut examples of polyphenism (Schwander et al. 2010).

Termites

Termites (Isoptera) are hemimetabolous insects and as such undergo an incomplete metamorphosis with well-developed larvae that in some respects resemble the adult insect (Noirot and Pasteels 1987). This has a profound effect on the division of labor, since in termite colonies all workers are at an immature stage of development and the only true adults are the primary reproductives. This contrasts strongly with the situation in the social Hymenoptera (bees, wasps and ants), where all work is performed by adult females and the larvae are legless, helpless grubs. Whereas in the social Hymenoptera the primary division is between workers and queens, in termites the situation is more complex. The queens and kings are the primary reproductives, also called alates since they form flying mating swarms prior to founding new colonies. Not all workers are destined to become adults; through a process of stationary or even regressive molts (in which features such as wing buds can be reabsorbed so that the termite resembles an earlier larval stage) workers may be held at an arrested and immature stage of development indefinitely. In the higher termites (Termitidae) there is an early canalization, by the second larval molt, between individuals destined to reach sexual maturity and those that will remain workers. In other termite families more fluidity remains and workers may under certain circumstances become reproductives (Noirot 1985). Workers lack wings, have a reduced thorax, are eyeless with a concomitant reduction of the optic lobes of the brain, and only have rudimentary sex organs. They do however have greater development of the head, mouthparts and gut compared to reproductives and they show many distinctive behaviors necessary to the function of the colony (Noirot 1969).

In addition to workers and primary reproductives, all termites have a distinct soldier caste, which are typically characterized by a massively developed head and mandibles. These individuals are so specialized that they are unable to feed themselves and must rely on workers to do so. The soldier caste is one endpoint of termite development but they are not true adults as, for example, their prothoracic glands are still present (a larval character). Termite soldiers pass through a characteristic pre-soldier or white soldier instar before becoming a full soldier, after which no further plasticity is possible. There may also be a number of secondary reproductives in the colony, which may succeed the colony founders (Vargo and Husseneder 2009). These neotenic reproductives resemble workers, except that they have the greatly enlarged abdomen containing the massive ovaries characteristic of primary queens.

There are therefore many kinds of phenotypic plasticity occurring within termite colonies, the mechanistic basis of which is still largely not understood. It has been long known that pheromones produced by queens inhibit the production of new reproductive individuals (reviewed in Wilson 1971). It was not until 2010 however, that Matsuura et al. identified a mixture of n-butyl-n-butyrate and 2-methyl-1-butanol as the active components of a volatile inhibitory pheromone that suppresses the differentiation of workers into new neotenic queens in *Reticulitermes speratus*. The same pheromone is also produced by eggs, where it acts as an attractant to workers and ensures that the eggs are taken care of. This dual role may provide a mechanism for ensuring the honest signaling of reproductive capability with a tight coupling between fertility and the inhibition of the formation of new reproductives.

There is evidence that juvenile hormone has a central role in the termite caste differentiation (Cornette et al. 2013, Hoffmann et al. this book). Implantation of corpora allata into workers causes them to develop into soldiers. This led to the development of a physiological model of termite caste differentiation by Nijhout and Wheeler (1982) in which sexual, non-sexual and soldier characteristics were governed by exposure to juvenile hormone (or the lack of it) during three successive sensitive periods in the prelude to molting. The development of the soldier phenotype was proposed to be caused by a sustained and high juvenile hormone titer. The development of full alate adults would be triggered by sustained low juvenile hormone titers during the entire critical period, similar to the situation that triggers the molt to adulthood in other insects. Stationary molts in workers would occur against a background of a declining juvenile titer whereas the converse pattern, a rising juvenile hormone titer, would induce the development of neotenic reproductives (Nijhout and Wheeler 1982).

Experimental data partially support this model: Soldiers can be induced through injection of juvenile hormone or its analogs into workers (Lelis and Everaerts 1993, Miura et al. 2003). Cornette et al. (2008) measured juvenile hormone titers in the large species *Hodotermopsis sjostedti*. They found a peak of juvenile hormone occurring during worker stationary molts and molts leading to the pre-soldier form. There was no juvenile hormone peak prior to the molt to alate adults, as predicted by the model. However, juvenile hormones rose in adult females during maturation, probably in relation to vitellogenesis and the general role of juvenile hormone as a gonadotropin in adults. By contrast, juvenile hormone titer was found to be surprisingly low in neotenic reproductives.

Hexamerins, hemolymph proteins that interact with juvenile hormone also have a significant role in regulating caste differentiation in termites (Zhou et al. 2006a, 2007). Hexamerins are amongst the most abundant proteins in insects, where they are typically described as storage proteins (Burmester and Scheller 1999). They also act as hormone binding proteins (Braun and Wyatt 1996). In termites, hexamerins are involved in a caste regulatory mechanism by inhibiting workers from molting into soldiers (Zhou et al. 2006a), possibly by sequestering juvenile hormone and making it unavailable to other tissues. Juvenile hormone is much more effective at inducing soldier differentiation after the silencing of hexamerin gene expression using RNAi (Zhou et al. 2006a). Juvenile hormone treatment significantly increases the expression of four genes encoding hemolymph proteins, including two vitellogenins and two hexamerins (Scharf et al. 2005, Cornette et al. 2013), possibly as part of a regulatory feedback mechanism. Finally, binding of juvenile hormone to hexamerins is suggested by the recognition of the termite hexamerins by an anti-juvenile hormone antibody (Zhou et al. 2006b).

Zhou and colleagues (2007) began the process of analyzing how hexamerins influence juvenile hormone-dependent gene expression in termites using RNAi of hexamerins coupled with quantification of gene expression. Hexamerin silencing resulted in significant alterations in the expression on 15 out of 17 morphogenesis-associated genes analyzed, and the effect of applying ectopic juvenile hormone caused broadly similar changes in gene expression. The most affected genes were a larval cuticle protein (LCP) and the translation factor *nanos*. Other affected genes included those encoding for four transcription/translation factors, seven signaling molecules, and two muscle troponin isoforms. Cornette et al. (2013) analyzed the effect of a juvenile hormone analog on gene expression in pseudergate workers of another species of termite, *H. sjostedti*, and found that its application caused a substantial increase in hexamerin expression and a decrease in the expression of several digestive enzymes and several other digestive system related proteins. Rapid growth, particularly of the head and mandibles is an essential part of the transition of workers to soldiers and a recent study (Hattori et al. 2013) has demonstrated a role for insulin/insulin-like growth factor hormone signaling in controlling allometric growth during soldier differentiation. *In situ* hybridization showed that an insulin receptor was strongly expressed in mandibular epithelial tissues and RNAi against this receptor disrupted soldier-specific morphogenesis.

Soldiers may themselves regulate their own numbers through feedback inhibition onto workers mediated by a soldier-produced pheromone (Tarver et al. 2010). The presence of live soldiers down-regulated the expression of *famet-2*, which is part of the mevalonate pathway necessary to synthesize juvenile hormone. This could in principle lead to reduced juvenile hormone production and hence decreased worker-to-soldier differentiation. The recent identification of a functional DNA methylation system in termites (Lo et al. 2012, Glastad et al. 2013) opens up an entirely new mechanistic pathway by which gene expression can be altered to produce new phenotypes. Juvenile hormone clearly has a key role in the complex regulation of caste and morphogenesis in termites, but it sits in the center of a regulatory network of other signaling and effector systems, most of which remain little understood.

Bees

The honeybees have been the most intensively studied social hymenopteran with regards to mechanisms of phenotypic plasticity. The primary division in bee colonies is between fertile queens, which are larger with highly developed ovaries, and workers which are smaller and have a rich behavioral repertoire connected with the maintenance of the hive, rearing brood and foraging. Workers undergo a shift in the tasks they perform as they age; as young adults they remain in the hive and tend to the larvae as nursemaids, as they age they leave the nest to become foragers, a high risk activity (Robinson 1992). Whilst the transition from nurse to forager is accompanied by many changes in behavior, brain structure and underlying neurochemistry (Robinson 1992, Schulz et al. 1992, Withers et al. 1993), this is part of a normal process of ontogenetic adult development rather than phenotypic plasticity. Schulz and Robinson (1999) showed that differences in the titers of dopamine, octopamine and serotonin in mushroom bodies of foragers and nurse bees were age-dependent, whereas in the antennal lobes, the differences were related to different tasks and therefore experience-dependent. Juvenile hormone titers increase with age in adult bees, so that foragers have higher levels than nurses. Treating younger bees with methoprene, a juvenile hormone analog, accelerates the onset of foraging (Bloch et al. 2002). The rate at which workers transition from nurses to foragers is plastic, however; a shortfall in foragers with a concomitant fall in the amount of food reaching the hive triggers the premature development of the forager phenotype.

The trigger for the formation of queens is diet. Larvae are fed on a mixture of pollen and a secretion from the mandibular glands of nurse bees known as royal jelly. It has been known for over a century that larvae fed small amounts of royal jelly become workers, while those that are fed on large amounts develop into queens (Haydak 1943, Shuel and Dixon 1960). Juvenile hormone titer during larval development has likewise been long known to correlate with caste: in larvae destined to become queens there is a peak during the fourth (out of five) larval stage, whereas a consistently low juvenile hormone titer is the hallmark of future workers (Wirtz and Beetsma 1972). This pattern contrasts with that of many other insects, in which higher levels of juvenile hormone result in more juvenile-like adult morphologies, as for example winglessness (Simpson et al. 2011).

It was not until the work of Kamakura (2011) that a breakthrough was made connecting an active agent in royal jelly to queen morphogenesis when he identified royalactin, a 57 kDa protein, as a substance that causes the significantly increases in body size and ovarian development characteristic of queens. Royalactin resembles epidermal growth factor, and remarkably is effective in both vertebrate cells and in *Drosophila,* where it exerts a similar phenotypic effect as in bees (Kamakura 2011). Royalactin acts through the epidermal growth factor receptor and through this has downstream actions on a number of effectors. These include S6K—which is activated by both phosphatidylinositol-dependent kinase 1 (PDK1) downstream of PI3K and target of rapamycin (TOR) downstream of PI3K/PDK1/Akt (Rintelen et al. 2001), which induce increased body size; mitogen-activated protein kinase (MAPK), which decreases developmental time; and also causes the previously identified increase in juvenile hormone titers, which promote ovary development.

Previous studies of caste-specific gene expression corroborate some of these findings (Evans and Wheeler 2001, Patel et al. 2007, Smith et al. 2008). Genes associated with metabolism and respiration are also upregulated in larvae destined to become queens. In larvae that become workers there is higher expression of storage proteins including hexamerins (Martins et al. 2010). Transcription of the hexamerin genes *hex110* and *hex70a* is greater in late larval stage incipient workers than in incipient queens, where it is inversely correlated with juvenile hormone titer. This is reminiscent of the situation in termites, and it has been suggested (Martins et al. 2010) that the development of queens occurs when juvenile hormone titer exceeds the binding capacity of hexamerins. Caste-specific differences in expression are also carried through into pupal and adult fat bodies.

Honey bees possess a functional DNA methylation system (Wang et al. 2006). Using RNAi to reduce the expression of the DNA methyltransferase 3 gene (*dnmt3*), responsible for *de novo* methylation in vertebrates, resulted in a strong bias towards production of adults with queen-like features (Kucharski et al. 2008). These results indicate that DNA methylation is critical for differentiation of the worker caste. An analysis using bisulfite sequencing to examine the epigenomes of queen and worker brain tissues revealed that approximately 69,000 cytosines out of a total of 60 million in the entire genome were found to be methylated, just 0.1% of the total (Lyko et al. 2010). Nearly all methylated cytosines were located in cytosine-guanine (CpG) dinucleotides and found in the exons of 5,854 genes which showed greater sequence conservation than non-methylated genes. Over 550 of these genes showed significant differentiation in methylation patterns between queen and workers, which may thus contribute to the developmental differentiation of queens and workers (Elango et al. 2009), though the mechanism by which this is achieved is still unknown.

Conclusions

Many of the examples of polyphenism discussed above do have changes in morphogenetic hormone signaling occurring at key points in developments. Locust phase change is perhaps the exception here as there appears to be rather little involvement of either juvenile hormone or ecdysteroids at the initiation of the process of change. This may be because the morphological change is relatively subtle in locusts and that hitherto the research focus has been on earlier changes in behavior preceding morphological change. In other cases the role of morphogenetic hormones is ambiguous or unclear, as in aphid wing-polyphenism, and generally as their role is analyzed in more detail their function as simple master switches becomes more questionable.

One issue that remains to be resolved is how changes in circulating hormone titers are translated into physiological mechanisms, and whether a titer is a physiologically meaningful way to consider the action of hormones rather than just a marker of difference. Indeed, *Methoprene-tolerant* (Met), a transcription factor of the basic helix–loop–helix Per-ARNT-Sim gene family has only fairy recently been confirmed as a juvenile hormone receptor (Wilson et al. 1986, Ashok et al. 1998, Charles et al. 2011, Kayukawa et al. 2012), so much remains to be discovered of the mode of action of morphogenetic hormones. In particular, how ecdysteroids and juvenile hormones interact to alter the pattern of gene expression still remains largely unknown. The discovery of functional DNA methylation systems in several insects with polyphenism: bees, termites, the pea aphid and two species of locusts, offers a possible way for gene expression to be silenced or otherwise altered, though how the methylation system interacts with hormonal or other biochemical effectors still remains to be established. It is interesting that, to date, evidence for functional DNA methylation systems have only been found in insects with polyphenism; *Drosophila, Anopheles* and *Tribolium* for example lack DNA methylation (Simpson et al. 2011).

New mass screening techniques of molecular change including genomic and proteomic approaches have begun to identify a plethora of systems affected by phenotypic plasticity, and it is becoming clearer that ecdysteroids and juvenile hormone exist in the center of extensive regulatory networks of signaling and effector systems. Establishing the chain of cause and effect remains a major challenge since the window of activation for many kinds of plasticity is temporally narrow; for example the 1–4 h serotonin peak responsible for gregarization in the Desert Locust (Anstey et al. 2009). Therefore, using an appropriate experimental design is essential.

Nevertheless, genomics approaches have been invaluable in gathering mass data with which to form hypotheses. Large scale screening has highlighted common proteins, for example hexamerins and heat shock proteins that appear to differ in expression in different morphs across disparate taxa and may have a mechanistic role in causing plasticity. Furthermore, the development of techniques such as RNA interference which have found wide application even in insects for which we don't have the full genome (Cullen 2012), are providing the tools to analyze the effects of individual genes on phenotypic plasticity.

In each of the model systems that have been considered only part of the pathway responsible for generating polyphenism has been revealed. Nowhere do we have the complete chain of events from environmental trigger, to detection of the trigger stimulus by the organism, through to the initiation of alternate developmental pathways through to the final implementation of the full phenotype. The black box of mechanism has only been partially opened, but several substantial breakthroughs in the last few years hold out the prospect for further rapid progress.

Acknowledgements

Stephen Rogers is supported by a Laureate Fellowship from the Australian Research Council to Professor Stephen Simpson. I thank Darron Cullen for reading the manuscript.

Keywords: phenotypic plasticity, seasonal polyphenism, *Bicyclus anynana*, locust, phase change, dispersal polyphenism, aphid, cricket, caste determination, termite, honey bee, juvenile hormone, ecdysteroid, gene expression

References

Angersbach, D. 1975. The direction of incident light and its perception in the control of pupal melanisation in *Pieris brassicae*. J. Insect Physiol. 21: 1691–1696.

Anstey, M.L., S.M. Rogers, S.R. Ott, M. Burrows and S.J. Simpson. 2009. Serotonin mediates behavioural gregarization underlying swarm formation in desert locusts. Science 323: 627–630.

Applebaum, S.W., E. Avisar and Y. Heifetz. 1997. Juvenile hormone and locust phase. Arch. Insect Biochem. Physiol. 35: 375–391.

Ashok, M.C. Turner and T.G. Wilson. 1998. Insect juvenile hormone resistance gene homology with the bHLH-PAS family of transcriptional regulators. Proc. Natl. Acad. Sci. USA. 95: 2761–2766.

Badisco, L., J. Huybrechts, G. Simonet, H. Verlinden, E. Marchal, R. Huybrechts, L. Schoofs, A. De Loof and J. Vanden Broeck. 2011. Transcriptome analysis of the desert locust central nervous system: production and annotation of a *Schistocerca gregaria* EST database. PLoS One 6: e17274.

Badisco, L., S.R. Ott, S.M. Rogers, T. Matheson, D. Knapen, L. Vergauwen, H. Verlinden, E. Marchal, M.R.J. Sheehy, M. Burrows and J. Vanden Broeck. 2012. Microarray-based transcriptomic analysis of differences between long-term gregarious and solitarious desert locusts. PLoS One 6: 11.

Ban, L., A. Scaloni, A. Brandazza, S. Angeli, L. Zhang, Y. Yan and P. Pelosi. 2003. Chemosensory proteins of *Locusta migratoria*. Insect Mol. Biol. 12: 125–134.

Beldade, P. and P.M. Brakefield. 2002. The genetics and evo–devo of butterfly wing patterns. Nature Rev. Genetics 3: 442–452.

Beldade, P., P.M. Brakefield and A.D. Long. 2002. Contribution of *Distal-less* to quantitative variation in butterfly eyespots. Nature 415: 315–318.

Berthier, K., M.P. Chapuis, S.J. Simpson, H.J. Ferenz, C.M.H. Kane, L. Kang, A. Lange, S.R. Ott, M.A.B. Ebbe, K.W. Rodenburg, S.M. Rogers, B. Torto, J. Vanden Broeck, J.J.A. van Loon and G.A. Sword. 2010. Laboratory populations as a resource for understanding the relationship between genotypes and phenotypes: A global case study in locusts. Adv. Insect Physiol. 39: 1–37.

Blackburn, L.M., S.R. Ott, T. Matheson, M. Burrow and S.M. Rogers. 2010. Motor neurone responses during a postural reflex in solitarious and gregarious desert locusts. J. Insect Physiol. 56: 902–910.

Bloch, G., D.E. Wheeler and G.E. Robinson. 2002. Endocrine influences on the organization of insect societies. pp. 195–235. *In*: D.W. Pfaff, A.P. Arnold, A.M. Etgen, S.E. Fahrbach and R.T. Rubin (eds.). Hormones, Brain and Behavior, Vol. 3. Academic Press, San Diego, CA.

Boerjan, B., F. Sas, U.R. Ernst, J. Tobback, F. Lemière, M.B. Vandegehuchte, C.R. Janssen, L. Badisco, E. Marchal, H. Verlinden, L. Schoofs and A. De Loof. 2011. Locust phase polyphenism: Does epigenetic precede endocrine regulation? Gen. Comp. Endocrinol. 173: 120–128.

Bouaichi, A., P. Roessingh and S.J. Simpson. 1995. An analysis of the behavioural effects of crowding and re-isolation on solitary-reared adult desert locusts (*Schistocerca gregaria*, Forskål) and their offspring. Physiol. Entomol. 20: 199–208.

Braendle, C., I. Friebe, M.C. Caillaud and D.L. Stern. 2005. Genetic variation for an aphid wing polyphenism is genetically linked to a naturally occurring wing polymorphism. Proc. R. Soc. Lond. Ser. B 272: 657–664.

Braendle, C., G. Davis, J.A. Brisson and D.L. Stern. 2006. Wing dimorphism in aphids. Heredity 97: 192–199.

Brakefield, P.M. and W.A. Frankino. 2009. Polyphenism in Lepidoptera: multidisciplinary approaches to studies of evolution and development. pp. 337–368. *In*: D.W. Whitman and T.N. Ananthakrishnan (eds.). Phenotypic Plasticity of Insects. Science Publishers, Enfield, New Hampshire.

Brakefield, P.M. and V. Mazzota. 1995. Matching field and laboratory environments: effect of neglecting daily temperature variation on insect reaction norms. J. Evol. Biol. 8: 559–573.

Brakefield, P.M., J. Pijpe and B.J. Zwaan. 2007. Developmental plasticity and acclimation both contribute to adaptive responses to alternative seasons of plenty and of stress in *Bicyclus* butterflies. J. Biosci. 32: 465–475.

Brakefield, P.M. and N. Reitsma. 1991. Phenotypic plasticity, seasonal climate and the population biology of *Bicyclus* butterflies (Satyridae) in Malawi. Ecol. Entomol. 16: 291–303.

Brakefield, P.M., J. Gates, D. Keys, F. Kesbeke, P.J. Wijngaarden, A. Montelro, V. French and S.B. Carroll. 1996. Development, plasticity and evolution of butterfly eyespot patterns. Nature 384: 236–242.

Braun, R.P. and G.W. Wyatt. 1996. Sequence of the hexameric juvenile hormone-binding protein from the hemolymph of *Locusta migratoria*. J. Biol. Chem. 6: 31756–31762.

Brisson, J.A., G.K. Davis and D.L. Stern. 2007. Shared genome-wide gene expression patterns underlying the wing polyphenism and polymorphism in the pea aphid (*Acyrthosiphon pisum*). Evol. Dev. 9: 338–346.

Brisson, J.A. 2010. Aphid wing dimorphisms: linking environmental and genetic control of trait variation. Phil. Trans. R. Soc. B 365: 605–616.

Brisson, J.A., A. Ishikawa and T. Miura. 2007. Wing development genes of the pea aphid and differential gene expression between winged and unwinged morphs. Insect Mol. Biol. 19: 63–73.

Brunetti, C.R., J.E. Selegue, A. Monteiro, F. French, P.M. Brakefield and S.B. Carroll. 2001. The generation and diversification of butterfly eyespot color patterns. Curr. Biol. 11: 1578–1585.

Burmester, T. and K. Scheller. 1999. Ligands and receptors: common theme in insect storage protein transport. Naturwissenschaften 86: 468–474.

Carroll, S.B., J. Gates, D.N. Keys, S.W. Paddock, G.E. Panganiban, J.E. Selegue and J.A. Williams. 1994. Pattern formation and eyespot determination in butterfly wings. Science 265: 109–114.

Charles, J.-P., T. Iwemab, V.C. Epa, K. Takaki, J. Rynes and M. Jindra. 2011. Ligand-binding properties of a juvenile hormone receptor, *Methoprene-tolerant*. Proc. Natl. Acad. Sci. USA 108: 21128–21133.

Chen, S., P. Yang, F. Jiang, Y. Wei, Z. Ma and L. Kang. 2010. *De novo* analysis of transcriptome dynamics in the migratory locust during the development of phase traits. PLoS One 5: e15633.

Collett, M., E. Despland, S.J. Simpson and D.C. Krakauer. 1998. The spatial scales of locust gregarisation. Proc. Natl. Acad. Sci. USA 95: 13052–13055.

Cornette, R., H. Gotoh, S. Koshikawa and T. Miura. 2008. Juvenile hormone titers and caste differentiation in the damp-wood termite *Hodotermopsis sjostedti* (Isoptera, Termopsidae). J. Insect Physiol. 54: 922–930.

Cornette, R., H. Yoshinobu, K. Shigeyuki and T. Miura. 2013. Differential gene expression in response to juvenile hormone analog treatment in the damp-wood termite *Hodotermopsis sjostedti* (Isoptera, Archotermopsidae). J. Insect Physiol. 59: 509–518.

Cullen, D.A. 2012. RNAi unravels the biology of the hemimetabolous and ametabolous Insects. Adv. Insect Physiol. 42: 37–72.

Cullen, D., G.A. Sword, T. Dodgson and S.J. Simpson. 2010. Behavioural phase change in the Australian plague locust, *Chortoicetes terminifera*, is triggered by tactile stimulation of the antennae. J. Insect Physiol. 56: 937–942.

Despland, E. 2001. Role of olfactory and visual cues in the attraction/repulsion responses to conspecifics by gregarious and solitarious desert locusts. J. Insect Behav. 14: 35–46.

Despland, E. and S.J. Simpson. 2000a. The role of food distribution and nutritional quality in behavioural phase change in the desert locust. Anim. Behav. 59: 643–652.

Despland, E. and S.J. Simpson. 2000b. Small-scale vegetation patterns in the parental environment influence the phase state of hatchlings of the desert locust. Physiol. Entomol. 25: 74–81.

Elango, N., B.G. Hunt, M.A.D. Goodisman and S.V. Yi. 2009. DNA methylation is widespread and associated with differential gene expression in castes of the honeybee, *Apis mellifera*. Proc. Natl. Acad. Sci. USA 106: 11206–11211.

Endo, K. and Y. Kamata. 1985. Hormonal control of seasonal morph determination in the small copper butterfly, *Lycaena phlaeas daimio* Seitz. J. Insect Physiol. 31: 701–706.

Evans, J.D. and D.E. Wheeler. 2001. Gene expression and the evolution of insect polyphenisms. BioEssays 23: 62–68.

French, V. 1997. Pattern formation in colour on butterfly wings. Curr. Opin. Genet. Dev. 7: 524–529.

Ghanim, M., A. Dombrovsky, B. Raccah and A. Sherman. 2006. A microarray approach identifies *ANT*, *OS-D* and *takeout*-like genes as differentially regulated in alate and apterous morphs of the green peach aphid *Myzus persicae* (Sulzer). Insect Biochem. Mol. Biol. 36: 857–868.

Glastad, K.M., B.G. Hunt and M.A.D. Goodisman. 2013. Evidence of a conserved functional role for DNA methylation in termites. Insect Mol. Biol. 22: 143–154.

Gray, L.J., G.A. Sword, M.L. Anstey, F.J. Clissold and S.J. Simpson. 2009. Behavioural phase polyphenism in the Australian plague locust (*Chortoicetes terminifera*). Biol. Lett. 5: 306–309.

Guo, W., X. Wang, Z. Ma, L. Xue, J. Han, D. Yu and L. Kang. 2011. *CSP* and *takeout* genes modulate the switch between attraction and repulsion during behavioral phase change in the migratory locust. PLoS Genetics 7: e1001291.

Guo, X., Z. Ma and L. Kang. 2013. Serotonin enhances solitariness in phase transition of the migratory locust. Front. Behav. Neurosci. 7: 129.

Hägele, B.F., F.-H. Wang, F. Sehnal and S.J. Simpson. 2004. Effects of crowding, isolation, and transfer from isolation to crowding on total ecdysteroid content of eggs in *Schistocerca gregaria*. J. Insect Physiol. 50: 621–628.

Hammock, B.D. 1985. Regulation of juvenile hormone titer: degradation. pp. 431–472. *In*: G.A. Kerkut and L.J. Gilbert (eds.). Comprehensive Insect Physiology, Biochemistry and Pharmacology. Pergamon Press, New York.

Hardie, J. and A.D. Lees. 1985. Endocrine control of polymorphism and polyphenism. pp. 441–490. *In*: G.A. Kerkut and L.J. Gilbert (eds.). Comprehensive Insect Physiology, Biochemistry and Pharmacology. Pergamon Press, New York.

Harrison, R.G. 1980. Dispersal polymorphisms in insects. Annu. Rev. Ecol. Syst. 11: 95–111.

Hattori, A., Y. Sugime, C. Sasa, H. Miyakawa, Y. Ishikawa, S. Miyazaki, Y. Okada, R. Cornette, L.C. Lavine, D.J. Emlen, S. Koshikawa and T. Miura. 2013. Soldier morphogenesis in the damp-wood termite is regulated by the insulin signaling pathway. J. Exp. Zool. (Mol. Dev. Evol.) 320B: 295–306.

Haydak, M.H. 1943. Larval food and development of castes in the honeybee. J. Econ. Entomol. 36: 778–792.

Hazel, W.N. 2002. The environmental and genetic control of seasonal polyphenism in larval color and its adaptive significance in a swallowtail butterfly. Evolution 56: 342–348.

Hazel, W.N. and D.A. West. 1979. Environmental control of pupal colour in swallowtail butterflies (Lepidoptera: Papilioninae): *Battus philenor* (L.) and *Papilio polyxenes* Fabr. Ecol. Entomol. 4: 393–400.

Hazell, S.P., D.M. Gwynn, S. Ceccarelli and M.D.E. Fellowes. 2005. Competition and dispersal in the pea aphid: clonal variation and correlations across traits. Ecol. Entomol. 30: 293–298.

Holloway, G.J., P.M. Brakefield, S. Kofman and J.J. Windig. 1992. An artificial diet for butterflies, including *Bicyclus* species, and its effect on development period, weight and wing pattern. J. Res. Lepidopt. 30: 121–128.

Hoste, B., S.J. Simpson, S. Tanaka, A. De Loof and M. Breuer. 2002. A comparison of phase-related shifts in behavior and morphometrics of an albino strain, deficient in [His(7)]-corazonin, and a normally colored *Locusta migratoria* strain. J. Insect Physiol. 48: 791–801.

Islam, M.S., P. Roessingh, S.J. Simpson and A.R. McCaffery. 1994a. Parental effects on the behaviour and colouration of nymphs of the desert locust *Schistocerca gregaria*. J. Insect Physiol. 40: 173–181.

Islam, M.S., P. Roessingh, S.J. Simpson and A.R. McCaffery. 1994b. Effects of population density experienced by parents during mating and oviposition on the phase of hatchling desert locusts, *Schistocerca gregaria*. Proc. R. Soc. Lond. Ser. B 257: 93–98.

Jacquin-Joly, E., R.G. Vogt, M.C. Francois, L. Nagnan and P. Meillour. 2001. Functional and expression pattern analysis of chemosensory proteins expressed in antennae and pheromonal gland of *Mamestra brassicae*. Chem. Senses 26: 833–844.

Jermy, T. 1987. The role of experience in the host plant selection of phytophagous insects. pp. 143–157. *In*: R.F. Chapman, E.A. Bernays and J.G. Stoffolano (eds.). Perspectives in Chemoreception and Behaviour. Springer-Verlag, New York.

Johnson, B. and P.R. Birks. 1960. Studies on wing polymorphism in aphids. I. The developmental process involved in the production of the different forms. Entomol. Exp. Appl. 3: 327–339.

Joly, L. 1954. Résultats d'implantations systématiques de corpora allata à de jeunes larves de *Locusta migratoria* L. C. R. Séanc. Soc. Biol. Paris 148: 579–583.

Joly, L., J. Hoffmann and P. Joly. 1977. Contrôle humoral de la différenciation phasaire chez *Locusta migratoria migratorioides* (R. & F.) (Orthoptères). Acrida 6: 33–42.

Kalmus, H. 1945. Correlations between flight and vision, and particularly between wings and ocelli in insects. Proc. R. Entomol. Soc. Lond. A 20: 84–96.

Kamakura, M. 2011. Royalactin induces queen differentiation in honeybees. Nature 473: 478–483.

Kandel, E.R. 2001. The molecular biology of memory storage: A dialogue between genes and synapses. Science 294: 1030–1038.

Kang, L., X. Chen, Y. Zhou, B. Liu, W. Zheng, R. Li, J. Wang and J. Yu. 2004. The analysis of large scale gene expression correlated to the phase changes of the migratory locust. Proc. Natl. Acad. Sci. USA 101: 17611–17615.

Kawada, K. 1987. Polymorphism and morph determination. pp. 255–266. *In*: A.K. Minks and P. Harrewijn (eds.). Aphids, their Biology, Natural Enemies and Control. Elsevier, Amsterdam.

Kayukawa, T., C. Minakuchi, T. Namiki, T. Togawa, M. Yoshiyama, M. Kamimura, K. Mita, S. Imanishi, M. Kiuchi, Y. Ishikawa and T. Shinoda. 2012. Transcriptional regulation of juvenile hormone-mediated induction of *Krüppel* homolog 1, a repressor of insect metamorphosis. Proc. Natl. Acad. Sci. USA 109: 11729–11734.

Kennedy, J.S. 1956. Phase transformation in locust biology. Biol. Rev. 31: 349–370.

Keys, D.N., D.L. Lewis, J.E. Selegue, B.J. Pearson, L.V. Goodrich, R.L. Johnson, J. Gates, M.P. Scott and S.B. Carroll. 1999. Recruitment of a *hedgehog* regulatory circuit in butterfly eyespot evolution. Science 283: 532–534.

Kingsolver, J.G. 1995. Viability selection on seasonal polyphenic traits: wing melanin pattern in Western White butterflies. Evolution 49: 932–941.

Kingsolver, J.G. 1996. Experimental manipulation of wing pattern and survival in Western White butterflies. Am. Nat. 147: 296–306.

Kitabayashi, A.N., T. Arai, T. Kubo and S. Natori. 1998. Molecular cloning of cDNA for p10, a novel protein that increases in the regenerating legs of *Periplaneta americana* (American cockroach). Insect Biochem. Mol. Biol. 28: 785–790.

Koch, P.B. 1992. Seasonal polyphenism in butterflies: a hormonally controlled phenomenon of pattern formation. Zool. Jb. Physiol. 96: 227–240.

Koch, P.B., P.M. Brakefield and F. Kesbeke. 1996. Ecdysteroids control eyespot size and wing color pattern in the polyphenic butterfly *Bicyclus anynana* (Lepidoptera: Satyridae). J. Insect Physiol. 42: 223–230.

Koch, P.B. and D. Bückmann. 1987. Hormonal control of seasonal morphs by the timing of ecdysteroid release in *Araschnia levana* L. (Nymphalidae). J. Insect Physiol. 33: 823–829.

Koch, P.B., U. Lorenz, P.M. Brakefield and R.H. ffrench-Constant. 2000. Butterfly wing pattern mutants: developmental heterochrony and co-ordinately regulated phenotypes. Dev. Genes Evol. 210: 536–544.

Kring, J.B. 1977. Structure of the eyes of the pea aphid, *Acyrthosiphon pisum*. Ann. Entomol. Soc. Am. 70: 855–860.

Kucharski, R., J. Maleszka, S. Foret and R. Maleszka. 2008. Nutritional control of reproductive status in honeybees via DNA methylation. Science 319: 1827–1830.

Kunert, G. and W. Weisser. 2005. The importance of antennae for pea aphid wing induction in presence of natural enemies. Bull. Entomol. Res. 95: 125–131.

Lambers, D.H.R. 1966. Polymorphism in Aphididae. Annu. Rev. Entomol. 11: 47–78.

Lees, A.D. 1977. Action of juvenile hormone mimics on the regulation of larval-adult and alary polymorphisms in aphids. Nature 267: 46–48.

Lelis, A. and C. Everaerts. 1993. Effects of juvenile hormone analogues upon soldier differentiation in the termite *Reticulitermes santonensis* (Rhinotermitidae, Heterotermitinae). J. Morphol. 217: 239–261.

Lester, R.L., C. Grach, M.P. Pener and S.J. Simpson. 2005. Stimuli inducing gregarious colouration and behaviour in nymphs of *Schistocerca gregaria*. J. Insect Physiol. 51: 737–747.

Lo, N., B. Li and B. Ujvari. 2012. DNA methylation in the termite *Coptotermes lacteus*. Insect. Soc. 59: 257–261.

Lyko, F., S. Foret, R. Kucharski, S. Wolf, C. Falckenhayn and R. Maleszka. 2010. The honey bee epigenomes: differential methylation of brain DNA in queens and workers. PLoS Biol. 8: e1000506.

Lyytinen, A., P.M. Brakefield, L. Lindström and J. Mappes. 2003. Does predation maintain eyespot plasticity in *Bicyclus anynana*? Proc. R. Soc. Lond. Ser. B 271: 279–283.

Ma, Z.Y., J. Yu and L. Kang. 2006. LocustDB: a relational database for the transcriptome and biology of the migratory locust (*Locusta migratoria*). BMC Genomics 7: 11.

Ma, Z., W. Guo, X. Guo, X. Wang, L. Xue, J. Han, D. Yu and L. Kang. 2011. Modulation of behavioral phase changes of the migratory locust by the catecholamine metabolic pathway. Proc. Natl. Acad. Sci. USA 108: 3882–3887.

McCaffery, A.R., S.J. Simpson, M.S. Islam and P. Roessingh. 1998. A gregarizing factor present in the egg pod foam of the desert locust *Schistocerca gregaria*. J. Exp. Biol. 201: 347–363.

McMillan, W.O., A. Monteiro and D.D. Kapan. 2002. Development and evolution on the wing. Trends Ecol. Evol. 17: 125–133.

Maeno, K. and S. Tanaka. 2004. Hormonal control of phase-related changes in the number of antennal sensilla in the desert locust, *Schistocerca gregaria*: possible involvement of [His7]-corazonin. J. Insect Physiol. 50: 855–865.

Martins, J., F. Nunes, A. Cristino, Z. Simões and M. Bitondi. 2010. The four hexamerin genes in the honey bee: structure, molecular evolution and function deduced from expression patterns in queens, workers and drones. BMC Mol. Biol. 11: 23.

Matsuura, K., C. Himuro, T. Yokoi, Y. Yamamoto, E.L. Vargo and L. Keller. 2010. Identification of a pheromone regulating caste differentiation in termites. Proc. Natl. Acad. Sci. USA 107: 12963–12968.

Meunier, N., Y.A. Belgacem and J.-R. Martin. 2007. Regulation of feeding behaviour and locomotor activity by *takeout* in *Drosophila*. J. Exp. Biol. 210: 1424–1434.

Miura, T., S. Koshikawa and T. Matsumoto. 2003. Winged presoldiers induced by a juvenile hormone analog in *Zootermopsis nevadensis*: implications for plasticity and evolution of caste differentiation in termites. J. Morphol. 257: 22–32.

Miyazaki, M. 1987. Forms and morphs of aphids. pp. 163–195. *In*: A.K. Minks and P. Harrewijn (eds.). Aphids, their Biology, Natural Enemies and Control. Elsevier, Amsterdam.

Mole, S. and A.J. Zera. 1994. Differential resource consumption obviates a potential flight-fecundity trade-off in the wing-dimorphic cricket *Gryllus rubens*. Oecologia 93: 121–127.

Monteiro, A.F., P.M. Brakefield and V. French. 1994. The evolutionary genetics and developmental basis of wing pattern variation in the butterfly *Bicyclus anynana*. Evolution 48: 1147–1157.

Monteiro, A., P.M. Brakefield and V. French. 1997. Butterfly eyespots: the genetics and development of the color rings. Evolution 51: 1207–1216.

Moran, N.A. 1992. The evolution of aphid life cycles. Annu. Rev. Ecol. Syst. 37: 321–348.

Müller, C.B., I.S. Williams and J. Hardie. 2001. The role of nutrition, crowding and interspecific interactions in the development of winged aphids. Ecol. Entomol. 26: 330–340.

Müller, U. and T.J. Carew. 1998. Serotonin induces temporally and mechanistically distinct phases of persistent PKA activity in *Aplysia* sensory neurons. Neuron 21:1423–1434.

Nijhout, H.F. 1985. The developmental physiology of color patterns in Lepidoptera. Adv. Insect Physiol. 18: 181–247.

Nijhout, H.F. 1994. Insect Hormones. Princeton University Press, Princeton.

Nijhout, H.F. and D.E. Wheeler. 1982. Juvenile hormone and the physiological basis of insect polymorphisms. Quart. Rev. Biol. 57: 109–133.

Nijhout, H.F. and G. Davidowitz. 2009. The developmental-physiological basis of phenotypic plasticity. pp. 589–608. *In*: D.W. Whitman and T.N. Ananthakrishnan (eds.). Phenotypic Plasticity of Insects. Science Publishers, Enfield, New Hampshire.

Noirot, C. 1969. Formation of castes in the higher termites. pp. 311–350. *In*: K. Krishna and F.M. Weesner (eds.). Biology of Termites, Vol. 1. Academic Press, New York.

Noirot, C. 1985. Pathways of caste development in lower termites. pp. 41–57. *In*: J.A.L. Watson, B.M. Okot-Kotber and C. Noirot (eds.). Caste Differentiation in Social Insects. Pergamon Press, London.

Noirot, C. and J.M. Pasteels. 1987. Ontogenetic development and evolution of the worker caste in termites. Experientia 43: 851–860.

Ott, S.R. and S.M. Rogers. 2010. Gregarious desert locusts have substantially larger brains with altered proportions compared with the solitarious phase. Proc. R. Soc. Lond. Ser. B 277: 3087–3096.

Ott, S.R., H. Verlinden, S.M. Rogers, C.H. Brighton, P.S. Quah, R.K. Vleugels, R. Verdonck and J. Vanden Broeck. 2012. A critical role for protein kinase A in the acquisition of gregarious behavior in the desert locust. Proc. Natl. Acad. Sci. USA 109: E381–E387.

Patel, A., M.K. Fondrk, O. Kaftanoglu, C. Emore, G. Hunt, K. Frederick and G.V. Amdam. 2007. The making of a queen: TOR pathway is a key player in diphenic caste development. PLoS One 2: e509.

Pener, M.P. 1991. Locust phase polymorphism and its endocrine relations. Adv. Insect Physiol. 23: 1–79.

Pener, M.P. and S.J. Simpson. 2009. Locust phase polyphenism: An update. Adv. Insect Physiol. 36: 1–286.

Reynolds, S. 2013. Postembryonic development. pp. 398–454. *In*: S.J. Simpson and A.E. Douglas (eds.). The Insects: Structure and Function, 5th Edition. Cambridge University Press, Cambridge, UK.

Rintelen, F., H. Stocker, G. Thomas and E. Hafen. 2001. PDK1 regulates growth through Akt and S6K in *Drosophila*. Proc. Natl. Acad. Sci. USA 98: 15020–15025.

Robinson, G.E. 1992. Regulation of division of labor in insect societies. Annu. Rev. Entomol. 37: 637–665.

Robinson, K.L., D. Tohidi-Esfahani, N. Lo, S.J. Simpson and G.A. Sword. 2011. Evidence for widespread genomic methylation in the migratory locust, *Locusta migratoria* (Orthoptera: Acrididae). PLoS One 6(12): e28167.

Roessingh, P., S.J. Simpson and S. James. 1993. Analysis of phase-related changes in behaviour of desert locust nymphs. Proc. Roy. Soc. Lond. Ser. B 252: 43–49.

Roessingh, P. and S.J. Simpson. 1994. The time-course of behavioural phase change in nymphs of the desert locust, *Schistocerca gregaria*. Physiol. Entomol. 19: 191–197.

Roessingh, P., A. Bouaichi and S.J. Simpson. 1998. Effects of sensory stimuli on the behavioural phase state of the desert locust, *Schistocerca gregaria*. J. Insect Physiol. 44: 883–893.

Roff, D.A. 1986. The evolution of wing dimorphism in insects. Evolution 40: 1009–1020.

Rogers, S.M., T. Matheson, E. Despland, T. Dodgson, M. Burrows and S.J. Simpson. 2003. Mechanosensory-induced behavioural gregarization in the desert locust *Schistocerca gregaria*. J. Exp. Biol. 206: 3991–4002.

Rogers, S.M., T. Matheson, K. Sasaki, K. Kendrick, S.J. Simpson and M. Burrows. 2004. Substantial changes in central nervous system neurotransmitters and neuromodulators accompany phase change in the locust. J. Exp. Biol. 207: 3603–3617.

Rogers, S.M. 2014. The neurobiology of a transformation from asocial to social life during swarm formation in desert locusts. New Front. Soc. Neurosci. Res. Perspect. Neurosci. 21: 11–23.

Rogers, S.M., D.A. Cullen, M.L. Anstey, M. Burrows, E. Despland, T. Dodgson, T. Matheson, S.R. Ott, K. Stettin, G.A. Sword and S.J. Simpson. 2014. Rapid behavioural gregarization in the desert locust, *Schistocerca gregaria* entails synchronous changes in both activity and attraction to conspecifics. J. Insect Physiol. 65: 9–26.

Rountree, D.B. and H.F. Nijhout. 1995. Hormonal control of seasonal polyphenism in *Precis coenia* (Lepidoptera, Nymphalidae). J. Insect Physiol. 41: 987–992.

Roussel, J.-P. 1993. Modification des taux d'ecdysteroides selon la phase chez *Locusta migratoria* L. Bull. Soc. Zool. Fr. 118: 367–373.

Sarov-Blat, L., W.V. So, L. Liu and M. Rosbash. 2000. The *Drosophila takeout* gene is a novel molecular link between circadian rhythms and feeding behavior. Cell 101: 647–656.

Sawada, H., M. Nakagoshi, R.K. Reinhardt, I. Ziegler and P.B. Koch. 2002. Hormonal control of GTP cyclohydrolase I gene expression and enzyme activity during color pattern development in wings of *Precis coenia*. Insect Biochem. Mol. Biol. 32: 609–615.

Scharf, M.E., C.R. Ratliff, D. Wu-Scharf, X. Zhou, B.R. Pittendrigh and G.W. Bennett. 2005. Effects of juvenile hormone III on *Reticulitermes flavipes*: changes in hemolymph protein composition and gene expression. Insect Biochem. Mol. Biol. 35: 207–215.

Schulz, D.J., A.B. Barron and G.E. Robinson. 2002. A role for octopamine in honey bee division of labor. Brain Behav. Evol. 60: 350–359.

Schulz, D.J. and G.E. Robinson. 1999. Biogenic amines and division of labor in honey bee colonies: behaviorally related changes in the antennal lobes and age-related changes in the mushroom bodies. J. Comp. Physiol. A 184: 481–488.

Schwander, T., N. Lo, M. Beekman, B.P. Oldroyd and L. Keller. 2010. Nature *vs* nurture in social insect caste determination. Trends Ecol. Evol. 25: 275–282.

Shuel, R.W. and S.E. Dixon. 1960. The early establishment of dimorphism in the female honeybee, *Apis mellifera* L. Insect. Soc. 7: 265–282.

Shapiro, A.M. 1976. Seasonal polyphenism. Evol. Biol. 9: 259–333.

Simpson, S.J., E. Despland, B.F. Hägele and T. Dodgson. 2001. Gregarious behaviour in desert locusts is evoked by touching their back legs. Proc. Natl. Acad. Sci. USA 98: 3895–3897.

Simpson, S.J., A.R. McCaffery and B.F. Hagele. 1999. A behavioural analysis of phase change in the desert locust. Biol. Rev. Cambridge Philos. Soc. 74: 461–480.

Simpson, S.J., D. Raubenheimer, S.T. Behmer, A. Whitworth and G.A. Wright. 2002. A comparison of nutritional regulation in solitarious- and gregarious-phase nymphs of the desert locust *Schistocerca gregaria*. J. Exp. Biol. 205: 121–129.

Simpson, S.J. and G.A. Sword. 2009. Phase polyphenism in locusts: mechanisms, population consequences, adaptive significance and evolution. pp. 147–190. *In*: D.W. Whitman and T.N. Ananthakrishnan (eds.). Phenotypic Plasticity of Insects. Science Publishers, Enfield, New Hampshire.

Simpson, S.J., G.A. Sword and N. Lo. 2011. Polyphenism in insects. Curr. Biol. 21: R738–R749.

Smith, C.R., A.L. Toth, A.V. Suarez and G.E. Robinson. 2008. Genetic and genomic analyses of the division of labour in insect societies. Nat. Rev. Genet. 9: 735–748.

Song, H. 2011. Density-dependent phase polyphenism in non model locusts: A mini review. Psyche 2011, Article ID 741769.

Stay, B., B.A.S. Bachman, J.A. Stoltzman, S.E. Fairbairn, C.G. Yu and S.S. Tobe. 1994. Factors affecting allatostatin release in a cockroach (*Diploptera punctata*): nerve section, juvenile hormone analog and ovary. J. Insect. Physiol. 40: 365–372.

Sutherland, O.R.W. 1969. The role of crowding in the production of winged forms by two strains of the pea aphid, *Acyrthosiphon pisum*. J. Insect Physiol. 15: 1385–1410.

Sword, G.A. 1999. Density-dependent warning coloration. Nature 397: 217.

Tanaka, S. 1994. Endocrine control of ovarian development and flight muscle histolysis in a wing dimorphic cricket *Modicogryllus confirmatus*. J. Insect Physiol. 40: 483–490.

Tarver, M.R., X. Zho and M.E. Scharf. 2010. Socio-environmental and endocrine influences on developmental and caste-regulatory gene expression in the eusocial termite *Reticulitermes flavipes*. BMC Mol. Biol. 11: 28.

Tawfik, A.I., A. Mat'hová, F. Sehnal and S.H. Ismail. 1996. Hemolymph ecdysteroids in the solitary and gregarious larvae of *Schistocerca gregaria*. Arch. Insect Biochem. Physiol. 31: 427–438.

Tawfik, A.I., S. Tanaka, A. De Loof, L. Schoofs, G. Baggerman, E. Waelkens, R. Derua, Y. Milner, Y. Yerushalmi and M.P. Pener. 1999a. Identification of the gregarization-associated dark-pigmentotropin in locusts through an albino mutant. Proc. Natl. Acad. Sci. USA 96: 7083–7087.

Tawfik, A.I., A. Vedrová and F. Sehnal. 1999b. Ecdysteroids during ovarian development and embryogenesis in solitary and gregarious *Schistocerca gregaria*. Arch. Insect Biochem. Physiol. 41: 134–143.

The International Aphid Genomics Consortium. 2010. Genome sequence of the pea aphid *Acrythosiphon pisum*. PLoS Biol. 8: e1000313.

Tsuji, H. and K. Kawada. 1987. Effects of starvation on life span and embryo development of four morphs of pea aphid (*Acyrthosiphon pisum* (Harris)). Jap. J. Appl. Entomol. Zool. 31: 36–40.

Uvarov, B.P. 1921. A revision of the genus *Locusta* L. (= *Pachytylus*, Fieb.), with a new theory as to periodicity and migrations of locusts. Bull. Entomol. Res. 12: 135–163.

Uvarov, B. 1966. Grasshoppers and Locusts. Vol. 1. Cambridge University Press, London, UK.

Uvarov, B. 1977. Grasshoppers and Locusts, Vol. 2. Centre for Overseas Pest Research, London, UK.

Vargo, E.L. and C. Husseneder. 2009. Biology of subterranean termites: Insights from molecular studies of *Reticulitermes* and *Coptotermes*. Annu. Rev. Entomol. 54: 379–403.

Walsh, T.K., J.A. Brisson, H. Robertson, K. Gordon, S. Jaubert-Possamai, D. Tagu and O.R. Edwards. 2010. A functional DNA methylation system in the pea aphid, *Acyrthosiphon pisum*. Insect Mol. Biol. 19: 215–228.

Wang, Y., M. Jorda, P.L. Jones, R. Maleszka, X. Ling, H.M. Robertson, C.A. Mizzen, M.A. Peinado and G.E. Robinson. 2006. Functional CpG methylation system in a social insect. Science 314: 645–647.

Wang, X., X. Fang, P. Yang, X. Jiang, F. Jiang, D. Zhao, B. Li, F. Cui, J. Wei, C. Ma, Y. Wang, J. He, Y. Luo, Z. Wang, X. Guo, W. Guo, X. Wang, Y. Zhang, M. Yang, S. Hao, B. Chen, Z. Ma, D. Yu, Z. Xiong, Y. Zhu, D. Fan, L. Han, B. Wang, Y. Chen, J. Wang, L. Yang, W. Zhao, Y. Feng, G. Chen, J. Lian, Q. Li, Z. Huang, X. Yao, N. Lv, G. Zhang, Y. Li, J. Wang, J. Wang, B. Zhu and L. Kang. 2014. The locust genome provides insight into swarm formation and long-distance flight. Nature Communications 5: 2957.

West-Eberhard, M.J. 2003. Developmental Plasticity and Evolution. Oxford University Press, USA.

Wiesel, G., S. Tappermann and A. Dorn. 1996. Effects of juvenile hormone and juvenile hormone analogues on the phase behaviour of *Schistocerca gregaria* and *Locusta migratoria*. J. Insect Physiol. 42: 385–395.

Wilson, E.O. 1971. The Insect Societies. Belknap Press, Cambridge, MA.

Wilson, I.D. and E.D. Morgan. 1978. Variations in ecdysteroid levels in 5th instar larvae of *Schistocerca gregaria* in gregarious and solitary phases. J. Insect Physiol. 24: 751–756.

Wilson, T.G. and J. Fabian. 1986. A *Drosophila melanogaster* mutant resistant to a chemical analog of juvenile hormone. Dev. Biol. 118: 190–201.

Wirtz, P. and J. Beetsma. 1972. Induction of caste differentiation in the honeybee (*Apis mellifera* L.) by juvenile hormone. Entomol. Exp. Appl. 15: 517–520.

Withers, G.S., S.E. Fahrbach and G.E. Robinson. 1993. Selective neuroanatomical plasticity and division of labor in the honeybee. Nature 364: 238–240.

Woltereck, R. 1909. Weitere experimentelle Untersuchungen über Artveränderung, speziell über das Wesen quantitativer Artenunterschiede bei Daphniden. Verh. Deutsch Zool. Gesell. 110–172.

Zera, A.J. 2009. Wing polymorphism in *Gryllus* (Orthoptera: Gryllidae): proximate endocrine, energetic and biochemical mechanisms underlying morph specialization for flight vs. reproduction. pp. 609–653. In: D.W. Whitman and T.N. Ananthakrishnan (eds.). Phenotypic Plasticity of Insects. Science Publishers, Enfield, New Hampshire.

Zera, A.J. and G. Cisper. 2001. Genetic and diurnal variation in the juvenile hormone titer in a wing-polymorphic cricket: implications for the evolution of life histories and dispersal. Physiol. Biochem. Zool. 74: 293–306.

Zera, A.J. and R.F. Denno. 1997. Physiology and ecology of dispersal polymorphism in crickets. Annu. Rev. Entomol. 42: 207–231.

Zera, A.J., L.G. Harshman and T.D. Williams. 2007. Evolutionary endocrinology: the developing synthesis between endocrinology and evolutionary genetic. Annu. Rev. Ecol. Evol. Syst. 38: 793–817.

Zera, A.J. and Y. Huang. 1999. Evolutionary endocrinology of juvenile hormone esterase: functional relationship with wing polymorphism in the cricket *Gryllus firmus*. Evolution 53: 837–847.

Zera, A.J., C. Strambi, K.C. Tiebel, A. Strambi and M.A. Rankin. 1989. Juvenile hormone and ecdysteroid titers during critical periods of wing morph determination in *Gryllus rubens*. J. Insect Physiol. 35: 501–511.

Zera, A.J. and K.C. Tiebel. 1988. Brachypterizing effect of group rearing, juvenile hormone-III, and methoprene on wing length development in the wing dimorphic cricket, *Gryllus rubens*. J. Insect Physiol. 34: 489–498.

Zera, A.J. and K.C. Tiebel. 1989. Differences in juvenile hormone esterase activity between presumptive macropterous and brachypterous *Gryllus rubens*: implications for the hormonal control of wing polymorphism. J. Insect Physiol. 35: 7–17.

Zhao, Z. and A.J. Zera. 2004. A morph-specific daily cycle in the rate of JH biosynthesis underlies a morph-specific daily cycle in the hemolymph JH titer in a wing-polymorphic cricket. J. Insect Physiol. 50: 965–973.

Zhou, X., F.M. Oi and M.E. Scharf. 2006a. Social exploitation of hexamerin: RNAi reveals a major caste-regulatory factor in termites. Proc. Natl. Acad. Sci. USA 103: 4499–4504.

Zhou, X.G., M.R. Tarver, G.W. Bennett, F. Oi and M.E. Scharf. 2006b. Two hexamerin genes from the termite *Reticulitermes flavipes*: sequence, expression, and proposed functions in caste regulation. Gene 376: 47–58.

Zhou, X., M.R. Tarver and M.E. Scharf. 2007. Hexamerin-based regulation of juvenile hormone-dependent gene expression underlies phenotypic plasticity in a social insect. Development 134: 601–610.

Zijlstra, W.G., M.J. Steigenga, P.M. Brakefield and B.J. Zwaan. 2003. Simultaneous selection on two components of life history in the butterfly *Bicyclus anynana*. Evolution 57: 1852–1862.

Zijlstra, W.G., M.J. Steigenga, P.B. Koch, B.J. Zwaan and P.M. Brakefield. 2004. Butterfly selected lines explore the hormonal basis of interactions between life histories and morphology. Am. Nat. 163: E76–E87.

2

Toxins, Defensive Compounds and Drugs from Insects

Konrad Dettner

Introduction

Arthropods and insects as largest groups of organisms with respect to species numbers or biomass contain an incredible number of biologically active low and high molecular compounds (Pietra 2002, Gronquist and Schroeder 2010). When taken by humans these compounds may have medicinal, intoxicating (venoms, toxins), performance enhancing or many other effects and therefore are called drugs. A large fraction of commercially available drugs represent natural products or represent derivatives of natural products (Dettner 2011), wherefore the search for such compounds is very important (Cragg et al. 2012, Tringali 2012). Very often these chemicals are toxic or may deter other animals and are therefore called allomones (defensive compounds) which are advantageous for the sender and disadvantageous for the receiver. With respect to predator prey interactions there exist various antipredator mechanisms in insects. Primary defenses are active before predators perceive prey (passive defenses such as mimicry, crypsis), secondary defenses only work after the predator has discovered its prey. As shown by Witz (1990) secondary defensive mechanisms are more important as compared with primary defensive mechanisms. Among secondary defense mechanisms chemical

Chair of Animal Ecology II, BayCEER, University of Bayreuth, 95440 Bayreuth, Germany.
Email: k.dettner@uni-bayreuth.de

defenses are most important as compared with mechanical defense, defensive stridulation or escape (Eisner 1970, Blum 1981, Eisner 2003, Eisner et al. 2005, Dossey 2010, Unkiewicz-Winiarczyk and Gromysz-Kalkowska 2012). These insect natural products usually are produced by the arthropods themselves (Bradshaw 1985, Morgan 2004), in many cases however such compounds originate from dietary plants or animals or might be even produced by symbiontic microorganisms (Pankewitz and Hilker 2008). In many cases the functional significance of these biologically active molecules within insects is unknown. In addition microorganisms isolated from insects may produce biologically active metabolites in the laboratory whereas these products can be never detected within the host insect. Table 1 compiles all insect orders according to Beutel et al. (2014) and their ways of chemical defense. However in this review the term insects includes also entognathous orders which are placed outside of insects by Beutel et al. (2014). Chemical defenses are obviously seldom in primitive and hemimetabolous insects, in contrast especially large holometabolous orders are characterized by a variety of very different chemical defensive mechanisms (Eisner 1970, 2003, Hilker and Meiners 2002, Eisner et al. 2005).

Primitive forms of chemical defenses within insects are represented by regurgitation (Fig. 1/2) and defecation (Fig. 1/3, 1/4 and Table 1; Eisner 1970, Weiss 2006). In many cases both regurgitates and fecal materials not only represent contents of fore- and hindguts but additionally contain behavior-modifying constituents which enhance deterrent and repellent effects of depleted materials. Defecation and regurgitation mechanisms are especially found in grasshoppers (Table 1), adult beetles, lepidopteran and hymenopteran larvae.

Many insect orders have toxic hemolymph which is often liberated from the interior of the body through reflex bleeding (= autohemorrhage, Fig. 1/9; Beauregard 1890, Eisner 1970, Evans and Schmidt 1990). Such ways to defend are found from Collembola (Table 1), Plecoptera, Orthoptera, Thysanoptera and Heteroptera. In holometabolan insects these ways to defend are distributed in Neuroptera, Coleoptera (Fig. 1/9), Lepidoptera, Diptera and Hymenoptera. With respect to egg stages of insects, toxic and behavior-modifying compounds are usually found within the egg or on the egg surface (Hinton 1981, Hilker and Meiners 2002). A lot of other taxa especially from holometabolous orders exhibit various kinds of exocrine glandular defenses (Fig. 1/1, 1/5, 1/6, 1/8). These exocrine glands are found within body cavites especially at exposed body parts and are sometimes eversible. When the insect developmental stage molts, these exocrine glands with their epicuticular linings must be replaced in the following developmental stage. Such secretions may also be collected in glandular hairs (Fig. 1/7), and subcuticular cavities as shown

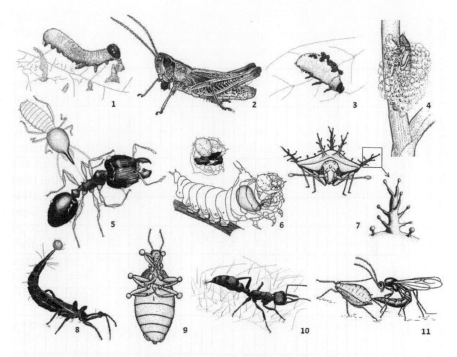

Figure 1. Various mechanisms of chemical defenses in insects (**1–11**).

1. Herbivorous *Stauronematus* larva (sawflies) depositing foamy pales from their salivary glands around its feeding place. The foam effectively deters ants (taken from a photograph).
2. Regurgitation behavior of a grasshopper (taken from a photograph).
3. Larva of *Leptinotarsa rubiginosa* (Chrysomelidae), the dorsal side of which is covered by feces (beetles; taken from a photograph).
4. Leaf-hopper larva of *Philaenus spumarius* hidden within foamy materials, mainly produced by Malpighian tubules (taken from a photograph).
5. Termite soldier attacking ant species *Pheidole megacephala* (with its sticky secretions from frontal gland). Drawing modified according to Spektrum der Wissenschaft 10 (1983).
6. Ventral prothoracic defensive gland of caterpillar of *Schizura unicornis* (Notodontidae). The inlet figure symbolizes the muscles which may help to aim the defensive spray (according to Whitman et al. 1990).
7. Larva of a tingid bug with its glandular hairs situated at body outgrowths (Scholze and Dettner 1992, unpublished).
8. Thysanopterid species depletes a secretion droplet from exocrine hindgut-glands at its abdominal tip (taken from a photograph).
9. *Meloe violacea* exhibiting thanatosis and reflex-bleeding (liberation of toxic cantharidin containing hemolymph) at femoral-tibian-articulations (according to Meixner 1935).
10. Australian primitive bulldog ant *Myrmecia piliventris* starting to introduce venom with her sting into a human skin (taken from a photograph).
11. Hymenopteran parasitoid *Lysiphlebus testaceipes* introducing her egg into its aphid host. Very often other components are simultaneously injected into the host insect (according to Berland 1951).

Table 1. Survey on insect orders according to Beutel et al. (2014). Apart from worldwide species numbers various strategies of chemical defense are indicated. Marks symbolizing certain defenses only indicate that this kind of defense is present in certain orders irrespective of species numbers.

	Species	chemical defense unknown	poisons bite/sting	hemo-lymph-toxin	reflex bleeding	regurgi-tation	defe-cation	defensive gland(s)	bacterial/ fungal toxins/ metabolites	defensive adhesive
Collembola (springtails)	8000			x	x			x		x
Protura (coneheads)	750							x		x
Diplura (dipluran, two-pronged bristletails)	1000	x?								
Archaeognatha (jumping bristletails)	500	x?								
Zygentoma (silverfish)	510	x?								
Ephemeroptera (mayflies)	3000	x								
Odonata (dragonflies, damselflies)	5600	x							x	
Plecoptera (stoneflies)	3500				x			x		
Dermaptera (earwigs)	2000							x		
Embioptera (webspinners)	360	x?								
Phasmatodea (stick- and leaf insects)	3030	x?						x		
Orthoptera (bush-crickets, crickets, grasshoppers, locusts)	22500			x	x	x	x		x	
Zoraptera (angel insects)	39	x								
Grylloblattodea (ice crawlers, rock crawlers, ice bugs)	32	x								
Mantophasmatodea (heel walkers, gladiators)	18	x								
Mantodea (praying mantises)	2300	x								
Blattodea incl. Isoptera (roaches incl. termites)	7600					x		x		x
Psocoptera (booklice, barklice and barklice)	5500	x?	x/	x						
Phthiraptera (true lice)	5000	x	x/							
Thysanoptera (thrips, fringe wings)	5822		x/				x	x		x
Auchenorrhyncha (cicadas, leaf-, plant-, frog- and treehoppers)	45000			x	x			x	x	
Sternorrhyncha (plantlice)	16400									x
Coleorrhyncha (moss bugs)	36	x								
Heteroptera (true bugs)	40000		x/	x	x			x	x	x
Neuroptera (netwinged insects)	6000		x/		x			x?	x	
Megaloptera (alderflies, dobsonflies, fishflies)	325	x				x				
Raphidioptera (snakeflies, camelneck flies)	235									
Coleoptera (beetles)	355000	x	x/	x	x	x	x	x	x	x
Strepsiptera (twisted-winged parasites)	600	x								
Trichoptera (caddisflies)	14500							x		
Lepidoptera (butterflies and moths)	175000			x	x	x	x	x		
Mecoptera (scorpionflies, hanging flies)	550	x?				x	x		x	
Siphonaptera (fleas)	2000		x/							
Diptera (true flies & midges)	154000	x?	x/	x	x	x	x	x	x	x
Hymenoptera (sawflies, woodwasps, bees, wasps, ants)	132000		x/x	x	x	x	x	x	x	x

in larvae of zygaenid moths (Franzl and Naumann 1985). Finally minute amounts of toxins and venoms may be specifically targeted and applied onto and within targets by stinging or biting (Schmidt 1982, Fitzgerald and Flood 2006). Whereas stinging was only realized in Hymenoptera (Table 1, Fig. 1/10, 1/11; Hermann 1984, Evans and Schmidt 1990), biting structures are more distributed in Phthiraptera, Heteroptera (Table 1), Neuroptera, Coleoptera, Diptera Siphonaptera, and Hymenoptera. Host immune reactions and various allergens were reviewed by Lehane (2005) and Richard and Ledford (2014). Moreover in various insect groups defensive secretions are depleted as adhesives (Fig. 1/5, 1/7), which is compiled in detail by Betz (2010).

Defensive compounds and toxins of arthropods and insects which represent secondary compounds show an extreme chemical and biosynthetic variation (Dettner 2010, Gronquist and Schroeder 2010). In this review fungicidal or bactericidal hemolymph peptides which are produced by the insects themselves are not regarded (see Wiesner and Vilcinskas 2011, Kastin 2013), in contrast microbial products from insect symbionts are discussed. Because of their enormous diversity insects are an interesting source for low molecular compounds but also for larger biomolecules (Blagbrough et al. 1992, Pemberton 1999, Vilcinskas and Gross 2005, Srivastava et al. 2009, Cherniack 2010, Lokeshwari and Shantibala 2010, Dossey 2011) and especially for unusual bacteria and fungi (Dowd 1992). In some cases especially aposematically colored and therefore often toxic insects represent promising candidates for bioprospecting (Helson et al. 2009, Dettner 2011, Vilcinskas 2011).

The following data usually are arranged according to different taxa within an insect order. Within an order sometimes the text was also arranged according to types of exocrine defensive glands.

Natural Compounds According to Insect Orders

There are various insect taxa as Strepsiptera, Raphidioptera, Coleorrhyncha, Phthiraptera, Mantodea, Mantophasmatodea, Grylloblattodea, Zoraptera, Embioptera, Ephemeroptera devoid of toxins, or any way of chemical defense (Table 1). Merely representatives of Phthiraptera (Anoplura) may cause skin lesions (Pediculosis Corporis) and allergic reactions in humans (Alexander 1984). Often these orders comprise only few species or may represent ectoparasites. As expected orders with high species numbers especially holometabolous insects such as Hymenoptera, Lepidoptera are characterized by various ways of chemical defense (see Eisner 1970, Evans and Schmidt 1990). Table 1 illustrates all orders which have few or many ways to defend chemically by gut contents, exocrine glands, hemolymph toxins, stings, bites or adhesives, in addition insect associated fungal or bacterial metabolites may be of importance.

The following chapters contain those insect orders, where the above mentioned chemical defenses, but also microbial toxins, are present. If there exist reviews concerning behavior modifying chemicals, drugs or toxins of certain insect taxa (e.g., Bettini 1978, Blum 1981, Schulz 2004, 2005, Mebs 2010, Vilcinskas 2011), these compilations are cited for further consultations.

Collembola (springtails)

The jumping ability and detachable hairs or scales represent important modi of defense in this large group of apterygote insects (worldwide 8,000 springtail species). However chemical defense may be of high significance not only in those species which are eyeless or have lost their jumping ability but are at least present in all poduromorph collembolans (Poduridae, Hypogastruridae, Neanuridae, Onychiuridae, Tullbergiidae) and few representatives of entomobryomorphs (e.g., Isotomidae). Some species such as genera *Piroides* or *Corynephra* (Dicyrtomidae) or certain representatives of Neanuridae (e.g., Uchidanurinae) show warning coloration or spine-like colored projections on body surface and some of them may possess club-like defensive glands at abdominal tip. Many species especially of Onychiuridae may release defensive compounds from secretory cells below pseudocells (integumental pores) or via toxic hemolymph by reflex bleeding. On molestation *Tetrodontophora bielanensis* (Onychiuridae) depletes 3 aromatic alkaloids 2,3-dimethoxypyrido[2,3-b] pyrazine, 3-isopropyl-2-methoxypyrido[2,3-b]pyrazine (Fig. 2), and 2-methoxy-4*H*-pyrido[2,3-b]pyrazine-3-one (Stransky et al. 1986, Dettner et al. 1996) which act as deterrents against carabid beetles. These compounds are mainly present in pseudocellar secretions but are also recorded in minor amounts from adult hemolymph; larvae only contain traces as was also recorded from *Onychiurus scotarius* and *O. circulans* (Nilsson and Bengtsson 2004). In addition there was identified the open chain tetraterpene lycopane (2,6,10,14,19,23,27,31-octamethyldotriacontane) apart from and in *Tetrodontophora*.

Of the 4 aromatics recorded from *Neanura muscorum* (Neanuridae: phenol, 2-aminophenol, 1,3-dimethoxybenzene, and 2,4-dimethoxyanilin), only 2 aminophenol had a deterrent activity against predatory mite *Pergamasus* (Messer et al. 2000) and was found as a chemotaxonomic marker for Neanurinae subfamily (Porco and Deharveng 2007). 1,3-Dimethoxybenzene was also identified as intraspecific acting alarm pheromone of *Neanura* (Messer et al. 1999), a behavior which was also described from other genera such as *Hypogastrura* (Hypogastruridae),

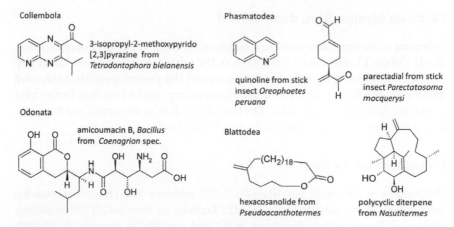

Figure 2. Defensive secretions and hemolymph toxins from Collembola, Phasmatodea and Blattodea. Amicoumacin B represents a microbial product from Odonata.

Folsomia (Isotomidae) and *Megalothorax* (Neelidae) (Purrington et al. 1991). In addition 1,3-dimethoxybenzene and 2,4-dimethoxyanyline have been found as major constituents in hemolymph of *Neanura* adults and in eggs, whereas exuviae of adults only contained small amounts of phenol (Messer et al. 1999). From *Ceratophysella denticulata* (Hypogastruridae; especially from intguments) there were identified 3-hydroxy-4,5-dimethoxy, and 4-hydroxy-3,5-dimethoxy benzoic acids, which represent effective deterrents against *Stenus* predators (Bitzer et al. 2004). Extracts of this species have also been found to act intraspecifically as alarm substances (Pfander and Zettel 2004). From *Podura aquatica* body surfaces finally the new hydrocarbon tetraterpene poduran, with an unusual tricyclo[6.2.0.0]decane system had been identified (Schulz et al. 1997). It seems interesting to denote, that unknown toxic hemolymph constituents of *Folsomia fimetaria* (Isotomuridae) negatively influence reproduction of predatory linyphiid spider *Erigone atra* (Marcussen et al. 1999).

Protura (coneheads)

The minute apterygote and entognathous Protura (750 species) possess huge paired abdominal defensive glands which open on the eigth tergite. When molested the species exude an acid gland material which hardens on contact with air and forms sticky threads (Janetschek 1970, Hansen et al. 2010). The secretion is insoluble in both water and ethanol and contains neutral glycoproteins (Francois and Dallai 1986).

Odonata (dragonflies, damselflies)

Odonata with their worldwide 5,600 species exhibit no chemical defense at all (Table 1). However *Bacillus*-species which were isolated from the gut of *Coenagrion* dragonfly larvae produced the pseudopeptide antibiotic amicoumacin B (Fig. 2) with anti-inflammatory, anti-ulcer and herbicidal activities (Gebhardt et al. 2002, Dettner 2011). It is unknown if gut bacteria of *Coenagrion* produce these compounds *in vivo*.

Plecoptera (stoneflies)

Plecoptera with their worldwide 3,500 species are characterized by terrestrial living adults and aquatic larvae. In two adult (*Pteronarcys proteus, Peltoperla maria*; Benfield 1974) and one larval species (*P. dorsata*; Moore and Williams 1990) there was recorded reflex bleeding. When disturbed, adults bleed from intersegmental membranes of coxo-trochanteral and tibio-femoral joints. Because there could be recorded neither low molecular compounds nor deterrent-effects of hemolymph it was suggested that the reflex bleeding of stoneflies which is accompanied by an audible popping sound has only a mechanical effect on potential target organisms (Stocks 2008). Zwick (1980) suggests that plecopteran hemolymph primarily acts as an adhesive. In larvae of *P. dorsata* reflex bleeding was observed for the first time in an aquatic insect after contact with benthic tactile crayfish predators (Moore and Williams 1990). In contrast the larva exhibited thanatosis when fish predators appeared.

Dermaptera (earwigs)

Most advanced species of the worldwide known nonparasitic 2,000 Dermaptera species possess abdominal pincers and one or two pairs of defensive gland reservoirs which open on the posterior margins of third and fourth abdominal tergites. The stinking quinoid secretions of earwigs may be discharged as droplets or as a spray which is improved by revolving the abdomen (Eisner 1960, Eisner et al. 2005). The European Earwig *Forficula auricularia* produces methyl-and ethyl-1,4-benzoquinones together with their corresponding hydroquinones (Schildknecht and Weis 1960), whereas the American species *Doru taeniatum* contains methyl-and 2,3-dimethyl-1,4-benzoquinones (as crystals) which are partly dissolved in an organic pentadecane-phase and in a co-occurring aqueous phase. Discharges always occur as quinoid solutions (Eisner et al. 2000). Whether the paired larval glands at the base of the pincers are true defensive glands (Vosseler 1890; defensive glands at the base of larval abdomen are not present) remains to be investigated.

Phasmatodea (stick- and leaf insects)

There have been described more than 3,000 species of Phasmida (stick and leaf insects) whose most important primary defense against predation is crypsis and mimicry. Secondary defenses include startle displays, aposematism, thanatosis, autotomy, sound production or ejection of an irritating spray from a pair of prothoracic exocrine glands (Eisner et al. 2005). Some of the phasmid species are the largest chemically defended insects in the world (Dossey 2010). Especially in males of the genus *Eurycantha* there are found long and hard spines on femora of hind legs and warning odors which originate from the abdominal tips (Seiler et al. 2000). Obviously the defense glands are present in all phasmid suborders. In certain species the secretion may be ejected a well directed beam of secretion which may reach 50 cm in some genera. The secretions represent effective deterrents against arthropods or birds but are also lachrymogenous, cause sneezing and their vapors are painfully irritating when inhaled (Dossey 2010, 2011). Representatives of the genera *Anisomorpha* (Eisner et al. 2005, Dossey 2010), *Autolyca* (Pavan 1975), *Graeffea* (Smith et al. 1979) and *Peruphasma* (McLeod et al. 2007) are characterized by cyclopentanoid monoterpenes as trans-, trans-iridodial, trans-, cis-iridodial, anisomorphal (= dolichodial), peruphasmal or nepetalactone, which are biosynthesized by the phasmids (Meinwald et al. 1966). *Megacrania alpheus* (Chow and Lin 1986, Dossey 2010) produces considerable amounts of actinidine and analogs (probably from iridoids) whereas *Oreophoetes peruana* contains an aqueous emulsion of pure quinoline (Fig. 2) within its defensive glands which represents both repellent and irritant (Eisner et al. 1997). The novel compound parectadial (Fig 2) was recently recorded from *Parectatosoma mocquerysi* from Madagaskar, the compound shows cytotoxic and cytostatic properties (Dossey 2010, 2011). In the defensive secretion of *Asceles glaber* there were described the two spiroketals (2S,6R)-(–)(*E*)-2-methyl-1,7-dioxaspiro[5.5] undecane and 2-ethyl-1,6-dioxaspiro[4.5]decane (Dossey et al. 2012). In contrast *Phyllium westwoodii* sprays its defensive secretion which contains a mixture of 3-isobutyl-2,5-dimethylpyrazine, 2,5-dimethyl-3-(2-methylbutyl)pyrazine and 2,5-dimethyl-3-(3-methylbutyl)pyrazine (Dossey et al. 2009). In *Sipyloidea sipylus* the defensive secretion contains diethyl ether, acetic acid, benzaldehyde, limonene and benzothiazole (Bouchard et al. 1997), whereas various other species produce methyl-1-hepten-3-one (Dossey 2010). As compared with other chemically defended species the secretion of this Asian species was extraordinary effective (Carlberg 1986).

Orthoptera (bush-crickets, crickets, grasshoppers, locusts)

Among Orthoptera with worldwide about 22,500 species a lot of primary and secondary defensive mechanisms such as crypsis, aposematism, jumping, biting, flight, stridulation or autotomy are realized (see Preston-Mafham 1990, Eisner et al. 2005). Employed chemical defenses are regurgitation (Fig. 1/2), anal discharges or defecation, glandular defensive secretions and internal toxins. Few species may also produce allergens (Richard and Ledford 2014). These various mechanisms were reviewed by Whitman (1990) for most representatives of the suborder Caelifera (grasshoppers). Here especially recent data on chemical defense of *Romalea/Taeniopoda*-grasshoppers, Tridactyloidea (pygmy mole crickets; Caelifera), the sequestration of alkaloids in grasshoppers and chemical defenses of suborder Ensifera are treated.

Both *Romalea* and *Taeniopoda*-species can discharge their defensive secretions from paired metathoracic spiracles respectively from metathoracic tracheal glands (Whitman et al. 1992). Moreover it was shown that adults of *R. guttata* secrete increased amounts of catechol and hydroquinone from their defensive glands when fed especially with *Nepeta*-diet (Snook et al. 1993). In contrast when it was reared on *Allium*-plants, *R. guttata* sequester sulphur volatiles. The defensive secretion in both genera also contained proteinase inhibitors which might act against entomopathogenic fungus (Polanowski et al. 1997). Hatle and Faragher (1998) demonstrated that slow movements may increase survivorship in frog encounters of chemically defended *R. guttata*-specimens. Finally defensive secretion of *T. eques* females may act as sex pheromone eliciting mating behavior in males (Whitman 2008). *R. guttata* was shown to sequester 2,5-dichlorophenol which originates from the herbicide 2,4-D (see Opitz and Müller 2009). In *Anacridium* a sequestration of gallic acid in the integument was reported (Opitz and Müller 2009).

Larvae and both sexes of adult Tridactylidae (more than 200 species; Caelifera) possess paired sternal glands which open laterally on the second abdominal sternum. The chemically unknown defensive secretion may effectively repel *Tetramorium*-ants and *Pardosa*-spiders (Moriya and Ichinose 1988, Moriya 1989). Obviously homologeous defensive glands are present in subterranean sandgropers *Cylindraustralia kochii* (Cylindrachetidae; Houston 2007).

Several grasshopper species are known to excrete or to sequester alkaloids from plants. *Melanoplus* can metabolize toxic pyrrolizidine alkaloids in the gut, yielding more polar metabolites (Ehmke et al. 1989). In addition certain grasshoppers as *Zonocerus elegans* are attracted to pure pyrrolizidine alkaloids such as heliotrine and may even ingest and harbouring them (Boppré et al. 1984). In *Schistocerca emarginata* degrees of distastefulness through regurgitation depend on the uptake of toxic plant material (Sword 2001).

Exocrine glands and allomones are also found among Ensifera (more than 10,000 species). When disturbed, females of Chinese and European mole-crickets (*Gryllotalpa*; Grylloidea) deplete mucoproteid-containing sticky secretions from their paired anal glands in order to disorientate predators. Since gland extracts are therapeutically used and may improve wound epithelialization and neovascularization in mammals it is of interest to elucidate the chemistry of these gland secretions (Zimmer et al. 2006).

Other partly eversible exocrine defensive glands are typical for representatives of Tettigoniidae (Tettigonioidea; more than 6,000 species) and may be restricted to the thoracic area (Mecopodinae: head/ thorax; Pseudophyllinae: procoxae; Rhaphidophoridae: intersegmental membranes), where insectivorous lizards usually seize their prey. A *Vestria*-species belonging to the tropical Katydids emits pyrrazines from odoriferous glands near its abdominal tips (Nickle et al. 1996). Other Tettigoniidae may effectively defend themselves by reflex bleeding. Especially in *Eugaster* hemolymph may be squirted up to 80 cm distance (velocity 2–3 m/sec) from bleeding pores at the base of the legs (Grzeschik 1969). In an African king cricket *Onosandrus* spec. it could be shown that defecation is not effective as deterrent against lizards, skinks and toads (Wolf et al. 2006).

Blattodea Inclusive Isoptera (roaches inclusive termites)

There exist various strategies by which cockroaches with their worldwide known more than 4,600 species may escape the attentions of predators. These range from stridulation, flying, quickly running away, coiling into a sphere, diving into water, swimming under water, burrowing, headstanding, to aposematism (warning coloration) and co-occurring aggregations. Sometimes cockroaches may be camouflaged and ressemble dead or green leafs, sometimes they may even mimic unpalatable lycid beetles (Preston-Mafham 1990). In addition cockroaches may cause allergic symptoms and bites in humans (Alexander 1984) and often may possess various exocrine glands where allomones are produced. These defensive mechanisms were reviewed 1978 by Roth and Alsop, by Brossut (1983), and partly by Eisner et al. (2005).

Various types of exocrine glands in males and females probably represent defensive glands, the internal surfaces of the gland reservoirs are often covered by setae which may increase evaporation of defensive secretions, when these reservoirs are everted (Beier 1974). In adults and larvae of some Blattidae the paired **laterocervical glands** produce UV-fluorescent secretions of unknown chemistry (Beier 1974).

Between abdominal spiracles of adults of the primitive Polyphagidae there are situated paired **pleural glands**, which emit complex defensive secretions. In adults of both sexes of *Therea petiveriana* the secretion of these brown eversible pouches serves as allomone and alarm pheromone and contains indole, nonanal, phenol, p-cresol, C_2-, C_4-, C_6-, C_7-, C_9-, C_{10}-C_{12}-carboxylic acids, various fatty acids and the main compounds N-3-methylbutylacetamide and N-3-methylbutylpropanamide (Farine et al. 2002).

Proteinaceous, slimy secretions which represent mechanically deterrents especially against ants are produced by exocrine **pygidial glands**, which are found on tergites 6, 7 and 10 of adult females and larvae of Blattidae and Blattellidae. There were found no active substances in the slime which suggests that it acts by virtue of its stickyness (Eisner et al. 2005). In larvae of *Megaloblatta blaberoides* the gray sticky proteinaceous secretion on the last abdominal segments and orange spots on its dorsum co-occur with a disturbance stridulation (Schal et al. 1982).

In Blaberidae there are found paired sac-like swellings in the spiracular trunks leading from the lateral trunk to the second spiracles. These swellings which are sometimes also associated with 6th and 7th spiracles represent gland reservoirs of the **tracheal glands** containing various p-quinones, phenols (2-ethylphenol; p-cresol), naphthole, methacrylic and tiglic acids, octanol and 2-decenal (see Brossut 1983), which are directed against arthropods, or pathogenic fungi and bacteria. Force for the ejection of secretion as a fine mist is provided by air from the internal tracheal air sacs (Eisner et al. 2005). Since the lining of the tracheal glands is shed after molting, larvae are defenseless at that time. In contrast freshly molted adult females are attractive for males and during subsequent copulations they may even derive profit from male gland-discharges since males are capable for chemical defense (Wyttenbach and Eisner 2001). In freshly molted *Diploptera*-adults Baldwin et al. (1990) showed that refilling of defensive glands with quinones was delayed which might be due to a competition for aromatic amino acids between cuticular and defensive quinone synthesis.

Larvae and adults of Blattidae possess unpaired **ventral** or **sternal glands** (sometimes with various lobules) which open between 6th and 7th sternites. Volatile secretions with 2-methylene butanal, E-2-hexenal, various short branched carboxylic acids or 2-acetyl-4-methylbutyrolactone (Brossut 1983) represent effective defensive secretions which may be sprayed and may repel predatory vertebrates and invertebrates. Nonvolatile fractions contain gluconic acid, gluconolactone and glucose of unknown biological activity. The organic fraction of this secretion, which may even be ejected was analyzed in *Eurycotis floridana* (Farine et al. 1997) and was shown to contain 40 compounds, especially 3-ethoxyhexanal, 3-hydroxyhexanal, ((*E*)-1-pentenyl)-4-propyl-1,3-dioxane, and 3-(*E*)-2-hexenoxyl-hexanal are unusual for insects. It was shown that the *Eurycotis*-secretion acts against small mammals however single individuals were capable of neutralizing this chemical defense by thrusting the roach's abdomen into the substrate (O'Connell and Reagle 2002). In other species of the genera *Periplaneta* and *Blatta* there were recorded 3-hydroxy-2-butanone (= acetoin), p-cresol, 2- and 4-ethylphenol, 4-vinylphenol, and indol (Takegawa and Takahashi 1990). In *E. floridana* it was additionally shown, that (*E*)-2-hexenal, (*E*)-2-hexenol and (*E*)-2-hexenoic acid are preferentially biosynthesized de novo from labelled acetate (Farine et al. 2000). The secretion proved to be an effective deterrent against mice, ants and other cockroaches (Turnbull and Fashing 2002).

Between 5th and 6th tergites in larvae and adults (both sexes) of Blattinae (Blattidae) there open **paired tergal glands**, the secretions of which represent allomones with 2- and 3-pentanone, 3-octanone, 2,3,4-methylcyclohexanone or α-terpineol. The toxic mixtures can be oozed out or spread from the body side being stimulated. In *Deropeltis wahlbergi* females and larvae assume a rigid stance just before ejection of secretion. In the fully wing males, wings are rotated upwards before secretion is emitted (Eisner et al. 2005). Residual defensive secretions on the abdominal tergites may fortify the repellency of the secretion for a longer time.

Because bioluminescence may also represent an effective chemical defensive system it should be noted that there were discovered luminescent cockroaches in the canopy forests of South america (Vršanský et al. 2012, see Oba, this book).

Female cockroaches deposit their eggs into an ootheca which is produced by female accessory glands. The protein arthropodin is hardened by tanning with ortho-quinones which are formed by oxidation of protocatechuic acid. It might be that calcium oxalate which can form up to 15% of the dry weight of the ootheca represents a chemical defensive agent (Wigglesworth 1974).

Termites with their worldwide 3,000 known species are today integrated within Blattodea. As social insects with high biomass values they have various enemies such as specialized mammals, ants but also so-called guests which may parasitize within the termite colonies. Therefore, representatives of Isoptera developed various defensive mechanisms. Especially their subterranean life and the development of the soldier caste represent effective defensive strategies. However worker termites are not defenseless. Mainly the soldiers, the percentage of which is between 4–9% in primitive termites and increases to 20–30% of individuals in highly developed families, are mechanically and especially chemically defended. Whereas chemical defensive compounds are found in salivary glands in lower termites, higher termites especially use their frontal glands (Fig. 1/5). The secretion of this unpaired gland may flow over a discharging gutter or groove to the mandibles or the labrum where it contacts the wounds. Very often the opening of the frontal gland is enlarged into a tube and the secretion can be squirted and distributed with a kind of brush. In various species salivary glands and especially frontal glands extend into the abdomen and may release their defensive secretion (together with gut contents) by autothysis (gland rupture) ("Kamikaze-soldiers") and abdominal dehiscence (intestinal rupture). In addition and especially in workers there was observed both defecation and regurgitation of defensive fluids in termites.

Many exocrine compounds and especially allomones of termites were compilated by Wheeler and Duffield (1988) and Laurent et al. (2005). Various aspects of chemical defense in termites were reported by Prestwich (1983, 1984), Howse (1984) or Quennedey (1984). Apart from some general remarks, here only recent data on termite chemical defense are compiled.

According to their functions termite defensive secretions are regarded as antihealants and irritants (especially mono- and sesquiterpenes, diterpenes, quinones), contact poisons (e.g., nitroalkenes, vinyl ketones, β-ketoaldehydes; Spanton and Prestwich 1982), alarm substances or glues (aqueous lipid-gylcoprotein-mucopolysaccharide).

As examples for aggressive termite gland products there is figured hexacosanolide (Fig. 2) from *Pseudoacanthotermes spiniger*. The compound from the salivary gland paralyzes and sometimes may kill aggressors (Laurent et al. 2005). Moreover Fig. 2 shows a polycyclic diterpene (Laurent et al. 2005) from the frontal glands of *Nasutitermes* and *Trinervitermes* soldiers. The secretions are irritating and usually act as glues.

Secretions from glands localized at base of mandibles and from cibarial gland may penetrate the wound inflicted by the mandibles and are considered as auxiliary defense organs. If tarsal glands might contribute to chemical defense has to be investigated. Chemical defensive secretions are produced in the paired salivary (non-volatile, aqueous secretions) and unpaired frontal glands (especially developed in Rhinotermitidae, Termitidae). The frontal gland secretions represent terpenoids (mono-, di- and sesquiterpenes: hydrocarbons and oxygenated products), acetate-derived compounds (quinones, macrocyclic lactones, alkanes, nitroalkenes, vinyl ketones, ketoaldehydes) and other chemicals from amino acid and/or carbohydrate metabolism (proteins, mucopolysacccharides) (Prestwich 1984). Ohta et al. (2007) discovered novel free ceramides as components of the frontal gland of *Coptotermes formosanus*.

Psocoptera (booklice, barkflies and barklice)

Psocids or barklice with about 5,500 species worldwide may produce human allergens, but have no distinct defensive organs (Dathe 2003, Parker 1982). Only certain larvae possess glandular hairs to which particles of algae, lichens or feces may adhere. It seems also possible that psocoptera may sequester aromatic compounds which are typical for lichens. Therefore especially these stages but also all stages of many other species are camouflaged, although few species may be brightly colored. Many species can produce silk from glands which are associated with the labium (see Neuenfeldt and Scheibel, this book). These constructs may protect from predators, desiccation or wetting. Howard and Lord (2003) reported that the booklice *Liposcelis bostrychophila* was unaffected by three entomopathogenic fungi of broad host range. They suggested that certain kinds of the cucicular lipids of these booklice such as C_{15}- and C_{16}-aldehydes, C_{16}- to C_{22}-amides and C_{16}- and C_{18}-fatty acids have fungistatic properties.

Thysanoptera (thrips, fringe wings)

Worldwide there are described more than 5,800 Thysanoptera (thrips)-species. Many of these minute insects with body lengths of 1–3 mm may cause considerable economic damage through sucking and by potential transfer of viruses, bacteria and fungi. Often Thysanoptera take up plant juices (suborder Terebrantia), however many species may also feed on fungal spores (Tubulifera: Phlaeothripidae) or represent true arthropod predators. In addition many thysanopteran species may cause skin lesions in humans due to thrip bites (Alexander 1984).

There exist various defense strategies of thrips-species against natural arthropod predators (e.g., ants, mites), ectoparasitic mites, internal parasites or entomopathogenic fungi. Most specimens escape by jumping and flying, they may feign death or may release a defensive secretion. Anal secretions have been described as sources for various allomones and alarm pheromones (e.g., perillene). When attacked scorpion-like thrips flex the tips of their abdomina, which represent true applicators due to their setae, towards aggressors and exude a droplet of anal fluid (Fig. 1/8; Howard et al. 1983, Tschuch et al. 2004). Obviously long setae cover the droplet, whereas short setae prevent spreading of the secretion (Tschuch et al. 2004). Apart from bactericidal secretions the chemical defense of thrips therefore acts either via odor (repellency) or via topical application (contact activity; deterrency). In some cases active compounds may spread over the surface of the aggressor (Moritz 2006). Gall-inhabiting species may use their volatile anal droplets as fumigants (Blum 1991). The anal defensive secretion may consist of one compound or even can contain up to 11 constituents. Concerning chemical defense in Thysanoptera there exist two compilations (Blum 1991, Moritz 2006).

Various saturated and unsaturated hydrocarbons very often represent major allomones in a considerable number of Thysanoptera species. From dodecane to heptadecane all saturated hydrocarbons are especially found in gall-inhabiting taxa. There were also recorded various alkenes such as (Z)-7-pentadecene, (Z)-8-heptadecene, (Z)-9-octadecene, (Z)-9-nonadecene, or nonadecadiene (Blum 1991, Suzuki et al. 2004, Moritz 2006). Phenole was identified from two genera *Arrhenothrips* and *Euryaplothrips* (Blum 1991, Moritz 2006). There were additionally recorded two aromatic aldehydes from Thysanoptera: 2-hydroxy-6-methyl-benzaldehyde (*Xylaplothrips*) and phenylacetaldehyde (*Arrhenothrips, Euryaplothrips*, Blum 1991, Moritz 2006). In 2 *Holothrips*-species there could be even identified 3-butanoyl-4-hydroxy-6-methyl-2H-pyran-2-one (Suzuki et al. 1993). Thrips species and especially representatives of Idalothripinae may also emit three naphthoquinones which are partly known as allelopathic agents: juglone (*Bactrothrips*), 7-methyl- and 2-methyl-juglone (*Ponticulothrips, Dolichothrips*; Blum 1991, Moritz 2006).

Among esters various acetates [C10, C12, C14, C16, C18, C20 acetates, dodecadienylacetate, (Z)-9-hexadienyl- and (Z)-11,19-eicosadienylacetate; Moritz 2006], together with hexadecylbutanoate are found in anal secretions of Thysanoptera (Blum 1981, 1991, Tschuch et al. 2008). Several of these esters which are sometimes classified as solvents seem to represent true repellents [e.g., (11Z)-11,19-eicosadienylacetate in *Suocerathrips linguis*; Howard et al. 1987, Teerling et al. 1993, Tschuch et al. 2002a,b, Suzuki et al. 2004]. Under natural temperature conditions solid esters such as stearyl acetate may be dissolved in the fluidal (11Z)-11,19-eicosadienylacetate (Tschuch et al. 2004, 2005). *Bagnalliella yuccae* contained 0.12 µg (larva) to 0.27 µg (adult) of γ-decalactone per specimen which is not present in leaves of the hostplant but probably synthesized in the hindgut (Howard et al. 1983, Haga et al. 1989). 4-Octadec-9-enolide was recorded from secretion of *Euryaplothrips* (Blum 1991). The aromatic lactone mellein was recorded from genus *Haplothrips* (Blum et al. 1992).

Among the various terpenes which were especially recorded from gall-inducing Phlaeothripidae, the furanomonoterpene perillene was found in various genera (*Arrhenothrips, Leeuwenia, Liothrips, Oidanothrips, Varshneyia, Xylaplothrips*, Blum 1981, Suzuki et al. 1986, 1988, Blum 1991, Moritz 2006). The perillene isomer 3-methyl (3-methyl-2-buten-1-yl)-furane (= rose furane) is present in *Arrhenothrips* and was recorded for the first time in animals. Other terpenes are the unstable dialdehyde β-acaridial, which was found in the genera *Gynaikothrips, Liothrips,* and *Varshneyia,* but is also known from mites (Suzuki et al. 1988, 1989). Further terpenes constitute β-myrcene (*Dolichothrips, Thilothrips*), dolichodial (e.g., *Callococcithrips, Leeuwenia*), α- and β-pinene (*Dolichothrips*), neral, geranial (*Eugynothrips*) and the sesquiterpene caryophyllene (*Dolichothrips*).

Apart from short chained carboxylic acids 2-methylbutanoic acid (*Varshneyia*) and 3-methylbutanoic acid (*Dinothrips*), there were found several long chained saturated (C8, C10, C12) and unsaturated acids [(*E*)-4-decenoic acid, (*Z*)-5- and (*E*)-3-dodecenoic acids, (*Z*)-5-tetradecenoic acid, dodecadienoic acid, and 5,8-tetradecadienoic acid]. The anal droplets of the eusocial gall-inducing species *Kladothrips intermedius* contained unsaturated wax esters, short chain fatty acids and 7-octenoic, 8-nonenoic- and 9-decenoic acids (De Facci et al. 2014).

Auchenorrhyncha (cicadas, leaf-, plant-, frog- and treehoppers)

Chemical defense mechanisms of Auchenorrhyncha were only recorded from few taxa as compared with 45,000 species worldwide known. From few species it is reported that they contain the toxic terpenanhydrid cantharidin (see Fig. 4; Hemp and Dettner 2001). This was proved by Feng et al. (1988) and Dettner (1997) for *Lycorma delicatula*. In the case of *Huechys sanguinea* the presence of cantharidin was suggested (Juanjie et al. 1995). It is unknown wether the terpenanhydrid is sequestered from animals or biosynthesized by the cicadas.

Moreover, reflex bleeding was reported from 44 species of Neotropical froghoppers from the tribes Tomaspidini and Ischnorhinini where adults are able to exude hemolymph from pretarsal pads by rupture of membranes (Peck 2000, Stocks 2008). The authors suggested that this behavior is defensive because many species have an aposematic coloration and may emit warning odors such as pyrazines (Guilford et al. 1987). These pyrazine odors significantly make visually conspicuous prey aversive to bird predators (Lindström et al. 2001). The role of the volatile mevalolactone which is emitted by *Psammotettix alienus* is remarkable (Alla et al. 2002). Finally from dried juvenile cicadas, which are used in Chinese medicine, there was registered the peptide cicadine with antifungal activity (Wang and Ng 2002).

An unusual behavior is reported from nymphs of spittlebugs (Mello et al. 1987). The larvae secrete a froth which surrounds their body and within which they are living and at the same time are protected from desiccation and from harmful environmental factors (Fig. 1/4). By Mello et al. (1987) it was shown that the froth contains ten polypeptides (most of them glycopeptides), acid proteoglycans. Del Campo et al. (2011) analyzed *Aphrophora*-froth and identified fatty-acid derived alcohols, γ-lactones, 1-monoacylglycerol, pinitol and poly-3-hydroxybutyrate. The froth repelled ants but showed no topical irritancy against cockroaches.

In contrast Auchenorrhyncha harbored a lot of symbiontic microorganisms with biologically active molecules. *Nilaparvata lugens* contains both *Bacillus*-species producing the decapeptide antibiotic polymyxin M_1 (Fig. 3) and unknown bacteria which produce the polyketide antibiotic andrimid (Fig. 3). Andrimid and similar compounds such as moiramides, which were isolated from various marine organisms and their bacteria represent broad spectrum antibacterials, which inhibit the fatty acid biosynthesis in bacteria (Dettner 2011). In the meantime the gene cluster responsible for andrimid-biosynthesis was identified in *Enterobacter* (Yu et al. 2005). Pyoluteorine (Fig. 3), a pyrrol-antibiotic and polyketide was isolated from *Sogatella*-plant hoppers (Delphacidae). It is also produced by *Pseudomonas*-species (Kenny et al. 1989). Finally from endosymbionts of a White-blacked planthopper species diacetylphloroglucinol (Fig. 3), an antimicrobial, fungicidal and phytotoxic polyketid was isolated (Kenny et al. 1989). Recently a gene cluster for synthesis of this interesting compound was identified in *Pseudomonas fluorescens* (see Dettner 2011).

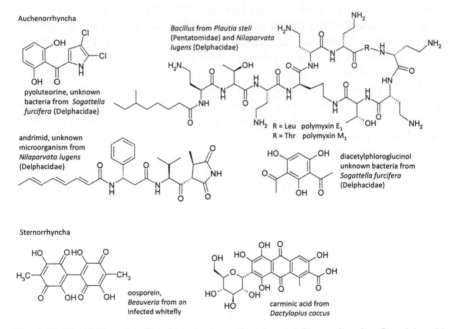

Figure 3. Microbial metabolites from Auchenorrhyncha and Sternorrhyncha. Carminic acid from *Dactylopius coccus* represents a hemolymph toxin.

Sternorrhyncha (plantlice)

From the worldwide known 16,400 species of plantlice various defense strategies are reported (Dixon 1997). Many taxa are known to sequester toxic compounds from plants. Important metabolites are pyrrolizidine alkaloids which may be sequestered by *Ceroplastes* (Coccidae) or *Aphis* (Aphididae) (see Opitz and Müller 2009). Moreover, sequestration of quinolizidine alkaloids was reported from the genera *Macrosiphum* and *Aphis* (Aphididae), whereas *Brevicoryne* and *Lipaphis* (Aphididae) may sequester glucosinolates (Opitz and Müller 2009). Sequestration of both toxicants protected the plantlice from attacks by carabid or ladybird beetles. Several plantlice may also sequester cardiac (*Aphis*) and iridoid glycosides (*Acyrthosiphon*) (Opitz and Müller 2009).

As reported by Eisner (1970, 2003) and Eisner et al. (2005) wax production may be an important defensive mechanism against plantlice predators.

A well-known important defensive mechanism is due to the red dye of various homopteran species. The most important compound is represented by carminic acid (Fig. 3) which may be found together with other compounds in the genera *Kermes*, *Dactylopius*, *Porphyrophora*, *Kerria*, *Lakshadia*, *Acantocuccus*, *Ceroplastes*, *Eriococcus* and *Tachardia*. The biological role of these deterrents has been thoroughly reviewed by Eisner (1970, 2003) and Eisner et al. (2005). Larvae from other predacous insects such as *Hyperaspis* coccinelids, *Leucopis* flies and *Laetilia* pyralid caterpillars ingest carminic acid from their plantlice prey, sequester the compound and emit the deterrent fluid when they are disturbed (Eisner et al. 1994). Chemical data concerning these insect pigments (kermes acid, flavokermes acid, lacain acids A–D, erythrolaccin, desoxyerythrolaccin, isoerythrolaccin, ceroalbinic acid, 7-hydroxyemodin and emodin) are found in Brown (1975) and Schweppe (1993). From *Beauveria* species (Ascomycota) which originated from an infected whitefly there was isolated the red toxic and antifungal dibenzoquinone pigment oosporein (Fig. 3; Eyal et al. 1994, Dettner 2011).

When attacked by predators many aphids release a cornicle secretion. On the one hand these secretions contain alarm pheromones that alert other members of the colony. In addion these cornicle secretion may also threaten an attacker (Moayeri et al. 2014). It was reported by Callow et al. (1973) that cornicle secretion contained triglycerides, especially with hexanoic, sorbic, myristic and palmitic acids. Obviosly the waxy material may be kept in a liquid state by a solvent which evaporates rapidly in the air which would explanate the rapid change of cornicle wax from the liquid to a chrystaline state. Another explanation would be that the liquid wax is in a supercooled state and that foreign material provides a seeding nucleus for the rapid crystallization (Edwards 1966).

Finally it is reported that larvae of genus *Aleurocanthus* posses certain dorsal spines which release through apical orifices sticky, viscous retentive droplets. Carver (1991) suggested that the droplets have a defensive function and also serve for gluing the exuviae of the previous instar to the glandular spines. The semiochemistry of aphids was recently reviewd by Pickett et al. (2013).

Heteroptera (true bugs)

There exist various reviews with respect to exocrinology and especially defensive chemistry of true bugs which comprise 40,000 species worldwide and possess a lot of exocrine structures including secretory hairs (Fig. 1/7). In 1978 (Dazzini Valcurone and Pavan 1978, Weatherston and Percy 1978a) and 1988 (Aldrich 1988) gave a complete survey on glands and chemical defensive systems of Heteroptera. The volatiles (including allomones) and especially pheromones of true bugs were reviewed by Millar (2005). Morphological details and glandular structures such as metathoracic scent glands, abdominal glands, ventral and sternal glands, brindley's glands, accessory glands, pronotal exocrine glands and secretory hairs (Fig. 1/7) were compiled by Staddon (1979) and Aldrich (1988). A survey of the glandular defensive compounds was presented by Aldrich (1988) and Millar (2005). It is interesting to denote that secretions from the brindley gland of Reduviidae have defensive functions (Aldrich 1988, Audino et al. 2007). Moreover there were identified unusual acetogenins which are produced by secretory hairs in larvae and adults of lace bugs (Tingidae) and are directed against bacteria, fungi, nematodes and other predators (see Millar 2005). Appart form Tingidae secretory hairs are also known from larvae of stilt bugs (Berytidae) and assesin bugs (Reduviidae). From the saliva of both selected Reduviidae and Cimicidae there have been isolated several anticoagulant and allergenic proteins (Richard and Ledford 2014). In addition bites of many heteropterans are painful (Alexander 1984, Schaefer and Panizzi 2000).

Heteroptera are also important with respect to sequestration of organic compounds which might be present in both hemolymph and exocrine glands. Several species as *Largus rufipennis* and *Neocoryphus bicrucis* may sequester pyrrolizidine alkaloids (see Opitz and Müller 2009). Moreover, Heteroptera can selectively take up and enrich (Opitz and Müller 2009) tropane alkaloids (*Acanthocoris*), cyanogenic glycosides, cyanolipids (*Leptocoris*, *Jadera*), glucosinolates (*Murgancia*), cardiac glycosides (*Oncopeltus*, *Caenocoris*, *Spilostethus*) and certain toxic phorbol esters (*Pachycoris*). In addition various heteropteran species (Lygaeidae, Miridae, Tingidae; see Hemp and Dettner 2001) pharmacophageously take up the toxic terpene anhydride cantharidin from meloid and oedemerid beetles (Dettner 1997, Hemp and Dettner 2001).

A lot of Heteroptera possess powerful venoms which are directed against both vertebrates (Schmidt 1982) and invertebrates such as insects (Sahayaraj and Vinothkanna 2011). Clinical reactions to bug bites are discussed by Alexander (1984). Further informations are compiled in Schaefer and Panizzi (2000). Several of these enzymes could be characterized chemically (giant water bugs Belostomatidae: Swart et al. 2006; Reduviidae: Sahayaraj and Vinothkanna 2011). Few species are also known to produce allergens (Richard and Ledford 2014).

Several heteropteran species produce unusual defensive compounds. For example representatives of aquatic Belostomatidae contain deoxycorticosterone, pregnenolone, progesterone in their cephalic glands (Lokensgard et al. 1993) and ressemble defensive secretions of dytiscid water beetles (Dettner 2014). There are several reports on unusual interactions between heteropteran defensive secretions and other arthropods. For example (*E*)-2-octenal or (*E*)-2-decenal may attract male crab spiders (Thomisidae) or kleptoparasitic flies (Milichiidae, Chloropidae) (Aldrich and Barros 1995). Furthermore a male *Satyrium*-butterfly was attracted to a female *Banasa*-stink bug (Moskowitz 2002). Finally *Trissolcus*-egg parasitoids (Scelionidae) are attracted to α,β-unsaturated aldehydes from heteropteran defensive glands (Mattiacci et al. 1993).

Heteroptera may also contain interesting microorganisms apart from true symbionts from mycetomes. A *Bacillus* species which was isolated from *Plautia stali* (Pentatomidae) produced the cyclic decapeptide antibiotic polymyxin E_1 (Fig. 3, Kenny et al. 1989).

Neuroptera (netwinged insects)

There exist about 6,000 neuropteran species worldwide (probably 10,000). Chemical defense, although actually poorly understood, seems widespread in all developmental stages of this holometabolous, mainly predatory insect order.

Eggs of many neuropteran species (Berothidae, Mantispidae, Chrysopidae) are usually deposited on the top of a slender stalk or egg pedicel which is formed from a gelatinous fluid from female accessory glands. In certain species such as *Ceraeochrysa smithi* the stalks are coated by droplets of an oily secretion produced by females, which contains oleic acid, butanal, decanal, pentadecanal, and isopropyl myristate (Eisner et al. 1996, Eisner et al. 2005). After hatching from the egg, young *Ceraeochrysa*-larvae avoid body contact with these ant repellents by ingesting the fluid (Eisner et al. 1996).

During prey capture neuropteran larvae (e.g., Osmylidae, Chrysopidae, Ascalaphidae, Hemerobiidae, Myrmeleontidae) inject toxic secretions as regurgitants into prey in order to paralyze and to kill it. In ant lion larvae (genus *Myrmeleon*; Myrmeleontidae) which feed on liquefied internal components of insect prey production of toxins was observed. These are derived from both *Myrmeleon*-larvae and their bacterial symbionts and they contribute to the prey's paralyzation and death. Larval toxins were identified as a larval ALMB-toxin (Yoshida et al. 1999) and as a paralytic 165-167 kDa polypeptide, which was even more active as a paralyzing agent than tetrodotoxin (Matsuda et al. 1995). There were also identified compounds produced by bacterial isolates cultured from larval fore- and midguts of *M. bore*, gut parts which are not connected with the hindgut. *Bacillus*- and *Enterobacter*-species from the insect larvae may produce a paralyzing toxin which is a homolog of GroEL, a protective heat-shock protein known as molecular chaperone (Yoshida et al. 2001). The symbiontic bacteria also synthesize an insecticidal sphingomyelinase C (34 kDa) which probably acts via its phospholipid-degrading activity (Nishiwaki et al. 2004). Finally a *Bacillus sphaericus*-isolate from the crops of *Myrmeleon bore* produced a novel insecticidal pore-forming 53 kDa-toxin, named sphaericolysin (Nishiwaki et al. 2007). The role of two low molecular compounds, two isoindoline alkaloids from *Myrmeleon bore* is unknown (Nakatani et al. 2006). It is known from east Africa that larvae of ant lions (Myrmeleontidae) are preferably collected by young girls, are brought into contact with the girls breasts and depleted secretions together with mechanically stimulations may stimulate the breast growth in adolescent girls (Kutalek and Kassa 2005). Until now it is unknown what kind of constituents in the regurgitant of larvae of ant lions may stimulate breast growth in the young girls (Dettner 2014).

From Australian Chrysopidae-larvae (which are usually camouflaged by cast skins or other materials) there were reported injuries to humans, because the sharp-pointed jaws are capable of piercing human skin (Southcott 1991). Effects for humans are local pain with erythema and a local papule, lasting few hours to a day.

Apart from these toxins, neuropteran larvae may show other means of defense. Chrysopid larvae very often cover themselves with remnants of their insect prey, trichomes of certain plants, lichens or even waxy filaments (Eisner et al. 2005). For locomotion or anchoring, chrysopid larvae of various genera (e.g., *Ceraeochrysa* = *Chrysopa*, *Chrysoperla*, *Eremochrysa*) often use anal droplets of a sticky proteinaceous secretion. In addition after molestations the larvae move their flexible abdomen in any direction and expose a fluid droplet of a defensive secretion to the aggressor (LaMunyon and Adams 1987). The secretion which is probably produced in the Malpighian tubules contains precursors of prepupal silk and is stored in the hindgut. It is extremely repulsive to ants and shows a paralyzation within 15 minutes. It cannot spread and quickly dries at the surface of ants especially the antennae. There the material can be quickly removed by cleansing behavior, because it is highly soluble in the saliva of ants (LaMunyon and Adams 1987). It is interesting to denote that larvae of the genus *Lomamyia* (Berothidae) which live subterraneously and feed on certain termites may release an aerosol from their abdominal tips which paralyzes their termite prey (Johnson and Hagen 1981). Finally larvae of *Mantispa uhleri* (Mantispidae; closely related with Berothidae) which live in spider cocoons and feed on spider eggs use unknown chemicals (aggressive allomones) in order to halt the development of spider eggs (Redborg 1983).

Adults of Neuroptera exhibit various mechanisms of chemical defense. Even the characteristic waxy secretions of adult Coniopterygidae seem to represent a true chemical and mechanical defense. Within genus *Semidalis* it was shown that apart from the eyes the total body was covered with waxy particles consisting of tetracosanoic acid (Nelson et al. 2003). However neuropteran adults often possess exocrine glands located in the pro- and metathorax and in the abdomina, which often are functionally and chemically unknown (Güsten 1996).

The paired prothoracic glands of Chrysopidae open at the frontal margin of the sometimes redly colored prothorax and obviously represent defensive glands (Güsten and Dettner 1992). In many species investigated 1-tridecene (Blum et al. 1973) and (Z)-4-tridecene (Güsten and Dettner 1992) represent the main constituents. Depending on species other compounds of the often noxious and skunklike odor are octanoic acid, skatole (= 3-methyl indole), terpenoids, hydrocarbons and amides (Güsten and Dettner 1992). There were observed no sexual dimorphisms, neither in gland morphology nor in secretion chemistry in 20 species (Güsten 1996). The most primitive species investigated of genus *Ninela* only synthesized the alkene which could indicate a spreading of regurgitants or feces together with the alkene (Güsten and Dettner 1992). As a whole, dendrograms revealing phylogeny of Chrysopidae based on both electrophoretic data and glandular chemistry were very similar (Güsten and Dettner 1992). Repellent effects of a chrysopid secretion were registered not only against mammals but also against ants (Blum et al. 1973). In the meantime it was shown that (Z)-4-tridecene elicited a significant EAD-response (single sensillum recordings) in *Chrysoperla carnea* and an avoidance behavior in predatory ants (Zhu et al. 2000). It was shown that also inodorous species (without skatole) possess the paired prothoracic glands (Güsten and Dettner 1992).

Also in Osmylidae paired eversible prothoracic gland vesicles have been described in both sexes which supports the homology of the structure in both families (Güsten 1996). On molestation of adults of *Osmylus* spec. a disagreeable odor of unknown chemistry is perceivable.

In the meantime there were detected further semiochemicals from *Chrysopa*-species together with EAD-responses. Antennae of both sexes of *Chrysopa oculata* responded to following three compounds from male abdomina: nonanal, nonanoic acid, and (1*R*, 2*S*, 5*R*, 8*R*)-iridodial. Both sexes of this species were especially attracted by (1*R*, 2*S*, 5*R*, 8*R*)-iridodial (Zhang et al. 2004, Chauhan et al. 2007). Even (1*R*, 4a*S*, 7*S*, 7a*R*)-nepetalactol (an aphid sex pheromone component) showed a weak attraction especially to males. Thoracic extracts of both sexes contained the antennal-stimulatory 1-tridecene and the EAD-inactive skatole (Zhang et al. 2004). Male specific epidermal glands in *C. oculata* have been described as potential pheromone glands (Zhang et al. 2004). This corresponds with the data of Güsten (1996) that abdominal glands in seven neuropteran families (including Chrysopidae) confined to males could represent pheromone glands.

In adults of Myrmeleontidae the paired metathoracic gland system was morphologically analyzed in 10 species from 6 tribes (Güsten 1998). It was also described that the setae-bearing knob (pilula axillaries) projecting from the base of male hindwings and fitting into the reservoir opening of the metathoracic gland serves for evaporation of the secretion (Güsten 1996). Besides species with smaller female glands, there are also species with female glands absent, species with glands lacking in both sexes and species with glands equally developed in both sexes. Therefore, it was suggested that the metathoracic glands in Myrmeleontidae were originally defensive glands and later developed separately in both sexes, that means from defensive secretions to species-specific attractants and aphrodisiacs (Güsten 1998).

Megaloptera (alderflies, dobsonflies, fishflies)

Among Megaloptera with their worldwide known 325 species larvae and prepupae of *Neohermes* can vomit a noxious-smelling liquid when disturbed (Smith 1970) and therefore exhibit a very pronounced avoidance reaction on encountering others. Adults, such as females of genus *Corydalus* also can bite (Parfin 1952).

Coleoptera (beetles)

Defensive substances and pheromones of beetles with their worldwide known 355,000 species as largest order of life on Earth have been reviewed recently (Dettner 1987, Francke and Dettner 2005, Laurent et al. 2005). All kind of chemical defenses (Table 1) are realized in Coleoptera, including defecation (Fig. 1/3) and reflex bleeding (hemolymph toxins, Fig. 1/9). In addition several species produce skin eruptions in mammals, which may be due to pederin from the genus *Paederus* (Staphylinidae), cantharidin from Meloidae and Oedemeridae or serrated hairs of larvae from Dermestidae (Alexander 1984). Other results from Coleoptera are mainly from Eisner (1970, 2003), Eisner et al. (2005), Weatherston and Percy (1978b), Blum (1981) or from Hilker and Meiners (2002).

Concerning their pygidial gland chemistry two representatives of the adephagous family Trachypachidae (*Trachypachus gibbsii, T. slevini*) seem closer to Geadephaga (aliphatic constituents: isopropyl methacrylate and –ethacrylate, 2- and 3-hexanone, isovaleric-, isobutyric-, 2-methylbutyric-, methacrylic-, ethacrylic- and nonanoic acids, (Z)-7- and (Z)-9-tricosene, (Z)-7- and (Z)-9-pentacosene, pentacosadiene) than to Hydradephaga (aromatics: 2-phenethylethanol, 2-phenethyl methacrylate- and ethacrylate) (Attygalle et al. 2004). Further data on pygidial glands of Carabidae were presented by Will et al. (2000). In *Chlaenius cordicollis* (Carabidae) the variability of the pygidial gland secretion was assessed zoogeographically with 3-methylphenol as main constituent, together with 2,5-dimethylphenol, 3-ethylphenol, 2,3-dimethylphenol and 3,4-dimethylphenol (Holliday et al. 2012). In addition representatives of Carabidae and Paussidae can eject blistering gland contents from their pygidial glands which may produce skin irritations (Alexander 1984).

Adult whirligig beetles (Gyrinidae) and predaceous diving beetles (Dytiscidae) respectively their defensive secretions with norsesquiterpens (gyrinids) and hormone-like steroids (Dytiscids) are used in order to stimulate breast growth in young girls in East Africa (Kutalek and Kassa 2005). For further discussions see Dettner (1985, 2014). One unusual steroid mirasorvone from the aposematically colored water beetle *Thermonectus marmoratus* is figured in Fig. 5. New data on the volatiles (3-methyl-1-butanal, 2-methyl-1-propanol, 3-methyl-1-butanol, 6-methyl-5-hepten-2-on) of pygidial glands of Gyrinidae were compiled by Ivarsson et al. (1996). The pygidial gland chemistry of Haliplidae with 3-hydroxy-octanoic and decanoic acids, phenylacetic acid, 4-hydroxyphenylacetic acid or phenyllactic acid was reviewed by Dettner and Böhner (2009). A complete compilation of pygidial and prothoracic defensive glands constituents of Dytiscidae with many steroids and other terpenes is presented by Dettner (2014).

The Staphylinidae or rove beetles produce a diverse array of defensive chemicals which was illustrated by Dettner (1987, 1993) and Francke and Dettner (2005). A detailed analysis of anal and oral secretions and head space of *Nicrophorus vespilloides* (carrion beetle) revealed 34 compounds which partly showed antimicrobial activities (Degenkolb et al. 2011). Within representatives of rove beetle subfamily Steninae there were found various new chemical structures from the paired anal defensive glands. Apart from stenusin and norstenusin there were identified three new pyridine alkaloids [especially 3-(2-methyl-1-propenyl)pyridine] and the unusual cicindeloine (Lusebrink et al. 2009, Müller et al. 2012). It is remarkable that the spreading potential and skimming behavior of selected Steninae varies considerably (Lang et al. 2012). In addition there were found intrageneric differences in the four stereoisomers of stenusin (the spreading alkaloid) in *Stenus*-species (Lusebrink et al. 2007). The ratio of the four stereoisomers (2′R,3R)-, (2′S,3R)-, (2′S,3S)-, and (2′R,3S)-stenusine varied depending on species respectively subgenera. The multifunctional role, ecological significance and evolution of the pygidial gland system in Steninae was reviewed by Schierling et al. (2013) and Schierling and Dettner (2013).

Concerning rove beetle genus *Paederus* (Paederinae) the transfer of endosymbionts of genus *Pseudomonas* from females to eggs was reported by Kador et al. (2011). The bacteria which are located in the female accessory glands and produce the cytotoxic polyketide pederin are smeared on the egg surface and must be taken by the hatching larvae. Pederin with the biosynthetically related pederone and pseudopederin are shown in Fig. 4. Many aspects on biological activities and medicinal chemistry of polyketids of pederin/mycalamide family are discussed by Mosey and Floreancig (2012).

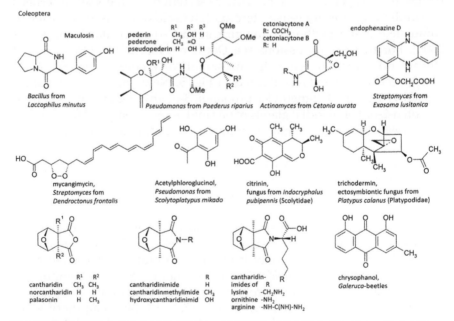

Figure 4. Bacterial and fungal (citrinin, trichodermin) metabolites from Coleoptera. Coleopteran hemolymph toxins are illustrated in the third row.

The semiochemistry of various beetle groups was recently studied in detail. These groups with defensive glands and allomones are Scarabaeoidea (Vuts et al. 2014), Erotylidae (Drilling and Dettner 2010, Drilling et al. 2013) and Coccinellidae (King and Meinwald 1996, Laurent et al. 2005). In Scarabaeidae there were reported repellents such as β-necrodol or butyl sorbate (Vuts et al. 2014). Moreover pygidial gland secretions of neotropical *Canthon*-dung rollers contain compounds such as geraniol, guaiacol or phenol which act as deterrents against predatory *Camponotus*-ants (Cortez et al. 2012). Remarkably *Canthon*-carrion beetles are attracted to defensive volatile secretions of diplopod species such as HCN, benzaldehyde or various benzoquinones (Bedoussac et al. 2007, Dettner 2010). Beetles of the genera *Calopteron* and *Lycus* (Lycidae) are well known to be chemically protected which seems mainly due to the novel octadeca-5E,7E-dien-9-ynoic acid (= lycidic acid). As warning odor they produce 2-methoxy-3-isopropylpyrazine (Eisner et al. 2008). Remarkably certain cerambycid beetle taxa such as *Elytroleptus* mimic these lycid beetles and feed on them (Eisner et al. 2008). Several defensive compounds and natural products of Coccinellidae have characteristic structures. This is illustrated in Fig. 5, where nitrogen containing molecules as epilachnene from *Epilachana varivestis*, psylloborine A from *Psyllobora 22-punctata*, chilocorine B from *Chilocorus cacti*, euphococcinine, a pyrrolidinooxazolidinine and two piperidines from *Epilachna* are seen. From *Harmonia axyridis* there have been reported various inhalant allergens (Richard and Ledford 2014). In addition this neozoic beetle is protected by alkaloids such as

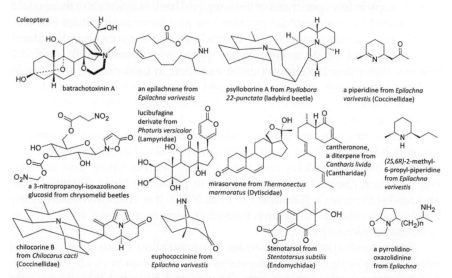

Figure 5. Hemolymph toxins and glandular defensive compounds (mirasorvone, 3-nitropropanoyl-isoxazolinone glucosid) from Coleoptera.

harmonine and (S)-3-hydroxypiperidin-2-one and contains various 2-alkyl-3-methoxypyrazines (Sloggett et al. 2011). Remarkably the harmonine-alkaloid titer per beetle is correlated with the percentage area of orange body color. From a phylogenetically primitive subfamily both the larval and pupal defensive compounds from secretory hairs have been described by Deyrup et al. (2014). Representatives of the genus *Delphastus* produce both isoprenoids such as (E,E)-2,6-diacetoxygermacr-1(10),4-diene and various polyketides such as catalypyrones H-J. Both types of compounds which differ from typical N-containing defensive compounds are biosynthesized endogenously (Deyrup et al. 2014). In *Hippodamia convergens* 2-isobutyl-3-methoxypyrazine which was interpreted as warning odor efficiently acted as aggregation pheromone (Wheeler and Cardé 2013). In *Stenotarsus subtilis*, a representative of handsome fungus beetles (Endomychidae, about 1,300 species) the sister group of Coccinellidae, there could be identified stenotarsol (Fig. 5; Laurent et al. 2005) with a new type of terpene skeleton. Obviously endomychid beetles are characterized by a unique chemical defense which is different from that of Coccinellidae (Laurent et al. 2005).

Larvae of certain fireflies (Lampyridae) are unique in that they are subaquatic and posses fork-shaped glands laterally from meso- and metathorax which can be everted on molestation. Simultaneously there was observed thanatosis, a glowing from paired abdominal glands and a pine oil-like odor which is partly due to the presence of terpinolene and γ-terpinene (Fu et al. 2007, Oba et al. 2011, Oba, this book). By capillary NMR-spectroscopy in few specimens of the lampyrid beetle *Lucidota atra* there could be identified 12 new steroidal pyrones (see also lucibufagine from *Photuris*, Fig. 5) and one compound which instead of a pyrone substituent contains a pentenoic acid amide moiety. As compared with known firefly steroidal pyrones these steroids show a great variation in oxidation of the steroid skeleton and mostly possess a *trans*-fused A-B ring system (Gronquist et al. 2005). Recently the presence of steroidal pyrones could also be confirmed for Eurasian species *Lampyris noctiluca* (Tyler et al. 2008).

Beetles of *Choresine pulchra* (Malachiidae) were shown to contain various batrachotoxins such as batrachotoxin, homobatrachotoxin, batrachotoxin A (and its *cis*-O- and *trans*-O-crotonates, and acetate; Fig. 5) which are probably transferred to passerine birds (*Pitohui*-birds) and render them toxic (Dumbacher et al. 2004). It is unknown whether the beetles biosynthesize their own batrachotoxins or get them from prey or symbiontic microorganisms.

From Australian jewel beetles (Buprestidae) there were isolated bitter tasting acylglycosides buprestin A and B (Brown et al. 1985) with a deterrent activity against ants (Moore and Brown 1985). There was subsequently developed a chemical and enzymatic approach to these buprestins A and B (Schramm et al. 2006). In addition there were identified various buprestins with altered O-6 acyl moieties in *Anthaxia hungarica* and *Chalcophora mariana* by incorporation of hydroxylated cinnamoyl residues (buprestin D, E, F) or substituted benzoyl moieties (buprestin G, G; Ryczek et al. 2009).

In the meantime there was analyzed the transcriptome of odoriferous defensive stink glands of *Tribolium castaneum* (Tenebrionidae; Li et al. 2013). In this pioneering study 29 genes (38%) presented strong visible phenotypes, while 67% genes showed alterations of at least one gland content. Three of this genes which showed quinone-less phenotypes were isolated, molecularly characterized, their expressions identified in both types of the secretory gland cells and their function determined. The authors also showed that quinoic secretions are neccesary to inhibit bacterial or fungal growth in the beetle cultivars. Phylogenetic analyses of this genes indicate that they have evolved independently and specifically for chemical defense in beetles.

The chemical defensive systems of chrysomelid beetles and their developmental stages are fascinating. Several important metabolites are compiled in Pasteels et al. (2004) and Laurent et al. (2005) and Fig. 5 (3-nitropropanoyl-isoxazolinone glucoside), however a thorough view is presented in a separate bookchapter by Burse and Boland (this book).

Recently it was shown that the defensive structures in larvae of Sermylini (Galerucinae) resemble exocrine defensive glands in Chrysomelinae (the putative sister group of Galerucinae) evolved independently (Bünnige and Hilker 2005, Bünnige et al. 2008). In *Agelastica alni*-larvae the sac-like cuticular invaginations near the abdominal spiracles obviously represent defensive structures (Bünnige and Hilker 1999).

When disturbed, larvae of cassidine chrysomelid beetles can aim and wave shields against attackers, which are made of cast skin and feces. In *Chelymorpha alternans* it was shown for the first time, that the chlorophyll catabolite phaeophorbide a exhibits a deterrent acitivity against ants. Shields of one beetle larva contained more than 100 µg of the porphyrinic compound (Vencl et al. 2009). The chemical ecology of longhorned beetles (Cerambycidae) was reviewed by Weatherston and Percy (1978b), Allison et al. (2004) and Francke and Dettner (2005). In addition there were identified various repellents and deterrents (Allison et al. 2004) such as (–)-germacrene D, conophthorin, hexanal, Z-3-hexenol, ethylbenzene, guaiacol, nonylaldehyde, p-dimethoxybenzene, myrcene, (*E*)-β-ocimene, α-cubebene, (*E*)-4,8-dimethylnona-1,3,7-triene, (*E*)-2-hexen-1-ol and (*E*)-2-hexenal.

Sequestration of plant secondary compounds by herbivorous beetles according to Opitz and Müller (2009) seems variable and there is a range from phenolic glycosides (larvae of *Chrysomela*, *Phratora*), prenylated aromatic compounds, cycasin (curculionid beetle *Rhopalotria*), pyrrolizidine alkaloids (chrysomelid beetles: *Platyphora, Longitarsus, Oreina; Chauliognathus* Cantharidae), cardiac glycosides (*Chrysochus*), cucurbitacines (*Diabrotica, Acalymma*), iridoidglycosides (*Dibolia*), to isoprenoids (*Platyphora* spec.). Keefover-Ring (2013) studied the fecal shields of larvae of tortoise species *Physonota* which feed on *Monarda* plants (Lamiaceae). It was evident that those plants subjected to herbivory emitted higher titers of volatiles, which are incorporated into the larval fecal shields of *Physonota* larvae. Attraction of the following species to the toxic animal derived cantharidin (including subsequent sequestration) was reported from Anthicidae (many species from genera *Acanthinus, Anthicus, Aulacoderus, Cordicomus, Cyclodinus, Endomia, Formicilla, Formicomus, Hirticomus, Mecynotarsus, Microhoria, Notoxus, Omonadus, Pseudoleptaleus, Pseudonotoxus, Sapintus, Tenuicomus, Vacusus, Trichananca, Tomoderus*), Endomychidae (*Aphorista, Danae, Lycoperdina, Xenomycetes*), Cleridae (*Cymatodera, Pallenothriocera*), Chrysomelidae (*Aristobrotica, Barombiella*), Pyrochroidae (many species from genera *Anisotria, Pedilus, Neopyrochroa, Pyrochroa, Schizotus*), and Staphylinidae (*Eusphalerum*) (Dettner et al. 1997, Hemp and Dettner 2001).

The polyketides chrysazin, chrysophanol (Fig. 4), dithranol and chrysarobin are found especially in eggs but also in other developmental stages of leaf beetles of Galerucinae. By treatment of the female beetles of *Galeruca tanaceti* with antimicrobial substances it could be shown that freshly laid eggs contained these polyketides, indicating that these compounds are produced by beetle enzymes and not by endosymbiontic microorganisms (Pankewitz et al. 2007a). The bioactive anthraquinone chrysophanol represents the first product of a polyketid synthase that is built up by more than one polyketid folding mode (Bringmann et al. 2006). In actinomycetes the cyclization follows mode S (from *Streptomyces*) in contrast in higher plants, fungi and insects it is formed via folding mode F (referring to fungi). In addition there was observed no increase of polyketid titer in freshly laid eggs as compared with freshly hatched larvae. Instead a significant decrease in total amounts of dithranol and chrysophanol from egg deposition in autumn to spring 5 months later was registered (Pankewitz and Hilker 2006). As already mentioned leaf beetles of the taxon Galerucini transfer antimicrobially active 1,8-dihydroxylated anthraquinones and anthrones into their eggs. It was shown that obligatory, cytoplasmatically inherited α-Proteobacteria, respectively certain genotypes of these bacteria may settle such eggs in spite of the presence of antimicrobics (Pankewitz et al. 2007b). Generally a polyketid synthases (PKS) in insects may have evolved from insect fatty acid synthases (FAS; Pankewitz and Hilker 2008).

Coleoptera host a lot of bacteria and fungi, which may produce interesting natural compounds in the laboratory. However the role of these constituents in their host remains unknown. The phytotoxic and cytotoxic diketopiperazine maculosin (Fig. 4) was isolated from *Bacillus pumilus* and from *Laccophilus minutus*-foregut (Gebhardt et al. 2002). Pederin, pederone, pseudopederin from *Pseudomonas* (Fig. 4) from *Paederus* represent important tumor-inhibiting compounds (see above). Their role in beetles as spider-deterrents was unequivocally proved (Kellner and Dettner 1996). Cetoniacytone A and B (Fig. 4) were released from an *Actinomyces*-species which was isolated from hind guts of *Cetonia aurata* (Schlörke et al. 2002). Both compounds show significant growth inhibition against hepatocellular carcinoma and breast adenocarcinoma. Further research in these *Cetonia*-strains resulted in isolation of a gene cluster responsible for biosynthesis of cetoniacytone A (Wu et al. 2009).

Endophenazines such as endophenazine D (Fig. 4) were produced from *Streptomyces annulatus* which were isolated from *Exosoma*-chrysomelid crops (Gebhardt et al. 2002). These compounds inhibited growth of *Botrytis* and showed herbicidal properties. Cantheronone (Fig. 5) represents an interesting diterpene enone which was isolated from total body extracts of *Cantharis livida* (see Gronquist and Schroeder 2010). Mycangimycin (Bode 2011) (Fig. 4), a polyene peroxide with strong antifungal activity was produced from a *Streptomyces* strain isolated from *Dendroctonus frontalis*. The polyketid metabolite acetylphloroglucinol (Fig. 4) was isolated from *Pseudomonas* from *Sciolytoplatus mikado*.Citrinine (Fig. 4), an antibiotic pentaketide was isolated from an unknown fungus from *Indocryphalus pubipennis* (Scolytidae) (Kenny et al. 1989). Finally trichodermin (Fig. 4) an important inhibitor of protein synthesis is produced by an ectosymbiontic fungus from *Platypus calanus* (Platypodidae) (Kenny et al. 1989).

Meloid and oedemerid beetles possess unusual terpenoid compounds apart from cantharidin (Fig. 4). In addition these Coleoptera may contain palasonin, cantharidinimide, cantharidinmethylimide, hydroxycantharidinimide, and cantharidinimides of lysine, ornithine, and arginine (Fig. 4). All these compounds are very interesting due to their antitumor activities, however, high toxicities due to inhibition of protein phosphatase PP2A prevents their medicinal utilization (Dettner 2011).

Trichoptera (caddisflies)

Chemical defensive systems seem to be present in various taxa of the worldwide known 14,500 species of caddisflies. Many larval caddisflies exhibit case-building from silk glands as a primary defense behavior. In addition when disturbed the aquatic larvae of the genera *Apatania* and *Apataniana* (Limnephilidae) release a defensive fluid from an eversible prothoracic gland which shows a paralyzing effect against small invertebrate predators (e.g., other trichopteran larvae such as *Rhyacophila* sp.) but seems ineffective against larger targets such as plecopteran larvae or fishes (Wagner et al. 1990). The secretion contains C_7 to C_{14}-carboxylic acids with the major constituents 5-octenoic-, octanoic-, decanoic-, 3-dodecenoic-, 3,5-dodecadienoic- and 3,5,7,11-tetradecatetrenoic acids (Wagner et al. 1990).

When adult caddisflies are handled they can cause allergic responses (asthma, dermatitis; Bowles 1992). In addition depending on species they may often emit a distinct odor which can be ascribed as sweet or foul skatolic. A pair of sac-like glands from the 5th abdominal sternum but also glands from the 4th segment represent the source of many volatiles such as p-cresol, indole, skatole (*Pycnopsyche*, Limnephilidae; Duffield et al. 1977), 6-methyl-nonan-3-one (Hesperophylax, Limnephilidae; Bjostadt et al. 1996), 2-heptan-2-one, heptan-2-ol, nonan-2-one, nonan-2-ol (*Rhyacophila*, Rhyacophilidae; Duffield 1981, Löfstedt et al. 1994), hexanoic- and octanoic acids, heptan-2-ol, acetophenone (*Rhyacophila*, Rhyacophilidae; Ansteeg and Dettner 1991, Löfstedt et al. 1994), 2-methylbutanoic acid (*Polycentropus*, Polycentropodidae; Ansteeg and Dettner 1991), 1-pentanol, 1-octanol, 3-methyl-2-heptanone, 2-phenylethanol (*Phryganea*, Phryganeidae; Ansteeg and Dettner 1991), heptane-2-one, (S)-heptan-2-ol, nonan-2-one, nonan-2-one (*Hydropsyche*, Hydropsychidae; Löfsted et al. 1994), (S)-nonan-2-ol (*Molanna*, Molannidae; Löfstedt et al. 2008), (S)-4-methyl-3-heptanone, (4S,6S)-4,6-dimethyl-3-octanone, (4S,6S)-4,6-dimethyl-3-nonanone (*Potamophylax, Glyphotaelius*, Limnephilidae; Bergmann et al. 2001), (1R,3S,5S,7S)-1-ethyl-3,5,7-trimethyl-2,8-dioxabicyclo[3.2.1] octane and 1,3-diethyl-4,6-dimethyl-2,7-dioxabicyclo[2.2.1] heptane (*Potamophylax, Glyphotaelius*, Limnephilidae; Bergmann et al. 2004).

As evidenced by field trapping experiments, in many cases these volatiles such as methylcarbinols and methylketones have pheromonal function and show antennal response in males (Löfstedt et al. 2008). However in other species chemically deviating compounds may represent defensive secretions which are emitted in larger amounts (Ivanov and Melnitsky 1999). Löfstedt et al. (2008) argued that pheromone components such as methylcarbinols and methylketones, which are produced in sternum glands of segment 5 may be a basal character in both Trichoptera and Lepidoptera which constitute the superorder Amphiesmenoptera. For example the short chain alcohols and corresponding methyl ketones identified from caddisflies are similar to the sex pheromone components identified from the archaic moth families Eriocraniidae and Nepticulidae. Several observations indicate that not all abdominal gland secretions in Trichoptera may represent pheromones: The 5th segment glands of certain caddisflies are developed in both sexes (although containing different compounds in males and females; Ansteeg and Dettner 1991). Moreover homologous abdominal glands may be present in both sexes of primitive lepidopteran Micropterigidae, but no pheromone communication at all was shown especially in this group (Kozlov and Zvereva 1999).

Lepidoptera (butterflies and moths)

Butterflies and moths with their 175,000 species worldwide are characterized by various systems to defend themselves chemically. Important reviews covering chemical defenses of all lepidopteran developmental stages are from Pavan and Valcurone Dazzini (1976), Brower (1984), Rothschild (1985) or Hallberg and Poppy (2003). As herbivorous insects they are at first important in sequestration of toxic plant chemicals (Trigo 2000, Nishida 2002, Opitz and Müller 2009). The authors report on natural compounds such as various acids, phenolic glycosides, cycasin, pyrrolizidin alkaloids, quinolizidine alkaloids, tropane alkaloids, amarillidaceae alkaloids, cyanogenic glycosides but also cardiac glycosides, iridoid glycosides or isoprenoids such as *Podocarpus* lactones. Chemical defense mechanisms of lepidopteran species were in detail reviewed by Delgado Quiroz (1978) and Dekker (1984). Valuable informations were also given by Eisner et al. (2005) and Laurent et al. (2005). Just as tenthredinid larvae (Hymenoptera) plant feeding lepidopteran larvae show many mechanisms of chemical defense as was reviewed by Bowers (1993), Hallberg and Poppy (2003), Vegliante and Hasenfuss (2012), Dyer and Smilanich (2012) and Greeney et al. (2012). As an example there are shown ventral prothoracic defensive glands of *Schizura unicornis* caterpillars (Fig. 1/6) which usually produce fromic acid mixed with various lipophiles (Weatherston et al. 1986). Within larvae there were found osmeteria, thoracic and abdominal defensive glands or oral secretions. In osmeteria of *Papilio glaucus* larvae the ecological, ontogenetic and genetic variations in gland chemistry was investigated by Frankfater et al. (2009). There was found a change from terpenoic constituents (directed against ants) to the production of 2-methylbutyric and isobutyric acids (directed against larger predators). Other larvae are characterized by gelatinous coatings which probably acts as deterrents to attack (Epstein et al. 1994) or possess cuticular storage chambers for cyanoglucoside-containing defensive secretions (Franzl and Naumann 1985). In several other families caterpillars are characterized by glandular tracheal nodes (Hings and Byers 1975) or even show sac-like glands (Thyrididae) secreting various deterrents ranging from mandelonitrile to benzaldehyde, benzoic acid, farnesene and 3-methylbutyl-3-methylbutanoate (Darling et al. 2001). Cossidae-larvae of genera *Cossus, Zeuzera* and *Xyleutes* possess large mandibular glands which contain strongly smelling defensive compounds such as dienols, trienols, acetades or 3-hydroxy acids (Pavan and Valcurone Dazzini 1976, Blum 1981). In the genus *Chilecomadia* Reyes-Garcia et al. (2011) identified (Z)-5,13-tetradecadienyl acetate, (Z)-5-tetradecenyl acetate and dodecyl acetate. However the first mentioned acetate seems to represent an aggregation pheromone. Within adult Lepidoptera Hallberg and Poppy (2003) report on tarsal and cervical defensive glands together with ventral hair pencils. In adults of nymphalid butterfly *Agraulis vanillae* there were identified abdominal defensive glands which produce 6-methyl-5-hepten-2-one, 6-methyl-5-hepten-2-ol esters, 1,15-hexadecanediol diacetate and 1,16-hexadecanediol diacetate (Ross et al. 2001). Important aspects of lepidopteran chemical defense are related to urticating hairs or spines of larvae or adults of Zygaenoidea, Bombycoidea, Notodontoidea or Noctuoidea. Various aspects including morphology, chemistry and medicinal aspects were reported by Weidner (1936), Alexander (1984), Deml and Dettner (1993, 1996, 1997, 2003, 2004), Aldrich et al. (1997), Deml (2001, 2003) and Battisti et al. (2011). Larvae of *Pieris rapae* caterpillars possess glandular hairs which bear oily droplets containing unsaturated lipids (mayolenes) which are derived from 11-hydroxylinolenic acid and act as deterrents against ants (Smedley et al. 2002). In some species every year there are numerous accidents in Southamerica because venoms of the *Lonomia* caterpillar cause disseminated and intravascular coagulation and a consumptive coagulopathy which can lead to a hemorrhagic syndrome (Arocha-Piñango et al. 1988, Carrijo-Carvalho and Chudzinski-Tavassi 2007).

Mecoptera (scorpionflies, hanging flies)

When molested representatives of Mecoptera (Scorpion Flies) with about 600 species fall to the ground if they are not camouflaged or show thanatosis (e.g., *Boreus hyemalis*). With their enlarged genital bulb especially males can also defend themselves through biting. Many species emit malodorous digestive fluid from the mouth when they are handled (Kaltenbach 1978).

Siphonaptera (fleas)

The worldwide known 2,000 species are almost all parasitic insects and seem to possess no chemical defensive systems. Salivary allergens of cat flies have been reported as 18 kDa-proteins (Richard and Ledford 2014). Generally fleas can produce skin lesions in humans (flea bites, papular urticaria, tungiasis; see Alexander 1984). Only in the unique endoparasitic larva of *Uropsylla tasmanica* which parasitizes Tiger cats and Tasmanian devils subdermally, there were identified mandibular glands. Their function remains unclear, however it was speculated that the secretion may contain anaesthetics to reduce irritations of the hosts (Williams 1986). Finally, from *Bacillus* species which was isolated from the tissue of adults of *Ceratophyllus* spec. the cyclic peptide and bactericidal bacitracin A (Fig. 6) was isolated (Gebhardt et al. 2002).

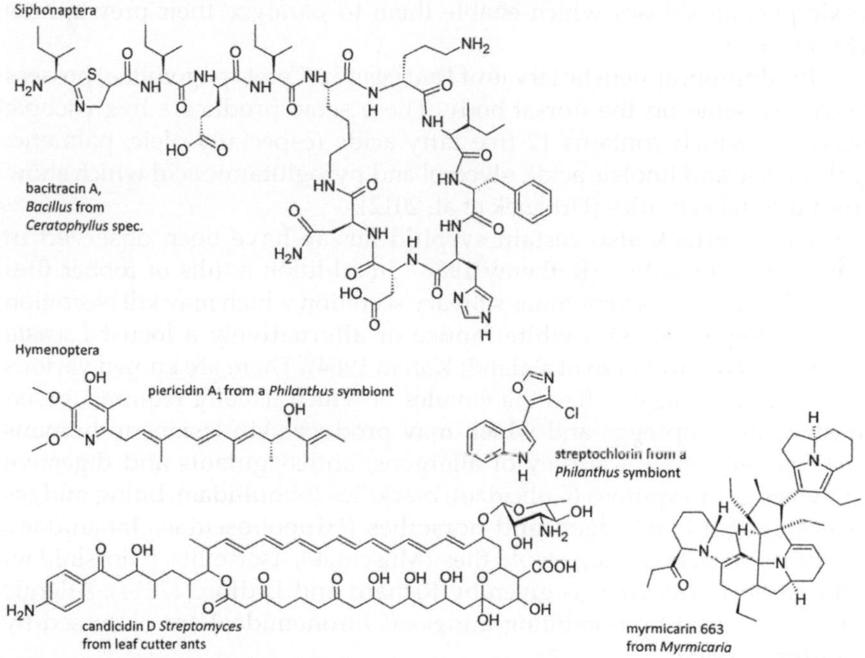

Figure 6. Microbial metabolite from a *Ceratophyllus* flea-species and compounds from Hymenoptera (piericidin A1 and streptochlorin from a *Philanthus* symbiont; candicidin D from a *Streptomyces* symbiont from leaf cutter ants and gland compound myrmicarin 663 from *Myrmicaria*).

Diptera (true flies and midges)

Although there exist a lot of dipteran species worldwide (about 154,000 species), chemical defenses as far as secondary defenses are concerned, seem to be very rare in this large insect order. Further examples of secondary defense include catalepsis, or rolling and dropping from the plant. In contrast primary defenses such as crypsis, translucence, or bird dropping resemblance may be found in various dipteran taxa (e.g., Rotheray 1986). Certainly careful bionomical investigations may reveal further allomones, toxic or venomous secretions especially in larvae of this large insect order.

Biologically active compounds were registered from pupae of the sandfly *Lutzomyia longipalpis* (Psychodidae). Both squashed pupae and hexane extracts of pupae were deterrent to *Lasius*-ants and their larvae (Dougherty and Hamilton 1996), however, chemical structures are not available. From salivary glands of larvae of predatory gall midges (Cecidomyiidae) such as *Aphidoletes aphidimyza* there could be isolated toxic phenoloxidases which enable them to paralyze their prey aphids (Mayr 1975).

In addition, apneustic larvae of *Forcipomyia* (Ceratopogonidae) possess secretory setae on the dorsal body. These setae produce a hygroscopic secretion which contains 12 free fatty acids (especially oleic, palmetic, palmitoleic and linoleic acid), glycerol and pyroglutamic acid which show antibacterial activities (Urbanek et al. 2012).

Under attack also certain syrphid larvae have been observed in emitting sticky saliva (Rotheray 1986). In addition adults of robber flies (Asilidae) possess a venomous salivary secretion which may kill (secretion form 1-2 specimens) a white mouse or alternatively a locust *Locusta migratoria* (1/2 to 1/126 of a gland; Kahan 1964). There are known various other hematophagous flies the females of which usually require a blood meal to develop eggs and which may produce skin lesions in humans (Alexander 1984). A survey of allergens, anticoagulants and digestive enzymes in mosquitoes (Culicidae), blackflies (Simuliidae), biting midges (Ceratopogonidae), deer- and horseflies (Hippoboscidae, Tabanidae), sandflies (Psychodidae), stable flies (Muscidae), Tsetseflies (Glossinidae) and *Stomoxys calcitrans* is given by Richard and Ledford (2014). Allergic diseases caused by non-biting midges (Chironomidae) are discussed by Cranston (1995).

Adults of many Sepsidae (swing flies) possess stink glands in both sexes which are associated with the rectum. It has been described that the aromatic smelling secretions may deter potential swing fly predators (Bristowe 1979). On the other hand, certain Sepsidae show aggregations, which might be evoked by such glandular secretions. Many volatiles are known to be produced in tephritid flies (Fletcher and Kitching 1995) and are usually reported to act as aggregation or sexual pheromones. In rectal ampullae of males of two gall-forming *Urophora*-species there has been identified 4-methyl-3Z,5-hexadienoic acid (Frenzel et al. 1990), a compound which seems to be also involved in the territorial defense between males.

Autogenous chemicals obviously are found in larvae of certain fungus-gnats (Mycetophilidae), which distribute considerable amounts of oxalic acid along the silken threads of their webs. The silk proteins contain large amounts of alanine, glycine, asparagic and glutaminic acid (Plachter 1979; see Neuenfeldt and Scheibel, this book). Droplets of this strong acid originate from larval salivary glands and are deposited on both trapping webs of larvae and pupal cocoons (Plachter 1979). Buston (1933) registered pH 1.8 when he tested the web fluid of *Platyura-* and *Ceroplatus*-species and found it toxic against oligochaet species. Pupal secretion of *Rhynchosciara* was also investigated by Terra and de Bianchi (1974). Oxalic acid may protect both larvae and pupae from aggressors (predators, fungi, bacteria) but also can be used by the larvae in order to trap small prey organisms. The origin of the larval oxalic acid remains unclear. However, it seems possible that the adults my take up free oxalic acid which was excreted by wood-rotting basidiomycets (Munir et al. 2001).

In Diptera there are known several cases of sequestration of defensive chemicals from animal origin. Adults of certain species of Anthomyiidae, Platystomatidae, and especially Ceratopogonidae are attracted to the toxic terpen anhydride cantharidin or naturally occurring cantharidin sources (e.g., dead meloid or oedemerid beetles; canthariphilous insects) and may ingest, detoxify and subsequently sequester this important protein phosphatase inhibitor (Frenzel and Dettner 1994, Dettner 1997).

In tephritid flies (Tephritidae) sequestration of phenylpropanoids such as methyl eugenol but also production of various behavior modifying compounds was reported by Fletcher and Kitching (1995) and Opitz and Müller (2009). Larvae of chamaemyiid flies (Chamaemyiidae) of genus *Leucopis* sequester carminic acid (Fig. 3: 1.12) from cochineal insects which constitute their prey. The larvae eject the compound with their rectal fluid in order to deter ants (Eisner et al. 1994).

It is important to note that bioluminescence for attracting prey and for defense was described in Mycetophilidae as *Arachnocampa, Orfelia* (= *Platyura*) or *Keroplatus* (Viviani et al. 2002, Oba this book). It is interesting, that the two bioluminescent systems in *Arachnocampa* and *Orfelia* are different. In *Orfelia* there was found a 140 kDa luciferase, and a luciferin-luciferase reaction, which is activated by dithiothreitol and ascorbic acid, whereas *Arachnocampa* hosts a 36 kDa luciferase, and is characterized by a system activated by adenosine triphosphate (ATP) (Viviani et al. 2002).

Apart from antimicrobial peptides which are found in the hemolymph of various dipteran larvae (e.g., rat-tailed maggot of *Eristalis tenax*; Vilcinskas 2011), blowfly larvae of *Lucilia seriata* may produce various biologically active constituents which are externalized by defecation or vomiting (Vilcinskas 2011, Joop and Vilcinskas this book). Larvae of various dipterans may produce cutaneous or furuncular myiasis by invasion of human skin (Alexander 1984). Moreover many dipteran species from extreme habitats probably harbour a considerable fraction of gut bacteria with interesting activities and microbial products (Kadavy et al. 2000).

Hymenoptera (sawflies, woodwasps, bees, wasps, ants)

With respect to the large insect order of Hymenoptera with worldwide 132,000 species there exist several reviews which cover defensive mechanisms in general (Bischoff 1927) and especially defensive compounds and toxins. A detailed paper on hymenopteran semiochemicals was given by Keeling et al. (2004). Various venoms of hymenopteran and non-hymenopteran species were reviewed by Schmidt (1982). Recent data also focus on defensive odours and venoms of endoparasitoid wasps (Quicke 1997, Asgari and Rivers 2011) and their role in host-parasite interactions (see Fig. 1/11, e.g., Völkl et al. 1994). There have been identified various low molecular volatile compounds from parasitic wasps such as skatole (Gomez et al. 2005), 6-methylhept-5-en-2-one, 3-hydroxy-3-methylbutan-2-one or alkyl spiroacetals (Davies and Madden 1985). Both gland morphological and chemical details of venoms of Apidae, Sphecidae, Pompilidae, Mutillidae, Bethylidae, Vespidae, Braconidae and Formicidae (see Fig. 1/10) are found in Bettini (1978) and Piek (1986), medicinal and pharmacological aspects were presented by Piek 1986 (Vespidae, Apidae, Formicidae), Levick et al. 2000 and Fitzgerald and Flood (2006). In addition allergenic reactions of humans to hymenopteran stings were discussed by Müller (1988), King and Spangfort (2000) or Klotz et al. (2009). Especially honeybee venoms were studied in detail (e.g., Dotimas and Hider 1987). With respect to allomones of Formicidae and other Hymenoptera recent chemical data are presented by Laurent et al. (2005). Formicidae and their venoms were especially reviewed by Numata and Ibuka (1987), Leclercq et al. (2000) and Hoffman (2010). The chemical ecology of bumblebees was summarized by Ayasse and Jarau (2014). As an example Fig. 6 shows the unique structure of one complex alkaloid myrmicarin 663 which has been isolated together with other air- and temperature-sensitive-alkaloids from the poison gland of a *Myrmicaria* species. Compilations covering exocrine compounds of Hymenoptera (e.g., pheromones, allomones) were given by Wheeler and Duffield (1988).

There exist a lot of data with respect to the herbivorous sawfly larvae (Tenthredinidae) which often feed exophytically on leaves. Consequently they are subject to heavy predation and must possess defensive mechanisms such as aposematic coloration, defensive glands, toxic hemolymph, diverticular sacs associated with the foregut or sequestration of toxins against natural enemies. Very often chemical defensive mechanisms in these larvae are accompanied by a group display and many larvae simultaneously rear their hind body. Herbivorous *Stauronematus* larvae even deposit foamy pales from their salivary glands around their feeding place (Fig. 1/1). The foam, which contains proteins, cholesterol and various fatty acids (Zhao et al. 2009) may effectively deters ants or larvae of armyworm *Pseudaletia separata*. Chemical data with respect to tenthredinid allomones were given by Laurent et al. (2005). Various aspects of biology and defensive behavior of tenthredinid larvae were presented by Schedl (1991), Codella and Raffa (1993), Boevé et al. (2000), Boevé (2006), Petre et al. (2007) and Boevé and Heilporn (2009). It is remarkable that some sawfly larvae are able to deplete small amounts of hemolymph during defense without damaging their integument (Boevé 2009). In addition certain sawfly larvae contain toxic peptides which may cause death of cattle and sheep. For example there could be detected a toxic octapeptide lophyrotomin from sawfly larvae *Lophyrotoma interrupta* (Daly et al. 1993).

Sometimes sequestrations of secondary compounds from plants are also reported from Hymenoptera. The tenthredinid specialist *Rhadinoceraea nodicornis* sequesters ceveratrum alkaloids from *Veratrum album* and may even convert them (see Opitz and Müller 2009). Moreover *Tenthredo grandis* sawfly larvae take up catalpol (iridoid glycoside) from their food. Other species such as *Athalia* spec. sequester neo-clerodane diterpenoids such as clerodendrin D, ajugachin A or athaliadiol from their food plants (see Opitz and Müller 2009) or sequester Furostanol saponins (*Monophadnus*, Prieto et al. 2007). Further tenthredinid species feeding on toxic plants may show reflex bleeding. In addition, hymenopteran parasites maybe attracted to toxic cantharidin (Fig. 4) from meloid and oedemerid beetles (Hemp and Dettner 2001). It is probable that these chemicals which are also present in canthariphilic insects serve as kairomones for these hymenopteran parasitoids. Therefore it seems probable that cantharidin is also sequestered by Braconidae (*Blacus, Periltius, Streplocera, Syrrhizus, Microtonus,* and *Melittobia*) and Diapriidae (gen. spec.; Hemp and Dettner 2001).

Many Hymenoptera are associated with symbiotic bacteria which produce interesting metabolites. One important example is represented by beewolf digger wasps (*Philanthus*) which cultivate specific symbiontic bacteria (*Streptomyces*) that are incorporated into the cocoon as protection against pathogens (Kroiss et al. 2010). It was reported that the bacteria produce a cocktail of nine antibiotics such as piericidin A_1, an inhibitor of NADH dehydrogenase (Fig. 6) or streptochlorin, which is active against several cancer cell lines (Fig. 6). Another chemically and biologically interesting macrolide is represented by candicidin (Fig. 6), which is used medicinally in the treatment of vulvovaginal candidiasis. *Acromyrmex*— leaf cutting ants are living in an obligate symbiosis with *Leucoagaricus*— fungi which they grow with harvested plant material. These symbiontic fungi serve as major food source of the ants. However certain pathogenic fungi such as *Escovopsis* can overcome the symbiotic fungus and therefore may destroy the whole colony. Now it was discovered (Haeder et al. 2009) that the ants protect *Streptomyces*-species which produce antibiotics such as candicin D in order to control these parasites. The role of defensive microbiological symbionts in Hymenoptera was reviewed by Kaltenpoth and Engl (2014).

Acknowledgements

Drawings and figures were kindly produced by E. Helldörfer (Bayreuth). Furthermore the help of S. Wagner (Bayreuth) is highly acknowledged. Finally I thank my colleague Prof. Dr. K.H. Hoffmann for his patience and help in improving this book chapter.

References

Aldrich, J.R. 1988. Chemical ecology of the heteroptera. Annu. Rev. Entomol. 33: 211–238.
Aldrich, J.R. and T.M. Barros. 1995. Chemical atraction of male crab spiders (Araneae, Thomisidae) and kleptoparasitic flies (Diptera, Milichiidae and Chloropidae). J. Arachnol. 23: 212–214.
Aldrich, J.R., P.W. Schaefer, J.E. Oliver, P. Puapoomchareon, C.-J. Lee and R.K. Vander Meer. 1997. Biochemistry of the exocrine secretion from gypsy moth caterpillars (Lepidoptera: Lymantriidae). Ann. Entomol. Soc. Am. 90: 75–82.
Alexander, J.O.D. 1984. Arthropods and Human Skin. Springer, Berlin.
Alla, S., C. Malosse, S. Cassel, P. Rollin and B. Frérot. 2002. La mévalonolactone: un composé volatil produit par *Psammotettix alienus* (Dhb). Compt. Rend. Biol. 325: 941–946.
Allison, J.D., J.H. Borden and S.J. Seybold. 2004. A review of the chemical ecology of the Cerambycidae (Coleoptera). Chemoecol. 14: 123–150.
Ansteeg, O. and K. Dettner. 1991. Chemistry and possible biological significance of secretions from a gland discharging at the 5th abdominal sternit of adult caddisflies (Trichoptera). Entomol. Gener. 15: 303–312.
Arocha-Pinango, C.L., N. Blumenfeld de Bosch, A.L. Nouel, A. Torres, J. Perales, M.E. Alonso, S. de Rodriguez, Z. Carvajal, A. Ojeda, M.L. Tasayco and W. Chitty. 1988. Fibrinolytic and procoagulant agents from a Saturnidae moth caterpillar. pp. 223–240. *In*: H.A.U. Pirkle (ed.). Hemostasis and Animal Venoms. CRC Press, New York.
Asgari, S. and D.B. Rivers. 2011. Venom proteins from endoparasitic wasps and their role in host-parasite interactions. Annu. Rev. Entomol. 56: 313–335.
Attygalle, A.B., X. Wu, J. Ruzicka, S. Rao, S. Garcia, K. Herath, J. Meinwald, D.R. Maddison and K.W. Will. 2004. Defensive chemical of two species of *Trachypachus* Motschulski. J. Chem. Ecol. 30: 577–588.
Audino, P.G., R.A. Alzogaray, C. Vassena, M. Mashuh, A. Fontán, P. Gatti, A. Martinez, F. Camps, A. Cork and E. Zerba. 2007. Volatile compounds secreted by Brindley's glands of adult *Triatoma infestans*: identification and biological activity of previously unidentified compounds. J. Vector Ecology 32: 75–82.
Ayasse, M. and S. Jarau. 2014. Chemical ecology of bumble bees. Annu. Rev. Entomol. 59: 299–319.
Baldwin, I.T., D.B. Dusenbery and T. Eisner. 1990. Squirting and refilling: Dynamics of p-benzoquinone production in defensive gland of *Diploptera punctata*. J. Chem. Ecol. 16: 2823–2834.
Battisti, A., G. Holm, B. Fagrell and S. Larsson. 2011. Urticating hairs in arthropods: Their nature and medical significance. Annu. Rev. Entomol. 56: 203–220.
Beauregard, H. 1890. Les Insectes Vésicants. F. Alcan, Paris.
Bedoussac, L., M.E. Favila and R.M. Lopez. 2007. Defensive volatile secretions of two diplopod species attract the carrion ball roller scarab *Canthon morsei* (Coleoptera: Scarabaeidae). Chemoecol. 17: 163–167.

Beier, M. 1974. Blattariae (Schaben). pp 1–127. In: Handbuch der Zoologie, 2. Aufl., Band 4, 2. Hälfte, 2. Teil, Lieferung 13, de Gruyter, Berlin.

Benfield, E.F. 1974. Authemorrhage in tow stoneflies (Plecoptera) and its effectiveness as a defense mechanism. Ann. Entomol. Soc. Am. 67: 739–742.

Bergmann, J., C. Löfstedt, V.D. Ivanov and W. Francke. 2001. Identification and assigment of the absolute configuration of biologically active methyl-branched ketones from limnephilid caddis flies. Eur. J. Org. Chem. 2001: 3175–3179.

Bergmann, J., C. Löfstedt, V.D. Ivanov and W. Francke. 2004. Identification and synthesis of new bicyclic acetals from caddisflies (Trichoptera). Tetrahedron Lett. 45: 3669–3672.

Berland, L. 1951. Superfamille des Ichneumonoidea. pp. 902–931. In: P.P. Grassé (ed.). Traité de Zoologie 10, Fasc. 2. Masson, Paris.

Bettini, S. 1978. Arthropod Venoms, Handbook of Experimental Pharmacology. Vol. 48. Springer, Berlin.

Betz, O. 2010. Adhesive exocrine glands in insects: Morphology, ultrastructure, and adhesive secretion. pp. 111–152. In: J. von Byern and I. Grunwald (eds.). Biological Adhesive Systems. Springer, Wien.

Beutel, R.G., F. Friedrich, S.Q. Ge and X.K. Yang. 2014. Insect Morphology and Phylogeny. De Gruyter, Berlin.

Bischoff, H. 1927. Biologie der Hymenopteren—Eine Naturgeschichte der Hautflügler. Biologische Studienbücher, Springer, Berlin.

Bitzer, C., G. Brasse, K. Dettner and S. Schulz. 2004. Benzoic acid derivatives in a hypogastrurid collembolan: temperature-dependent formation and biological significance as deterrents. J. Chem. Ecol. 30: 1591–1602.

Bjostadt, L.B., D.K. Jewett and D.L. Brigham. 1996. Sex pheromone of caddisfly *Hesperophylax occidentalis* (Banks) (Trichoptera: Limnephilidae). J. Chem. Ecol. 22: 103–121.

Blagbrough, I.S., P.T.H. Brackley, M. Bruce, B.W. Bycroft, A.J. Mather, S. Millington, H.L. Sudan and P.N.R. Usherwood. 1992. Arthropod toxins as leads for novel insecticides: an assessment of polyamine amides as glutamate antagonists. Toxicon 30: 303–322.

Blum, M.S. 1981. Chemical Defenses of Arthropods. Academic Press, New York.

Blum, M.S. 1991. Chemical ecology of the Thysanoptera. pp. 95–112. In: B.L. Parker, L. Bruce, M. Skinner and T. Lewis (eds.). Towards Understanding Thysanoptera (Gen. Tech. Rep. NE-147). U.S. Department of Agriculture, Forest Service, Northeastern Forest Experiment Station.

Blum, M.S., J.B. Wallace and H.M. Fales. 1973. Skatole and tridecene: Identification and possible role in a chrysopid secretion. Insect Biochem. 3: 353–357.

Blum, M.S., R. Foottit and H.M. Fales. 1992. Defensive chemistry and function of the anal exudate of the thrips *Haplothrips leucanthemi*. Comp. Biochem. Physiol. 102C: 209–211.

Bode, H.P. 2011. Insect-associated microorganisms as a source for novel secondary metabolites with therapeutic potential. pp. 77–93. In: A. Vilcinskas (ed.). Insect Biotechnology, Series: Biologically-Inspired Systems, Vol. 2. Springer, Dordrecht.

Boevé, J.-L. 2006. Differing patterns of chemcially-mediated defence strategies in Nematinae versus Phymatocerini larvae (Hymenoptera: Tenthredinidae). pp. 63–71. In: S.M. Blank, S. Schmidt and A. Taeger (eds.). Recent Sawfly Research: Synthesis and Prospects. Goecke and Evers, Keltern.

Boevé, J.-L. 2009. Easily damaged integument of some sawflies (Hymenoptera) is part of a defence strategy against predators. pp. 31–43. In: S.N. Gorb (ed.). Functional Surfaces in Biology, Vol. 1, Springer, Berlin.

Boevé, J.-L. and S. Heilporn. 2009. Secretion of the ventral glands in *Craesus* sawfly larvae. Biochem. System. Ecol. 36: 836–841.

Boevé, J.-L., S. Heilporn, K. Dettner and W. Francke. 2000. The secretion of the ventral glands in *Cladius*, *Priophorus* and *Trichiocampus* sawfly larvae. Biochem. System. Ecol. 28: 857–864.

Boppré, M., U. Seibt and W. Wickler. 1984. Pharmacophagy in grasshoppers. Entomol. Exp. Appl. 35: 115–117.

Bouchard, P., C.C. Hsiung and V.A. Yaylayan. 1997. Chemical analysis of defense secretions of *Sipyloidea sipylus* and their potential use as repellents against rats. J. Chem. Ecol. 23: 2049– 2057.

Bowers, M.D. 1993. Aposematic caterpillars: Life-styles of the warningly colored and unpalatable. pp. 331–371. *In*: N.E. Stamp and T.M. Casey (eds.). Caterpillars: Ecological and Evolutionary Constraints on Foraging. Chapman, New York.

Bowles, D.E. 1992. The medical importance of caddisflies. Braueria 19: 24.

Bradshaw, J.W.S. 1985. Insect Natural Products - Compounds derived from Acetate, Shikimate and Amino Acids. pp. 655–703. *In*: G.A. Kerkut and L.I. Gilbert (eds.). Comprehensive Insect Physiology, Biochemistry and Pharmacology, Vol. 11, Pergamon Press, Oxford.

Bringmann, G., T.F. Noll, T.A.M. Gulder, M. Grüne, M. Dreyer, C. Wilde, F. Pankewitz, M. Hilker, G.D. Payne, A.L. Jones, M. Goodfellow and H.-P. Fiedler. 2006. Different polyketide folding modes converge to an identical molecular architecture. Nature Chem. Biol. 2: 429–433.

Bristowe, W.S. 1979. The mysterious swarms of sepsid flies and their unpalatability to spiders. Proc. Trans. Brit. Entomol. Nat. Hist. Soc. 12: 16–19.

Brossut, R. 1993. Allomonal secretions in cockroaches. J. Chem. Ecol. 9: 143–158.

Brower, L.P. 1984. Chemical defense in butterflies. pp. 109–134. *In*: R.I. Vane-Wright and P.R.Ackery (eds.). The Biology of Butterfl ies. Academic Press, London.

Brown, K.S. 1975. The chemistry of aphids and scale insects. Chem. Soc. Rev. 4: 263–288.

Brown, W.V., A.J. Jones, M.J. Lacey and B.P. Moore. 1985. Chemistry of buprestins A and B. Bitter principles of jewel beetles (Coleoptera: Buprestidae). Austral. J. Chem. 38: 197–206.

Bünnige, M. and M. Hilker. 1999. Larval exocrine glands in the galerucine *Agelastica alni* L. (Coleoptera: Chrysomelidae): their morphology and possible functions. Chemoecology 9: 55–62.

Bünnige, M. and M. Hilker. 2005. Do "glanduliferous" larvae of Galerucinae (Coleoptera: Chrysomelidae) possess defensive glands? A scanning electron microscopic study. Zoomorphology 124: 111–119.

Bünnige, M., M. Hilker and S. Dobler. 2008. Convergent evolution of chemical defence in Galerucine larvae. Biol. J. Linn. Soc. 93: 165–175.

Buston, H.W. 1933. Note on the chemical nature from the webs of larvae of *Platyura* and *Ceroplatus*. Trans. Roy. Entomol. Soc. Lond. 81: 90–92.

Callow, R.K., A.R. Greenway and D.C. Griffiths. 1973. Chemistry of the secretion from the cornicles of various species of aphids. J. Insect Physiol. 19: 737–748.

Carlberg, U. 1986. Chemical defence in *Sipyloidea sipylus* (Westwood) (Insecta: Phasmida). Zool. Anz. 217: 31–38.

Carrijo-Carvalho, L.C. and A.M. Chudzinski-Tavassi. 2007. The venom of the *Lonomia* caterpillar: An overview. Toxicon 49: 741–757.

Carver, M. 1991. Secretory spines in immatures of *Aleurocanthus* Quaintance and Baker (Hemiptera: Aleyrodidae). J. Austral. Entomol. Soc. 30: 265–266.

Chauhan, K.R., V. Levi, Q.H. Zhang and J.R. Aldrich. 2007. Female goldeneyed lacewings (Neuroptera: Chrysopidae) approach but seldom enter traps baited with the male-produced compound iridodial. J. Econ. Entomol. 100: 1751–1755.

Cherniack, E.P. 2010. Bugs as drugs, Part 1: Insects. The "new" alternative medicine for the 21st century? Altern. Med. Rev. 15: 124–135.

Chow, Y.S. and Y.M. Lin. 1986. Actinidine, a defensive secretion of stick insect, *Megacrania alpheus* Westwood (Orthoptera: Phasmatidae). J. Entomol. Sci. 21: 97–101.

Codella, S.G.J. and K.F. Raffa. 1993. Defense strategies of folivorous sawflies. pp. 261–294. *In*: M.R. Wagner and K.F. Raffa (eds.). Sawfly Life History Adaptations to Woody Plants. Academic Press, San Diego.

Cortez, V., M.E. Favila, J.R. Verdu and A.J. Ortiz. 2012. Behavioral and antenal electrophysiological responses of a predator ant to the pygidial gland secretions of two species of neotropic dung-roller beetles. Chemoecology 22: 29–38.

Cragg, G.M., D.G.I. Kingston and D.J. Newman (eds.). 2012. Anticancer Agents from Natural Products, 2nd Ed. CRC Press, Boca Raton.

Cranston, P.S. 1995. Medical Significance. pp. 365–384. *In*: P. Armitage, P.S. Cranston, L.C.V. Pinder (eds.). The Chironomidae. Chapman and Hall, London.

Daly, N.L., A.R. Atkins and R. Smith. 1993. Solution structure of the toxic octapeptide, lophyrotomin. Int. J. Pept. Protein Res. 42: 366–371.

Dathe, H.H. 2003. Wirbellose Tiere 5. Teil Insecta (2. Aufl.). Lehrbuch der Speziellen Zoologie. Spektrum, Heidelberg.

Darling, D.C., F.C. Schroeder, J. Meinwald, M. Eisner and T. Eisner. 2001. Production of cyanogenic secretion by a thyridid caterpillar (*Calindoea trifascialis*, Thyrididae, Lepidoptera). Naturwissenschaften 88: 306–309.

Davies, N.W. and J.L. Madden. 1985. Mandibular gland secretion of two parasitoid wasps (Hymenoptera: Ichneumonidae), J. Chem. Ecol. 11: 1115–1127.

Dazzini Valcurone, M. and M. Pavan. 1978. Scent glands and defensive secretions of Rhynchota. Pubblicazioni Dell'Istituto Di Entomologia Dell'Universita Di Pavia 5: 1–46.

De Facci, M., H.-L. Wang, J.K. Yuvarai, I.A.N. Dublon, G.P. Svensson, T.W. Chapman and O. Anderbrant. 2014. Chemical composition of anal droplets of the eusocial gall-inducing thrips *Kladothrips intermedius*. Chemoecology 24: 85–94.

Degenkolb, T., R.A. Düring and A. Vilcinskas. 2011. Secondary metabolites released by the burring beetle *Nicrophorus vespilloides*: chemical analyses and possible ecological functions. J. Chem. Ecol. 37: 724–735.

Dekker, M. 1984. Biology and venoms of Lepidoptera. pp. 291–330. *In*: A.T. Tu (ed.). Handbook of Natural Toxins, Vol. 2. M. Dekker, New York.

Del Campo, M.L., J.T. King and M.R. Gronquist. 2011. Defensive and chemical characterization of the froth produced by the cecropid *Aphrophora cribrata*. Chemoecology 21: 1–8.

Delgado Quiroz, A. 1978. Venoms of Lepidoptera. pp. 555–612. *In*: S. Bettini (ed.). Arthropod Venoms. Springer, Berlin.

Deml, R. 2001. Secondary compounds in caterpillars of four moth families (Noctuoidea, Bombycoidea) are partly identical. Nota lepid. 24: 65–76.

Deml, R. 2003. Pyrrolidonyl and Pyridyl Alkaloids in *Lymantria dispar*. Z. Naturforsch. 58c: 860–866.

Deml, R. and K. Dettner. 1993. Biogenic amines and phenolics characterize the defensive secreticon of saturniid caterpillars (Lepidoptera: Saturniidae): a comparative study. J. Comp. Physiol. B 163: 123–132.

Deml, R. and K. Dettner. 1996. "Balloon hairs" of gipsy moth larvae (Lep., Lymantriidae): morphology and comparative chemistry. Comp. Biochem. Physiol. 112B: 673–681.

Deml, R. and K. Dettner. 1997. Chemical defence of emperor moths and tussock moths (Lepidoptera: Saturniidae, Lymantriidae). Entomologia Gener. 21: 225–251.

Deml, R. and K. Dettner. 2003. Comparative morphology and secretion chemistry of the scoli in caterpillars of *Hyalophora cecropia*. Naturwissenschaften 90: 460–463.

Deml, R. and K. Dettner. 2004. Defensive potential of the colourful scoli and haemolymph from caterpillars of *Hyalophora cecropia* (LINNAEUS 1758) (Lepidoptera: Saturniidae). Entomol. Z. 114: 23–26.

Dettner, K. 1985. Ecological and phylogenetic significance of defensive compounds from pygidial glands of Hydradephaga (Coleoptera). Proc. Acad. Natl. Sci. Philadelphia 137: 156–171.

Dettner, K. 1987. Chemosystematics and evolution of beetle chemical defenses. Annu. Rev. Entomol. 32: 17–48.

Dettner, K. 1993. Defensive secretions and exocrine glands in free-living staphylinid beetles —Their bearing on phylogeny (Coleoptera: Staphylinidae). Biochem. System. Ecol. 21: 143–162.

Dettner, K. 1997. Inter- and intraspecific transfer of toxic insect compound cantharidin. pp. 115–145. *In*: K. Dettner, G. Bauer and W. Völkl (eds.). Vertical Food Web Interactions. Springer. Berlin.

Dettner, K. 2010. Chemical defense and toxins of lower terrestrial and freshwater animals. pp. 387–410. *In*: L. Mander and H.W. Lui (eds.). Comprehensive Natural Products II -Chemistry and Biology, Vol. 4. Elsevier, Oxford.

Dettner, K. 2011. Potential pharmaceuticals from insects and their co-occurring microorganisms. pp. 95–119. *In*: A. Vilcinskas (ed.). Insect Biotechnology, Series: Biologically-Inspired Systems, Vol. 2. Springer, Dordrecht.

Dettner, K. 2014. Chemical ecology and biochemistry of Dytiscidae. pp. 235–306. *In*: D.A. Yee (ed.). Ecology, Systematics, and the Natural History of Predaceous Diving Beetles (Coleoptera: Dytiscidae). Springer, Berlin.

Dettner, K. and M. Böhner. 2009. Die Pygidialdrüse der Wassertreter (Coleoptera: Haliplidae): Morphologie, Chemie, Funktion und phylogenetische Bedeutung. Contr. Nat. Hist. 12: 437–460.

Dettner, K., G. Bauer and W. Völkl. 1997. Evolutionary patterns and driving forces in vertical food-web interactions. pp. 337–377. *In*: K. Dettner, G. Bauer and W. Völkl (eds.). Vertical Food Web Interactions. Springer Verlag, Berlin.

Dettner, K., A. Scheuerlein, P. Fabian, S. Schulz and W. Francke. 1996. Chemical defense of giant springtail *Tetrodontophora bielanensis* (WAGA) (Insecta: Collembola). J. Chem. Ecol. 22: 1051–1074.

Deyrup, S.T., L.E. Eckman, E.E. Lucadamo, P.H. McCarthy, J.C. Knapp and J.R. Smedley. 2014. Antipredator activity and endogenous biosynthesis of defensive secretion in larval and pupal *Delphastus catalinae* (Horn) (Coleoptera: Coccinelidae). Chemoecology 24: 145–157.

Dixon, A.F.G. 1997. Aphid Ecology, 2nd edition. Springer, Berlin.

Dossey, A.T. 2010. Insects and their chemical weaponry: New potential for drug discovery. Nat. Prod. Rep. 27: 1737–1757.

Dossey, A.T. 2011. Chemical defenses of insects: A rich resource for chemical biology in the tropics. pp. 27–57. *In*: J.M. Vivanco and T. Weir (eds.). Chemical Biology of the Tropics – An Interdisciplinary Approach. Springer, Heidelberg.

Dossey, A.T., M. Gottardo, J.M. Whitaker, W.R. Roush and A.S. Edison. 2009. Alkyldimethylpyrazines in the defensive spray of *Phyllium westwoodii*: A first for order Phasmatodea. J. Chem. Ecol. 35: 861–870.

Dossey, A.T., J.M. Whitaker, M.C.A. Dancel, R.K. Vander Meer, U.R. Bernier, M. Gottardo and W.R. Roush. 2012. Defensive spiroketals from *Asceles glaber* (Phasmatodea): absolute configuration and effects on ants and mosquitoes. J. Chem. Ecol. 38: 1105–1115.

Dotimas, E.M. and R.C. Hider. 1987. Honeybee venom. Bee World 68: 51–70.

Dougherty, M.J. and J.G.C. Hamilton. 1996. A biologically-active compound from pupae of the sandfly *Lutzomyia longipalpis* (Diptera: Psychodidae) and its possible role in defense. Bull. Entomol. Res. 86: 11–16.

Dowd, P.F. 1992. Insect fungal symbionts: a promising source of detoxifying enzymes. J. Industr. Microbiol. 9: 149–161.

Drilling, K. and K. Dettner. 2010. First insights into the chemical defensive system of the erotylid beetle, *Tritoma bipustulata*. Chemoecology 20: 243–253.

Drilling, K., K. Dettner and K.-D. Klass. 2013. The distribution and evolution of exocrine compound glands in Erotylinae (Insecta: Coleoptera: Erotylidae). Ann. Soc. Entomol. France (N.S.) 49: 36–52.

Duffield, R.M. 1981. 2-Nonanol in the exocrine secretion of the Nearctic caddisfly *Rhyacophila fuscula* (Walker) (Rhyacophilidae: Trichoptera). Proc. Entomol. Soc. Wash. 83: 60–63.

Duffield, R.M., M.S. Blum, J.B. Wallace, H.-A. Lloyd and F.-E. Regnier. 1977. Chemistry of the defensive secretion of the caddisfly *Pycnopsyche scabripennis* (Trichoptera: Limnephilidae). J. Chem. Ecol. 3: 649–656.

Dumbacher, J.P., A. Wako, S.R. Derrickson, A. Samuelson, T.F. Spande and J.W. Daly. 2004. Melyrid beetles (*Choresine*): A putative source for the batrachotoxin alkaloids found in poison-dart frogs and toxic passerine birds. Proc.Natl. Acad. Sci. USA 101: 15857–15860.

Dyer, H.F.G.L.A. and A.M. Smilanich. 2012. Feeding by lepidopteran larvae is dangerous: A review of caterpillars' chemical, physiological, and behavioral defenses against natural enemies. Invertebr. Survival J. 9: 7–34.

Edwards, J.S. 1966. Defense by smear: supercoiling in the cornicle wax of Aphids. Nature 5044: 73–74.

Ehmke, A., P. Proksch, L. Witte, T. Hartmann and M.B. Isman. 1989. Fate of ingested pyrrolizidine alkaloid N-oxide in the grasshopper *Melanoplus sanguinipes*. Naturwissenschaften 76: 27–29.

Eisner, T. 1960. Defense mechanisms of arthropods. II. The chemical and mechanical weapons of an earwig. Psyche 67: 62–70.

Eisner, T. 1970. Chemical defense against predation in arthropods. pp. 157–217. *In*: E. Sondheimer and J.B. Simeone (eds.). Chemical Ecology. Academic Press, New York.

Eisner, T. 2003. For Love of Insects. Belknap Press of Harvard University Press, Cambridge.

Eisner, T., R. Ziegler, L. McCormick, M. Eisner, E.R. Hoebeke and J. Meinwald. 1994. Defensive use of an acquired substance (carminic acid) by predaceous insect larvae. Experientia 50: 610–615.

Eisner, T., A.B. Attygalle, W.E. Conner, M. Eisner, E. MacLeod and J. Meinwald. 1996. Chemical egg defense in a green lacewing (*Ceraeochrysa smithi*). Proc. Natl. Acad. Sci. USA 93: 3280–3283.

Eisner, T., M. Eisner and M. Siegler. 2005. Secret Weapons. Belknap Press, Cambridge.

Eisner, T., R.C. Morgan, A.B. Attygalle, S.R. Smedley, K.B. Herath and J. Meinwald. 1997. Defensive production of quinoline by a phasmid insect (*Oreophoetes peruana*). J. Exp. Biol. 200: 2493–2500.

Eisner, T., C. Rossini and M. Eisner. 2000. Chemical defense of an earwig (*Doru taeniatum*). Chemoecology 10: 8–87.

Eisner, T., F.C. Schroeder, N. Snyder, J.B. Grant, D.J. Aneshansley, D. Utterback, J. Meinwald and M. Eisner. 2008. Defensive chemistry of lycid beetles and of mimetic carambycid beetles that feed on them. Chemoecology 18: 109–119.

Epstein, M., S.R. Smedley and T. Eisner. 1994. Sticky integumental coating of a dalcerid caterpillar: a deterrent to ants. J. Lepidopt. Soc. 48: 381–386.

Evans, D.L. and J.O. Schmidt. 1990. Insect Defenses. State University of New York Press, Albany.

Eyal, J., A. Mabud, K.L. Fischbein, J.F. Walter, L.S. Osborne and Z. Landa. 1994. Assessment of *Beauveria bassiana* Nov. EO-1 strain, which produces a red pigment for microbial control. Appl. Biochem. Biotechnol. 44: 65–80.

Farine, J.-P., C. Everaerts, J.-L. Le Quere, E. Semon, R. Henry and R. Brossut. 1997. The defensive secretion of *Eurycotis floridana* (Dictyoptera, Blattidae, Polyzosteriinae): Chemical identification and evidence of an alarm function. Insect Biochem. Mol. Biol. 27: 577–586.

Farine, J.P., C. Everaets, D. Abed and R. Brossut. 2000. Production, regeneration and biochemical precursors of the major components of the defensive secretion of *Eurycotis floridana* (Dictyoptera, Polyzosteriinae). Insect Biochem. Mol. Biol. 30: 601–608.

Farine, J.-P., E. Semon, C. Everaerts, D. Abed, P. Grandcolas and R. Brossut. 2002. Defensive secretion of *Therea petiveriana*: chemical identification and evidence of an alarm function. J. Chem. Ecol. 28: 1629–1640.

Feng, Y., M. Jianqi, L. Zhongren and G. Tianpeng. 1988. A preliminary investigation on the cantharidin resources of Shaanxi province. Acta Univ. Sept. Occid. Agric. 16: 23–28.

Fitzgerald, K.T. and A.A. Flood. 2006. Hymenopteran stings. Clin. Tech. Small Animal Practice 21: 194–204.

Fletcher, M.T. and W. Kitching. 1995. Chemistry of fruit flies. Chem. Rev. 95: 789–828.

Francke, W. and K. Dettner. 2005. Chemical signalling in beetles. Top. Curr. Chem. 240: 85–166.

Francois, J. and R. Dallai. 1986. Les glandes abdominals des Protures. pp. 273–280. *In*: R. Dallai (ed.). 2nd Internat. Seminar on Apterygota. Siena.

Frankfater, C., M.R. Tellez and M. Slattery. 2009. The scent of alarm: ontogenetic and genetic variation in the osmeterial chemistry of *Papilio glaucus* (Papilionidae) caterpillars. Chemoecology 19: 81–96.

Franzl, S. and C.M. Naumann. 1985. Cuticular cavities: Storage chambers for cyanoglucoside-containing defensive secretions in larvae of a zygaenid moth. Tissue Cell 17: 267–278.

Frenzel, M., K. Dettner, W. Boland and P. Erbes. 1990. Identification and biological significance of 4-metyl-3Z,5-hexadienoic acid produced by males of the gall-forming tephritids *Urophora cardui* (L.) and *Urophora stylata* (Fab.) (Diptera: Tephritidae). Experientia 46: 542–547.

Frenzel, M. and K. Dettner. 1994. Quantitation of cantharidin in canthariphilic ceratopogonids (Diptera: Ceratopogonidae), Anthomyiids (Diptera) and cantharidin-producing Oedemerids (Coleoptera). J. Chem. Ecol. 20: 1795–1812.

Fu, X., F.V. Vencl, O. Nobuyoshi, V.B. Meyer-Rochow, C. Lei and Z. Zhang. 2007. Structure and function of the eversible glands of the aquatic firefly *Luciola leii* (Coleoptera: Lampyridae). Chemoecology 17: 117–124.

Gebhardt, K., J. Schimana, P. Krastel, K. Dettner, J. Rheinheimer, A. Zeeck and H.-P. Fiedler. 2002. Endophenazines A–D, new phenazine antibiotics from the arthropod associated endosymbiont *Streptomyces anulatus*. I. Taxonomiy, fermentation, isolation and biological activities. J. Antibiotics 55: 794–800.

Gomez, J., J.F. Barrera, J.C. Rojas, J. Macias-Samano, J.P. Liedo, L. Cruz-Lopez and M.H. Badii. 2005. Volatile compounds released by disturbed females of *Cephalonomia stephanoderis* (Hymenoptera: Bethylidae): A parasitoid of the coffee berry borer *Hypothenemus hampei* (Coleoptera: Scolytidae). Florida Entomol. 88: 180–187.

Greeney, H.F., L.A. Dyer and A.M. Smilanich. 2012. Feeding by lepidopteran larvae is dangerous: A review of caterpillars' chemical, physiological, morphological and behavioral defenses against natura enemies. Invertebr. Survival J. 9: 7–34.

Gronquist, M., J. Meinwald, T. Eisner and F. Schroeder. 2005. Exploring uncharted terrain in nature's structure space using capillary NMR spectroscopy: 13 steroids from 50 fireflies. J. Am. Chem. Soc. 127: 10810–10811.

Gronquist, M. and F.C. Schroeder. 2010. Insect natural products. pp. 67–107. *In*: L. Mander and H.W. Liu (eds.). Comprehensive Natural Products II Chemistry Biology, Vol. 2. Elsevier, Amsterdam.

Grzeschik, K.H. 1969. On the systematics, biololgy and ethology of *Eugaster* SERVILLE (Orthoptera, Tettigoniidae). Forma et Functio 1: 111–144.

Güsten, R. 1996. A review of epidermal glands in the order Neuroptera (Insecta). pp. 129–146. *In*: M. Canard, H. Aspöck and M.W. Mansell (eds.). Pure and Applied Research in Neuropterology. Proc. 5th Int. Symp. Neuropterol. Cairo, Egypt, 1994.

Güsten, R. 1998. The morphology of the metathoracic gland system in the Myrmeleontidae (Neuroptera): a preliminary overview. Acta Zool. Fennica 209: 121–127.

Güsten, R. and K. Dettner. 1992. The prothoracic gland of the Chrysopidae (Neuropteroidea: Planipennia). Proc. 4. Europ. Congr. Entomol., XIII. SIEEC, Gödöllö 60–65.

Guilford, T., C. Nicol, M. Rothschild and B. Moore. 1987. The biological roles of pyrazines: evidence for a warning odour function. Biol. J. Linn. Soc. 31: 113–128.

Haeder, S., R. Wirth, H. Herz and D. Spiteller. 2009. Candicidin-producing *Streptomyces* support leaf-cutting ants to protect their fungus garden against the pathogenic fungus *Escovopsis*. Proc. Natl. Acad. Sci. USA 106: 4742–4746.

Haga, K., T. Suzuki, S. Kodama and Y. Kuwahara. 1989. Secretion of thrips, III. Secretion of acid-emitting thrips, *Holothrips japonicus* (Thysanoptera: Phlaeothripidae). Appl. Entomol. Zool. 24: 242–244.

Hallberg, E. and G. Poppy. 2003. Exocrine Glands: Chemical communication and chemical defense. pp. 361–375. *In*: Handbook of Zoology, Lepidopetera Vol. II, De Gruyter, Berlin.

Hansen, J.A., E.C. Bernard and J.K. Moulton. 2010. A defensive behaviour of *Acerentulus confinis* (Berlese) (Protura, Acerentomidae). Proc. Entomol. Soc. Wash. 112: 43–46.

Hatle, J.D. and S.G. Faragher. 1998. Slow movement increases the survivorship of a chemically defended grasshopper in predatory encounters. Oecologia 115: 260–267.

Helson, J.E., L.C. Todd, T. Johns, A. Aiello and D.M. Windsor. 2009. Ecological and evolutionary bioprospecting: using aposematic insects as guides to rainforest plants active against disease. Front. Ecol. Environm. 7: 130–134.

Hemp, C. and K. Dettner. 2001. Compilation of canthariphilous insects. Beitr. Entomol. 51: 231–245.

Hermann, H.R. 1984. Defensive Mechanisms in Social Insects. Praeger, New York.

Hilker, M. and T. Meiners (eds.). 2002. Chemoecology of Insect Eggs and Egg Deposition. Blackwell, Berlin.

Hings, C.F. and J.R. Byers. 1975. A New glandular organ in some toxic caterpillars. Experientia 31: 965–967.

Hinton, H.E. 1981. Biology of Insect Eggs, Vol. 1–3, Pergamon, Oxford.

Hoffman, D.R. 2010. Ant venoms. Curr. Opin. Allergy Clin. Immunol. 10: 1–5.

Holliday, A.E., N.J. Holliday, T.M. Mattingly and K.N. Naccarato. 2012. Defensive secretion of the carabid beetle *Chlaenius cordicollis*: chemical components and their geographic patterns of variation. J. Chem. Ecol. 38: 278–286.

Houston, T.F. 2007. Oberservations on the biology and immature stages of the sandgroper *Cylindraustralia kochii* (Saussure), with notes on some congeners (Orthoptera: Cylindrauchetidae). Rec. West. Austral. Museum 23: 219–234.

Howard R.W. and J.C. Lord. 2003. Cuticular lipids of the stored food pest, *Liposcelis bostrychophila*: hydrocarbons, aldehydes, fatty acids and fatty amides. J. Chem. Ecol. 29: 597–609.

Howard, D.F., M.S. Blum and H.M. Fales. 1983. Defense in thrips: forbidden fruitiness of a lactone. Science 220: 335–336.

Howard, D.F., M.S. Blum, T.H. Jones, H.M. Fales and M.D. Tomalski. 1987. Defensive function and chemistry of the anal exudate of the Cuban laurel thrips *Gynaikothrips ficorum* (MARCHAL). Phytophaga 1: 163–170.

Howse, P.E. 1984. Sociochemicals of termites. pp. 475–519. *In*: W.J. Bell and R.T. Cardé (eds.) Chemical Ecology of Insects. Chapman and Hall, London.

Ivanov, V.D. and S.I. Melnitsky. 1999. The structure of the sternal pheromone glands in caddis flies (Trichoptera). Entomologicheskoe Obozrenie 78: 505–526.

Ivarsson, P., B.-I. Henrikson and J.A.E. Stenson. 1996. Volatile substances in the pygidial secretion of gyrinid beetles (Coleoptera: Gyrinidae). Chemoecology 7: 191–193.

Janetschek, H. 1970. Protura (Beintastler). pp. 1–72. *In*: J.G. Helmcke, D. Stark and H. Wermuth (eds.). Handbuch der Zoologie, Vol. 4. De Gruyter, Berlin.

Johnson, J.B. and K.S. Hagen. 1981. A neuropteran larva uses an allomone to attack termites. Nature 289: 506–507.

Juanjie, T., Z. Jouwei, W. Shuyong, D. Zengji and Z. Xhuanxian. 1995. Investigation on the natural resources und utilization of the Chinese medicinal beetles—Meloidae. Acta Entomol. Sin. 38: 324–331.

Kadavy, D.R., J.M. Hornby, T. Haverkost and K.W. Nickerson. 2000. Natural antibiotic resistance of bacteria isolated from larvae of the oil fly *Helaeomyia petrolei*. Appl. Environm. Microbiol. 4615–4619.

Kador, M., M.A. Horn and K. Dettner. 2011. Novel oligonucleotide probes for *in situ* detection of pederin-producing endosymbionts of *Paederus riparius* rove beetles (Coleoptera: Staphylinidae). FEMS Microbiol. Lett. 319: 73–81.

Kahan, D. 1964. The toxic effect of the bite and the proteolytic activity of the saliva and stomach contets of the robber flies (Diptera Asilidae). Israel J. Zool. 13: 47–57.

Kaltenbach, A. 1978. Mecoptera (Schnabelhafte, Schnabelfliegen). pp. 1–111. *In*: J.-G. Helmcke, D. Starck and H. Wermuth (eds.). Handbuch der Zoologie, IV. Band, 2. Hälfte (Insecta) (M. Beier, ed.) 2. Teil Spezielles, Band 28. De Gruyter, Berlin.

Kaltenpoth, M. and T. Engl. 2014. Defensive microbial symbionts in Hymenoptera. Functional Ecol. 28: 315–327.

Kastin, A. 2013. Handbook of Biologically Active Peptides, 2nd edition. Elsevier, San Diego.

Keefover-Ring, K. 2013. Making scents of defense: do fecal shields and herbivore-caused volatiles match host plant chemical profiles? Chemoecology 23: 1–11.

Keeling, C.I., E. Plettner and K.N. Slessor. 2004. Hymenopteran semiochemicals. pp. 133–177. *In*: S. Schulz (ed.). The Chemistry of Pheromones and other Semiochemicals, Vol. I. Springer, Berlin.

Kellner, R.L.L. and K. Dettner. 1996. Differential efficacy of toxic pederin in deterring potential arthropod predators of *Paederus* offspring (Coleoptera: Staphylinidae). Oecologia 107: 293–300.

Kenny, P.T.M., S.Y. Tamura, A. Fredenhagen, Y. Naya, K. Nakanishi, K. Nishiyama, M. Suguira, H. Kita and H. Komura. 1989. Symbiotic microorganisms of insects: A potential new source for biologically active substances. Pest. Sci. 27: 117–131.

King, A.G. and J. Meinwald. 1996. Review of the defensive chemistry of coccinellids. Chem. Rev. 96: 1105–1122.

King, T.P. and M.D. Spangfort. 2000. Structure and biology of stinging insect venom allergens. Int. Arch. Allergy Immunol. 123: 99–106.

Klotz, J.H., S. Klotz and J.L. Pinnas. 2009. Animal bites and stings with anaphylactic potential. J. Emerg. Med. 36: 148–156.

Kozlov, M.V. and E.L. Zvereva. 1999. A failed attempt to demonstrate pheromone communication in archaic moths of the genus *Sabatinca* Walker (Lepidoptera, Micropterigidae). Ecol. Lett. 2: 215–218.

Kroiss, J., M. Kaltenpoth, B. Schneider, M.-G. Schwinger, C. Hertweck, R.K. Maddula, E. Strohm and A. Svatoš. 2010. Symbiotic streptomycetes provide antibiotic combination prophylaxis for wasp offspring. Nature Chem. Biol. 6: 261–263.

Kutalek, R. and A. Kassa. 2005. The use of Gyrinids and Dytiscids for stimulating breast growth in East Africa. J. Ethnobiol. 25: 115–128.

LaMunyon, C.W. and P.A. Adams. 1987. Use and effect of an anal defensive secretion in larval Chrysopidae (Neuroptera). Ann. Entomol. Soc. Am. 80: 804–808.

Lang, C., K. Seifert and K. Dettner. 2012. Skimming behaviour and spreading potential of *Stenus* species and *Dianous coerulescens* (Coleoptera: Staphylinidae). Naturwissenschaften 99: 937–947.

Laurent, P., J.C. Braekman and D. Daloze. 2005. Insect chemical defense. Top. Curr. Chem. 240: 167–230.

Leclercq, S., J.C. Braekman, D. Daloze and J.M. Pasteels. 2000. The defensive chemistry of ants. Progr. Chem. Org. Nat. Prod. 79: 115–229.

Lehane, M.J. 2005. The Biology of Blood-sucking in Insects. 2nd edition, Cambridge Univ. Press, Cambridge.

Levick, N.R., J.O. Schmidt, J. Harrison, G.S. Smith and K.D. Winkel. 2000. Review of bee and wasp sting injuries in Australia and the USA. pp. 437–447. *In*: A.D. Austin, M. Dowton (eds.). Hymenopetra, Evolution, Biodiversity and Biological Control. CSIRO-Publishing, Melbourne.

Li, J., S. Lehmann, B. Weißbecker, I.O. Naharros, S. Schütz, G. Joop and E.A. Wimmer. 2013. Odoriferous defensive stink gland transcriptome to identify novel genes necessary for quinone synthesis in the red flour beetle *Tribolium castaneum*. Plos Genetics 9: 1–18.

Lindström, L., C. Rowe and T. Guilford. 2001. Pyrazine odour makes visually conspicuous prey aversive. Proc. Roy. Soc. Lond. Serie B 268: 159–162.

Löfstedt, C., B.S. Hansson, E. Petterson, P. Valeur and A. Richards. 1994. Pheromonal secretions from glands on the 5th abdominal sternite of hydropsychid and rhyacophilid caddisflies (Trichoptera). J. Chem. Ecol. 20: 153–170.

Löfstedt, C., J. Bergmann, W. Francke, E. Jirle, B.S. Hansson and V.D. Ivanov. 2008. Identification of a sexpheromone produced by sternal glands in females of the caddisfliy *Molanna angustata* Curtis. J. Chem. Ecol. 34: 220–228.

Lokensgard, J., R.L. Smith, T. Eisner and J. Meinwald. 1993. Pregnanes from defensive glands of a belostomatid bug. Experientia 49: 175–176.

Lokeshwari, R.K. and T. Shantibala. 2010. A review on the fascinating world of insect resources: Reasons for thoughts. Psyche 1–11.

Lusebrink, I., D. Burkhardt, T. Gedig, K. Dettner, A. Mosandl and K. Seifert. 2007. Intrageneric differences in the four stereoisomers of stenusine in the rove beetles genus, *Stenus* (Coleoptera, Staphylinidae). Naturwissenschaften 94: 143–147.

Lusebrink, I., K. Dettner, A. Schierling, T. Müller, C. Daolio, B. Schneider, J. Schmidt and K. Seifert. 2009. Newpyridine alkaloids from rove beetles from the Genus *Stenus* (Coleoptera: Staphylinidae). Zeitschr. Naturforsch. 64c: 271–278.

Marcussen, B.M., J.A. Axelsen and S. Toft. 1999. The value of two Collembola species as food for a linyphiid spider. Entomol. Exp. Appl. 92: 29–36.

Matsuda, K., H. Suzuki, F. Nakanishi, K. Shio, K. Komai and K. Nishimura. 1995. Purification and characterization of a paralytic polypeptide from larvae of *Myrmeleon bore*. Biochem. Biophys. Res. Commun. 215: 167–171.

Mattiacci, L., S.B. Vinson, H.J. Williams, J.R. Aldrich and F. Bin. 1993. A long-range attractant kairomone for egg parasitoid *Trissolcus basalis*, isolated from defensive secretion of its host, *Nezara viridula*. J. Chem. Ecol. 19: 1167–1181.

Mayr, L. 1975. Untersuchungen zur Funktion der Speicheldrüsen räuberischer Gallmückenlarven (*Aphidoletes aphidimyza* Rond.) Z. ang. Ent. 77: 270–273.

McLeod, M.P., A.T. Dossey and M.K. Ahmed. 2007. Application of attenuated total reflection infrared spectroscopy in the study of *Peruphasma schultei* defensive secretion. Spectroscopy 21: 169–176.

Mebs, D. 2010. Gifttiere, 3. Ed. Wissenschaftliche Verlagsgesellschaft, Stuttgart.

Meinwald, J., G.M. Happ, J. Labours and T. Eisner. 1966. Cyclopentanoid terpene biosynthesis in a phasmid insect and in catmint. Science 151: 79–80.

Meixner, J. 1935. Achte Überordnung der Pterygogenea: Coleopteroidea 1037–1382, Textfig. pp. 1157–1423. *In*: Th. Krumbach (Herausgeber). Handbuch der Zoologie, gegr. v. W. Kükenthal , 4. Band, 2. Hälfte, 1. Teil, Insecta 2. Walter de Gruyter & Co., Berlin.

Mello, M.L.S., E.R. Pimentel, A.T. Yamada and A. Storopoli-Neto. 1987. Composition and structure of the froth of the spittlebug, *Deoiss*p. Insect Biochem. 17: 49–502.

Messer, C., K. Dettner, S. Schulz and W. Francke. 1999. Phenolic compounds in *Neanura muscorum* (Collembola, Neanuridae) and the role of 1,3-dimethoxybenzene as an alarm substance. Pedobiologia 43: 174–182.

Messer, C., J. Walther, K. Dettner and S. Schulz. 2000. Chemical deterrents in podurid Collembola. Pedobiologia 44: 210–220.

Millar, J.G. 2005. Pheromones of true bugs. pp. 37–84. *In*: S. Schulz (ed.). The Chemistry of Pheromones and Other Semiochemicals, Vol. II. Springer, Berlin.

Moayeri, H.R.S., A. Rasekh and A. Enkegaard. 2014. Influence of cornicle droplet secretions of the cabbage aphid, *Brevicoryne brassicae*, on parasitism behavior of naïve and experienced *Diaeretiella rapae*. Insect Sci. 21: 56–64.

Moore, B.P. and W.V. Brown. 1985. The buprestins: bitter principles of jewel beetles (Coleoptera: Buprestidae). J. Austral. Entomol. Soc. 24: 81–85.

Moore, K.A. and D.D. Williams. 1990. Novel strategies in the complex defensive repertoire of a stonefly (*Pteronarcys dorsata*) nymph. Oikos 57: 49–56.

Morgan, E.D. 2004. Biosynthesis in Insects. Royal Society of Chemistry, Cambridge.

Moritz, G. 2006. Thripse. pp. 1–384. Westarp, Hohenwarsleben.

Moriya, N. 1989. Morphology and histology of the scent glands of the pigmy mole cricket *Tridactylus japonicus* DE HAAN (Orthoptera: Tridactylidae). Appl. Entomol. Zool. 24: 161–168.

Moriya, N. and T. Ichinose. 1988. Function of the scent from the pigmy mole cricket *Tridactylus japonicus* DE HAAN (Orthoptera: Tridactylidae). Appl. Entomol. Zool. 23: 321–328.

Mosey, R.A. and P.E. Floreancig. 2012. Isolation, biological activity, synthesis, and medicinal chemistry of the pederin/mycalamide family of natural products. Nat. Prod. Report 29: 980–995.

Moskowitz, D.P. 2002. An unusual interaction between a banded hairstreak butterfly *Satyrium calanus* (Lycaenidae) and a stink bug *Banasa dimiata* (Pentatomidae). Entomol. News 113: 183–186.

Müller, T., M. Göhl, I. Lusebrink, K. Dettner and K. Seifert. 2012. Cicindeloine from *Stenus cicindeloides*—isolation, structure elucidation and total synthesis. Eur. J. Org. Chem. 12: 2323–2330.

Müller, U.R. 1988. Insektenstichallergie. Gustav Fischer, Stuttgart.

Munir, E., J.J. Yoon, T. Tokimatsu, T. Hattori and M. Shimada. 2001. A physiological role for oxalic acid biosynthesis in the wood-rotting basidiomycet *Formitopsis palustris*. Proc. Natl. Acad. Sci. USA 98: 11126–11130.

Nakatani, T., E. Nishimura and N. Noda. 2006. Two isoindoline alkaloids from the crude drug, the ant lion (the larvae of Myrmeleontidae species). J. Nat. Med. 60: 261–263.

Nelson, D.R., T.P. Freeman, J.S. Buckner, K.A. Hoelmer, C.G. Jackson and J.R. Hagler. 2003. Characterization of the cuticular surface wax pores and the waxy particles of the dustywing, *Semidalis flinti* (Neuroptera: Coniopterygidae). Comp. Biochem. Mol. Biol. 136: 343–356.

Nickle, D.A., J.L. Castner, S.R. Smedley, A.B. Attygalle, J. Meinwald and T. Eisner. 1996. Glandular pyrrazine emission by a tropical Katydid: an example of chemical aposematism? (Orthoptera: Tettigoniidae: Copiphorinae: *Vestria* Stål). J. Orthopt. Res. 5: 221–223.

Nilsson, E. and G. Bengtsson. 2004. Endogenous free fatty acids repel and attract Collembola. J. Chem. Ecol. 30: 1431–1433.

Nishida, R. 2002. Sequestration of defensive substances from plants by Lepidoptera. Annu. Rev. Entomol. 47: 57–92.

Nishiwaki, H., K. Ito, K. Otsuki, H. Yamamoto, K. Komai and K. Matsuda. 2004. Purification and functional characterization of insecticidal sphingomyelinase C produced by *Bacillus cereus*. Eur. J. Biochem. 271: 601–606.

Nishiwaki, H., K. Nakashima, C. Ishida, T. Kawamura and K. Matsuda. 2007. Cloning, functional characterization, and mode of action of a novel insecticidal pore-forming toxin, sphaericolysin, produced by *Bacillus sphaericus*. Appl. Environm. Microbiol. 73: 340–3411.

Numata, A. and T. Ibuka. 1987. Alkaloids from ants and other insects. The Alkaloids 31: 193–315.

Oba, Y., M.A. Branham and T. Fukatsu. 2011. The terrestrial bioluminscent animals of Japan. Zool. Sci. 28: 771–789.

O' Connell, T.J. and N.Z. Reagle. 2002. Is the chemical defense of *Eurycotis floridana* a deterrent to small mammal predators? Florida Sci. 65: 245–249.

Ohta, M., F. Matsuura, G. Henderson and R.A. Laine. 2007. Novel free ceramides as components of the soldier defense gland of the Formosan subterranean termite (*Coptotermes formosanus*). J. Lip. Res. 48: 656–664.

Opitz, S.E.W. and C. Müller. 2009. Plant chemistry and insect sequestration. Chemoecology 19: 117–154.

Pankewitz, F. and M. Hilker. 2006. Defensive components in insect eggs: are anthraquinones produced during egg development? J. Chem. Ecol. 32: 2067–2072.

Pankewitz, F. and M. Hilker. 2008. Polyketides in insects: ecological role of these widespread chemicals and evolutionary aspects of their biogenesis. Biol. Rev. 83: 209–226.

Pankewitz, F., A. Zöllmer, Y. Gräser and M. Hilker. 2007a. Anthraquinones as defensive compounds in eggs of Galerucini leaf beetles: biosynthesis by the beetles? Arch. Insect Biochem. Physiol. 66: 98–108.

Pankewitz, F., A. Zöllmer, M. Hilker and Y. Gräser. 2007b. Presence of *Wolbachia* in insect eggs containing antimicrobially active anthraquinones. Microbial Ecol. 54: 713–721.

Parfin, S. 1952. Notes on the bionomics of *Corydalus cornutus* (Linné), *Chauliodes rastricornis* Rambur, *C. pectinicornis* (Linné) and *Neohermes* sp. Am. Midland Naturalist 47: 426–434.

Parker, S.P. 1982. Synopsis and Classification of Living Organisms, Vol. 2. McGraw-Hill, New York.

Pasteels, J.M., D. Daloze, J.-C. de Biseau, A. Termonia and D.M. Windsor. 2004. Patterns in host-plant association and defensive toxins produced by neotropical chrysomeline beetles. pp. 669–676. *In*: P. Jolivet, J.A. Santiago-Blay and M. Schmitt (eds.). New Developments in the Biology of Chrysomelidae. SPB Academic Publishing, The Hague.

Pavan, M. 1975. Gli iridoidi negli insetti. Pubblicationes Instituto Entomologia Agraria Universita di Pavia 2: 1–49.

Pavan, M. and M. Valcurone Dazzini. 1976. Sostanze di difesa dei lepidotteri. Publicationes Instituto Entomologia Agraria Universita di Pavia 3: 1–23

Peck, D.C. 2000. Reflex bleeding in froghoppers (Homoptera: Cercopidae): variation on behavior and taxonomic distribution. Ann. Entomol. Soc. Am. 93: 1186–1194.

Pemberton, R.W. 1999. Insects and other arthropods used as drugs in Korean traditional medicine. J. Ethnopharmacol. 65: 207–216.

Petre, C.A., C. Detrain and J.-L. Boevé. 2007. Anti-predator defence mechanisms in sawfly larvae of *Arge* (Hymenoptera, Argidae). J. Insect Physiol. 53: 668–675.

Pfander, I. and J. Zettel. 2004. Chemical communication in *Ceratophysella sigillata* (Collembola: Hypogastruridae): intraspecific reaction to alarm substances. Pedobiologia 48: 575–580.

Pickett, J.A., R.K. Allemann and M.A. Birkett. 2013. The semiochemistry of aphids. Nat. Prod. Rep. 30: 1277–1283.

Piek, T. (ed.). 1986. Venoms of the Hymenoptera-Biochemical, Pharmacological and Behavioural Aspects. Academic Press, London.

Pietra, F. 2002. Biodiversity and Natural Product Diversity. Pergamon Press, Amsterdam.

Plachter, H. 1979. Zur Kenntnis der Präimaginalstadien der Pilzmücken. Teil I: Gespinnstbau. Zool. Jahrb. Anat. 101: 168–266.

Polanowski, A., M.S. Blum, D.W. Whitman and J. Travis. 1997. Proteinase inhibitors in the nonvenomous defensive secretion of grasshoppers: antiproteolytic range and possible significance. Comp. Biochem. Physiol. 117B: 525–529.

Porco, D. and L. Deharveng. 2007. 1,3-Dimethoxybenzene, a chemotaxonomic marker for the Neanurinae subfamily (Collembola). Biochem. Syst. Ecol. 35: 160–161.

Preston-Mafham, K. 1990. Grasshoppers and Mantids of the World. Facts on File, New York.

Prestwich, G.D. 1983. Chemical systematics of termite exocrine secretions. Annu. Rev. Ecol. Syst. 14: 287–311.

Prestwich, G.D. 1984. Defense mechanisms of termites. Annu. Rev. Entomol. 29: 201–232.

Prieto, J.M., U. Schaffner, A. Barker, A. Braca, T. Siciliano and J.-L. Boevé. 2007. Sequestration of furostanol saponins by *Monophadnus* sawfly larvae. J. Chem. Ecol. 33: 513–524.

Purrington, F.F., P.A. Kendall, J.E. Bater and B.R. Stinner. 1991. Alarm pheromone in a gregarious poduromorph collembolan (Collembola: Hypogastruridae). Great Lakes Entomol. 24: 75–78.

Quennedey, A. 1984. Morphology and ultrastructure of termite defense glands. pp. 151–200. *In*: H.R. Hermann (ed.). Defensive Mechanisms in Social Insects. Praeger, New York.

Quicke, D.L.J. 1997. Parasitic Wasps. Chapman and Hall, London.

Redborg, K.E. 1983. A mantispid larva can preserve its spider egg prey: evidence for an aggressive allomone. Oecologia 58: 230–231.

Reyes-Garcia, L., M. Fernando Flores and J. Bergmann. 2011. Biological activity of the larval secretion of *Chilecomadia valdiviana*. J. Chem. Ecol. 37: 1137–1142.

Richard, F.L. and D.K. Ledford. 2014. Allergens and Allergen Immunotherapy, 5th edition. CRC Press, Taylor and Francis, Boca Raton.

Ross, G.N., H.M. Fales, H.A. Lloyd, T. Jones, E.A. Sokoloski, K. Marshall-Batty and M.S. Blum. 2001. Novel chemistry of abdominal defensive glands of nymphalid butterfly *Agraulis vanillae*. J. Chem. Ecol. 27: 1219–1228.

Roth, L.M. and D.W. Alsop. 1978. Toxins of Blattaria. pp. 465–487. *In*: S. Bettini (ed.). Handbook of Experimental Pharmacology, Vol. 48. Springer, Berlin.

Rotheray, G.E. 1986. Colour, shape and defence in aphidophagous syrphid larvae (Diptera). Zool. J. Linn. Soc. 88: 201–216.

Rotschild, M. 1985. British aposematic Lepidoptera. pp. 9–62. *In*: J. Heath and A.M. Emmet (eds.). The Moths and Butterflies of Great Britain and Ireland. Harley, Essex.

Ryczek, S., K. Dettner and C. Unverzagt. 2009. Synthesis of buprestins D, E, F, G and H; structural confirmation and biological testing of acyl glucoses from jewel beetles (Coleoptera: Buprestidae). Bioorg. Med. Chem. 17: 1187–1192.

Sahayaraj, K. and A. Vinothkanna. 2011. Insecticidal activity of venomous saliva from *Rhynocoris fuscipes* (Reduviidae) against *Spodoptera litura* and *Helicoverpa armigera* by microinjection and oral administration. J. Ven. Anim. Toxins including Tropical Diseases 17: 486–490.

Schaefer, C.W. and A.R. Panizzi (eds.). 2000. Heteroptera of Economic Importance. CRC Press, Boca Raton.

Schal, C., J. Fraser and W.J. Bell. 1982. Disturbance stridulation and chemical defence in nymphs of the tropical cockroach *Megaloblatta blaberoides*. J. Insect Physiol. 28: 541–552.

Schedl, W. 1991. Hymenoptera, Unterordnung Symphyta. Handbuch der Zoologie; Band IV: Arthropoda: Insecta. M. Fischer (ed.). de Gruyter, Berlin.

Schierling, A. and K. Dettner. 2013. The pygidial defense gland system of the Steninae (Coleoptera, Staphylinidae): Morphology, ultrastructure and evolution. Arthrop. Struct. Dev. 42: 197–208.

Schierling, A., K. Seifert, S. Sinterhauf, J.B. Rieß, J.C. Rupprecht and K. Dettner. 2013. The multifunctional pygidial gland secretion of the Steninae (Coleoptera: Staphylinidae): ecological significance and evolution. Chemoecology 23: 45–57.

Schildknecht, H. and K.H. Weis. 1960. VI. Mitteilung über Insektenabwehrstoffe: Zur Kenntnis des Pygidialdrüsen-Sekretes vom gemeinen Ohrwurm, *Forficula auricularia*. Zeitschr. Naturforschung 15b: 755–757.

Schlörke, O., P. Krastel, I. Müller and I. Usón. 2002. Structure and biosynthesis of cetoniacytone A, a cytotoxic aminocarba sugar produced by an endosymbiontic *Actenomyces*. J. Antibiot. 55: 635–642.

Schmidt, J.O. 1982. Biochemistry of insect venoms. Annu. Rev. Entomol. 27: 339–368.

Schramm, S., K. Dettner and C. Unverzagt. 2006. Chemical and enzymatic synthesis of buprestin A and B–bitter acylglucosides from Australian jewel beetles (Coleoptera: Buprestidae). Tetrahedron Lett. 47: 7741–7743.

Schulz, S. 2004. The Chemistry of Pheromones and Other Semiochemicals I. Springer, Berlin.

Schulz, S. 2005. The Chemistry of Pheromones and Other Semiochemicals II. Springer, Berlin.

Schulz, S., C. Messer and K. Dettner. 1997. Poduran, an unusual tetraterpen from the springtail *Podura aquatica*. Tetrahedron Lett. 38: 2077–2080.

Schweppe, H. 1993. Handbuch der Naturfarbstoffe. Ecomed, Landsberg.

Seiler, C., S. Bradler and R. Koch. 2000. Phasmiden. bede-Verlag, Ruhmannsfelden.

Sloggett, J.J., A. Magro, F.J. Verheggen, J.-L. Hemptinne, W.D. Hutchinson and E.W. Riddick. 2011. The chemical ecology of *Harmonia axyridis*. BioControl 56: 643–661.

Smedley, S.R., F.C. Schroeder, D.B. Weibel, J. Meinwald, K.A. Lafleur, J.A. Renwick, R. Rutowski and T. Eisner. 2002. Mayolenes: Labile defensive lipids from the glandular hairs of a caterpillar (*Pieris rapae*). Proc. Natl. Acad. Sci. USA 99: 6822–6827.

Smith, E.L. 1970. Biology and structure of the Dobsonfly, *Neohermes californicus* (Walker) (Megaloptera: Corydalidae). Pan-Pacific Entomol. 46: 142–150.

Smith, R.M., J.J. Brophy, G.W.K. Cavill and N.W. Davies. 1979. Iridodials and nepetalactone in the defensive secretion of coconut stick insects *Graeffea crouani*. J. Chem. Ecol. 5: 727–735.

Snook, M.E., M.S. Blum, D.W. Whitman, R.F. Arrendale, C.E. Costello and J.S. Harwood. 1993. Caffeoyltartronic acid from catnip (*Nepeta cataria*): A precursor for catechol in lubber grasshoppers (*Romalea guttata*) defensive secretions. J. Chem. Ecol. 19: 1957–1966.

Southcott, R.V. 1991. Injuries from larval Neuroptera. Med. J. Australia 154: 329–332.

Spanton, S.G. and G.D. Prestwich. 1982. Chemical defense and self-defense. Tetrahedron 38: 1921–1930.

Srivastava, S.K., N. Babu and H. Pandey. 2009. Traditional insect bioprospecting—As human food and medicine. Indian J. Traditional Knowledge 8: 485–494.

Staddon, B.W. 1979. The scent glands of Heteroptera. Adv. Insect Physiol. 14: 351–418.

Stocks, I. 2008. Reflex bleeding (Autohemorrhage). pp. 3132–3139. *In*: J.L. Capinera (ed.). Encyclopedia of Entomology, 2nd ed. Springer, Berlin.

Stransky, K., M. Psoky and M. Streibl. 1986. Lipid compounds from the extract of springtail *Tetrodontophora bielanensis* (Waga). Collect. Czech Chem. Commun. 51: 948–955.

Suzuki, T., K. Haga and Y. Kuwahara. 1986. Anal secretion of thrips. I. Identification of perillene from *Leeuwenia pasanii* (Thysanoptera: Plaeothripidae). Appl. Entomol. Zool. 21: 461–466.

Suzuki, T., K. Haga, S. Kodama, K. Watanabe and Y. Kuwahara. 1988. Secretion of thrips, II. Secretion of the three gall-inhabiting thrips (Thysanoptera: Phlaeothripidae). Appl. Entomol. Zool. 23: 291–297.

Suzuki, T., K. Haga, W.S. Leal, S. Kodama and Y. Kuwahara. 1989. Secretion of Thrips, IV: Identification of β-Acaridial from Three Gall-Forming Thrips (Thysanoptera: Phlaeothripidae). Appl. Entomol. Zool. 24: 222–228.

Suzuki, T., K. Haga, M. Izuno, S. Matsuiama and Y. Kuwahara. 1993. Secretion of thrips, VII. Identification of 3-butanoyl-4-hydroxy-6-methyl-2H-pyran-2-one from *Holothrips japonicus* and *H. hagai* (Thysanoptera: Phlaeothripidae). Appl. Entomol. Zool. 28: 108–112.

Suzuki, T., K. Haga, T. Tsutsumi and S. Matsuyama. 2004. Analysis of anal secretions from Phlaeothripine thrips. J. Chem. Ecol. 30: 409–423.

Swart, C.C., L.E. Deaton and B.E. Felgenhauer. 2006. The salivary gland and salivary enzymes of the giant waterbugs (Heteroptera, Belostomatidae). Comp. Biochem. Physiol. 145A: 114–122.

Sword, G.A. 2001. Tasty on the outside, but toxic in the middle: grasshopper regurgitation and host plant-mediated toxicity to a vertebrate predator. Oecologia 128: 416–421.

Takegawa, H. and S. Takahashi. 1990. Allomonal secretions in six species of the genera *Periplaneta* and *Blatta* (Dictyoptera: Blattidae). Jap. J. Environm. Entomol. Zool. 2: 123–127.

Teerling, C.R., D.R. Gillespie and J.H. Borden. 1993. Utilization of western flower thrips alarm pheromone as a prey-finding kairomone by predators. Can. Entomol. 125: 431–437.

Terra, W.R. and A.G. de Bianchi. 1974. Chemical composition of the cocoon of the fly, *Rhychosciara americana*. Insect Biochem. 4: 173–183.

Trigo, J.R. 2000. The chemistry of antipredator defense by secondary compounds in neotropical Lepidoptera: Facts, perspectives and caveats. J. Braz. Chem. Soc. 11: 551–561.

Tringali, C. (ed.). 2012. Bioactive Compounds from Natural Sources. CRC Press, Boca Raton.

Tschuch, G., P. Lindemann and G. Moritz. 2002a. Chemical defence in thrips. pp. 277–278. *In*: R. Marullo and L. Mound (eds.). Thrips and Tospoviruses: Proceedings of the 7th International Symposium on Thysanoptera. Calabria, Italy.

Tschuch, G., Lindemann and G. Moritz. 2002b. Chemical defence in the thrips *Suocerathrips linguis* Mound and Marullo 1994 (Phlaeothripidae, Thysanoptera, Insecta). Zoology 105(Suppl. V): 99.

Tschuch, G., G. Kießling, C. Engel, P. Lindemann and G. Moritz. 2004. Chemische Abwehr bei Thysanopteren, Mitteilungen der Deutschen Gesellschaft für Allgemeine und Angewandte Entomologie. 14: 183–186.

Tschuch, G., P. Lindemann, A. Niesen, R. Csuk and G. Moritz. 2005. A novel long-chained acetate in the defensive secretion of thrips. J. Chem. Ecol. 31: 1555–1565.

Tschuch, G., P. Lindemann and G. Moritz. 2008. Anunexpected mixture of substances in the defensive secretions of the tubuliferan thrips, *Callococcithrips fuscipennis* (Moulton). J. Chem. Ecol. 34: 742–747.

Turnbull, M.W. and N.J. Fashing. 2002. Efficacy of the ventgreal abdominal secretion of the cockroach *Eurycotis floridana* (Blattaria: Blattidae) as a defensive allomone. J. Insect Behav. 15: 369–384.

Tyler, J., W. McKinnon, G.A. Lord and P.J. Hilton. 2008. A defensive steroidal pyrone inthe glow-worm *Lampyris noctiluca* L. (Coleoptera: Lampyridae). Physiol. Entomol. 33: 167–170.

Unkiewicz-Winiarczyk, A. and K. Gromysz-Kalkowska. 2012. Ethological defence mechanisms in insects. III. Chemical defence. Annales Universitates Mariae Curie-Sklodowska Lublin—Polonia, LXVII, 2, Sectio C: 63–74.

Urbanek, A., R. Szadziewski, P. Stepnowski, J. Boros-Majewska, I. Gabriel, M. Dawgul, W. Kamysz, D. Sosnowska and M. Golebiowski. 2012. Composition and antimicrobial activity of fatty acids detected in the hygroscopic secretion collected from the secretory setae of larvae of the biting midge *Forcipomyia nigra* (Diptera: Ceratopogonidae). J. Insect Physiol. 58: 1265–1276.

Vegliante, F. and I. Hasenfuss. 2012. Morphology and diversity of exocrine glands in lepidopteran larvae. Annu. Rev. Entomol. 57: 187–204.

Vencl, F.V., N.E. Gómez, K. Ploß and W. Boland. 2009. The chlorophyll catabolite, pheophorbide a, confers predation resistance in a larval tortoise beetle shield defense. J. Chem. Ecol. 35: 281–288.

Vilcinskas, A. 2011. Insect Biotechnology. Springer, Dordrecht.

Vilcinskas, A. and J. Gross. 2005. Drugs from bugs: the use of insects as a valuable source of transgenes with potential in modern plant protection strategies. J. Pest. Sci. 78: 187–191.

Viviani, V.R., J.W. Hastings and T. Wilson. 2002. Two bioluminescent diptera: the North American *Orfelia fultoni* and the Australian *Arachnocampa flava*. Similar niche, different bioluminescence systems. Photochem. Photobiol. 75: 22–27.

Völkl, W., G. Hübner and K. Dettner. 1994. Interactions between *Alloxysta brevis* (Hymenoptera, Cynipoidea, Alloxystidae) and honeydew collecting ants: How an aphid hyperparasitoid overcomes ant aggression by chemical defense. J. Chem. Ecol. 20: 2901–2905.

Vosseler, J. 1890. Die Stinkdrüsen der Forficuliden. Arch. Mikroskop. Anatomie 36: 565–578.

Vršanský, P., D. Chorvát, I. Fritzsche, M. Hain and R. Ševčik. 2012. Light-mimicking cockroaches indicate tertiary origin of recent terrestrial luminescence. Naturwissenschaften 99: 739–749.

Vuts, J., Z. Imrei, M.A. Birkett, J.A. Pickett, C.M. Woodcock and N. Tóth. 2014. Semiochemistry of the Scarabaeoidea. J. Chem. Ecol. 40: 190–210.

Wagner, R., M. Aurich, E. Reder and H.J. Veith. 1990. Defensive secretion from larvae of *Apatania fimbriata* (Pictet) (Trichoptera: Limnephilidae). Chemoecology 1: 96–104.

Wang, H. and T.B. Ng. 2002. Isolation of cicadin, a novel and potent antifungal peptide from dried juvenile cicadas. Peptides 23: 7–11.

Weatherston, J. and J.E. Percy. 1978a. Venoms of Rhynchota (Hemiptera). pp. 489–509. *In*: S. Bettini (ed.). Arthropod Venoms. Handbuch der Experimentellen Pharmakologie, Vol. 48. Springer, Berlin.

Weatherston, J. and J.E. Percy. 1978b. Venoms of Coleoptera. pp. 511–554. *In*: S. Bettini (ed.). Arthropod Venoms. Handbuch der Experimentellen Pharmakologie, Vol. 48. Springer, Berlin.

Weatherston, I., J.A. MacDonald, D. Miller, G. Riere, J.E. Percy-Cunningham and M.H. Benn. 1986. Ultrastructure of exocrine prothoracic gland of *Datana ministra* (Druby) (Lepidoptera: Notodontidae) and the nature of its secretion. J. Chem. Ecol. 12: 2039–2050.

Weidner, H. 1936. Beitrag zu einer Monographie der Raupen mit Gifthaaren. Zeitschr. Angew. Entomol. 23: 432–484.

Weiss, M.R. 2006. Defecation behavior and ecology of insects. Annu. Rev. Entomol. 51: 635–661.

Wheeler, C.H. and R.T. Cardé. 2013. Defensive allomones function as aggregation pheromones in diaposing ladybird beetles *Hippodamia convergens*. J. Chem. Ecol. 39: 723–732.

Wheeler, J.W. and R.M. Duffield. 1988. Pheromones of Hymenoptera and Isoptera. pp. 59–206. *In*: E.D. Morgan and N. Bhushan Mandava (eds.). CRC Handbook of Natural Pesticides, Vol. IV, Pheromones, Part B. CRC Press, Boca Raton.

Whitman, D.W. 1990. Grasshopper chemical communication. pp. 357–392. *In*: R.F. Chapman and A. Joern (eds.). The Biology of Grasshoppers. Wiley, New York.

Whitman, D.W. 2008. Grasshopper sexual pheromone: a component of the defensive secretion in *Taeniopoda eques*. Physiol. Entomol. 7: 111–115.

Whitman, D.W., M.S. Blum and D.W. Alsop. 1990. Allomones: chemicals for defense. pp. 288–351. *In*: D.L. Evans and J.O. Schmidt (eds.). Insect Defenses. State University of New York Press, Albany.

Whitman, D.W., C.G. Jones and M.S. Blum. 1992. Defensive secretion production in lubber grasshoppers (Orthoptera: Romaleidae): Influence of age, sex, diet, and discharge frequency. Ann. Entomol. Soc. Am. 85: 96–102.

Wiesner, J. and A. Vilcinskas. 2011. Therapeutic potential of antimicrobial peptides from insects. pp. 29–65. *In*: A. Vilcinskas (ed.). Insect Biotechnology. Springer, Dordrecht.

Wigglesworth, V.B. 1974. The Principles of Insect Physiology. Chapman & Hall, London.

Will, K.W., A.B. Attygalle and K. Herath. 2000. New defensive chemical data for ground beetles (Coleoptera: Carabidae): interpretations in a phylogenetic framework. Biol. J. Linn. Soc. 71: 459–481.

Williams, B. 1986. Mandibular glands in the endoparasitic larva of *Uropsylla tasmanica* Rothschild (Siphonaptera: Pygiopsyllidae). Int. J. Insect Morphol. Embryol. 15: 263–268.

Witz, B.W. 1990. Antipredator mechanisms in arthropods: A twenty year literature survey. Flor. Entomol. 73: 71–99.

Wolf, S., H. Brettschneider and P.W. Bateman. 2006. The predator defence system of an African king cricket (Orthoptera: Anostostomatidae): does it help to stink? Afr. Zool. 41: 75–80.

Wu, X., P.M. Flatt, H. Xu and T. Mahmud. 2009. Biosynthetic gene cluster of Cetoniacytone A, an unusual aminocyclitol from the endosymbiotic bacterium *Actinomyces* sp. Lu 9419. Chem. Bio. Chem. 10: 304–314.

Wyttenbach, R. and T. Eisner. 2001. Use of defensive glands during mating in a cockroach (*Diploptera punctata*). Chemoecology 11: 25–28.

Yoshida, N., H. Sugama, S. Gotoh, K. Matsuda, K. Nishimura and K. Komai. 1999. Detection of ALMB-toxin in the larval body of *Myrmeleon bore* by anti-N-terminus peptide antibodies. Biosci. Biotechnol. Biochem. 63: 232–234.

Yoshida, N., K. Oeda, E. Watanabe, T. Mikami, Y. Fukita, K. Nishimura, K. Komai and K. Matsuda. 2001. Chaperonin turned insect toxin. Nature 411: 44.

Yu, X.-P., J.-L. Zhu, X.-P. Yao, S.-C. He, H.-N. Huang, W.-L. Chen, Y.-H. Hu and D.-B. Li. 2005. Identification of *anrI* gene, a homology of *admM* of andrimid biosynthetic gene cluster related to the antangonistic activity of *Enterobacter cloacae* B8. World J. Gastrenterol. 11: 6152–6158.

Zhang, Q.H., K.R. Chauhan, E.F. Erbe, A.R. Vellore and J.R. Aldrich. 2004. Semiochemistry of the goldeneyed lacewing *Chrysopa oculata*: Attraction of males to a male-produced pheromone. J. Chem. Ecol. 30: 1849–1870.

Zhao, X., Q. Meng, L. Yu and M. Li. 2009. Composition and bioactivity of secretion from *Stauronematus compressicornis* (Fabricius). For. Stud. China 11: 122–126.

Zhu, J., R.C. Unelius, K.-C. Park, S.A. Ochieng, J.J. Obrycki and T.C. Baker. 2000. Identification of (Z)-4-tridecene from defensive secretion of grenn lacewing *Chrysoperla carnea*. J. Chem. Ecol. 26: 2421–2434.

Zimmer, M.M., J. Frank, J.H. Barker and H. Becker. 2006. Effect of extracts from the Chinese and European mole cricket on wound epithelialization and neovascularization: *in vivo* studies in the hairless mouse ear wound model. Wound Repair Regeneration 14: 142–151.

Zwick, P. 1980. Plecoptera, Steinfliegen. Handbuch der Zoologie, II. Bd. 2. Hälfte, 2. Teil, 7. Part. de Gruyter, Berlin.

3

Insect Bioluminescence in the Post-Molecular Biology Era

Yuichi Oba

Introduction

This chapter provides an overview of recent advances in the study of luminous insects. Right after publishing a comprehensive review entitled 'Environmental Aspects of Insect Bioluminescence' by K.H. Hoffmann (Hoffmann 1984), beetle luciferase genes were isolated for the first time from the North American firefly *Photinus pyralis* in 1985 and 1987 (de Wet et al. 1985, 1987), and then from both the Japanese firefly *Luciola cruciata* in 1989 (Masuda et al. 1989, Tatsumi et al. 1989) and the Jamaican luminous click beetle *Pyrophorus plagiophthalmus* in 1989 (Wood et al. 1989). Subsequently, cDNAs for beetle luciferase have been cloned from over 30 species in Lampyridae, Phengodidae, Rhagophthalmidae, and Elateridae, and these genes are currently utilized in various biotechnological applications, such as reporter assay and DNA sequencing methodology. Accumulating nucleotide and amino acid sequences for beetle luciferases promoted functional and evolutional investigations on bioluminescence in beetles. The earliest molecular phylogenetic analysis of fireflies was conducted using 16S ribosomal DNA in 1997 (Suzuki 1997). After that, several analyses using other molecular markers and more taxa have been performed, and the results have been made available for further discussion

Graduate School of Bioagricultural Sciences, Nagoya University, Nagoya 464-8601, Japan.
Email: oba@agr.nagoya-u.ac.jp

on the evolution of bioluminescence. In contrast to the remarkable progress with luminous beetles, however, molecular studies on bioluminescent insects other than beetles are still limited. Luciferase (or photoprotein) genes have yet to be identified from any luminous springtails or fungus gnats, including even the well-known cave glowworm *Arachnocampa*.

Bioluminescent Insects Except Coleoptera

Insects, the most diversified class of animals, include large numbers of luminous species. Based on Herring's list 'bioluminescence in living organisms' (Herring 1987) and revisions by Haddock et al. (2010) and Oba and Schultz (2014), bioluminescent insects are found in 144 genera, which is 19% of all genera, and in 67% of the Arthropod genera that contain luminous species. Luminous insect species are currently found in three orders: Collembola, Diptera, and Coleoptera (Oba et al. 2011).

Collembola (Springtails)

The order Collembola currently includes 33 families and over 8,000 described species worldwide (Janssens and Christiansen 2011), but only a small number of luminous species are recognized in the families Neanuridae and Onychiuridae (Harvey 1952). The luminescences and taxonomic statuses of luminous springtails have not been reexamined after Harvey's book (Harvey 1952), except for the observation and DNA barcoding of *Lobella* sp. (near *sauteri*) from Japan (Oba et al. 2011). This scarlet red species is 1–3 mm long and found under a coppice litter, emitted a continuous weak green light for some seconds from abdominal tubercles by mechanical stimulation. The luminescence was not secretory (Oba et al. 2011).

It should be noted that Collembola is currently classified as an Entognatha of subphylum Hexapoda (insects, in a traditional sense) as well as an Insecta (true insects, another member of Hexapoda), but the monophyly of Hexapoda has long been uncertain. Recent exhaustive molecular analyses, however, strongly support the Entognatha as the sister to Insecta, thereby indicating that Hexapoda is monophyletic (Regier et al. 2010, Sasaki et al. 2013).

Diptera (Fungus Gnats)

The order Diptera in Insecta includes about 150 families and over 150,000 species (Yeates et al. 2007), but only a small number of luminous species are recognized in a single family Keroplatidae (Oba et al. 2011, and references therein). Three genera, *Arachnocampa*, *Keroplatus*, and *Orfelia*, certainly contain luminous species, and an unknown keroplatid (?) luminous species was also recorded in Fiji (Harvey 1952) and the highlands of New Guinea (Lloyd 1978).

The luminescence of *Arachnocampa* and *Orfelia* in larvae attracts prey (Meyer-Rochow 2007). The luminescence of female *Arachnocampa* pupa has been considered to male attraction, but Broadley (2012) suggested by infrared video monitoring that pupal luminescence has no function in New Zealand *A. luminosa* but is instead carried over from the larval stage (Broadley 2012). Luminescence intended for prey sometimes attracts predators instead. The cave opilionid harvestman *Megalopsalis tumida* and *Hendea myersi* possess prominent eyes to detect the luminescence of the major prey, the larval *A. luminosa* (Meyer-Rochow and Liddle 1988). The larva of *Keroplatus* is a spore feeder (Matile 1997), and Sivinski (1998) speculated that the function of the luminescence might be to repel negatively phototropic enemies or to serve as an aposematic signal, but the details are still unknown. The sheet web of *K. nipponicus* larva is sticky and very acidic, as it is in some luminous and nonluminous fungus gnats (Matile 1997), accordingly, ants and other invertebrates never intrude into the web (Oba et al. 2011). These observations may support the aposematic display hypothesis by Sivinski (1998). Phylogenetic analysis using morphological codes indicated that the ancestral state of Keroplatidae was predaceous and that spore feeders appeared secondarily in this lineage (Matile 1997). Considering this evolutionary scenario, we speculate that the luminescence of *Keroplatus* is a kind of vestigial trait. Indeed, it seems that the luminescence of *K. nipponicus* larvae is too faint to deter enemies (Oba et al. 2011). Molecular phylogenetic analysis of Keroplatidae should be conducted for further elucidation.

Molecular phylogenetic analysis of the genus *Arachnocampa* using mitochondrial COII and 16S genes showed that Australian *A. flava* is subclustered into allopatric geographic groups (Baker et al. 2008). Together with morphological keys, five new species were described (Baker 2010). It has been known that cave populations show markedly reduced pigmentation and tend to make longer snares than rainforest populations. However, molecular data supported allopatric clustering regardless of habitat. Therefore, Baker et al. (2008) considered that the morphological differences between the cave dwellers and epigeans would be facultative adaptations to microclimatic conditions (Baker et al. 2008).

Biochemical studies on luminous fungus gnats are still very limited. Viviani et al. (2002) reported that the luciferin-luciferase reaction was reproduced for both *A. flava* and *Orfelia fultoni*, but their biochemistries were markedly different: there were no cross-reactions of enzymes or substrates between their systems; ATP was required for the luminescence reaction in *Arachnocampa* but not in *Orfelia*. Luminescence in *K. nipponicus* was negative for both the luciferin-luciferase reaction and the ATP effect (Haneda 1957, Oba et al. 2011). Molecular analysis of the bioluminescence of fungus gnats has not been reported in the literature to date. Sivinski (1998) suggested that morphological differences of light organs indicate independent evolution of bioluminescence systems among *Arachnocampa*, *Orfelia*, and *Keroplatus*.

Hemiptera and Blattodea—Doubtful Luminescence

The elongated head of the lantern fly *Fulgora laternaria* in South America had long been believed to be luminous (Harvey 1952), but this has become very doubtful (Goemans 2006). Recently, luminescence was claimed to occur in some South American cockroaches *Lucihormetica* (Blattodea) (Zompro and Fritzsche 1999, Vršanský et al. 2012), but the veracity of these reports is highly dubious (Merritt 2013).

Coleoptera (Beetles)

The order Coleoptera is the most diversified group of Insecta; it currently includes 176 families and 386,500 species (Slipinski et al. 2011). Luminescent species are recognized in only four families of single superfamily Elateroidea, Elateridae, Phengodidae, Rhagophthalmidae, and Lampyridae, except for some uncertain records (Oba 2009). The luminosity of larval *Omalisus fontisbellaquaei* (Omalisidae, Elateroidea) (Bertkau 1891) has been questioned (Burakowski 1988, M.A. Branham, personal communication). The luminosity of Telegeusidae (Elateroidea) (Barber 1908, Crowson 1972) is doubtful (Branham and Wenzel 2001, M.A. Branham, personal communication). The luminescence of a pair of pronotal 'light organs' of the adult *Balgus schnusei* (Costa 1984), a species that has now been assigned to the Thylacosterninae of the Elateridae (Costa et al. 2010), has not been confirmed by later observation. The luminescence of larval *Xantholinus* sp. (Staphylinidae) from Brazil (Costa et al. 1986) was recently reaffirmed (Rosa 2010). The larvae emitted green light on the dorsal median line from the prothorax to the eighth abdominal segment (Rosa 2010). To confirm the true luminosity, it is necessary to investigate the light source for excluding the possibility of luminous bacterial infection or luminescence from luminous organisms ingested, such as luminous click beetles and phengodid glowworms.

Phylogeny of Elateroidea and Evolution of Bioluminescence in Beetles

Superfamily Elateroidea currently includes 18 extant families (Slipinski et al. 2011). It is important to note here that all known species of Lampyridae, Phengodidae, and Rhagophthalmidae are luminous at least in the larval stages (Branham and Wenzel 2003, Branham 2010, Costa and Zaragoza-Caballero 2010). Phylogenetic relationships of the families in Elateroidea have been studied in the context of the evolution of bioluminescence, based on larval morphological characteristics (Crowson 1972, Pototskaja 1983, Beutel 1995, Branham and Wenzel 2001), but their cladograms were inconsistent with each other. In 2007, two molecular analyses of Elateroidea were reported (Bocakova et al. 2007, Sagegami-Oba et al. 2007a), but the relationships among the families are still not fully resolved. More recent molecular analysis, based on the mitochondrial genome sequences, strongly supported a sister relationship between Lampyridae and Rhagophthalmidae (Timmermans et al. 2010, Timmermans and Vogler 2012). Although their analyses included only 10 elateroid species from seven families (taxa in Phengodidae were not included), this result suggests the monophyly of Elateroidea and a single origin of the luminescence in Lampyridae and Rhagophthalmidae. An alternative approach, based on observation of embryonic development, also suggested an association between Lampyridae and Rhagophthalmidae (Kobayashi et al. 2002). Further studies using mitochondrial genome analysis of more elateroid families will be helpful for understanding the evolution and origin of bioluminescence in beetles.

Phylogeny of Elateridae and Evolution of Bioluminescence in Click Beetles

The family Elateridae includes about 10,000 species (Slipinski et al. 2011), which are widespread throughout the world, and Costa et al. (2010) subdivided it into 17 subfamilies. Of those, approximately 200 luminous species are recorded from tropical and subtropical regions of the Americas and some small Melanesian islands (Costa 1975, Costa et al. 2010). For instance, the tropical American *Pyrophorus noctilucus* is considered the largest (~30 mm) and brightest bioluminescent insect (Harvey and Stevens 1928, Levy 1998). All luminous click beetles belong to the tribe Pyrophorini of the subfamily Agrypninae, with the single exception of *Campyloxenus pyrothorax* (from Chile) belonging in the subfamily Campyloxeninae (Costa et al. 2010). Regarding the status of *Campyloxenus*,

Stibick (1979) suggested a clear relationship between Pyrophorini and Campyloxeninae (Stibick 1979). Phylogenetic analyses have been conducted using larval and/or adult morphologies (Hyslop 1917, Ôhira 1962, Dolin 1978, Stibick 1979, Ôhira 2013), but the topologies of these cladograms were inconsistent with each other. Molecular phylogenetic analysis of Elateridae was conducted, for the first time, using nuclear 28S fragments (Sagegami-Oba et al. 2007b). This result showed that luminous taxa are deeply nested within nonluminous species, and so those authors concluded that bioluminescence might have arisen independently in Elateridae (Oba and Sagegami-Oba 2007, Sagegami-Oba et al. 2007b). This conclusion was supported by more recent phylogenetic analyses using nuclear and mitochondrial genes (Kundrata and Bocak 2011) and morphological characteristics (Douglas 2011).

Relationship between Phengodidae and Rhagophthalmidae

The families Phengodidae and Rhagophthalmidae include about 250 and 30 species, respectively (Slipinski et al. 2011). The distributions of phengodid and rhagophthalmid species are geographically separated into the New World and the Old World, respectively (Lawrence 1982). These two families have been considered closely related, and sometimes have been combined into Phengodidae as subfamilies Phengodinae and Rhagophthalminae (Crowson 1972, Lawrence and Newton 1995). However, recent phylogenetic analyses raised a question about this traditional view. Based on the morphologies of male adults, Branham and Wenzel (2001) suggested that Phengodidae is not the sister group of Rhagophthalmidae, but instead is closely related to the genus *Stenocladius* of Lampyridae. Meanwhile, molecular analyses using mitochondrial 16S fragments supported the close relationship between the genera *Rhagophthalmus* (Rhagophthalmidae) and *Stenocladius* (Suzuki 1997, Stanger-Hall et al. 2007, South et al. 2011). Besides, molecular analysis using whole mitochondrial genomes supported that *Rhagophthalmus* is related more closely to *Pyrophorus* (Elateridae) than to *Pyrocoelia* (Lampyridae) (Arnoldi et al. 2007). However, molecular phylogenetic analyses based on nuclear ribosomal DNAs (Bocakova et al. 2007, Sagegami-Oba et al. 2007a) strongly support the sister relationship between Phengodidae and Rhagophthalmidae, and this result fits into the classical taxonomic system (Lawrence and Newton 1995). Together with strong support for the close relationship between Rhagophthalmidae and Lampyridae as described above, these results suggest a single origin of the luminescence in Lampyridae, Phengodidae, and Rhagophthalmidae.

Phylogeny of Lampyridae and Evolution of Bioluminescence in Fireflies

The family Lampyridae includes about 2,200 species worldwide (Slipinski et al. 2011), and currently consists of four subfamilies: Psilocladinae, Lampyrinae, Luciolinae, and Photurinae (Lawrence et al. 2010a). All known species are luminous at least in the larval stages, while only a portion of adult lampyrids are luminous (Branham 2010). Phylogenetic relationships of the species in Lampyridae have been studied in the context of the evolution of bioluminescence traits, based on male adult morphologies (Branham and Wenzel 2001, 2003), mitochondrial 16S genes (Suzuki 1997, Li et al. 2006), nuclear 18S genes (Sagegami-Oba et al. 2007a), and a combination of mitochondrial and nuclear genes (Bocakova et al. 2007, Stanger-Hall et al. 2007). Some of these results were questioned for the monophyly of Lampyridae (Suzuki 1997, Branham and Wenzel 2001, 2003, Li et al. 2006, Stanger-Hall et al. 2007). For example, morphological analysis using male adults reproduced a unique cladogram that placed the species in the subfamily Ototretinae at the outgroup of Lampyridae (Branham and Wenzel 2001, 2003). Based on this result, Lawrence et al. (2010a,b) considered the genera *Stenocladius* and *Drilaster* (both are formerly in Ototretinae of Lampyridae) as 'Elateriformia Incertae Sedis'. Molecular analyses based primarily on mitochondrial genes suggested that *Rhagophthalmus* is included within Lampyridae (Suzuki 1997, Li et al. 2006, Stanger-Hall et al. 2007). On the other hand, the analyses primarily based on nuclear genes (Bocakova et al. 2007, Sagegami-Oba et al. 2007a) strongly supported the monophyly of Lampyridae (*Rhagophthalmus* is not included). On the basis of these results, together with the conclusions in sections above, we proposed that bioluminescence in Coleoptera arose independently in two lineages: Elateridae and Lampyridae-Phengodidae-Rhagophthalmidae.

The status of Ototretinae defined by Crowson (1972) has been questioned by morphological analysis (Branham and Wenzel 2001, 2003) and some molecular analyses (Suzuki 1997, Li et al. 2006, Stanger-Hall et al. 2007). However, recent molecular analyses primarily based on nuclear genes reproduced the monophyly of Ototretinae (Bocakova et al. 2007, Sagegami-Oba et al. 2007a). More recent revision of adult morphology reconfirmed the status of Ototretinae (Janisova and Bocakova 2013).

On the basis of morphological and molecular phylogenetic analyses, several authors have discussed the evolution of traits in Lampyridae, such as gain or loss of luminescence, flight ability, and glow/flash (Suzuki 1997, Branham and Wenzel 2001, 2003, Stanger-Hall et al. 2007, South et al. 2011). However, the bootstrap support values on the nodes of the trees were relatively low when the molecular analysis was performed primarily on the basis of mitochondrial 16S and/or COI gene fragments. In my own opinion, we should discuss the history of fireflies on the basis of more robust tree having high bootstrap support values on the nodes by using nuclear genes and/or mitochondrial genomes.

Beetle Luciferase Genes

Since 1985, beetle luciferase genes have been isolated in over 30 species of luminous beetles (see Oba et al. 2013a). It is noteworthy that cloning efforts had started to aim the biotechnological application from the first (de Wet et al. 1985). However, the continual accumulation of sequenced genes has allowed structural-functional and evolutionary studies on beetle luciferase.

Molecular Cloning and Application

The firefly luciferase gene was cloned (de Wet et al. 1985) and the cDNA and genomic DNA sequences were determined (de Wet et al. 1987) for the first time from the North American firefly *Photinus pyralis*. The open reading frame encoded 550 amino acids, with the peroxisome targeting signal (Gould et al. 1989) at the C-terminus: the peroxisomal localization of firefly luciferase in the lantern was confirmed using immunocryoelectron microscopy (Keller et al. 1987). The recombinant proteins, expressed in bacteria *E. coli* (de Wet et al. 1985), plant cells (Ow et al. 1986), mammalian cells (de Wet et al. 1987), and yeast *S. cerevisiae* (Tatsumi et al. 1988), possessed bioluminescence activity upon the addition of firefly luciferin (beetle luciferin) and ATP. These findings strongly demonstrated the extensive applicability of the firefly luciferase gene, particularly for use as a reporter gene. Since then, beetle bioluminescence has used in every field of life science as a biotool. A second beetle luciferase gene was isolated from the Japanese Genji-firefly *Luciola cruciata* (Masuda et al. 1989, Tatsumi et al. 1989). The open reading frame encoded 548 amino acids, and the amino acid sequence identity to *P. pyralis* was 67%. The first click beetle luciferase genes were isolated from the Jamaican fire beetle *Pyrophorus plagiophthalmus* (Wood et al. 1989). They found four different types of genes from a single species, and the recombinant proteins catalyzed different colors of luminescence: green, yellow-green, yellow, and orange (Wood et al. 1989). The amino acid identities among click beetle luciferases were 95–99%, and that to *P. pyralis* luciferase was 48% (Wood et al. 1989). Phengodid and rhagophthalmid luciferase genes were isolated from the North American *Phengodes* sp. (Gruber et al. 1997) and the Japanese *R. ohbai* (Ohmiya et al. 2000), respectively. Now we know that beetle luciferases from all four families containing luminous species shared significant amino acid identity (>46%) with few gaps. Incidentally, this fact might oppose the hypothesis that the bioluminescence of luminous click beetles is independent of those of other luminous beetles (see above). We discuss this point further below.

To date, beetle luciferases isolated from luminous beetles have been used for various kinds of biotechnological tools. Perhaps the most noteworthy application for firefly luciferase is 'pyrosequencing', which is a DNA sequencing technology built on the real-time monitoring of DNA synthesis by bioluminescence (Ahmadian et al. 2006). The idea of pyrosequencing was proposed in 1987, the year the first firefly luciferase gene was identified (Nyrén 1987); recently it was established as a next-generation sequencing technology. Project 'JIM', a whole-genome sequencing project of James Watson, was completed using pyrosequencing (Wheeler et al. 2008). A draft sequence of the Neandertal genome was also completed with the assistance of pyrosequencing (Green et al. 2010).

Firefly luciferase genes have been isolated from nocturnal fireflies, whose adults emit bright light for sexual communication, in the subfamilies Lampyridae, Luciolinae, and Photurinae. Recently, luciferase genes were also isolated from diurnal fireflies, whose adults are nonluminous or faintly luminescent and mainly use pheromones for communication. The Asian firefly *Lucidina biplagiata* (Lampyrinae) is diurnal and nonluminous in the adult stage. The luciferase gene isolated from this species was most similar to that of the nocturnal firefly *P. pyralis* (Lampyrinae), and the luminescence activity of the recombinant protein was comparable to that of those in nocturnal fireflies *P. pyralis* and *L. cruciata* (Oba et al. 2010a). To understand the nonluminosity of the adult *L. biplagiata*, the luciferase and luciferin contents were compared to those in the bright luminescent species *L. cruciata* in adults. The results showed that luciferase and luciferin in *L. biplagiata* were about 0.1% of those in *L. cruciata* (Oba et al. 2008a, 2010a). Thus, the nonluminosity of *L. biplagiata* in adults is explained by low levels of both luciferase and luciferin contents. The lampyrid species in the subfamilies Psilocladinae and Ototretinae are diurnal and nonluminescent or weakly luminescent in adults, and may use pheromones as a major communication tool. We isolated the luciferase genes from *Cyphonocerus ruficollis* (Psilocladinae), *Drilaster axillaris* (Ototretinae), and *Stenocladius azumai azumai* (Ototretinae) (Oba et al. 2012). The amino acid sequence identities of these three luciferases were higher to each other (81–89%) than to other firefly luciferases previously isolated (60–81%). Phylogenetic analysis (Fig. 1) showed that these luciferases belong together in the clade of luciferases in Lampyridae (Oba et al. 2012). These results support the assumption that the species in *Drilaster* and *Stenocladius* are closely related to each other in the family Lampyridae.

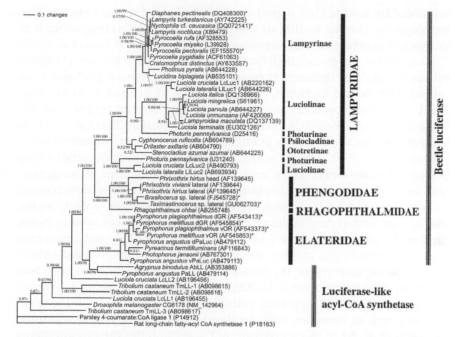

Figure 1. Bayesian phylogenetic tree of beetle luciferases and homologues based on the deduced amino acid sequences. The genes for which only partial sequences were reported are not included. Multiple amino acid sequence alignment was performed using the L-INS-i strategy of Mafft 7 (Kato et al. 2002) with a BLOSUM62 matrix. The Bayesian inference was carried out using MrBayes 3.2.2 (Huelsenbeck and Ronquist 2001) under mixed models. We conducted two simultaneous chains for 5,000,000 generations, sampled trees every 100 cycles, discarded the first 12,500 trees, and used majority rule consensus on the remaining trees to determine the clade posterior probabilities. The posterior probabilities/neighbor-joining bootstrap values (%) from 1,000 replicates calculated using ClustalW 2.1 are indicated on the nodes. Asterisks indicate the luciferase genes for which luminescence activities of the recombinant protein were not examined in the references. Luciferases in *Phengodes* sp. (Gruber et al. 1997), *Macrolampis* sp. (Viviani et al. 2005), *Amydetes fanestratus* (Viviani et al. 2011), *Fulgeochlizus bruchi* (Amaral et al. 2012), and 'protoluciferase' in *Zophobas morio* (Viviani et al. 2009) could not be included in this analysis because there has been no deposition in GenBank.

Structure and Function of Beetle Luciferase

Amino acid sequence similarity for beetle luciferases have been found in plant 4-coumarate:CoA ligase (34%, Schröder 1989) and in other adenylate-forming enzymes (>20%, Toh 1991), indicating that the protein structures are essentially homologous to each other (Chothia and Lesk 1986). The X-ray crystal structures of firefly luciferase were determined for the first time using the unliganded recombinant *P. pyralis* luciferase, and the results showed that firefly luciferase consisted two distinct domains separated by a wide cleft (Conti et al. 1996). Each domain structure was topologically similar to that of the phenylalanine-activating subunit of gramicidin synthetase 1, an adenylate-forming enzyme (Conti et al. 1997).

Luciferase genes isolated from various luminous beetles exhibited different colors of luminescence when the recombinant proteins were mixed with beetle luciferin, Mg^{2+}, and ATP at optimum weak basic conditions: the spectral maxima were from 536 nm (green) for the Panamanian luminous click beetle *Pyrophorus angustus* (Oba et al. 2010b) to 622 nm (red) for the Brazilian phengodid railroad worm *Phrixothrix hirtus* (Viviani et al. 1999). Amino acid comparisons revealed the relationship between luminescence color and the primary structure of luciferase, and several color-shifted mutant luciferases have been constructed by site-directed and random mutagenesis. However, a convincing mechanism underlying bioluminescence color differences among beetle luciferases remains elusive (see Branchini et al. 2007, and references therein). Interestingly, most of the mutant proteins exhibited red-shifted luminescence; conversely, the mutant for blue-shifted luminescence was relatively difficult to design. In 2006, the crystal structures of *L. cruciata* luciferase, wild-type emitting yellow-green and single-point mutant emitting red, in complex with luciferyl-AMP analog was determined (Nakatsu et al. 2006). A comparison of these catalytic core environments suggested that the molecular rigidity of the excited state oxyluciferin in the catalytic domain of luciferase controlled the color (Nakatsu et al. 2006).

Beetle bioluminescence is remarkable for its strong light with no detectable heat, and its efficiency was once expected to be almost 100% (Harvey 1952). In 1959, the quantum yield, the absolute number of light quanta emitted per luciferin molecule oxidized, of firefly bioluminescence was measured using purified and crystallized luciferin and luciferase from *P. pyralis* (Seliger and McElroy 1959). The results of 39 independent measurements showed an average value of 0.88 ± 0.25 at pH 7.6. Recently, the quantum yield was reexamined using firefly luciferase purified from *P. pyralis* and synthetic D-form-pure beetle luciferin, and the value was calculated to be 0.41 ± 0.074 at pH 8.5 (Ando et al. 2008). The quantum yield values were also measured using recombinant proteins of luciferase from Lampyridae, Phengodidae, and Elateridae, and the results showed that the value for elaterid *Pyrearinus termitilluminans* green-emitting luciferase (λmax = 539 nm) was highest (0.61) whereas that for phengodid *P. hirtus* red-emitting luciferase (λmax = 625 nm) was lowest (0.15) (Niwa et al. 2010). The value for the recombinant *P. pyralis* luciferase (0.45) was almost the same as that for naturally purified *P. pyralis* luciferase (0.48, Ando et al. 2008). These values are lower than formerly expected but comparable to those of other bioluminescent systems, such as 0.30 for marine ostracod

Vargula hilgendorfii, 0.11–0.18 for jellyfish *Aequorea aequorea*, and 0.10–0.16 for luminous bacteria (Shimomura 2006).

Phylogeny and Evolution of Beetle Luciferase

Molecular phylogenetic analyses revealed that luciferases in Phengodidae and Rhagophthalmidae are grouped together, and they are more closely related to those in Lampyridae than to those in Elateridae (Fig. 1). This result is consistent with recent molecular phylogenetic analyses of nuclear DNA and mitochondrial genomes described above.

It has been suggested that some luminous beetles may possess two or more luciferase genes: e.g., the North American firefly *Photuris pennsylvanica* (Ye et al. 1997), the Jamaican luminous click beetle *P. plagiophthalmus* (Wood et al. 1989), and the Brazilian phengodid railroad worm *P. hirtus* (Viviani et al. 1999, Arnoldi et al. 2010). However, the multiplicity of luciferase genes in these species was not fully elucidated, because two or more specimens were used for gene isolation in these studies.

In luminous click beetles, two or more different luciferase genes were separately isolated from single specimens of dorsal (pronotal) and ventral (abdominal) light organs in the Jamaican *P. plagiophthalmus* (Stolz et al. 2003), Dominican *Pyrophorus mellifluus* (Feder and Velez 2009), and Panamanian *P. angustus* (Oba et al. 2010b). Two luciferase isoforms, dPaLuc and vPaLuc, isolated from *P. angustus* were predominantly expressed in the adult dorsal and ventral light organs, and their recombinant proteins exhibited different luminescence colors: green (λmax = 536 nm) and yellow (λmax = 566 nm), respectively (Oba et al. 2010b). In Lampyridae, Day et al. (2009) clearly suggested the multiplicity of the luciferase gene in several lampyrid beetles by genomic PCR analysis using single specimen for each species (Day et al. 2009). The duality of the luciferase gene was demonstrated in Japanese *L. cruciata* (Oba et al. 2010c) and then in Japanese *Luciola lateralis* (Oba et al. 2013a) using a single specimen for gene isolation. In contrast to that of click beetle luciferases expressed in different light organs in adult (Oba et al. 2010b), the luciferase isoforms in fireflies were expressed in different life stages (Oba et al. 2010c, 2013). Two luciferase isoforms, LlLuc1 and LlLuc2, were isolated from *L. lateralis*: LlLuc1 was predominantly expressed in larval, pupal and adult lanterns, while LlLuc2 was expressed in eggs and the pupal whole body (Oba et al. 2013a). The luminescence colors of the recombinant proteins of LlLuc1 and LlLuc2 were yellow (λmax = 550 nm) and green (λmax = 539 nm), respectively, suggesting that LlLuc1 is responsible for the on/off-capable yellowish luminescence of larval, pupal and adult lanterns, and also suggesting that LlLuc2 is responsible for the continuous dim greenish glow of eggs and whole pupal body (Oba et al. 2013a). It should be expected that two different luciferase isoforms are present in the phengodid beetles, and were used for head and lateral lanterns of larviform adult females, respectively (Arnoldi et al. 2010). Larviform adult females in Rhagophthalmidae possess two different types of lanterns, a large abdominal lantern and small lateral lanterns, but the multiplicity of the luciferase gene was not examined.

Phylogenetic trees of luciferase genes indicated that gene duplication of the luciferase occurred independently in at least three lineages: Lampyridae, Phengodidae, and Elateridae (Fig. 1). Arnoldi et al. (2010) expected that the gene duplication of luciferase in Phengodidae occurred within the lineage of a tribe (Mastinocerini), because head lanterns were observed in species of the tribe Mastinocerini but not in the tribe Phengodini (Phengodidae) or Rhagophthalmidae. In luminous click beetles, although the multiplicity of the luciferase was elucidated in only a single genus, *Pyrophorus*, the color difference between dorsal and ventral lanterns was also observed in other genera, such as *Pyrearinus*, *Opselater*, and *Hapsodrilus* (Bechara 1988). Thus, the gene duplication of luciferase in Elateridae may have occurred before the divergence of the tribe Pyrophorini. In a phylogenetic tree (Fig. 1), click beetle luciferases derived from the dorsal lantern are not grouped together, nor are those from the ventral lantern. This is explained by the intergenic recombination between dorsal and ventral luciferases (Stolz et al. 2003, Feder and Velez 2009). In Lampyridae, on the other hand, the phylogenetic tree showed that gene duplication of luciferase occurred at the most basal level before the divergence of subfamilies, indicating the duality (or multiplicity) of luciferase in all lampyrid species. This view might be supported by the fact that two or more luciferase-like gene fragments were amplified by genomic PCR from the lampyrid species in the subfamilies Lampyrinae, Luciolinae, and Photurinae (Day et al. 2009).

Origin of Beetle Luciferase

The bioluminescence reaction in beetles is an oxygenation of beetle luciferin catalyzed by beetle luciferase in the presence of ATP, Mg^{2+}, and oxygen. The reaction occurs in two steps (Fig. 2A): first, the carboxyl group of beetle luciferin is adenylated by luciferase to give luciferyl-AMP; second, luciferase oxygenates luciferyl-AMP with O_2 to give the excited state of oxyluciferin, CO_2, and AMP; accompanied by the excited oxyluciferin to the ground state, visible light is elicited (Wood 1995). On the basis of the facts that the first step of the reaction is exactly analogous to the activation reactions of fatty acids and other carboxylic acids, and that luminescence is enhanced by the addition of CoA, the resemblance

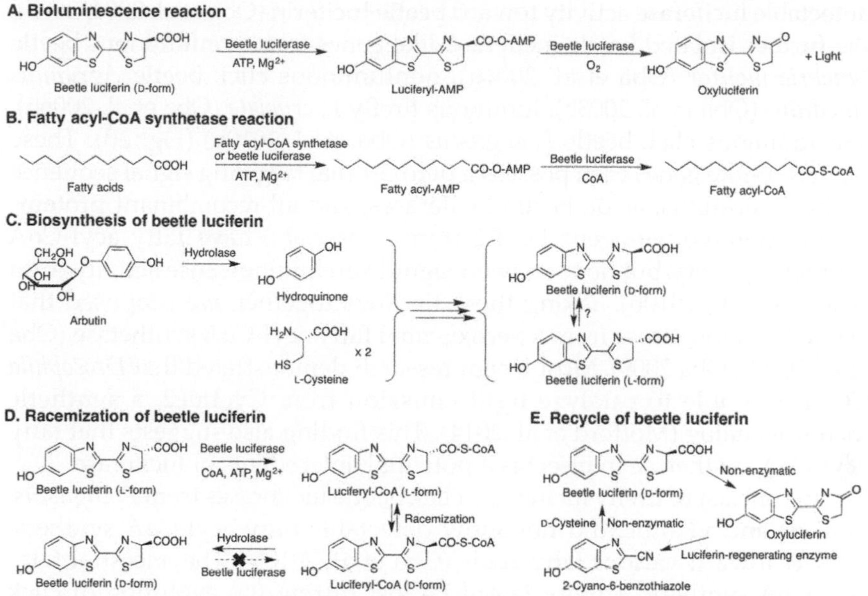

A. Bioluminescence reaction

Beetle luciferin (D-form)   Luciferyl-AMP   Oxyluciferin

B. Fatty acyl-CoA synthetase reaction

Fatty acids   Fatty acyl-AMP   Fatty acyl-CoA

C. Biosynthesis of beetle luciferin

Arbutin   Hydroquinone   Beetle luciferin (D-form)

L-Cysteine   Beetle luciferin (L-form)

D. Racemization of beetle luciferin

Beetle luciferin (L-form)   Luciferyl-CoA (L-form)

Beetle luciferin (D-form)   Luciferyl-CoA (D-form)

E. Recycle of beetle luciferin

Beetle luciferin (D-form)   Oxyluciferin

2-Cyano-6-benzothiazole

Figure 2. Schemes of biochemistry in luminous beetles. (A) Bioluminescence reaction catalyzed by beetle luciferases. (B) Fatty acyl CoA synthetic reaction catalyzed by fatty acyl-CoA synthetase in all organisms and firefly luciferases. (C) Biosynthetic process of beetle luciferin (after Oba et al. 2013b). Hydrolase (ß-glucosidase) and biosynthetic enzyme(s) are not identified. (D) Racemization process of beetle luciferin (after Niwa et al. 2006). Hydrolase (acyl-CoA thioesterase) is not identified. (E) Recycling process of beetle luciferin (after Gomi et al. 2001).

between beetle luciferase and acyl-CoA synthetase had been pointed out (McElroy et al. 1967) prior to the molecular cloning of luciferase. After beetle luciferases and several acyl-CoA synthetases were cloned, the significant similarity between their primary sequences was revealed. This finding indicated that beetle luciferase evolved from an acyl-CoA synthetase, such as 4-coumarate:CoA ligase, fatty acyl-CoA synthetase, and acetyl-CoA synthetase (Wood 1995).

In 2003, the author and colleagues found that firefly luciferase is a bifunctional enzyme, possessing not only luminescence activity toward beetle luciferin (Fig. 2A), but also fatty acyl-CoA synthetic activity toward various medium- and long-chain fatty acids (Oba et al. 2003). For example, lauric acid (C12:0) is efficiently converted to lauroyl-CoA by firefly luciferase in the presence of ATP, Mg^{2+}, and CoA (Oba et al. 2005) (Fig. 2B). We isolated and characterized CG6178, the most similar gene to beetle luciferase (41%) in the *Drosophila melanogaster* genome (Oba et al. 2004). The recombinant protein CG6178 had fatty acyl-CoA synthetic activity toward various medium- and long-chain fatty acids, but did not show detectable luciferase activity toward beetle luciferin (Oba et al. 2004, 2005). We further isolated beetle luciferase-like genes from nonluminous beetle *Tenebrio molitor* (Oba et al. 2006a), nonluminous click beetle *Agrypnus binodulus* (Oba et al. 2008b), luminous firefly *L. cruciata* (Oba et al. 2006b), and luminous click beetle *P. angustus* (Oba et al. 2010b) (Fig. 2B). These luciferase-like genes each possess a peroxisomal targeting signal sequence at the C-terminus, as do beetle luciferases, and all recombinant proteins (except homologous gene LcLL2 from *L. cruciata*) have fatty acyl-CoA synthetic activity, but none showed significant luminescence activity (Oba et al. 2006a,b, 2010b). Taking these findings together, we proposed that beetle luciferase arose from a peroxisomal fatty acyl-CoA synthetase (Oba et al. 2006b, Oba 2009). Most recent research demonstrated that *Drosophila* CG6178 is able to catalyze light emission from CycLuc2, a synthetic luciferin analog (Mofford et al. 2014). This finding also suggests that fatty acyl-CoA synthetase in insect is a potential antecedents of luciferase.

In contrast to firefly luciferase, click beetle luciferases from *P. angustus*, dPaLuc and vPaLuc, did not show detectable fatty acyl-CoA synthetic activity toward various fatty acids (Oba et al. 2010b). The ancestral fatty acyl-CoA synthetic activity might be lost during the evolution of click beetle luciferase. The acyl-CoA synthetic activity of the luciferases in Phengodidae and Rhagophthalmidae has not been examined. Day et al. (2009) reported that the orthologous gene of luciferase was not detected from the species in the family Cantharidae (Elateroidea) by genomic PCR analysis using degenerate primers. The 'nonluminous' families in Elateroidea might have lost the ancestral luciferase gene during the course of their evolution.

Recently, the author and colleagues demonstrated that fatty acyl-CoA synthetases in nonluminous click beetle *A. binodulus* (AbLL) and *Drosophila* (CG6178) acquired significant luminescence activities by site-directed mutagenesis (Oba et al. 2009, Inouye 2010). The mutation sites were selected from the residues interacting with the luciferin molecule elucidated by the crystal structure analysis of firefly luciferase (Nakatsu et al. 2006). As a result, wild-type AbLL, having luminescence so insignificant it was nearly at the background level, gained over 250-fold higher luminescence by the mutation of three residues: C343S/G344T/L345S. *Drosophila* CG6178 also gained about 3.5-fold higher luminescence by single-point mutation (L346S). Although the luminescence activity of the triple-point mutant of AbLL was only 0.001% that of native firefly luciferase, these results support the hypothesis that ancestral luciferase was a fatty acyl-CoA synthetase. Afterward, Prado et al. (2011) reported similar mutagenesis analysis using a luciferase-like gene isolated from the nonluminous beetle *Zophobas morio*. Although the original enzymatic function of this gene product was not examined, those authors suggested that the recombinant protein had significant luminescence activity when mixed with beetle luciferin and ATP (Prado et al. 2011). They also showed that single-point mutants at I327, which corresponds to C343 in AbLL, of *Z. morio* luciferase-like gene (I327S and I327T), exhibited 1.3- to 2.3-fold higher luminescence than the wild type (Prado et al. 2011). With these results taken together, it seems that the evolution of luciferase may have been readily triggered by small numbers of mutations for the ancestral enzyme, and this will support the hypothesis that the beetle luciferases arose independently in two lineages by parallel evolution (Oba 2009). However, if so, then the biosynthetic ability of beetle luciferin also must have evolved independently in these two lineages, because beetle luciferin in all luminous beetles is structurally identical.

Biosynthesis and Recycling of Beetle Luciferin

Elucidation of the origin of beetle luciferin is a key to understand the evolution of beetle bioluminescence, but it is still enigmatic (Day et al. 2004). Beetle luciferin, (*S*)-2-(6′-hydroxy-2′-benzothiazolyl)-2-thiazoline-4-carboxylic acid, is a unique natural product consisting of a benzothiazole unit attached to a thiazole carboxylic acid unit (Fig. 2); a benzothiazole unit is rare in nature, and the carbon center of the carboxylic acid in the thiazole unit has L-configuration, which is unusual in natural amino acids and their derivatives (Day et al. 2004). Luminous species in Lampyridae and Elateridae contain beetle luciferin in the range of pmol to hundreds of nmol per specimen, but no luciferin was detected in the nonluminous elateroid species in Cantharidae, Lycidae, and Elateridae (Oba et al. 2008a). Thus it is expected that beetle luciferin is present only in luminous beetles and not in their relatives.

The biosynthesis of beetle luciferin has been examined by injecting or feeding ^{14}C radioisotope-labeled compounds into living fireflies and luminous click beetles (Okada et al. 1976, McCapra and Razavi 1976). These results revealed that the biosynthetic component of the benzothiazole unit is *p*-benzoquinone (or its equivalent, hydroquinone) (Okada et al. 1976), and that of the thiazole unit is cysteine (McCapra and Razavi 1976, Colepicolo et al. 1988). However, no other component of the benzothiazole unit has been identified, and the stereochemistry of cysteine used for biosynthesis has not been solved. Recently, the author and colleagues demonstrated the full components of beetle luciferin biosynthesis by injecting stable isotope-labeled compounds into the adult lantern of the living *L. lateralis* (Oba et al. 2013b). The results showed that D-form beetle luciferin is biosynthesized in the lantern of the adult firefly from two L-cysteine molecules and hydroquinone (Fig. 2C): the benzothiazole unit is biosynthesized from one hydroquinone and one

L-cysteine, accompanied by decarboxylation of L-cysteine; and the D-form thiazole unit can be biosynthesized from L-cysteine (Oba et al. 2013b). In insects, the endogenous 'more toxic' hydroquinone has been identified from defensive secretions of some tenebrionid beetles (Chadha et al. 1961, Gnanasunderam et al. 1985, Holliday et al. 2009) and from a cockroach (Roth and Stay 1958). In fireflies, on the other hand, we could not detect the presence of hydroquinone in the extract of adult *L. lateralis*; instead we detected arbutin, a glycosylated 'less toxic' form of hydroquinone (Oba et al. 2013b). In the tenebrionid beetle *Eleodes longicollis*, it has been known that hydroquinone is stored in secretory cells as arbutin and is occasionally secreted to defensive glands in where hydroquinone is released by glucosidase and used for defensive substances (Happ 1968). Perhaps fireflies also store hydroquinone for the biosynthesis of firefly luciferin as its less-toxic form, arbutin, to avoid self-poisoning. To elucidate the biosynthesis mechanism and origin of beetle luciferin, further studies would be necessary, including the identification of the biosynthetic intermediate and isolation of the biosynthetic enzyme(s) using both fireflies and luminous click beetles.

Beetle luciferin has one chiral center, and only D-form luciferin is utilized for the luminescence reaction. In fireflies, most beetle luciferin present in the lantern is D-form, but a small amount of L-form is also detected (Niwa et al. 2006, Oba et al. 2013b). Niwa et al. (2006) showed that D-form beetle luciferin can be converted enzymatically from L-form beetle luciferin in the firefly lantern. They proposed that L-form luciferyl-CoA is generated from L-form luciferin by firefly luciferase, after which racemization occurs followed by D-form luciferyl-CoA hydrolyzation by an unidentified enzyme to give D-form luciferin, during the *de novo* biosynthesis and/or recycling process (Niwa et al. 2006) (Fig. 2D).

Okada et al. (1974) proposed that oxyluciferin, the reaction product of the luciferase reaction, is recycled to regenerate beetle luciferin in the firefly lantern via 2-cyano-6-hydrobenzothiazole. Actually, 2-cyano-6-hydroxybenzothiazole is generated from oxyluciferin in neutral buffer (Okada et al. 1974), and condenses with cysteine to yield beetle luciferin in pH 8 buffer at room temperature (White et al. 1963). Gomi et al. (2001) reported the isolation of a 'luciferin-regenerating enzyme' gene from lanterns of the firefly *P. pyralis*. They suggested that this enzyme catalyzes the first step of the luciferin recycling process, oxyluciferin to 2-cyano-6-hydroxybenzothiazole, while the second step, 2-cyano-6-hydroxybenzothiazole to beetle luciferin, proceeds non-enzymatically in the presence of D-cysteine (Gomi et al. 2001) (Fig. 2E). The luciferin-regenerating enzyme is approximately 34 kDa protein having significant similarity (~39%) to anterior fat body protein in the flesh fly (Gomi et al. 2002). Homologous genes were also isolated from the fireflies *L. cruciata* and *L. lateralis,* and the enzymatic activity of the recombinant proteins was confirmed *in vitro* (Gomi et al. 2002). However, further study will be necessary to elucidate the involvement of luciferin-regenerating enzyme for the recycling system of luciferin in luminous beetles *in vivo*.

Biogeography of Fireflies

The Japanese Genji firefly, *L. cruciata,* contains two ecotypes exhibiting 'fast-flash' and 'slow-flash' patterns: an adult male in the fast-flash ecotype emits repeatedly at approximately 2-sec intervals when flying to search for mates, while an adult male in the slow-flash ecotype emits at approximately 4-sec intervals (Kanda 1935, Ohba 1984, 2001). It has been reported that females of slow-flash populations tend to prefer slow-flash mates, and *vice versa* (Tamura et al. 2005). Molecular phylogenetic analysis of the *L. cruciata* populations using mitochondrial COII fragments revealed that the fast-flash ecotype distributed in Western Japan is the ancestral state and that the slow-flash ecotype distributed in Eastern Japan is a derived state, with a few discordances in the populations near the ecotype boundary areas (Suzuki et al. 2002). The discordances should be explained by introgressive hybridization between two ecotypes (Suzuki et al. 2002).

The firefly *Pyrocoelia rufa* is distributed widely in Korea and China, as well as on the Tsushima Islands in Japan. The female adult is flightless, with reduced elytra and the absence of a hindwing, and thus is expected to be a poor disperser. The adults of both sexes are luminous, and the male and female communicate using both luminescence and pheromone (Ohba 1983). Molecular analysis of Korean and Tsushima populations using mitochondrial COI fragments showed that the haplotypes in the Jeju Island population are clearly separate from those on the Korean mainland and on the Tsushima Islands (Lee at al. 2003), suggesting the necessity of reexamining the taxonomic status including morphologies and luminescent behavior. Indeed, the populations on Jeju Island were formerly classified as an independent species, *P. saishutonis* (Lee et al. 2003).

In the Ryukyu archipelago of Japan, two closely related species in Luciolinae, *Curtos costipennis* and *C. okinawanus,* are allopatrically distributed. These two species are easily distinguishable from the color of elytra, but interestingly, the distribution of the latter species occupies the middle of the distribution range of the former species (Ohba and Goto 1993). Molecular phylogenetic analysis of the populations of these two species using nuclear and mitochondrial DNAs revealed that *C. costipennis* is paraphyletic and nests the clade of *C. okinawanus* (Muraji et al. 2012). From this result, those authors concluded the necessity of reevaluating the taxonomic status of these two species (Muraji et al. 2012).

NO Synthase and Opsin in Fireflies

Except for the phylogenetic analyses and luciferase characterization (see above), there are very few molecular studies on luminous beetles. In this final section, the author highlights two such topics, NO synthase and opsin genes in fireflies.

It has been suggested that nitric oxide (NO) and NO synthase are involved in the neuronal flash on/off control by regulating the O_2 supply to the firefly lantern (Trimmer et al. 2001). Ohtsuki et al. (2008) isolated NO synthases from the Japanese fireflies *L. cruciata* and *L. lateralis*, and compared them to those from other insects. However, positively selected sites were not detected in these firefly NO synthases through the amino acid sequences (Ohtsuki et al. 2008).

It has been suggested that visual sensitivity in luminous beetles is optimized to accept their species-specific luminescence spectra (Lall et al. 2009). Electroretinogram analysis revealed that the compound eyes of the firefly *L. cruciata* show two spectral sensitivity peaks corresponding to ultraviolet (360 nm) and long (560 nm) wavelengths, and the latter matches the spectral maximum of the luminescence emission (Eguchi et al. 1984, Sakayori et al. 1992). Recently, the author and colleague isolated ultraviolet-sensitive and long-wavelength-sensitive opsin genes from the firefly *L. cruciata* (Oba and Kainuma 2009). Quantitative PCR showed that the mRNA level of long-wavelength-sensitive opsin in female adults is higher at night than in the day, peaking at 20:00, the time when luminescence behavior was most active. In adult males, dial changes in the expression of both opsins were not significant. These results suggest that the expression level of 'bioluminescence-sensitive' opsin in female *L. cruciata* is linked to mating behavior (Oba and Kainuma 2009c).

Conclusion

Studies on bioluminescent insects have focused largely on beetles, especially fireflies. After molecular cloning of firefly luciferase was first accomplished in 1985 and 1987, the study of insect bioluminescence has further concentrated on the mechanisms underlying luminescence in beetles. As a result, many questions regarding beetle luminescence systems have been answered. Conversely, it seems there has been a lack of focus on ecological studies of insect bioluminescence. However, the outcomes of extensive studies on beetle luciferase have yielded various biotechnological tools, especially reporter assays and next-generation sequencing techniques. The author thinks that these new technologies will shed new light on insect bioluminescence, including ecological aspects, not only in luminous beetles, but also in luminous fungus gnats and springtails in the near future.

Keywords: Bioluminescence, Coleoptera, evolution, fungus gnat, luciferase, luciferin, Lampyridae, Phengodidae, Rhagophthalmidae, Elateridae, firefly

References

Ahmadian, A., M. Ehn and S. Hober. 2006. Pyrosequencing: History, biochemistry and future. Clin. Chim. Acta 363: 83–94.

Amaral, D.T., R.A. Prado and V.R. Viviani. 2012. Luciferase from *Fulgeochlizus bruchi* (Coleoptera: Elateridae), a Brazilian click-beetle with a single abdominal lantern: molecular evolution, biological function and comparison with other-beetle luciferases. Photochem. Photobiol. Sci. 11: 1259–1267.

Ando, Y., K. Niwa, N. Yamada, T. Enomoto, T. Irie, H. Kubota, Y. Ohmiya and H. Akiyama. 2008. Firefly bioluminescence quantum yield and colour change by pH-sensitive green emission. Nature Photonics 2: 44–47.

Arnoldi, F.G.C., K. Ogoh, Y. Ohmiya and V.R. Viviani. 2007. Mitochondrial genome sequence of the Brazilian luminescent click beetle *Pyrophorus divergens* (Coleoptera: Elateridae): Mitochondrial genes utility to investigate the evolutionary history of Coleoptera and its bioluminescence. Gene 405: 1–9.

Arnoldi, F.G.C., A.J. da Silva Neto and V.R. Viviani. 2010. Molecular insights on the evolution of the lateral and head lantern luciferases and bioluminescence colors in Mastinocerini railroad-worms (Coleoptera: Phengodidae). Photochem. Photobiol. Sci. 9: 87–92.

Baker, C.H., G.C. Graham, K.D. Scott, S.L. Cameron, D.K. Yeates and D.J. Merritt. 2008. Distribution and phylogenetic relationships of Australian glow-worms *Arachnocampa* (Diptera: Keroplatidae). Mol. Phylogenet. Evol. 48: 506–514.

Baker, C.H. 2010. A new subgenus and five new species of Australian glow-worms (Diptera: Keroplatidae: *Arachnocampa*). Mem. Queensland Mus. Nature 55: 11–43.

Barber, H.S. 1908. The glow-worm *Astraptor*. Proc. Washington Ent. Soc. 9: 41–43.

Bechara, E.J.H. 1988. Luminescent elaterid beetles: Biochemical, biological, and ecological aspects. pp. 123–178. *In*: A.L. Baumstark (ed.). Advances in Oxygenated Processes, Vol. 1. JAI Press, London.

Bertkau, P. 1891. Beschreibung der Larve und des Weibchens von *Homalisus suturalis*. Dtsch. Ent. Zeitschr. 1: 37–42 + 1 plt.

Beutel, R.G. 1995. Phylogenetic analysis of Elateriformia (Coleoptera: Polyphaga) based on larval characters. J. Zoo. Syst. Evol. Res. 33: 145–171.

Bocakova, M., L. Bocak, T. Hunt, M. Teräväinen and A.P. Vogler. 2007. Molecular phylogenetics of Elateriformia (Coleoptera): evolution of bioluminescence and neoteny. Cladistics 23: 477–496.

Branchini, B.R., D.M. Ablamsky, J.M. Rosenman, L. Uzasci, T.L. Southworth and M. Zimmer. 2007. Synergistic mutations produce blue-shifted bioluminescence in firefly luciferase. Biochemistry 46: 13847–13855.

Branham, M.A. and J.W. Wenzel. 2001. The evolution of bioluminescence in cantharoids (Coleoptera: Elateroidea). Flor. Entomol. 84: 565–586.

Branham, M.A. and J.W. Wenzel. 2003. The origin of photic behavior and the evolution of sexual communication in fireflies (Coleoptera: Lampyridae). Cladistics 19: 1–22.

Branham, M.A. 2010. Lampyridae Latreille, 1817. pp. 141–149. *In*: R.A.B. Leschen, R.G. Beutel and J.F. Lawrence (eds.). Handbook of Zoology, Vol. IV, Arthropoda: Insecta, Teilband 39, Coleoptera, Beetles. Vol. 2: Morphology and Systematics. Walter de Gruyter, Berlin.

Broadley, R.A. 2012. Notes on pupal behaviour, eclosion, mate attraction, copulation and predation of the New Zealand glowworm *Arachnocampa luminosa* (Skuse) (Diptera: Keroplatidae), at Waitomo. Entomol. Soc. New Zealand 35: 1–9.

Burakowski, B. 1988. Observations on the larval morphology and biology of *Omalisus fontisbellaquei* Fourcroy (Coleoptera, Homalisidae). Bull. Entomol. Pologne. 58: 571–574.

Chadha, M.S., T. Eisner and J. Meinwald. 1961. Defence mechanisms of arthropods–IV: Para-benzoquinones in the secretion of *Eleodes longicollis* Lec. (Coleoptera: Tenebrionidae). J. Insect Physiol. 7: 46–50.

Chothia, C. and A.M. Lesk. 1986. The relation between the divergence of sequence and structure in protein. EMBO J. 5: 823–826.

Colepicolo, P., D. Pagni and E.J.H. Bechara. 1988. Luciferin biosynthesis in larval *Pyrearinus termitilluminans* (Coleoptera: Elateridae). Comp. Biochem. Physiol. B 91: 143–147.

Conti, E., N.P. Franks and P. Brick. 1996. Crystal structure of firefly luciferase throws light on a superfamily of adenylate-forming enzymes. Structure 4: 278–298.

Conti, E., T. Stachelhaus, M.A. Marahiel and P. Brick. 1997. Structural basis for the activation of phenylalanine in the non-ribosomal biosynthesis of gramicidin S. EMBO J. 16: 4174–4183.

Costa, C. 1975. Systematics and evolution of the tribes Pyrophorini and Heligmini, with description of Campyloxeninae, new subfamily (Coleoptera, Elateridae). Arq. Zool. S. Paulo 26: 49–190.

Costa, C. 1984. Note on the bioluminescence of *Balgus schnusei* (Heller, 1974) (Trixagidae, Coleoptera). Revta Bras. Ent. 28: 397–398.

Costa, C., S.A. Vanin and P. Colepicolo Neto. 1986. Larvae of Neotropical Coleoptera. XIV. First record of bioluminescence in the family Staphylinidae (Xantholinini). Revta Bras. Ent. 30: 101–104.

Costa, C., J.F. Lawrence and S.P. Rosa. 2010. Elateridae Leach, 1815. pp. 75–103. *In*: R.A.B. Leschen, R.G. Beutel and J.F. Lawrence (eds.). Handbook of Zoology, Vol. IV, Arthropoda: Insecta, Teilband 39, Coleoptera, Beetles. Vol. 2: Morphology and Systematics. Walter de Gruyter, Berlin.

Costa, C. and S. Zaragoza-Caballero. 2010. Phengodidae LeConte, 1861. pp. 126–135. *In*: R.A.B. Leschen, R.G. Beutel and J.F. Lawrence (eds.). Handbook of Zoology, Vol. IV, Arthropoda: Insecta, Teilband 39, Coleoptera, Beetles. Vol. 2: Morphology and Systematics. Walter de Gruyter, Berlin.

Crowson, R.A. 1972. A review of the classification of Cantharoidea (Coleoptera), with the definition of two new families, Cneoglossidae and Omethidae. Rev. Univ. Madrid 21: 35–77.

Day, J.C., L.C. Tisi and M.J. Bailey. 2004. Evolution of beetle bioluminescence: the origin of beetle luciferin. Luminescence 19: 8–20.

Day, J.C., T.I. Goodall and M.J. Bailey. 2009. The evolution of the adenylate-forming protein family in beetles: Multiple luciferase gene paralogues in fireflies and glowworms. Mol. Phylogenet. Evol. 50: 93–101.

de Wet, J.R., K.V. Wood, D.R. Helinski and M. DeLuca. 1985. Cloning of firefly luciferase cDNA and the expression of active luciferase in *Escherichia coli*. Proc. Natl. Acad. Sci. USA 82: 7870–7873.

de Wet, J.R., K.V. Wood, M. DeLuca, D.R. Helinski and S. Subramani. 1987. Firefly luciferase gene: Structure and expression in mammalian cells. Mol. Cell. Biol. 7: 725–737.

Dolin, V.G. 1978. Phylogeny of click beetles (Coleoptera, Elateridae). Vestnik Zoologii 3: 3–12.

Douglas, H. 2011. Phylogenetic relationships of Elateridae inferred from adult morphology, with special reference to the position of Cardiophorinae. Zootaxa 2900: 1–45.

Eguchi, E., A. Nemoto, V.B. Meyer-Rochow and N. Ohba. 1984. A comparative study of spectral sensitivity curves in three diurnal and eight nocturnal species of Japanese fireflies. J. Insect Physiol. 30: 607–612.

Feder, J.L. and S. Velez. 2009. Intergenic exchange, geographic isolation, and the evolution of bioluminescent color for *Pyrophorus* click beetles. Evolution 63: 1203–1216.

Gnanasunderam, C., H. Young and R. Hutchins. 1985. Defensive secretions of New Zealand tenebrionids: V. Presence of methyl ketones in *Uloma tenebrionoides* (Coleoptera: Tenebrionidae). J. Chem. Ecol. 11: 465–472.

Goemans, G. 2006. The Fulgoridae (Hemiptera, Fulgoromorpha) of Guatemala. pp. 337–344. *In*: E.B. Cano (ed.). Biodiversidad de Guatemala, Vol. 1. Pub. Univ. del Vall de Guatemala, Guatemala.

Gomi, K. and N. Kajiyama. 2001. Oxyluciferin, a luminescence product of firefly luciferase, is enzymatically regenerated into luciferin. J. Biol. Chem. 276: 36508–36513.

Gomi, K., K. Hirokawa and N. Kajiyama. 2002. Molecular cloning and expression of the cDNAs encoding luciferin-regenerating enzyme from *Luciola cruciata* and *Luciola lateralis*. Gene 294: 157–166.

Green, R.E., J. Krause, A.W. Briggs, T. Maricic, U. Stenzel, M. Kircher, N. Patterson, H. Li, W. Zhai, M.S.-Y. Fritz, N.F. Hansen, E.Y. Durand, A.-S, Malaspinas, J.D. Jensen, T. Marques-Bonet, C. Alkan, K. Prüfer, M. Meyer, H.A. Burbano, J.M. Good, R. Schultz, A. Aximu-Petri, A. Butthof, B. Höber, B. Höffner, M. Siegemund, A. Weihmann, C. Nusbaum, E.S. Lander, C. Russ, N. Novod, J. Affourtit, M. Egholm, C. Verna, P. Rudan, D. Brajkovic, Ž. Kucan, I. Gušic, V.B. Doronichev, L.V. Golovanova, C. Lalueza-Fox, M. de la Rasilla, J. Fortea, A. Rosas, R.W. Schmitz, P.L.F. Johnson, E.E. Eichler, D. Falush, E. Birney, J.C. Mullikin, M. Slatkin, R. Nielsen, J. Kelso, M. Lachmann, D. Reich and S. Pääbo. 2010. A draft sequence of the Neandertal genome. Science 328: 710–722.

Gruber, M.G., G.D. Kutuzova and K.V. Wood. 1997. Cloning and expression of a *Phengodes* luciferase. pp. 244–247. *In*: J.W. Hastings, L.J. Kricka and P. Stanley (eds.). Bioluminescence and Chemiluminescence: Molecular Reporting with Photons. John Wiley & Sons, Chichester.

Gould, S.J., G-.A. Keller, N. Hosken, J. Wilkinson and S. Subramani. 1989. A conserved tripeptide sorts proteins to peroxisomes. J. Cell Biol. 108: 1657–1664.

Haddock, S.H.D., M.A. Moline and J.F. Case. 2010. Bioluminescence in the sea. Annu. Rev. Mar. Sci. 2: 443–493.

Haneda, Y. 1957. Luminous insects of Hachijo Island, Japan. Sci. Rept. Yokosuka City Mus. 2: 24–27.

Happ, G.M. 1968. Quinone and hydrocarbon production in the defensive glands of *Eleodes longicollis* and *Tribolium castaneum* (Coleoptera, Tenebrionidae). J. Insect Physiol. 14: 1821–1837.

Harvey, E.N. and K.P. Stevens. 1928. The brightness of the light of the West Indian elaterid beetle, *Pyrophorus*. J. Gen. Physiol. 12: 269–272.

Harvey, E.N. 1952. Bioluminescence. Academic Press, New York.

Herring, P.J. 1987. Systematic distribution of bioluminescence in living organisms. J. Biolumin. Chemilumin. 1: 147–163.

Hoffmann, K.H. 1984. Environmental aspects of insect bioluminescence. pp. 225–245. *In*: K.H. Hoffmann (ed.). Environmental Physiology and Biochemistry of Insects. Springer, Berlin.

Holliday, A.E., F.M. Walker, E.D. Brodie III and V.A. Formica. 2009. Differences in defensive volatiles of the forked fungus beetle, *Bolitotherus cornutus*, living on two species of fungus. Chem. Ecol. 35: 1302–1308.

Huelsenbeck, J.P. and F. Ronquist. 2001. MRBAYES: Bayesian inference of phylogenetic trees. Bioinformatics 17: 754–755.

Hyslop, J.A. 1917. The phylogeny of the Elateridae based on larval characters. Ann. Ent. Soc. Amer. Columbus 10: 241–263.

Inouye, S. 2010. Firefly luciferase: an adenylate-forming enzyme for multicatalytic functions. Cell. Mol. Life Sci. 67: 387–404.

Janisova, K. and M. Bocakova. 2013. Revision of the subfamily Ototretinae (Coleoptera: Lampyridae). Zool. Anz. 252: 1–19.

Janssens, F. and K.A. Christiansen. 2011. Class Collembola Lubbock, 1870. pp. 192–194. *In*: Z.-Q. Zhang (ed.). Animal Biodiversity: An Outline of Higher-level Classification and Survey of Taxonomic Richness. Magnolia Press, Auckland.

Kanda, S. 1935. Firefly (Hotaru). Maruzen, Tokyo.

Kato, K., K. Misawa, K. Kuma and T. Miyata. 2002. MAFFT: a novel method for rapid multiple sequence alignment based on fast Fourier transform. Nucleic Acid Res. 30: 3059–3066.

Keller, G.-A., S. Gould, M. DeLuca and S. Subramani. 1987. Firefly luciferase is targeted to peroxisomes in mammalian cells. Proc. Natl. Acad. Sci. USA 84: 3264–3268.

Kobayashi, Y., H. Suzuki and N. Ohba. 2002. Embryogenesis of the glowworm *Rhagophthalmus ohbai* Wittmer (Insecta: Coleoptera, Rhagophthalmidae), with emphasis on the germ rudiment formation. J. Morphol. 253: 1–9.

Kundrata, R. and L. Bocak. 2011. The phylogeny and limits of Elateridae (Insecta, Coleoptera): is there a common tendency of click beetles to soft-bodiedness and neoteny? Zool. Scr. 40: 364–378.

Lall, A.B., T.W. Cronin, E.J.H. Bechara, C. Costa and V.R. Viviani. 2009. Visual ecology of bioluminescent beetles: Visual spectral mechanisms and the colors of optical signaling in Coleoptera, Elateroidea: Lampyridae, Elateridae and Phengodidae. pp. 201–228. *In*: V.B. Meyer-Rochow (ed.). Bioluminescence in Focus—A Collection of Illuminating Essays. Research Signpost, Kerala.

Lawrence, J.F. 1982. Coleoptera. pp. 482–553. *In*: S.P. Parker (ed.). Synopsis and Classification of Living Organisms. McGraw-Hill, New York.

Lawrence, J.F. and A.F. Newton Jr. 1995. Families and subfamilies of Coleoptera (with selected genera, notes, references and data on family-group names). pp. 779–1006. *In*: J. Pakaluk and S.A. Slipinski (eds.). Biology, Phylogeny, and Classification of Coleoptera: Papers Celebrating the 80th Birthday of Roy A. Crowson. Mus. Inst. Zool. PAN, Warszawa.

Lawrence, J.F., R.G. Beutel, R.A.B. Leschen and A. Ślipiński. 2010a. Changes in classification and list of families and subfamilies. pp. 1–7. *In*: R.A.B. Leschen, R.G. Beutel and J.F. Lawrence (eds.). Handbook of Zoology, Vol. IV, Arthropoda: Insecta, Teilband 39, Coleoptera, Beetles. Vol 2: Morphology and Systematics. Walter de Gruyter, Berlin.

Lawrence, J.F., I. Kawashima and M.A. Branham. 2010b. Elateriformia *Incertae Sedis*. pp. 162–177. *In*: R.A.B. Leschen, R.G. Beutel and J.F. Lawrence (eds.). Handbook of Zoology, Vol. IV, Arthropoda: Insecta, Teilband 39, Coleoptera, Beetles. Vol. 2: Morphology and Systematics. Walter de Gruyter, Berlin.

Lee, S.-C., J.-S. Bae, I. Kim, H. Suzuki, S.-R. Kim, J.-G. Kim, K.-Y. Kim, W.-J. Yang, S.-M. Lee, H.-D. Sohn and B.-R. Jin. 2003. Mitochondrial DNA sequence-based population genetic structure of the firefly, *Pyrocoelia rufa* (Coleoptera: Lampyridae). Biochem. Genet. 41: 427–452.

Levy, H.C. 1998. Greatest bioluminescence. pp. 72–73. *In*: T.J. Walker (ed.). Book of Insect Records. University of Florida, Florida.

Li, X., S. Yang, M. Xie and X. Liang. 2006. Phylogeny of fireflies (Coleoptera: Lampyridae) inferred from mitochondrial 16S ribosomal DNA, with reference to morphological and ethological traits. Prog. Nat. Sci. 16: 817–826.

Lloyd, J.E. 1978. Insect bioluminescence. pp. 241–272. *In*: P.J. Herring (ed.). Bioluminescence in Action. Academic Press, New York.

Masuda, T., H. Tatsumi and E. Nakano. 1989. Cloning and sequence analysis of cDNA for luciferase of a Japanese firefly, *Luciola cruciata*. Gene 77: 265–270.

Matile, L. 1997. Phylogeny and evolution of the larval diet in the Sciaroidea (Diptera, Bibionomorpha) since the Mesozoic. pp. 273–303. *In*: P. Grandcolas (ed.). The Origin of Biodiversity in Insects: Phylogenetic Tests of Evolutionary Scenarios. Mém. Mus. Natn. Hist. Nat. 173, Paris.

McCapra, F. and Z. Razavi. 1976. Biosynthesis of luciferin in *Pyrophorus pellucens*. J.C.S. Chem. Commun. 153–154.

McElroy, W.D., M. DeLuca and J. Travis. 1967. Molecular uniformity in biological catalyses: The enzymes concerned with firefly luciferin, amino acid, and fatty acid utilization are compared. Science 157: 150–160.

Merritt, D.J. 2013. Standards of evidence for bioluminescence in cockroaches. Naturwissenschaften 100: 697–698.

Meyer-Rochow, V.B. and A.R. Liddle. 1988. Structure and function of the eyes of two species of opilionid from New Zealand glow-worm caves (*Megalopsalis tumida*: Palpatores, and *Hendea myersicavernicola*: Laniatores). Proc. R. Soc. Lond. B 233: 293–319.

Meyer-Rochow, V.B. 2007. Glowworms: a review of *Arachnocampa* spp. and kin. Luminescence 22: 251–256.

Mofford, D.M., G.R. Reddy and S.C. Miller. 2014. Latent luciferase activity in the fruit fly revealed by a synthetic luciferin. Proc. Natl. Acad. Sci. USA 111: 4443–4448.

Muraji, M., N. Arakaki and S. Tanizaki. 2012. Evolutionary relationship between two firefly species, *Curtos costipennis* and *C. okinawanus* (Coleoptera, Lampyridae), in the Ryukyu Islands of Japan revealed by the mitochondrial and nuclear DNA sequences. Sci. World J. 653013.

Nakatsu, T., S. Ichiyama, J. Hiratake, A. Saldanha, N. Kobashi, K. Sakata and H. Kato. 2006. Structural basis for the spectral difference in luciferase bioluminescence. Nature 440: 372–376.

Niwa, K., M. Nakamura and Y. Ohmiya. 2006. Stereoisomeric bio-inversion key to biosynthesis of firefly ᴅ-luciferin. FEBS Lett. 580: 5283–5287.

Niwa, K., Y. Ichino, S. Kumata, Y. Nakajima, Y. Hiraishi, D. Kato, V.R. Viviani and Y. Ohmiya. 2010. Quantum yields and kinetics of the firefly bioluminescence reaction of beetle luciferases. Photochem. Photobiol. 86: 1046–1049.

Nyrén, P. 1987. Enzymatic method for continuous monitoring of DNA polymerase activity. Anal. Biochem. 167: 235–238.

Oba, Y., M. Ojika and S. Inouye. 2003. Firefly luciferase is bifunctional enzyme: ATP-dependent monooxygenase and a long chain fatty acyl-CoA synthetase. FEBS Lett. 540: 251–254.

Oba, Y., M. Ojika and S. Inouye. 2004. Characterization of *CG6178* gene product with high sequence similarity to firefly luciferase in *Drosophila melanogaster*. Gene 329: 137–145.

Oba, Y., M. Sato, M. Ojika and S. Inouye. 2005. Enzymatic and genetic characterization of firefly luciferase and *Drosophila CG6178* as a fatty acyl-CoA synthetase. Biosci. Biotechnol. Biochem. 69: 819–828.

Oba, Y., M. Sato and S. Inouye. 2006a. Cloning and characterization of the homologous genes of firefly luciferase in the mealworm beetle, *Tenebrio molitor*. Insect Mol. Biol. 15: 293–299.

Oba, Y., M. Sato, Y. Ohta and S. Inouye. 2006b. Identification of paralogous gene of firefly luciferase in the Japanese firefly, *Luciola cruciata*. Gene 368: 53–60.

Oba, Y. and R. Sagegami-Oba. 2007. Phylogeny of Elateridae inferred from molecular analysis. Konchu to Shizen 42: 30–33.

Oba, Y., T. Shintani, T. Nakamura, M. Ojika and S. Inouye. 2008a. Determination of the luciferin contents in luminous and non-luminous beetles. Biosci. Biotechnol. Biochem. 72: 1384–1387.

Oba, Y., K. Iida, M. Ojika and S. Inouye. 2008b. Orthologous gene of beetle luciferase in non-luminous click beetle, *Agrypnus binodulus* (Elateridae), encodes a fatty acyl-CoA synthetase. Gene 407: 169–175.

Oba, Y. 2009. On the origin of beetle luminescence. pp. 277–290. *In*: V.B. Meyer-Rochow (ed.). Bioluminescence in Focus—A Collection of Illuminating Essays. Research Signpost, Kerala.

Oba, Y., K. Iida and S. Inouye. 2009. Functional conversion of fatty acyl-CoA synthetase to firefly luciferase by site-directed mutagenesis: A key substitution responsible for luminescence activity. FEBS Lett. 583: 2004–2008.

Oba, Y. and T. Kainuma. 2009. Diel changes in the expression of long wavelength-sensitive and ultraviolet-sensitive opsin genes in the Japanese firefly, *Luciola cruciata*. Gene 436: 66–70.

Oba, Y., M. Furuhashi and S. Inouye. 2010a. Identification of a functional luciferase gene in the non-luminous diurnal firefly, *Lucidina biplagiata*. Insect Mol. Biol. 19: 737–743.

Oba, Y., M. Kumazaki and S. Inouye. 2010b. Characterization of luciferases and its paralogue in the Panamanian luminous click beetle *Pyrophorus angustus*: A click beetle luciferase lacks the fatty acyl-CoA synthetic activity. Gene 452: 1–6.

Oba, Y., N. Mori, M. Yoshida and S. Inouye. 2010c. Identification and characterization of a luciferase isotype in the Japanese firefly, *Luciola cruciata*, involving in the dim glow of firefly eggs. Biochemistry 49: 10788–10795.

Oba, Y., M.A. Branham and T. Fukatsu. 2011. The terrestrial bioluminescent animals of Japan. Zool. Sci. 28: 771–789.

Oba, Y., M. Yoshida, T. Shintani, M. Furuhashi and S. Inouye. 2012. Firefly luciferase genes from the subfamilies Psilocladinae and Ototretinae (Lampyridae, Coleoptera). Comp. Biochem. Biophys. B 161: 110–116.

Oba, Y., M. Furuhashi, M. Bessho, S. Sagawa, H. Ikeya and S. Inouye. 2013a. Bioluminescence of a firefly pupa: involvement of a luciferase isotype in the dim glow of pupae and eggs in the Japanese firefly, *Luciola lateralis*. Photochem. Photobiol. Sci. 12: 854–863.

Oba, Y., N. Yoshida, S. Kanie, M. Ojika and S. Inouye. 2013b. Biosynthesis of firefly luciferin in adult lantern: Decarboxylation of L-cysteine is a key step for benzothiazole ring formation in firefly luciferin synthesis. PLoS One 8: e84023.

Oba, Y. and D.T. Schultz. 2014. Eco-Evo Bioluminescence on Land and in the Sea. pp. 3–36. *In*: G. Thouand and R. Marks (eds.). Bioluminescence: Fundamentals and Applications in Biotechnology Vol. 1. Advances in Biochemical Engineering/Biotechnology. 144. Springer, Heidelberg.

Ohba, N. 1983. Studies on the communication system of Japanese fireflies. Sci. Rept. Yokosuka City Mus. 30: 1–62 + 6 plt.

Ohba, N. 1984. Synchronous flashing in the Japanese firefly, *Luciola cruciata* (Coleoptera: Lampyridae). Sci. Rept. Yokosuka City Mus. 32: 23–32 + 8 plt.

Ohba, N. and Y. Goto. 1993. Geographical variation on the morphology and behavior of *Curtos costipennis* and *C. okinawana* (Coleoptera: Lampyridae) in the Southwestern Islands. Sci. Rept. Yokosuka City Mus. 41: 1–14.

Ohba, N. 2001. Geographical variation, morphology and flash pattern of the firefly, *Luciola cruciata* (Coleoptera: Lampyridae). Sci. Rept. Yokosuka City Mus. 48: 45–89.

Ôhira, H. 1962. Morphological and taxonomic study on the larvae of Elateridae in Japan (Coleoptera). Entomol. Lab. Aichi Gakugei Univ., Okazaki.

Ôhira, H. 2013. Illustrated key to click beetles of Japan. pp. 227–251. *In*: Jpn. Soc. Environ. Entomol. Zool. (ed.). An Illustrated Guide to Identify Insects. Bunkyo Shuppan, Osaka.

Ohmiya, Y., M. Sumiya, V.R. Viviani and N. Ohba. 2000. Comparative aspects of a luciferase molecule form the Japanese luminous beetle, *Rhagophthalmus ohbai*. Sci. Rept. Yokosuka City Mus. 47: 31–38.

Ohtsuki, H., J. Yokoyama, N. Ohba, Y. Ohmiya and M. Kawata. 2008. Nitric oxide synthase (NOS) in the Japanese fireflies *Luciola lateralis* and *Luciola cruciata*. Arch. Insect Biochem. Physiol. 69: 176–188.

Okada, K., H. Iio, I. Kubota and T. Goto. 1974. Firefly bioluminescence III. Conversion of oxyluciferin to luciferin in firefly. Tetrahedron Lett. 15: 2771–2774.

Okada, K., H. Iio and T. Goto. 1976. Biosynthesis of firefly luciferin. Probable formation of benzothiazole from *p*-benzoquinone and cysteine. J.C.S. Chem. Commun. 1976: 32.

Ow, D.W., K.V. Wood, M. DeLuca, J.R. de Wet, D.R. Helinski and S.H. Howell. 1986. Transient and stable expression of the firefly luciferase gene in plant cells and transgenic plants. Science 234: 856–859.

Pototskaja, V.A. 1983. Phylogenetic links and composition of the superfamily Cantharoidea (Coleoptera) based on study of larval characters. Rev. Entomol. URSS 62: 549–554.

Prado, R.A., J.A. Barbosa, Y. Ohmiya and V.R. Viviani. 2011. Structural evolution of luciferase activity in *Zophobas* mealworm AMP/CoA-ligase (protoluciferase) through site-directed mutagenesis of the luciferin binding site. Photochem. Photobiol. Sci. 10: 1226–1232.

Regier, J.C., J.W. Shultz, A. Zwick, A. Hussey, B. Ball, R. Wetzer, J.W. Martin and C.W. Cunningham. 2010. Arthropod relationships revealed by phylogenomic analysis of nuclear protein-coding sequences. Nature 463: 1079–1084.

Rosa, S.P. 2010. Second record of bioluminescence in larvae of *Xantholinus* Dejean (Staphylinidae, Xantholinini) from Brazil. Revta Bras. Ent. 54: 147–148.

Roth, L.M. and B. Stay. 1958. The occurrence of para-quinones in some arthropods, with emphasis on the quinone-secreting tracheal glands of *Diploptera punctata* (Blattaria). J. Insect Physiol. 1: 305–318.

Sakayori, M., T. Hariyama, Y. Tsukahara and A. Terakita. 1992. The distribution of the visual pigments in the compound eyes of fireflies. Ouyou Jouhougaku Kenkyu Nenpou 17: 1–15.

Sagegami-Oba, R., N. Takahashi and Y. Oba. 2007a. The evolutionary process of bioluminescence and aposematism in cantharoid beetles (Coleoptera: Elateroidea) inferred by the analysis of 18S ribosomal DNA. Gene 400: 104–113.

Sagegami-Oba, R., Y. Oba and H. Ôhira. 2007b. Phylogenetic relationships of click beetles (Coleoptera: Elateridae) inferred from 28S ribosomal DNA: Insights into the evolution of bioluminescence in Elateridae. Mol. Phylogenet. Evol. 42: 410–421.

Sasaki, G., K. Ishiwata, R. Machida, T. Miyata and Z.-H. Su. 2013. Molecular phylogenetic analyses support the monophyly of Hexapoda and suggest the paraphyly of Entognatha. BMC Evol. Biol. 13: 236.

Schröder, J. 1989. Protein sequence homology between plant 4-coumarate: CoA ligase and firefly luciferase. Nucleic Acids Res. 17: 460.

Seliger, H.H. and W.D. McElroy. 1959. Quantum yield in the oxidation of firefly luciferin. Biochem. Biophys. Res. Commun. 1: 21–24.

Shimomura, O. 2006. Bioluminescence: Chemical Principles and Methods. World Scientific, Singapore.

Sivinski, J.M. 1998. Phototropism, bioluminescence, and the Diptera. Flor. Entomol. 81: 282–292.

Slipinski, S.A., R.A.B. Leschen and J.F. Lawrence. 2011. Order Coleoptera Linnaeus, 1758. pp. 203–208. *In*: Z.-Q. Zhang (ed.). Animal Biodiversity: An Outline of Higher-level Classification and Survey of Taxonomic Richness. Magnolia Press, Auckland.

South, A., K. Stanger-Hall, M.-L. Jeng and S.M. Lewis. 2011. Correlated evolution of female neoteny and flightlessness with male spermatophore production in fireflies (Coleoptera: Lampyridae). Evolution 65: 1099–1113.

Stanger-Hall, K.F., J.E. Lloyd and D.M. Hillis. 2007. Phylogeny of North American fireflies (Coleoptera: Lampyridae): Implications for the evolution of light signals. Mol. Phylogenet. Evol. 45: 33–49.

Stibick, J.N.L. 1979. Classification of the Elateridae (Coleoptera): Relationships and classification of the subfamilies and tribes. Pacific Insects 20: 145–186.

Stolz, U., S. Velez, K.V. Wood, M. Wood and J.L. Feder. 2003. Darwinian natural selection for orange bioluminescent color in a Jamaican click beetle. Proc. Natl. Acad. Sci. USA 100: 14955–14959.

Suzuki, H. 1997. Molecular phylogenetic studies of Japanese fireflies and their mating systems. Tokyo Metro. Univ. Bull. Natl. Hist. 3: 1–53.

Suzuki, H., Y. Sato and N. Ohba. 2002. Gene diversity and geographic differentiation in mitochondrial DNA of the Genji firefly, *Luciola cruciata* (Coleoptera: Lampyridae). Mol. Phylogenet. Evol. 22: 193–205.

Tamura, M., J. Yokoyama, N. Ohba and M. Kawata. 2005. Geographic differences in flash intervals and pre-mating isolation between populations of the Genji firefly, *Luciola cruciata*. Ecol. Entomol. 30: 241–245.

Tatsumi, H., T. Masuda and E. Nakano. 1988. Synthesis of enzymatically active firefly luciferase in yeast. Agric. Biol. Chem. 52: 1123–1127.

Tatsumi, H., T. Masuda, N. Kajiyama and E. Nakano. 1989. Luciferase cDNA from Japanese firefly, *Luciola cruciata*: Cloning, structure and expression in *Escherichia coli*. J. Biolumin. Chemilumin. 3: 75–78.

Timmermans, M.J.T.N., S. Dodsworth, C.L. Culverwell, L. Bocak, D. Ahrens, D.T.J. Littlewood, J. Pons and A.P. Vogler. 2010. Why barcodes? High-throughput multiplex sequencing of mitochondrial genomes for molecular systematics. Nucleic Acids Res. 38: e197.

Timmermans, M.J.T.N. and A.P. Vogler. 2012. Phylogenetically informative rearrangements in mitochondrial genomes of Coleoptera, and monophyly of aquatic elateriform beetles (Dryopoidea). Mol. Phylogenet. Evol. 63: 299–304.

Toh, H. 1991. Sequence analysis of firefly luciferase family reveals a conservative sequence motif. Protein Seq. Data Anal. 4: 111–117.

Trimmer, B.A., J.R. Aprille, D.M. Dudzinski, C.J. Lagace, S.M. Lewis, T. Michel, S. Qazi and R.M. Zayas. 2001. Nitric oxide and the control of firefly flashing. Science 292: 2486–2488.

Viviani, V.R., E.J.H. Bechara and Y. Ohmiya. 1999. Cloning, sequence analysis, and expression of active *Phrixothrix* railroad-worms luciferases: Relationship between bioluminescence spectra and primary structures. Biochemistry 38: 8271–8279.

Viviani, V.R., J.W. Hastings and T. Wilson. 2002. Two bioluminescent Diptera: The North American *Orfelia fultoni* and the Australian *Arachnocampa flava*. Similar niche, different bioluminescence systems. Photochem. Photobiol. 75: 22–27.

Viviani, V.R., T.L. Oehlmeyer, F.G.C. Arnoldi and M.R. Brochetto-Braga. 2005. A new firefly luciferase with bimodal spectrum: Identification of structural determinants of spectral pH-sensitivity in firefly luciferases. Photochem. Photobiol. 81: 843–848.

Viviani, V.R., D. Amaral, R. Prado and F.G.C. Arnoldi. 2011. A new blue-shifted luciferase from the Brazilian *Amydetes fanestratus* (Coleoptera: Lampyridae) firefly: molecular evolution and structural/functional properties. Photochem. Photobiol. Sci. 10: 1879–1886.

Viviani, V.R., R.A. Prado, F.C.G. Arnoldi and F.C. Abdalla. 2009. An ancestral luciferase in the Malpighi tubules of a non-bioluminescent beetle. Photochem. Photobiol. Sci. 8: 56–71.

Vršanský, P., D. Chorvát, I. Fritzsche, M. Hain and R. Ševčík. 2012. Light-mimicking cockroaches indicate Tertiary origin of recent terrestrial luminescence. Naturwissenschaften 99: 739–749.

Wheeler, D.A., M. Srinivasan, M. Egholm, Y. Shen, L. Chen, A. McGuire, W. He, Y.-J. Chen, V. Makhijani, G.T. Roth, X. Gomes, K. Tartaro, F. Niazi, C.L. Turcotte, G.P. Irzyk, J.R. Lupski, C. Chinault, X.-Z. Song, Y. Liu, Y. Yuan, L. Nazareth, X. Qin, D.M. Muzny, M. Margulies, G.M. Weinstock, R.A. Gibbs and J.M. Rothberg. 2008. The complete genome of an individual by massively parallel DNA sequencing. Nature 452: 872–876.

White, E.H., F. McCapra and G.F. Field. 1963. The structure and synthesis of firefly luciferin. J. Am. Chem. Soc. 85: 337–343.

Wood, K.V., Y.A. Lam, H.H. Seliger and W.D. McElroy. 1989. Complementary DNA coding click beetle luciferases can elicit bioluminescence of different colors. Science 244: 701–702.

Wood, K.V. 1995. The chemical mechanism and evolutionary development of beetle bioluminescence. Photochem. Photobiol. 62: 662–673.

Ye, L., L.M. Buck, H.J. Schaeffer and F.R. Leach. 1997. Cloning and sequencing of a cDNA for firefly luciferase from *Photuris pennsylvanica*. Biochim. Biophys. Acta 1339: 39–52.

Yeates, D.K., B.M. Wiegmann, G.W. Courtney, R. Meier, C. Lambkin and T. Page. 2007. Phylogeny and systematics of Diptera: Two decades of progress and prospects. Zootaxa 1668: 565–590.

Zompro, O. and I. Fritzsche. 1999. *Lucihormetica fenestrata* n. gen., n. sp., the first record of luminescence in an orthopteroid insect (Dictyoptera: Blaberidae: Blaberinae: Brachycolini). Amazoniana 15: 211–219.

4

A Glance on the Role of miRNAs in Insect Life

Sassan Asgari[a],* *and Mazhar Hussain*[b]

Introduction

The biological community has witnessed the fascinating expanding world of microRNAs (miRNAs) during the last decade. The knowledge created from miRNA research has now established the fact that nearly all cellular pathways from development to oncogenesis are regulated directly or indirectly by miRNAs in almost all eukaryotic organisms. More than half of all human protein coding genes are under the control of miRNA regulation. Our understanding of their biogenesis and functionality is building up and it has been realized that they are produced and act in several different ways, but still there are many unknowns and mechanisms yet to be discovered. Recent investigations on the discovery of miRNAs in several species of insects have opened up new vistas of studying insect life from miRNA perspectives (Chawla and Sokol 2011, Bellés et al. 2012 Asgari 2013, Lucas and Raikhel 2013). In this chapter, we discuss recent research findings on miRNA biogenesis and evolution in general and their role in aspects of insect biology. However, in several sections, relevant examples from non-insect models are mentioned, which are useful in comparative understanding of miRNA biogenesis and function.

School of Biological Sciences, The University of Queensland, Brisbane, QLD 4072, Australia.
[a] Email: s.asgari@uq.edu.au
[b] Email: m.hussein1@uq.edu.au
* Corresponding author

miRNA Biogenesis

Where do miRNAs originate from in the genome? Localization and transcription of miRNAs

miRNAs are derived from several different locations in the genome from coding as well as non-coding regions (ncRNA) (Rodrigueza et al. 2004). In mammalian cells, more than 70 percent of miRNAs come from introns of largely protein coding genes, while others are located in ncRNA regions. miRNAs have been found to originate from transcription of both sense and antisense strands of DNA (Finnegan and Pasquinelli 2013). In *Caenorhabditis elegans*, 30 precent of miRNA genes are on the antisense strand overlapping protein-coding genes (Martinez et al. 2008). A comprehensive analysis on genomic location of miRNAs have been conducted in *Drosophila* that reveal emergence of several miRNAs from antisense strand overlapping protein coding as well as non-coding genes (Berezikov et al. 2011). A large number of miRNAs are transcribed by RNA polymerase II with the exception of those that are located closer to Alu repeats, which are transcribed by RNA polymerase III (Borchert et al. 2006). miRNAs can be transcribed from the promoter of their host gene as well as from their own distinct/independent promoter. For this reason, several reports have revealed both coordinated and independent transcription of miRNAs from their host genes. In humans, one-third of intronic miRNAs are transcribed as independent units (Corcoran et al. 2009, Monteys et al. 2010). Multiple miRNAs can also be transcribed as one long transcriptional unit called a 'cluster' (Lee et al. 2002).

How are miRNAs generated in the cell? *The canonical pathway where RNase III proteins collaborate*

So far, most of the animal miRNAs have been reported to be produced through a canonical pathway controlled by RNase III enzymes. In the nucleus, biogenesis of miRNA begins with a normal transcription process mainly by RNA polymerase II that produces a variable length (100 to many thousands of nucleotides, nt) primary miRNA (pri-miRNA) possessing hairpin(s) (Kim et al. 2009) (Fig. 1). pri-miRNAs are further processed at the sites of hairpins by a microprocessor that contains an RNase III enzyme, Drosha, and its cofactor DGCR8/Pasha that results in the production of normally a ~70 nt precursor miRNA (pre-miRNA). In the cytoplasm, another RNase III enzyme, Dicer, processes pre-miRNA to a ~22 nt miRNA-5p:3p duplex. Until recently, the duplex was referred to as miRNA: miRNA* indicating miR-X-3p and miR-X-5p strands. In this chapter, we may still refer to the strands of some miRNAs as miR-X

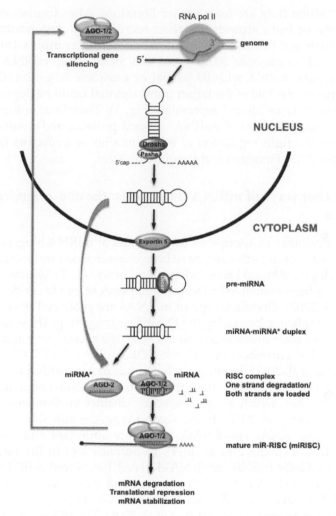

Figure 1. miRNA biogenesis. miRNA genes are transcribed by RNA polymerase II in the nucleus containing a 5'cap and a 3' polyA tail. The primary miRNA (pri-miRNA) transcript contains one or more stem-loop structures, which is processed by an RNase III type enzyme, Drosha, in connection with Pasha. This leads to the production of a ~70 nt hairpin structure called the precursor miRNA (pre-miRNA). Pre-miRNA is then transferred into the cytoplasm by Exportin-5, where the hairpin head from the stem-loop is removed by Dicer-1 forming a short miRNA-5p:3p duplex (also referred to as miRNA:miRNA* duplex) which becomes incorporated into the RISC complex. Ago proteins, Ago1 or Ago2, constitute the main component of the complex. One of the strands (passenger) is normally degraded or could be loaded into Ago2 as mature miRNA. The mature miRNA-RISC complex (miRISC) interacts with the target sequences in the mRNA target. The outcome of the interaction is variable and includes translational repression, mRNA degradation and in certain instances target upregulation. miRISC could also be transported into the nucleus where it binds to the promoter region of the target gene and induces transcriptional gene silencing (TGS).

or miR-X* when they are from earlier literature. After formation of the duplex, one or both strands are then recruited by Argonaute (Ago) proteins forming the miRNA-induced silencing complex (miRISC) (Carthew and Sontheimer 2009, Siomi and Siomi 2010). miRNA guides RISC to specific mRNA with its partial or complete target-binding site that determines the fate of the target transcript that could be degradation, stabilization or translational repression (Fig. 1). Therefore, in this whole process of manifestation of miRNA, several proteins are involved that work under the tight regulation of other proteins or co-factors (Ameres and Zamore 2013, Finnegan and Pasquinelli 2013).

Several other ways of miRNA biogenesis: *the non-canonical pathways*

Recently, a number of alternative mechanisms of miRNA biogenesis, so-called non-canonical pathways, have been characterized and include both Drosha-independent and Dicer-independent processes. Biogenesis of a few miRNAs has been shown to be Drosha-independent, but Dicer-dependent (Ruby et al. 2007). One such type of miRNAs are produced from hairpin introns after splicing, bypassing the Drosha cleavage step. These are called 'mirtrons' and were discovered first in *Drosophila melanogaster* and later in *C. elegans* and in a number of vertebrates (Okamura et al. 2007, Chong et al. 2010, Chung et al. 2011). In mirtrons, both ends of pre-miRNA are defined by splicing, but in few cases either the 5′ or the 3′ hairpin end matches the splice donor sites, which are processed by distinct mechanisms (Glazov et al. 2008, Flynt et al. 2010). In the second type, the miRNAs come from small nucleolar RNAs (snoRNAs), which are abundant and conserved antisense RNAs (Babiarz et al. 2011). In humans and in the flagellated protozoan *Giardia lamblia*, snoRNA-derived functional miRNAs have been discovered (Ender et al. 2008, Saraiya and Wang 2008). In the third type, miRNAs are derived from transfer RNA (tRNA) first found in mouse embryonic stem cells (Babiarz et al. 2008). The tRNA-Ile/miR-1983 precursor was predicted to fold into a typical tRNA cloverleaf cleaved by RNase P at the 5′ end and by tRNase Z at the 3′ end. Generation of miRNAs encoded by murine γ-herpes virus 68 (MHV68) was reported to be based on tRNase Z cleavage at the 5′end of the pre-miRNA hairpins (Bogerd et al. 2010). In the fourth type, miRNAs originate from the RNAi pathway and are well documented in *D. melanogaster* (Chung et al. 2008, Czech et al. 2008). Dicer-2 cleaves endogenous siRNAs (endo-siRNAs) from transposable elements which are mostly loaded into Ago2, but a few loaded into Ago1 can function as miRNAs (Okamura et al. 2008).

Recent research on vertebrate miR-451 has led to another way of miRNA biogenesis that is Dicer-independent and Ago2-dependent (Cheloufiet et al. 2010, Yang and Lai 2011). The biogenesis of mature miR-451 starts from normal Drosha/DGCR8 cleavage of pri-miR-451 into a short pre-miRNA with only ~18 bp of duplex stem, which is too short to serve as a substrate for Dicer. In this case, pre-miRNA is loaded directly into Ago2 and is sliced on its 3' hairpin arm, as guided by the 5' end of the hairpin, yielding a 30 nt RNA. Subsequently, the small RNA is subjected to a 3' resection activity that trims ~7 nt to produce a 23 nt miR-451. Interestingly, if this pre-miRNA is loaded into Ago1 protein (non-slicing), it cannot be processed into the mature miRNA. The processing of miR-451 was also recently reported in *D. melanogaster*, where pre-miRNA when loaded into Ago1 was resected into mature miR-451 and this maturation was strongly enhanced by knocking down Dicer-1 (Yang et al. 2013). However, after loading into Ago2, pre-miR-451 could not produce mature miRNA resulting in a 30 nt small RNA referred to as Ago-cleaved pre-miRNA (ac-pre-miRNA). The resection of ac-pre-miRNA in Ago2 was only successful after depletion of Hen1 methyltransferase (Yang et al. 2013).

A few examples of other types of non-canonical biogenesis come from virus-encoded miRNAs. For example, several miRNAs are produced by a retrovirus, Bovine leukemia virus (BLV), based on DNA polymerase III transcription *in vivo* as well as *in vitro* (Rosewick et al. 2013). The products of these transcripts were too small to be recognized by Drosha; therefore, these were directly processed by Dicer-1 to mature miRNAs.

The group of proteins that regulate processing and functioning of miRNAs

Following is a brief description of the major proteins and their regulators involved in miRNA biogenesis and function.

The microprocessor complex-*the primary processors*

The microprocessor is a complex of the RNase III enzyme Drosha and DiGeorge syndrome critical region gene 8 (DGCR8) or Partner of Drosha (Pasha). Drosha is a nuclear localized protein that catalyzes the initial processing of pri-miRNAs to pre-miRNAs (Lee et al. 2004). Drosha cleaves dsRNA non-specifically with the help of DGCR8 in humans and Pasha in invertebrates for specificity and proper positioning of ribonuclease domain. DGCR8/Pasha recognizes dsRNA-ssRNA junctions and positions the Drosha ribonuclease domain 11 nt, or one turn of the dsRNA helix, away (Han et al. 2006, Jinek and Doudna 2009). The structure of the microprocessor cleavage product is more important than its sequence and possesses a stem-loop containing a terminal loop, an upper stem, miRNA-5p:3p (miRNA: miRNA*) region and a lower stem with phosphorylated 5' ends and 3' dinucleotide overhangs (Voinnet 2009).

The amount of microprocessor in the cell is critical for regulation of the primary processing and interestingly Drosha and DGCR8 regulate each other at the post-transcriptional level resulting in auto-regulation of the microprocessor complex (Macias et al. 2010). This auto-regulation guarantees homeostasis of miRNA levels. Several modifying proteins and co-factors also regulate the microprocessor activity. Inefficient Drosha cleavage was observed in pri-miR-122 in which adenosine to inosine (A-to-I) editing was done in its sequence by ADAR (adenosine deaminase that acts on RNA) in human and mouse brain tissues (Luciano et al. 2004). A number of other proteins such as p53, p68, p72 and SMAD bind to the microprocessor and enhance processing of certain miRNAs (Suzuki et al. 2009, Lane and Levine 2010, Blahna and Hata 2012). SMAD proteins and p68 can also enhance miRNA levels by binding to the promoters of miRNA genes in a tissue-specific manner. Transforming growth factor β (TGF β) induces the expression of miR-216 and miR-217 in glomerular meningeal cells and represses the expression of miR-24 in myoblasts (Sun et al. 2008). Other examples of DNA/RNA binding proteins involved in induction/repression of subsets of miRNAs after binding to microprocessors are breast cancer 1 (BRCA1), hnRNP A1, KH-type splicing regulatory protein (KSRP), TAR DNA-binding protein-43 (TDP-43) and LIN-28 (Newman et al. 2008, Michlewski et al. 2010, Briata et al. 2011, Lightfoot et al. 2011, Kawahara and Mieda-Sato 2012, Kawai and Amano 2012). In *D. melanogaster*, genome-wide analyses were conducted to find out target transcripts of the Drosha-Pasha/DGCR8 complex. Based on Drosha knockdown strategy and by RNA PolII chromatin immunoprecipitation followed by *D. melanogaster* tiling microarrays most of the identified pri-miRNAs seemed extremely large (>10 kb). Along with well-defined miRNAs, more than a 100 additional RNAs not annotated as miRNAs were also found to be under Drosha control and are likely to be direct targets of Drosha (Kadener et al. 2009).

Exportin-5-*the exporter and protector*

For Dicer processing, pre-miRNA must be exported to the cytoplasm (Lee et al. 2002). Exportin-5 (Expo-5) is responsible for this transport and like other nuclear receptors it works with the cooperation of the co-factor Ran-GTP (Bohnsack et al. 2004, Lund et al. 2004). A general decline in miRNA levels was observed when Expo-5 was knocked down in human embryonic kidney 293 T cells (Yi et al. 2003) leading to the idea that Expo-5 also protects pre-miRNA from exonucleases, therefore, the protein must be vital for proper biogenesis of miRNA. Another interesting aspect of Expo-5 was revealed recently that its knockdown leads to decrease in Dicer protein levels with enhanced accumulation of Dicer mRNA in the nucleus (Bennasser et al. 2011). Thus, it was suggested that Expo-5 interacts with Dicer mRNA and transports it into the nucleus.

Dicer-*the perfect reaper*

Dicer enzymes cleave dsRNA to 21–27 nt small RNAs (siRNAs and miRNAs) which are then used for gene silencing (Carthew and Sontheimer 2009). Dicer acts as a molecular ruler that produces small RNAs by recognizing the end of a dsRNA substrate and cleaving a set distance away. The PAZ domain in Dicer facilitates its binding to the end of the dsRNA and the connector helix is the structural element that sets the measuring distance from the dsRNA end to the cleavage site (Lingel et al. 2003, Yan et al. 2003). A mechanistic logic behind generation of 21–27 nt small RNAs by Dicer has been shown. There is a distance of 65 Å in between the 3′-overhang-binding pocket of the PAZ domain and the active site of the RNase IIIa domain, which corresponds to the length of a 25 nt RNA duplex (Lingel et al. 2004, Ma et al. 2004). Another study suggests that Dicer measures from the monophosphorylated 5′ end of the precursor to position the RNase III domains for intramolecular dimerization that cleaves 21–24 nt away from the end (Park et al. 2011).

Recently, it was shown in *D. melanogaster* and mammals that the Dicer partner protein Loquacious (loqs) tunes the length of the mature miRNAs (Fukunaga et al. 2012). The investigators produced flies expressing all the three distinct Loqs-PA, Loqs-PB, or Loqs-PD and found that Loqs-PB tunes Dicer-1 cleavage of pre-miR-307a that results in a longer miRNA isoform with a distinct seed sequence and target specificity. The longer form of miR-307a represses glycerol kinase and taranis mRNA expression. Kinetic studies of human Dicer showed that the DEXD/H-boxATPase domain has an auto-inhibitory function, since cleavage rate is enhanced upon removal of this domain (Ma et al. 2008). In contrast, ATPase domain is vital for Dicer function in *D. melanogaster* and *C. elegans* because ATP promotes dsRNA processing by *D. melanogaster* Dicer-2 and *C. elegans* Dicer-1 (Nykanen et al. 2001). Only one Dicer that produces all types of small RNAs has been reported from humans and *C. elegans*, and two from insects (Dicer-1, usually for miRNA biosynthesis, and Dicer-2 mostly for siRNA). Several proteins have been reported that interact and modify Dicers activities. Some proteins such as HIV-1 TAR RNA binding protein (TRBP), protein activator of PKR (PACT), Loqs and RBM3 stimulate the processing of pre-miRNA through distinct mechanisms (Forstemann et al.

2005, Koscianska et al. 2011). Other proteins with Dicer repressive activity are monocyte chemoattractant protein 1-induced protein 1 (MCPIP1) and Lin 28 (Suzuki et al. 2011, Van Wynsberghe et al. 2011).

Argonaute (Ago)-*the most versatile player of the group*

The Ago superfamily can be divided mainly into Piwi type that binds piRNAs and the Ago type that is specific to miRNAs and siRNAs (Brennecke et al. 2007). Ago proteins are the central part of RISC and all small RNAs of 20–30 nt in length are believed to require one or more Ago proteins for their biogenesis. The double stranded small RNAs ultimately are loaded into RISC and proceed to unwinding duplex strands, culminating in binding of the guide strand to Ago protein (Hutvagner and Simard 2008). The guide strand then guides Ago proteins to base pair with mRNA complementary sequences resulting in mRNA cleavage or degradation. Ago proteins are multi-domain containing the N-terminal, PAZ, middle (MID) and PIWI domains (Cerutti et al. 2000). After the discovery that PIWI domain has RNaseH activity, it was found that Ago proteins possess the slicer activity of the RISC. RISC-catalyzed RNA cleavage yields a 5' RNA product with a free 3' OH and a 3' RNA product with a 5' P group (Song et al. 2004).

In humans, four Ago proteins (Ago 1–4) have been well characterized that are involved in the miRNA pathway, and it was recently shown that miRNAs are randomly sorted to individual Ago proteins (Wang et al. 2012a). In *Drosophila*, there are five Ago proteins; Ago1–3, Piwi, and Aub, in which Ago1 and Ago2 are involved in miRNA and RNAi pathways, respectively (Tomari et al. 2007, Ghildiyal et al. 2010). Recent investigations have highlighted insects' Ago2 association with miRNA pathways (Ghildiyal et al. 2010, Hussain et al. 2013a, Rubio et al. 2013). Moreover, only Ago2 in mammals and both Ago1 and Ago2 in flies possess slicer activity and can also be associated with the biosynthesis of miRNA bypassing the Dicer-1 processing step (Okamura et al. 2004). In *C. elegans*, 27 Ago proteins have been reported with ALG-1/2 (Ago-like genes 1 and 2) dedicated for miRNA pathway (Yigit et al. 2006). In *Schizosaccharomyces pombe*, Ago1's slicer activity is required for transcriptional silencing (Irvine et al. 2006). Two models for target recognition and cleavage by Ago have been developed (Filipowicz 2005, Tomari and Zamore 2005). The fixed-end model suggests that both ends of the guide RNA bind to Ago during slicing and limits base-pairing to less than one helical turn (11 base pairs). This model probably presents the idea that base pairing is limited to the seed region of the guide RNA. However, the other two-state model suggests that the target binds to the seed region of the guide strand, which leads to base pairing towards the 3' end of the guide strand.

Ago proteins may also provide stability to associated miRNAs by protecting them against exonucleolytic degradation by XRN-1 and XRN-2 as shown in *C. elegans* (Chatterjee and Grosshans 2009). Moreover, knockdown of Ago1 and Ago2 led to significant decreases in miRNA levels (Hussain et al. 2013a) and over expression of any of the four Ago proteins in human cells results in global increase in miRNA levels (Winter and Diederichs 2011). A new homeostatic mechanism has been revealed that regulate the abundance of Ago1 and Ago2 proteins in relation to the status of miRNA biogenesis in *D. melanogaster* and mammalian systems (Smibert et al. 2013). Accordingly, in *D. melanogaster,* both mutations of multiple miRNA factors and ectopic transcription of miRNAs affect Ago1 protein levels. Ago proteins are intracellularly localized to both the cytoplasm and the nucleus. Cytoplasmic Ago-miRNA complexes are mainly involved in cleavage or translational repression of target mRNAs, while the nuclear ones are engaged in transcriptional gene silencing, methylation, chromatin remodeling, and splicing (Ameyar-Zazoua et al. 2012).

miRNAs loaded into Ago1 can also undergo 3′ end trimming by exoribonucleases. It was observed in *D. melanogaster* that more than a quarter of all miRNAs undergo 3′ end trimming by the 3′-to-5′ exoribonuclease Nibbler/Knabber (CG9247), a member of the DEDD family of exonucleases (Han et al. 2011). miRNA trimming is crucial for proper functioning as RNAi-mediated silencing of nibbler causes developmental defects. Trimming of miRNA 3′ ends occurs after the removal of the passenger strand from pre-RISC and may be the final step in RISC assembly, ultimately enhancing target mRNA repression (Han et al. 2011).

How do miRNAs regulate gene expression: miRNA targets mRNA

The miRNA-5p:3p duplex is short-lived and upon loading in Ago proteins is rapidly unwound to make single strands (Czech and Hannon 2011). One of the strands guides Ago to the complementary mRNA, and a major determinant of this binding is the seed region (nucleotides 2–8 nt from the 5′ end of the miRNA) resulting in mRNA degradation or stabilization, and protein repression or activation (Place et al. 2008, Bartel 2009, Bruno et al. 2011). A number of recent studies in diverse organisms have changed the canonical approach towards miRNA-mRNA interaction that was established during the last decade. The followings are the canonical rules and their new modifications:

1) Functional miRNA target sites are localized in the 3′UTRs of protein-coding genes in relatively unstructured regions (Grimson et al. 2007).

At present, several target sites have been identified in the 5'UTR and coding region as well as introns of protein coding genes and even in the non-coding antisense RNA (Grey et al. 2010, Hafner et al. 2010, Rossi 2011, Reczko et al. 2012, Helwak et al. 2013, Hussain et al. 2013b, Meng et al. 2013). All of these interactions were experimentally validated. A detailed analysis on miRNA targets in the human genome revealed 42.6 percent in the coding region, 23.4 percent in the 3'UTR and 3.8 percent in the 5'UTR in the total targets obtained through CLASH (cross linking, ligation, and sequencing of hybrids) technique (Kudla et al. 2011, Helwak et al. 2013).

2) The seed region of miRNA with Watson-Crick pairing is indispensable for miRNA-mRNA interaction (Fabian et al. 2010). Now, there are several lines of evidence based on solid experimentation where functional targets have been shown to have multiple mismatches and wobbles in the seed region. About 15 percent of target genes against human miR-124 have G bulges in miRNA-target binding at nucleotide positions 5–6 of miRNA (Chi et al. 2012). Based on a human miRNA interactome analysis, 18 percent of miRNA-mRNA interactions involve complementarity with the miRNAs' 3' end with little complementarity at the 5' end (Helwak et al. 2013). Another type of interaction is "centered pairing" in which 11 consecutive Watson-Crick base pairs between the target and positions 4–14 or 5–15 of miRNA are involved (Shin et al. 2010).

3) miRNA-mRNA interaction results in mRNA degradation, cleavage or translation repression through several different mechanisms (Baek et al. 2008, Selbach et al. 2008, Guo et al. 2010). miRNAs are also linked to the induction or stabilization of mRNA and activation of protein translation (Vasudevan et al. 2007, Orom et al. 2008, Place et al. 2008, Bruno et al. 2011, Hussain et al. 2011). The mechanism of miRNA-mediated mRNA stability and translation enhancement was recently revealed in the case of the well-known miR-122 interaction with Hepatitis C virus (HCV) which enhances replication of the virus (Conrad et al. 2013). The investigators found that miR-122 recruits Ago2 to 5'UTR of HCV, which is important for induction of translation. Ago2 alone, without miR-122, was unable to mediate the function. This miR-122-Ago2 complex was also found to protect viral RNA from RNA decay. Another example of mechanism of induction is miR-128, which targets transcripts of nonsense mediated decay machinery leading to increases in the expression levels of genes involved in brain development (Bruno et al. 2011).

Target mRNA can regulate miRNA expression in both degradation as well as stability ways. Target-miRNA interaction with strong complementarity may lead to tailing or trimming at the 3′ end of a miRNA resulting in degradation of that miRNA. However, in some cases this interaction may provide stability to miRNA by rescuing it from exoribonuclease XRN1, as shown in mouse (Chatterjee and Grosshans 2009).

Insect miRNAs: *a Field Still in its Infancy with the Hope of Rapid Growth*

During the last few years with the advent of small RNA deep sequencing and reductions in its cost, miRNAs have been explored in several insect species including 12 *Drosophila* species (Stark et al. 2007), three mosquitoes (*Anopheles gambiae, Aedes albopictus* and *Culex quinquefasciatus*) (Winter et al. 2007, Skalsky et al. 2010), four hymenopterans (*Apis mellifera, Nasonia vitripennis, N. giraulti,* and *N. longicornis*) (Weaver et al. 2007, Sathyamurthy and Swamy 2010), five lepidopterans (*Bombyx mori, Manduca sexta, Heliconius melpomene, Plutella xylostella* and *Spodoptera frugiperda*) (He et al. 2008, Consortium 2012, Jagadeeswaran et al. 2010), one coleopteran (*Tribolium castaneum*) (Luo et al. 2008, Singh and Nagaraju 2008) and one hemipteran (*Acyrthosiphon pisum*) (Legeai et al.). miRNAs in *Locusta migratoria* (Wei et al. 2009), *Blattella germanica* (Cristino et al. 2011), *Bemisia tabaci* (Guo et al. 2013b), *Mayetiola destructor* (gall midge)(Khajuria et al. 2013) and *Panonychus citri* (citrus red mite) (Liu et al. 2013) have also been documented. In the coming years, their functional analyses will boost our knowledge in insect science. Here, recent work on miRNAs associated with multiple aspects of insects' life will be discussed. These functions are summarized in Table 1.

Metamorphosis: *miRNAs May Control Transition from Egg to Adult*

Endocrine regulated molts are the hallmarks of insect development (Yamanaka et al. 2013). Ecdysone is the master regulator among several known key hormones and neuropeptides associated with developmental transitions. Ecdysone is produced in prothoracic glands and released into the hemolymph where it is converted to the active form called 20-hydroxyecdysone (20E) by a P450 monooxygenase (Rewitz et al. 2006, Huang et al. 2008). It is well documented that 20E drives molting as it binds to its nuclear receptor leading to onset of cascades of gene expression. Ecdysone receptors (EcRs) are expressed in *Drosophila* in three isoforms; EcR-A, Ecr-B1 and EcR-B2 that possess sequence variability at the N-terminus and conservation at the C-terminus (Talbot et al. 1993). During metamorphosis, the EcR-A is mainly expressed in imaginal discs

Table 1. Insect miRNAs with experimentally validated functions.

Function	miRNAs	Insect species	References
Metamorphosis			
Ecdysone (20E) interaction	miR-100, miR-125, Let-7 regulate remodeling of neuro-musculature	*D. melanogaster*	(Chawla and Sokol 2012)
	miR-14 suppresses EcR	*D. melanogaster*	(Varghese and Cohen 2007)
	miR-8 depletion enhance U shaped (USH) in insulin pathway	*D. melanogaster*	(Jin et al. 2012)
Molting	miR-14 enhances EcR	*H. armigera*	(Jayachandran et al. 2013)
	miR-281 downregulates EcR-B	*B. mori*	(Jiang et al. 2013)
	miR-275	*Ae. aegypti*	(Bryant et al. 2010)
	miR-8-5p and miR-2a-3p downregulates Tr2 and PAGM genes in chitin biosynthesis	*N. lugens*	(Chen et al. 2013b)
Apoptosis			
Regulation of proapoptotic genes	Bantam and miR-2 suppresses *hid1*	*D. melanogaster*	(Brennecke et al. 2003, Stark et al. 2005)
	miR-6 and miR-2/13 family regulate caspase Drice	*D. melanogaster*	(Leaman et al. 2005, Xu et al. 2003)
	miR-6 and miR-11 regulate rpr, hid, grim and skl	*D. melanogaster*	(Ge et al. 2012)
	miR-263a and miR-263b downregulates hid	*D. melanogaster*	(Hilgers et al. 2010)
Hippo pathway	miR-2 family	*D. melanogaster*	(Zhang and Cohen 2013)
Transcription factor	miR-9 downregulates dLMO	*D. melanogaster*	(Bejarano et al. 2010, Biryukova et al. 2009)
Reproduction			
Oogenesis	miR-308 regulates dMyc in negative feedback mechanism	*D. melanogaster*	(Daneshvar et al. 2013)
	miR-184 regulates DPP receptor Saxophone, gurken transport factor K10 and Tramtracks69	*D. melanogaster*	(Iovino et al. 2009)
	miR-1a-3p downregulates vitelline membrane protein, BmVMP23	*B. mori*	(Chen et al. 2013a)
	miR-275 depletion leads to abnormal egg development	*Ae. aegypti*	(Bryant et al. 2010)

Table 1. contd....

Table 1. contd.

Function	miRNAs	Insect species	References
Spermatogenesis	miR-7 downregulates Bag-of-marbles (Bam)	*D. melanogaster*	(Pek et al. 2009)
Maternal inheritance	miR-34 detection in embryos before zygotic transcription	*D. melanogaster*	(Soni et al. 2013)
Pheromone production	miR-124 regulates *transformer*	*D. melanogaster*	(Weng et al. 2013)
Interaction with pathogens/ parasites			
Bacterial	miR-2940 regulates expression of metalloprotease m41 ftsh and DNA (cytosine-5) methyltransferases (AaDnmt2) in interaction with *Wolbachia,wMelPop strain*	*Ae. aegypti*	(Hussain et al. 2011, Zhang et al. 2013a)
	miR-12 regulates MCM6 and MCT1	*Ae. aegypti*	(Osei-Amo et al. 2012)
Viral	miR-375 regulation of REL1 and cactus genes in interaction with DENV-2	*Ae. aegypti*	(Hussain et al. 2013b)
	miR-7 targets WSSV gene (wsv447)	*M. japonicus*	(Huang and Zhang 2012)
	miR-8 targets BmNPV genes	*B. mori*	(Singh et al. 2012)
	miR-24 downregulates ascovirus genes	*H. zea*	(Hussain and Asgari 2010)
Parasitization	miR-275 downregulates carboxypeptidase in interaction with *D. semiclausum*	*P. xylostella*	(Etebari et al. 2013) (Etebari and Asgari 2013)
	miR-8 positively regulates serpin27		
miRNAs encoded by insect viruses			
DNA viruses	HvAV-miR-1 target viral DNA polymerase	*H. zea*	(Hussain et al. 2008)
	HzNV-1pag1 miRNA downregulates viral early gene hhi 1	*H. zea*	(Wu et al. 2011)
	bmnpv-miR-1 regulates cellular Exportin-5 cofactor Ran	*H. zea*	(Singh et al. 2012)
	SGIV miR-homoHSV targets viral TNF-alpha (LITAF)-likefactor gene (ORF136R)	*Fathead minnow cells*	(Guo et al. 2013a)
RNA viruses	WNV$_{KUN}$ KUN-miR-1 upregulates cellular GATA-4	*Ae. aegypti/ albopictus*	(Hussain et al. 2012)
	DENV-vsRNA-5 downregulates viral NS1	*Ae. aegypti*	(Hussain and Asgari 2014)

and functions in neuronal remodeling, while EcR-B is expressed in larval tissues that undergo apoptosis (Davis et al. 2005, Ronald et al. 2013).

Recently, positive as well as negative interactions between several cellular miRNAs and 20E have been revealed. In *D. melanogaster*, 20E interaction with its receptor directly activates transcription of the Let-7 complex, which encodes miR-100, miR-125 and let-7 that regulate target genes involved in remodeling of neuromusculature during larval to adult transition (Chawla and Sokol 2012). In *B. germanica*, expression of the Let-7 complex was also found to significantly increase in the insect's wing pads in the last nymphal instar around the molting peak of 20E (Rubio and Bellés 2013). Depletion of let-7 and miR-100, but not miR-125, was shown to interrupt wing development in the insect. In another report, a feedback loop mechanism in *D. melanogaster* was found where miR-14 suppresses EcR expression, while 20E-EcR interaction downregulates miR-14 expression (Varghese and Cohen 2007). Interaction of miR-14 and EcR was also recently reported in *Helicoverpa armigera*, where miR-14 targets EcR and enhances its transcript levels both *in vivo* and *in vitro* (Jayachandran et al. 2013). Similarly, in *B. mori*, miR-281 downregulates EcR-B, while its own expression was suppressed upon 20E treatment (Jiang et al. 2013). Thus, miR-281 was found to be involved in *B. mori* developmental regulation in Malpighian tubules.

In *Ae. aegypti* mosquitoes, the expression level of miR-275 was elevated when 20E was induced *in vitro* in incubated fat bodies from non-blood fed mosquitoes (Bryant et al. 2010). Recently, a study conducted in pre-metamorphic nymphal instar (N5) and metamorphic nymphal instar of *B. germanica* using high throughput sequencing revealed differential expression of three and 37 miRNAs in N5 and N6, respectively (Rubio et al. 2012). Further investigation confirmed higher expression of miR-252-3p in N5 as compared to N6, while upregulation of Let-7, miR-100 and miR-125 was found only in N6. Similarly, 20E induction led to increase in Let-7, miR-100 and miR-125 and decrease in miR-252-3p. The same research group had previously reported the importance of miRNAs in the metamorphosis of *B. germanica* by knocking down Dicer-1 using RNAi at the penultimate instar (N5). There was normal molting to the 6th instar, while extra molt was produced in the 7th instar with functional prothoracic glands instead of the adult (Gomez-Orte and Bellés 2009).

The synthesis and degradation of chitin is important in the molting process in insects. It was recently found in the planthopper *Nilaparvata lugens* that conserved miRNAs miR-8-5p and miR-2a-3p target and downregulate genes involved in the chitin biosynthesis pathway; membrane-bound trehalase (Tre-2) and phosphoacetylglucosamine mutase (PAGM) (Chen et al. 2013b). Interestingly, in response to 20E the levels of miR-8-5p and miR-2a-3p were decreased and the levels of Tre-2 and PAGM were boosted. Moreover, feeding of miR-2a-3p mimics resulted in reduced chitin content and defective molting leading to significant reduction in the survival rate. Thus, miR-8-5p and miR-2a-3p act as molecular links that tune the chitin biosynthesis pathway in response to 20E signaling. In *D. melanogaster*, miR-8 is downregulated by the induction of 20E resulting in an increase of U shaped (USH), a p13 kinase inhibitor in insulin pathway to control the body size (Jin et al. 2012). Diapause was also shown to be regulated by small RNAs in the flesh fly *Sarcophaga bullata*. These flies can be switched from direct-development to diapause-destined when embryos and 1st instar larvae are exposed to short-day length. Along with other proteins associated with small RNA synthesis, Ago1 and loquacious were upregulated during pupal diapause, indicating possible involvement of miRNAs in maintaining diapause (Reynolds et al. 2013).

Another study based on microarray analysis of *P. xylostella* miRNAs showed their temporal expressions at different developmental stages (Liang et al. 2013). In eggs, nine miRNAs from four families (miR-71, miR-11, miR-279 and miR-92) were highly expressed possibly associated with regulation during embryogenesis and metamorphosis of *P. xylostella*. While the investigators found upregulation of 16 miRNAs during larval stage (from 1st to 4th instar) and higher expressions of four novel miRNAs (PC-5p-3972, PC-5p-13964, PC-5p-3130 and PC-5p-81253) in pupae. Further, differentially expressed miRNAs were grouped into seven groups based on their induction at different developmental stages (Liang et al. 2013).

In the gall midge, using deep sequencing and homology-based prediction, 611 putative miRNAs were identified including 89 known and 530 new or putative miRNAs (Khajuria et al. 2013). The expression levels of several miRNAs were varied at different larval instars such as mde-miR-1-3p and mde-miR-8-3p at the 1st instar, which were expressed at very high levels, whereas mde-miR-iab-4-5p and mde-miR-2a-5p levels were undetectable. Moreover, there was downregulation of mde-miR-10-5p, mde-miR-1000-5p, and mde-miR-190-5p in 3-day-old larvae as compared to that in 1-day-old larvae; while mde-miR-305-5p, mde-miR-9c-5p, and mde-miR-965-3p were upregulated in 3-day-old larvae. Similarly, differential expression of miRNAs at different developmental stages was observed in citrus red mite (Liu et al. 2013). Based on deep sequencing data of small RNAs, the researchers found upregulation of most of the known miRNAs in larvae and nymphs as compared to embryo and adult. The levels of miR-1 was very high in all stages and peaked in nymphs, while miR-315 was downregulated in later developmental stages and in contrast miR-34 gradually increased later.

Apoptosis: *miRNAs as Master Regulator of Cell Death*

In *D. melanogaster*, proapoptotic genes are targets of several miRNAs. For example, bantam and miR-2 were found to repress expression of head involution defective (hid) gene (Brennecke et al. 2003, Stark et al. 2005). Other reported examples of miRNAs regulating apoptosis are miR-14, that potentially regulates apoptotic effector caspase Drice, and miR-6 and miR-2/13 family that repress several proapoptotic genes such as *hid, grim, reaper* (*rpr*) and *sickle* (*skl*) (Xu et al. 2003, Leaman et al. 2005). Recently, another research group studied regulation of the proapoptotic genes, *rpr, hid, grim* and *skl* by miR-6 and miR-11 (Ge et al. 2012). The mutant flies that lacked both miR-6 and miR-11 showed embryonic lethality and defects in the central nervous system. Zhang and Cohen demonstrated the interaction between Hippo pathway, p53 and miR-2 in the regulation of apoptosis (Zhang and Cohen 2013). Their findings revealed that Hippo pathway functions in association with Yorkie protein and p53 to control *rpr* gene expression. Yorkie further mediates *rpr* expression levels through regulation of members of miR-2 miRNA family to prevent apoptosis (Zhang and Cohen 2013). Another interesting study indicated that *hid* gene is also downregulated by both miR-263a and miR-263b in *Drosophila* bristle progenitors and prevents apoptosis. Further analysis on miR-263a/b deletion mutants reveals that loss of bristles is irregular, which shed light on the significance of these miRNAs in warranting tight regulation of the patterning process by mediating survival of these cells with specific functions (Hilgers et al. 2010). Further, during wing development in *D. melanogaster*, miR-9 downregulates a transcription factor dLMO and inhibits apoptosis (Biryukova et al. 2009, Bejarano et al. 2010).

Reproduction: *Hormonal Manipulations with miRNAs*

In insects, various aspects of reproduction, including vitellogenesis, oocyte maturation and ovarian growth are mainly regulated by juvenile hormone (JH) (Riddiford 2012). In *Ae. aegypti*, JH controls oocyte development by inducing fat body competence for vitellogenin (Vg) synthesis, while 20E regulates Vg expression after blood feeding (Raikhel et al. 2002, Shin et al. 2012). Similarly, in *D. melanogaster*, *T. castaneum*, lepidopteran insects, migratory locust and cockroaches, either JH or JH with 20E has been demonstrated to be involved in vitellogenesis, oocyte maturation and ovarian development (Carney and Bender 2000, Cruz et al. 2003, Bellés 2004, Telfer 2009, Sheng et al. 2011). Previously, it was shown that Vg synthesis in the adult female fat body of locust is inducible by treatment with methoprene, an active JH analog (JHA) (Wyatt et al. 1992). Recently, a study on differential expression of miRNAs in JH-deprived and JHA-exposed fat bodies from *L. migratoria* showed that a total of 83 miRNAs were upregulated and 60 were downregulated by methoprene treatment (Song et al. 2013). Using quantitative RT-PCR (qRT-PCR), the investigators further demonstrated that methoprene significantly upregulated levels of miR-71, miR-8, lm-miRNA-35, lm-miR'-57, lm-miRNA-1 and lm-miRNA-12 and downregulated miR-184, miR-7a, miR-2, lm-miRNA-1, lm-miRNA-2, lm-miRNA-6, lm-miRNA-33 and lm-miRNA-58. Interestingly, miR-14, miR-13a and miR-100 showed a tendency to decline at 6–12 h and then elevate again at 24–48 h post-JHA induction. However, the researchers did not experimentally confirm whether any of these miRNAs were involved in regulating vitellogenesis, but knocking down Ago1 in the fat body had drastic effects on Vg transcripts, follicular epithelium development, terminal oocyte maturation and ovarian growth. In another study, Ago1 along with Dicer-1, Pasha and Drosha were found involved in oocyte formation and germ line cell division in *D. melanogaster* (Azzam et al. 2012). Ago1 protein was enriched in the oocytes and the cytoplasm of follicle cells and multiple Ago1 mutant alleles showed only eight nurse cells in the egg chambers without an oocyte, which was also observed in Dicer-1, Pasha and Drosha mutants.

Myc is a transcription factor associated with cell proliferation and growth. In *D. melanogaster*, Myc is required for oogenesis and larval growth and its absence results in retardation and lethality (Maines et al. 2004). The interaction of dMyc and miR-308 in *D. melanogaster* has recently been shown. dMyc increases expression of miR-308 by binding to its locus and this higher level of miRNA destabilize dMyc mRNA as well as protein (Daneshvar et al. 2013). This negative feedback loop is important to maintain the balance of dMyc resulting in lower apoptotic activity and suppression of lethality. Further, the negative effect of higher expression levels of miR-308 on dMyc target genes such as *fibrillarin* was confirmed (Daneshvar et al. 2013).

In *B. mori*, expression of a vitelline membrane protein, BmVMP23, was found to be downregulated by bmo-miR-1a-3p (Chen et al. 2013a). In the ovaries of the wild type females, the expression level of bmo-miR-1a-3p was in contrast to that of BmVMP23. Specific miRNA-mRNA interaction was shown through the use of a reporter construct in which 3'UTR of BmVMP23 was cloned downstream to the *luciferase* gene whose expression was repressed in co-transfection experiments *in vitro* using miRNA mimics (Chen et al. 2013a). In another report, Bryant and colleagues revealed that miR-275 is essential for female mosquitoes as depletion of this miRNA led to severe deficiencies in blood digestion, fluid excretion and egg development (Bryant et al. 2010). Tanaka and Piulachs investigated the correlation between Dicer-1 and oocyte development in *B. germanica* (Tanaka and Piulachs 2012). First, they found that expression levels of *B. germanica* Dicer-1 (BgDCR1) was the highest at the onset of the 6th nymphal instar that went on decreasing as the basal oocyte matured, and reached minimum values in the adult just prior to oviposition. RNAi knockdown of BgDCR1 at the 6th nymphal instar or in the newly emerged adults led to the adults' failure to oviposit by affecting follicular epithelium development. Although the investigators showed downregulation of several miRNAs in BgDCR1 RNAi females, which miRNA was specifically involved is yet to be determined. In a similar study in *N. lugens* (Brown planthopper), RNAi silencing of Dicer-1 resulted in badly malformed oocytes in which follicular cells were not developed as normal (Zhang et al. 2013b).

In *D. melanogaster*, suppression of miR-184 leads to reduced egg production and defects in early embryogenesis (Iovino et al. 2009). The function of miR-184 is exerted through controlling three known key regulators of development: decapentaplegic (DPP) receptor Saxophone which regulates GSC differentiation, the gurken transport factor K10 which regulates dorso-ventral patterning of the egg shell, and the transcriptional repressor Tramtracks69 which regulates antero-posterior patterning of the blastoderm. Involvement of miRNAs in spermatogenesis in male *Drosophila* has also been shown. miR-7 targets Bag-of-marbles (Bam) mRNA in the 3'UTR and downregulates its expression, which subsequently impedes differentiation of germline stem cells (GSC) into primary spermatocytes in testes (Pek et al. 2009).

Recently, maternal inheritance of miR-34 was reported in *D. melanogaster* embryos as higher abundance of this miRNA was found in embryos before zygotic transcription. The processing of dme-miR-34 was revealed to be carried out by the maternal Dicer-1 (Soni et al. 2013). This suggests the role of miRNAs in insect development as part of maternal factors deposited into the eggs during oogenesis.

Gender-specific miRNAs: *miRNAs as Determinant of Sex*

A recent report highlighted the significance of miRNAs in sexual behavior in *D. melanogaster* (Weng et al. 2013). It was found that miR-124 is required for normal production of sex pheromone in male flies. The miRNA negatively regulates expression of the female form of *transformer*, which is a sex-specific splicing factor, hence, controls male sexual differentiation and behavior. miR-124 mutant male flies had more tendency to mate with other males, and wild type females showed less interest in the mutant male flies as compared to the wild type male flies when they were given a choice.

Variations in miRNA expression levels among males and females have been found in the cockroach *Eupolyphaga sinensis* (Wu et al. 2013b). Based on small RNA deep sequencing data, it was found that 45 miRNAs exhibited differential expression in both sexes at 4th instar larvae. In *E. sinensis*, striking morphological differences between the sexes (adult females are bigger in body size and wingless while adult males have smaller bodies and have forewings) makes it easy to distinguish them apart. Moreover, females have comparatively longer life span and larvae have to go through 9–11 instars to achieve sexual maturity, while only seven instars occur in male nymphs. Since differences in maturity and metamorphosis between both sexes exist, the investigators found downregulation of metamorphosis related miRNAs (miR-9c, miR-275 and miR-276) only in males. Four other miRNAs, miR-14 and bantam were expressed many folds higher in females, while miR-8 and miR-315 were upregulated in males. However, all of these findings from this work were based on deep sequencing results and qRT-PCR data and experimental target analyses would help in understanding the biology of sexes at the miRNA level.

A couple of other studies demonstrated the gender specific miRNA expression in *T. castaneum* and brown planthopper *N. lugens*. In *T. castaneum*, seven and four different miRNAs were shown highly expressed in females and males, respectively (Freitak et al. 2012). In *N. lugens*, miRNAs that were only expressed in females were miR-30d, miR-144* and miR-22, while miR-263, miR-9a and let-7b were highly expressed. In male adults, miR-317, miR-87, miR-277 and miR-34 were highly expressed (Chen et al. 2012).

Defence Against Pathogens: *miRNAs at the Heart of Resistance*

The role of miRNAs in defense and immunity against pathogens has been explored in vertebrates (O'Connell et al. 2010). Several studies from different organisms have revealed changes in miRNA expression levels upon immune activation (Delic et al. 2012, Yang et al. 2012). In mammals, immune system consists of adaptive and innate responses that include utilization of antibodies and non-specific components, respectively (Litman et al. 2010). Insects lack adaptive immunity, but they have evolved a competent innate immunity that includes production of anti-microbial peptides (AMPs), melanization, phagocytosis, encapsulation and coagulation of pathogens (Lemaitre and Hoffmann 2007). In order to counter fungal, bacterial and viral pathogens, insects use a variety of signaling pathways such as Toll and Imd (that regulate AMPs expression), JNK cascade and JAK-STAT pathways (Dostert et al. 2005, Lazzaro 2008, Bond and Foley 2009). In *D. melanogaster*, 73 miRNAs were identified as putative miRNAs involved in immune regulation based on their potential targets in genes involved in immune pathways (Fullaond and Lee 2012). The investigators found that seven miRNAs (miR-12, 31b, 33, 283, 304, 1003 and 1016) that target host genes are mainly expressed in immune-related tissues. In honeybees *A. mellifera*, differential expression of cellular miRNAs upon immune activation was studied in response to infection by the Gram-negative bacterium *Serratia marcescens* or the Gram-positive bacterium *Micrococcus luteus* (Lourenco et al. 2013). In *S. marcescens* infected bees, among the downregulated miRNAs were ame-bantam, miR-12, miR-34, miR-184, miR-278, miR-375, miR-989 and miR-1175, while some were upregulated such as ame-let-7, miR-2, miR-13a, miR-92a. The genes associated with immune system in honeybees could be potential targets of these miRNAs.

A number of papers have recently shown the involvement of cellular miRNAs in the host-*Wolbachia* wMelPop strain interactions in *Ae. aegypti* (Hussain et al. 2011, Osei-Amo et al. 2012, Zhang et al. 2013a). Functional analyses of upregulated miRNAs in *Wolbachia*-infected miRNAs (aae-miR-2940 and aae-miR-12) were conducted in these studies. First, it was found that aae-miR-2940 targets the 3'UTR region of the mosquito's metalloprotease m41 ftsh transcripts and increases its transcript levels. Silencing of the metalloprotease gene in both *Wolbachia*-infected cells and adult mosquitoes led to a significant reduction in *Wolbachia* density, as did inhibition of the miR-2940 in mosquitoes (Hussain et al. 2011). In a separate study, aae-miR-2940 was found to downregulate the expression of the DNA (cytosine-5) methyltransferase (AaDnmt2) (Zhang et al. 2013a). Ectopic expression of AaDnmt2 in mosquito cells led to inhibition of *Wolbachia* replication, but significantly promoted replication of Dengue virus, suggesting a miRNA based blocking of Dengue virus replication in *Wolbachia*-infected mosquitoes. Function of another *Wolbachia* induced cellular miRNA, aae-miR-12, was confirmed in regulation of two target genes; DNA replication licensing factor MCM6 and monocarboxylate transporter MCT1 (Osei-Amo et al. 2012).

In most hematophagous insects, such as mosquitoes, blood feeding is a significant event as it is required for egg development and there have been published reports on differential expression of a large number of genes as well as miRNAs following blood uptake (Bryant et al. 2010, Hussain et al. 2013b). The upregulated miRNAs in blood fed (BF) mosquitoes can regulate defense genes. One such miRNA, aae-miR-375, was exclusively detected in BF mosquitoes that could regulate six different genes involved in development and immunity at the post-transcriptional level. In these experiments, interaction of aae-miR-375 with two immune related genes, *cactus* and *REL1*, resulted in up and downregulation of their transcripts, respectively (Hussain et al. 2013b). As a consequence, aae-miR-375 was shown to enhance Dengue virus genomic RNA levels in mosquito cells by suppressing cactus, which is an inhibitor of activation of immune effector genes, such as cecropin, that has been shown to have anti-Dengue virus activity (Luplertlop et al. 2012). In another study in *An. gambiae*, female midguts carrying *Plasmodium berghei* showed downregulation of some miRNAs (aga-miR-34, miR-1174 and miR-1175) and upregulation of only aga-miR-989 (Winter et al. 2007).

Differential expression of cellular miRNAs in *B. mori* midgut upon *B. mori* cytoplasmic polyhedrosis virus (BmCPV) was investigated using deep sequencing of infected larvae at 72 and 96 h post-infection. In total, 58 miRNAs were found with significant up or downregulation in virus-infected larvae as compared to non-infected larvae (Wu et al. 2013a). Putative target genes against six differentially expressed miRNAs, bmo-miR-275, miR-14, miR-1a, N-50, N-46 and N-45, were associated with response to stimulus and immune system process based on gene ontology (GO) analysis. Moreover, significant increases in miR-275 levels upon BmCPV infection at 96 h could provide a basic explanation for the stunted growth in virus-infected *B. mori*, given miR-275 has been shown to regulate developmental processes during metamorphosis (Liu et al. 2010). In addition, BmCPV infection also led to downregulation of miR-1 and miR-274 at 96 h, which suggests a possible viral strategy to achieve successful infection. Global miRNA change was also observed in *S. frugiperda* (Sf9) cells upon infection by *Autographa californica* multiple nucleopolyhedrovirus(AcMNPV). There were several upregulated miRNAs (miR-184, miR-998 and miR-10) at 24 h, the expression of some of which was eventually downregulated at 72 h post-infection. In addition, 13 novel *S. frugiperda* miRNAs were identified in this study, three of which showed upregulation during virus infection (Mehrabadi et al. 2013). In another report, changes in miRNA levels were also observed upon virus infection where White spot syndrome virus (WSSV) upregulated 25 and downregulates six miRNAs in the shrimp *Marsupenaeus japonicus* (Huang et al. 2012). Another example is downregulation of miR-989 and upregulation of miR-92 in *C. quinquefasciatus* infected with West Nile virus (WNV) (Skalsky et al. 2010). A recent analysis of host miRNAs modulated by Dengue virus infection (2, 4 and 9 dpi) revealed differential expression of 35 miRNAs. In this analysis, miR-5119-5p, miR-34-3p, miR-87-5p and miR-988-5p were upregulated (Campbell et al. 2013).

Cellular miRNAs have also been shown to target viral genes (Hussain and Asgari 2010, Huang and Zhang 2012, Singh et al. 2012). In *B. mori*, several targets were found in *BmNPV* genes against miR-8 and inhibition of this miRNA resulted in 8-fold increase in viral load in the fat bodies of infected larvae (Singh et al. 2012). Moreover, in *Helicoverpa zea*, miR-24 was able to downregulate ascovirus *Heliothis virescens* (HvAV) genes; *DNA dependent RNA polymerase II RPC2* (*DdRP*; ORF64) and *DdRP β* subunits (*DdRP*; ORF82) (Hussain and Asgari 2010). In the shrimp *M. japonicus*, miR-7 targets were predicted in *WSSV* gene (*wsv447*) that is associated with early virus replication and proliferation (Huang and Zhang 2012).

In understanding the role of miRNAs in insect-virus interactions, another milestone was achieved with the discoveries of miRNAs encoded by viral genomes that target viral as well as host mRNAs. Among insect viruses, virus-encoded miRNAs have now been reported from both DNA and RNA viruses. From insect DNA viruses, the first report was from HvAV-3e encoding a miRNA, HvAV miR-1, which was found to target the virus's own DNA polymerase resulting in autoregulation of virus replication in host cells (Hussain et al. 2008). During latent infection, *H. zea* nudivirus (HzNV-1) expresses miRNAs from pag1 transcript, which downregulate viral early gene *hhi1*, that permits the virus to enter the latent infection stage in cell culture (Wu et al. 2011). BmNPV produces miR-1 which downregulates miRNA nuclear Exportin-5 cofactor Ran resulting in the cessation of the export of cellular pre-miRNAs into the cytoplasm (Singh et al. 2012). This results in an overall decrease in the cellular miRNAs abundance. Another viral miRNA, miR-homoHSV, that is encoded by Singapore grouper irido virus (SGIV), targets SGIV *pro-apoptotic lipopolysaccharide-induced TNF-alpha (LITAF)-like factor* gene (ORF136R) and attenuates SGIV-induced cell death (Guo et al. 2013a). Therefore, this miRNA serves as a feedback regulator of cell death during viral infection.

Recently, our group has discovered functional miRNAs encoded by insect arboviruses, West Nile Virus (WNV) and Dengue virus (DENV-2). First, in Kunjin strain of WNV (WNV$_{KUN}$), KUN-miR-1, was found to be derived from the 3' SL stem-loop located at the end of the 3'UTR of the virus in mosquito cells (Hussain et al. 2012). The mature KUN-miR-1 detection was also confirmed in transfections of plasmids carrying WNV$_{KUN}$ pre-miRNA (3'SL) as well as Semliki Forest virus (SFV) RNA replicon. Silencing of *Dicer-1* but not *Dicer-2* led to a reduction in the KUN-miR-1 level. Further, when a synthetic inhibitor of KUN-miR-1 was transfected into mosquito cells, replication of viral RNA was significantly reduced. Using cloning and bioinformatics approaches, the cellular GATA4 mRNA was identified as a target for KUN-miR-1 and this interaction of miRNA-mRNA results in an increase in GATA-4 mRNA levels. Depletion of GATA4 mRNA by RNAi led to a significant reduction in virus RNA replication, while a KUN-miR-1 RNA mimic enhanced replication of a mutant WNV$_{KUN}$ virus producing reduced amounts of KUN-miR-1, suggesting that GATA4-induction via KUN-miR-1 plays an important role in virus replication. In the other example, functional miRNAs or viral small RNAs (vsRNAs) encoded by DENV-2 in virus-infected *Ae. aegypti* mosquitoes and cell lines were discovered (Hussain and Asgari 2014). Initially, six vsRNAs, with candidate stem-loop structures in the 5' and 3'UTRs of the viral genomic RNA were found in deep sequencing data. Synthetic RNA inhibitors were used against these vsRNAs with the result that inhibition of only DENV-vsRNA-5 led to significant increases in viral replication. Further, silencing of RNAi/miRNA pathways associated proteins showed that Ago2, and to some extent Dicer-1 and Ago-1, is mainly involved in DENV-vsRNA-5 biogenesis. Interestingly, significant impact of synthetic mimic and inhibitor of DENV-vsRNA-5 on DENV RNA levels revealed DENV-vsRNA-5's role in virus autoregulation by targeting the virus genome at the NS1 gene region. The results revealed that DENV is able to encode functional vsRNAs, and one of those, which resembles miRNAs, specifically targets a viral gene, opening an avenue for possible utilization of the small RNA to limit DENV replication.

A couple of papers have also revealed the involvement of host miRNAs in response to parasitization. Differential expression of miRNAs was studied in the larvae of *Lymantria dispar* in response to *Glyptapanteles flavicoxis* wasp parasitization (Gundersen-Rindal and Pedroni 2010). In this study, significant upregulation of miR-1, miR-184 and miR-277 and downregulation of Let-7 and miR-279 in larval hemocytes were detected. Notably, expression of these miRNAs was tissue specific as was confirmed by RT-qPCR. In the fat bodies, expression levels of miR-277 and miR-279 were very high, while in the hemolymph miR-279 was downregulated and miR-1 upregulation was detected in all tissue types analyzed. Small RNA deep sequencing for miRNA expression profile of Diamondback moth *P. xylostella* was conducted in association with parasitization by *Diadegma semiclausum*. Combining the deep sequencing data and bioinformatics, 235 miRNAs were identified from *P. xylostella*, 96 of which were differentially expressed after parasitization (Etebari et al. 2013). Bantam*, miR-8, miR-184 and miR-281* were significantly downregulated and miR-279b and miR-2944b* were highly induced in parasitized larvae. Interestingly, high copy numbers and differential expression of several miRNA passenger strands (miRNA*) were also identified, which suggested their potential roles in the host-parasitoid interaction (Etebari et al. 2013). In a follow-up study, downregulation of miR-8 in *P. xylostella* after parasitization was shown to correlate with the induction of immune related genes, such as the anti-microbial peptide gloverin (Etebari and Asgari 2013). Experiments showed that miR-8 positively regulates Serpin27, an inhibitor of immune activation. As a consequence, due to decline in miR-8 levels after parasitization, Serpin27 levels also decline allowing activation of the Toll pathway and prophenoloxidase involved in melanization. Although this immune activation mediated by miR-8 is believed to be part of the host immune response following parasitization, parasitoid wasps interfere with various aspects of host immunity by introducing immunosuppressive factors into the host, such as venom and polydnaviruses (Asgari and Rivers 2011).

Differential Expression of miRNAs and Blood Feeding: *miRNAs Red Hot in Blood*

When a female mosquito takes a blood meal, a sequence of signals that originate from several tissues leads to alterations in gene expression. These signals include vertebrate insulin and increased amino acid levels in the hemolymph, peptide hormones from the gut and the central nervous system and ecdysteroids from the ovaries (Brown and Cao 2001, Hansen et al. 2004). These signals may be potential candidates for inducing expression of miRNAs, which in turn regulate the expression of multiple genes. In blood-feeding arthropods, such as mosquitoes and ticks, changes in the abundance of a number of miRNAs following blood uptake have been demonstrated. In *Ae. aegypti*, the expression levels of aae-miR-375, -2940*, -125, -317, -14, -1 and bantam were significantly induced in blood fed (BF) mosquitoes when compared to non-blood fed (NBF) mosquitoes (Hussain et al. 2013b). The analyses of expression levels of the six different potential target genes for aae-miR-375 in miRNA mimic transfected mosquitoes showed significant upregulation of *cactus, DEAD box ATP dependent RNA helicase, prohibitin* and *kinesin* and downregulation of *REL1* and a hypothetical protein (Hussain et al. 2013b). In another study, miR-275 was shown significantly induced in BF *Ae. aegypti* mosquitoes. Knockdown of this miRNA by injection of the synthetic antagomir of the miRNA into female mosquitoes led to severe defects in blood digestion, fluid excretion, and egg development (Bryant et al. 2010).

Using small RNA deep sequencing, differential expression of a number of miRNAs in BF and NBF tick (*Haemaphysalis longicornis*) salivary glands was shown (Zhou et al. 2013). After blood feeding, 162 known miRNAs were upregulated and six of them, miR-1810, miR-2138, miR-2140, miR-425*, miR-429, and miR-516*, were highly induced. However, 231 miRNAs were also downregulated after blood feeding including miR-2941-1*, miR-10-5p, miR-2973, miR-1183, miR-4006b-5p, and miR-881.

miRNAs in Cold and Hot Temperatures Stresses: *A friend in need*

Insects have evolved strategies to resist stresses, such as low temperatures and decreased food source. Physiological and molecular basis of this resistance are characterized by reduced metabolism, reversible protein phosphorylation (RPP), histone modification and protein SUMOylation

(Storey and Storey 2013). Recently, a couple of studies reported differential expression of miRNAs (cryomiRs) in response to cold temperature stresses in insects. In goldenrod gall fly, *Eurosta solidaginis*, miRNA expression was analyzed in control (+5°C) and chilled (−15°C) larvae (Courteau et al. 2012). Comparatively, in frozen larvae the levels of miR-11, miR-276, miR-71, miR-3742, miR-277-3p, miR-2543b and miR-34 were significantly decreased whereas miR-284, miR-3791-5p and miR-92c-3p were upregulated. The potential target genes for miR-277-3p and miR-284 were found to be involved in translation and the Krebs cycle. In another study, a comparative analysis was conducted on expression levels of miR-1 and miR-34 in freeze tolerant *E. solidaginis* and freeze-avoiding *Epiblema scudderiana* at +5 and −15°C that showed elevated levels of miR-1 only in frozen *E. solidaginis* with no effect on miR-34 in both insects (Lyons et al. 2013).

Differential expression levels of miRNAs in response to heat shock were studied in *T. castaneum* (Freitak et al. 2012). Based on microarray analysis, the researchers found that exposure of *T. castaneum* to mild heat shock induced six and repressed 40 miRNAs. In the same study, insects were exposed to long stress-like starvation that changed miRNA levels with upregulation of five and downregulation of 22 miRNAs. Interestingly, as a result of overall stresses, levels of stress-induced miRNAs in females were much higher than in males. In a separate transcriptome analysis, several genes were found to be responsive to stresses in *T. castaneum* (Altincicek et al. 2013). It would be interesting to find out if stress-induced miRNAs are associated with the regulation of differentially altered genes.

Insect miRNA Evolution: *A Tendency to Evolve and Flourish*

There are several views on the evolution of miRNA genes such as 1) they originate from duplication and nucleotide substitution of old miRNA genes (Tanzer and Stadler 2004, Zhuo et al. 2013), 2) come from inverted terminal repeats of transposable elements (TEs) (Piriyapongsa and Jordan 2007), or 3) from random hairpin structures in intronic or intergenic regions (Bentwich et al. 2005, Felippes et al. 2008). One study showed that gains or losses of miRNA genes frequently occur during the evolution of 12 *Drosophila* species. Accordingly, new miRNA genes mainly originate from random hairpin structures in intronic or intergenic regions as well as duplication of miRNAs genes (Nozawa et al. 2010, Mohammed et al. 2014). Moreover, more recent miRNA genes in *Drosophila* species are less conserved than old ones. Two studies have highlighted the role of shifting in the hairpin precursor 5′ or 3′ arms or in the seed region that may affect a number of miRNAs conserved between *D. melanogaster* and *T. castaneum* during insect miRNA evolution (Marco et al. 2010). It was found that bias toward the 5' or 3'arm usage, for a given miRNA that is finally processed into mature miRNA, can change the target profile and therefore function of a given miRNA (Marco et al. 2010), while seed shifting (changes in the 5' end of miRNA) might also significantly alter targets of mature miRNAs.

A recent study has revealed the evolution of bi-directional transcription at the iab-4/iab-8 miRNA locus based on computational analysis and *in vivo* assays in *D. melanogaster* and *T. castaneum* (Hui et al. 2013). The iab-4 locus produces two primary transcripts that fold into two distinct hairpin precursors, named dme-iab-4 and dme-iab-8, yielding four mature miRNAs that regulate translation of the Hox genes *Ubx* and *abd-A*. The investigators found differences in targeting of host genes in both insects, since in the fly iab-8 targets *abd-A* gene, while in *T. castaneum* both iab-4 and iab-8 can target this gene. There were also differences in the relative abundance of these miRNAs in the two insects. Iab-4 locus, containing a highly conserved hairpin sequence, was identified in small RNAseq data from *D. melanogaster*, *T. castaneum*, *L. migratoria* and *A. mellifera*, and mature miRNAs from this locus were detected in all of them (Hui et al. 2013). These miRNAs were also identified from a hemimetabolous insect, the pea aphid *A. pisum*.

The rate of evolution of genes associated with miRNA pathways can also shape the miRNAs evolutionary journey. Recent evidence suggests duplication events in *Dicer* and *Ago* genes in insects. In *D. melanogaster*, two distinct dicers have been reported; Dicer-1 for miRNA and Dicer-2 for siRNA production (Ye and Liu 2008). Similarly, multiple *Ago* genes have been found in *D. melanogaster*, *Culex pipiens* and *Ae. aegypti* (Campbell et al. 2008, Kim et al. 2009). Interestingly, expansion of these genes was observed in the pea aphid with the existence of two copies of *Dicer-1* (*Apidcr-1a* and *Apidcr-1b*), two copies of *Ago1* (*Apiago-1a* and *Apiago-1b*) and four copies of *Pasha* (*Apipasha-1* to *Apipasha-4*) (Jaubert-Possamai et al. 2010, Ortiz-Rivas et al. 2012). These studies present molecular analyses that show the simultaneous duplications of *Ago1* and *Dicer-1* and one copy of each duplicated gene; *ago-1b* and *dcr-1b*, had undergone a higher evolutionary rate and relaxation of selection, while the other two, *ago-1a* and *dcr-1a*, were conserved and subject to strong purifying selection (Ortiz-Rivas et al. 2012). Further analysis confirmed positive selection for *ago-1b* leading to new robust roles. Expression analyses of *dcr-1b* and *ago-1b* through RT-qPCR in different reproductive morphs of the pea aphid revealed different levels of gene expression in parthenogenetic and sexual aphids, suggesting a role for these genes in regulation of sexual polyphenism (Ortiz-Rivas et al. 2012). In the silkworm *B. mori*, four Ago proteins have been identified, named Ago1, Ago2, Ago3 and Siwi. Ago1 is involved in miRNA-mediated translational repression; Ago2 participates in double-stranded RNA (dsRNA)-induced RNAi, whereas Ago3 and Siwi are associated with Piwi-interacting RNA (piRNA) pathway (Wang et al. 2012b, Zhu et al. 2013).

In another study, *Ae. aegypti* exo-siRNA and miRNA pathway genes were found to be undergoing rapid, positive and diversifying selection (Bernhardt et al. 2012). The researchers studied comparative patterns of molecular evolution among the RNAi (*Dicer-2, Ago2, R2D2*) and miRNA pathways genes (*Dicer-1, Ago1, R3D1*) and revealed that the miRNA pathway genes do not evolve as rapidly as the genome average. This study was based on mosquitoes collected from six distinct geographic populations. The ratio of replacement to silent amino acid substitutions was higher in RNAi genes as compared to miRNA pathway genes. An extensive expansion of miRNA genes has recently been reported in gall midge (Liang et al. 2013). The expanded gene families include PC-3P-59454 with 9 genes, PC-5p-61169 with 10 genes, PC-3p-19591 with 11 genes, PC-3p-58746 with 13 genes, PC-3p-36826 with 14 genes, PC-5p-66343 with 22 genes, PC-3p-47103 with 24 genes, PC-5p-39989 with 28 genes, PC-5p-57811 with 60 genes, PC-3p-54311 with 66 genes, and PC-5p-67443 with 91 genes.

Concluding Remarks and Prospects

miRNAs have emerged as important regulators of gene expression, in most cases fine tuning their targets' transcript levels. As more research is conducted in this area, we continuously find new non-canonical mechanisms for production of miRNAs, alternative ways of their interactions with target sequences and their functions. The number of publications on insect miRNA functions has significantly increased in the last five years adding to our scant knowledge on their role in insects' biology. Many of these reports have focused on exploring the differential abundance of miRNAs under particular conditions/treatments (infection, morphs, developmental stage, feeding, stress, etc.). There is now a need to utilize this wealth of information to explore the role of miRNAs in insect function. Currently, many tools are available for the study of miRNA functions. In particular, keeping up-to-date with advances in miRNA research in vertebrates is essential considering major advancements that have been made in both basic and applied aspects of miRNA biology in vertebrates. The outcomes should help us to firstly gain a better understanding of the basic role of these small non-coding RNAs in the biology of insects and secondly pave the way to exploit them for the control of medically and agriculturally important pests or limit transmission of insect-borne plant and animal pathogens. This is certainly an exciting period!

Acknowledgements

We would like to thank Dr. Xavier Bellés from Institut de Biologia Evolutiva (CSIC-UPF) in Spain for critically reading this manuscript. We also would like to acknowledge the Australian Research Council (DP110102112 and DE120101512) and the National Health and Medical Research Council (APP1027110 and APP1062983) for providing funding, the outcomes of which are mentioned in this chapter.

Keywords: Argonaute, miRISC, Dicer, Drosha, Exprotin-5, Seed region, Apoptosis, Ecdysone, *Wolbachia*, AcMNPV, Ascovirus, Nudivirus, Dengue virus, West Nile virus, GATA4

References

Altincicek, B., A. Elashry, N. Guz, F.M.W. Grundler, A. Vilcinskas and H. Dehne. 2013. Next generation sequencing based transcriptome analysis of septic-injury responsive genes in the beetle *Tribolium castaneum*. PLoS One 8: e52004.

Ameres, S.L. and P.D. Zamore. 2013. Diversifying microRNA sequence and function. Nat. Rev. Mol. Cell Biol. 14: 475–488.

Ameyar-Zazoua, M., C. Rachez, M. Souidi, P. Robin, L. Fritsch, R. Young, N. Morozova, R. Fenouil, N. Descostes, J.-C. Andrau, J. Mathieu, A. Hamiche, S. Ait-Si-Ali, C. Muchardt, E. Batsché and A. Harel-Bellan. 2012. Argonaute proteins couple chromatin silencing to alternative splicing. Nat. Struct. Mol. Biol. 19: 998–1004.

Asgari, S. 2013. MicroRNA functions in insects. Insect Biochem. Mol. Biol. 43: 388–397.

Asgari, S. and D.B. Rivers. 2011. Venom proteins from endoparasitoid wasps and their role in host-parasite interactions. Annu. Rev. Entomol. 56: 313–335.

Azzam, G., P. Smibert, E.C. Lai and J.L. Liu. 2012. *Drosophila* Argonaute 1 and its miRNA biogenesis partners are required for oocyte formation and germline cell division. Dev. Biol. 365: 384–394.

Babiarz, J.E., R. Hsu, C. Melton, M. Thomas, E.M. Ullian and R. Blelloch. 2011. A role for noncanonical microRNAs in the mammalian brain revealed by phenotypic differences in Dgcr8 versus Dicer1 knockouts and small RNA sequencing. RNA 17: 1489–1501.

Babiarz, J.E., J.G. Ruby, Y. Wang, D.P. Bartel and R. Blelloch. 2008. Mouse ES cells express endogenous shRNAs, siRNAs, and other Microprocessor-independent, Dicer-dependent small RNAs. Genes Dev. 22: 2773–2785.

Baek, D., J. Ville´n, C. Shin, F.D. Camargo, S.P. Gygi and D.P. Bartel. 2008. The impact of microRNAs on protein output. Nature 455: 64–71.

Bartel, D.P. 2009. MicroRNAs: target recognition and regulatory functions. Cell 136: 215–233.

Bejarano, F., P. Smibert and E.C. Lai. 2010. miR-9a prevents apoptosis during wing development by repressing *Drosophila* LIM-only. Dev. Biol. 338: 63–73.

Bellés, X. 2004. Vitellogenesis directed by juvenile hormone. pp. 157–197. *In*: A.S. Raikhel (ed.). Reproduction Biology of Invertebrates: Progress in Vitellogenesis. Science Publishers Inc., Boca Raton.

Bellés, X., A.S. Cristino, E.D. Tanaka, M. Rubio and M.D. Piulachs. 2012. Insect microRNAs: From molecular mechanisms to biological roles. pp. 30–56. *In*: L.I. Gilbert (ed.). Insect Molecular Biology and Biochemistry. Elsevier Academic Press, Amsterdam.

Bennasser, Y., C. Chable-Bessia, R. Triboulet, D. Gibbings, C. Gwizdek, C. Dargemont, E.J. Kremer, O. Voinnet and M. Benkirane. 2011. Competition for XPO5 binding between Dicer mRNA, pre-miRNA and viral RNA regulates human Dicer levels. Nat. Struct. Mol. Biol. 18: 323–327.

Bentwich, I., A. Avniel, Y. Karov, R. Aharonov, S. Gilad, O. Barad, A. Barzilai, P. Einat, U. Einav, E. Meiri, E. Sharon, Y. Spector and Z. Bentwich. 2005. Identification of hundreds of conserved and nonconserved human microRNAs. Nat. Genetics 37: 766–770.

Berezikov, E., N. Robine, A. Samsonova, J.O. Westholm, A. Naqvi, J.-H. Hung, K. Okamura, Q. Dai, D. Bortolamiol-Becet, R. Martin, Y. Zhao, P.D. Zamore, G.J. Hannon, M.A. Marra, Z. Weng, N. Perrimon and E.C. Lai. 2011. Deep annotation of *Drosophila melanogaster* microRNAs yields insights into their processing, modification, and emergence. Genome Res. 21: 203–215.

Bernhardt, S.A., M.P. Simmons, K.E. Olson, B.J. Beaty, C.D. Blair and W.C. Black. 2012. Rapid intraspecific evolution of miRNA and siRNA genes in the mosquito *Aedes aegypti*. PLoS One 7: e44198.

Biryukova, I., J. Asmar, H. Abdesselem and P. Heitzler. 2009. *Drosophila* miR-9a regulates wing development via fine-tuning expression of the LIM only factor dLMO. Dev. Biol. 327: 486–496.

Blahna, M.T. and A. Hata. 2012. Smad-mediated regulation of microRNA biosynthesis. FEBS Lett. 586: 1906–1912.

Bogerd, H.P., H.W. Karnowski, X. Cai, J. Shin, M. Pohlers and B.R. Cullen. 2010. A mammalian herpesvirus uses noncanonical expression and processing mechanisms to generate viral microRNAs. Mol. Cell 37: 135–142.

Bohnsack, M.T., K. Czaplinski and D. Gorlich. 2004. Exportin 5 is a RanGTP-dependent dsRNA-binding protein that mediates nuclear export of pre-miRNAs. RNA 10: 185–191.

Bond, D. and E. Foley. 2009. A quantitative RNAi screen for JNK modifiers identifies Pvr as a novel regulator of *Drosophila* immune signaling. PLoS Pathog. 5: e1000655.

Borchert, G.M., W. Lanier and B.L. Davidson. 2006. RNA polymerase III transcribes human microRNAs. Nat. Struct. Mol. Biol. 13: 1097–1010.

Brennecke, J., A.A. Aravin, A. Stark, M. Dus, M. Kellis, R. Sachidanandam and G.J. Hannon. 2007. Discrete small RNA-generating loci as master regulators of transposon activity in *Drosophila*. Cell 128: 1089–1103.

Brennecke, J., D.R. Hipfner, A. Stark, R.B. Russell and S.M. Cohen. 2003. Bantam encodes a developmentally regulated microRNA that controls cell proliferation and regulates the proapoptotic gene hid in *Drosophila*. Cell 113: 25–36.

Briata, P., C.Y. Chen, M. Giovarelli, M. Pasero, M. Trabucchi, A. Ramos and R. Gherzi. 2011. KSRP, many functions for a single protein. Front. Biosci. 16: 1787–1796.

Brown, M.R. and C. Cao. 2001. Distribution of ovary ecdysteroidogenic hormone I in the nervous system and gut of mosquitoes. J. Insect Sci. 1: 3.

Bruno, I.G., R. Karam, L. Huang, A. Bhardwaj, C.H. Lou, E.Y. Shum, H.W. Song, M.A. Corbett, W.D. Gifford, J. Gecz, S.L. Pfaff and M.F. Wilkinson. 2011. Identification of a microRNA that activates gene expression by repressing nonsense-mediated RNA decay. Mol. Cell 42: 500–510.

Bryant, B., W. Macdonald and A.S. Raikhel. 2010. MicroRNA miR-275 is indispensable for blood digestion and egg development in the mosquito *Aedes aegypti*. Proc. Natl. Acad. Sci. USA 107: 22391–22398.

Campbell, C.L., W.C. Black IV, A.M. Hess and B.D. Foy. 2008. Comparative genomics of small RNA regulatory pathway components in vector mosquitoes. BMC Genomics 9: 425.

Campbell, C.L., T. Harrison, A.M. Hess and G.D. Ebel. 2013. MicroRNA levels are modulated in *Aedes aegypti* after exposure to Dengue-2. Insect Mol. Biol. 23: 132–139.

Carney, G.E. and M. Bender. 2000. The *Drosophila* ecdysone receptor (EcR) gene is required maternally for normal oogenesis. Genetics 154: 1203–1211.

Carthew, R.W. and E.J. Sontheimer. 2009. Origins and mechanisms of miRNAs and siRNAs. Cell 136: 642–655.

Cerutti, L., N. Mian and A. Bateman. 2000. Domains in gene silencing and cell differentiation proteins: The novel PAZ domain and redefinition of the Piwi domain. Trends Biochem. Sci. 10: 481–482.

Chatterjee, S. and H. Grosshans. 2009. Active turnover modulates mature microRNA activity in *Caenorhabditis elegans*. Nature 461: 546–549.

Chawla, G. and N.S. Sokol. 2011. MicroRNAs in *Drosophila* development. Int. Rev. Cell. Mol. Biol. 286: 1–65.

Chawla, G. and N.S. Sokol. 2012. Hormonal activation of let-7-C microRNAs via EcR is required for adult *Drosophila melanogaster* morphology and function. Development 139: 1788–1797.

Cheloufiet, S., C.O. Dos Santos, M.M. Chong and G.J. Hannon. 2010. A dicer-independent miRNA biogenesis pathway that requires Ago catalysis. Nature 465: 584–589.

Chen, A., D. Xia, Z. Qiu, P. Gao, S. Tang, X. Shen, F. Zhu and Q. Zhao. 2013a. Expression of a vitelline membrane protein, BmVMP23, is repressed by bmo-miR-1a-3p in silkworm, *Bombyx mori*. FEBS Lett. 587: 970–975.

Chen, J., Z. Liang, Y. Liang, R. Pang and W. Zhang. 2013b. Conserved microRNAs miR-8-5p and miR-2a-3p modulate chitin biosynthesis in response to 20-hydroxyecdysone signaling in the brown planthopper, *Nilaparvata lugens*. Insect Biochem. Mol. Biol. 43: 839–848.

Chen, Q., L. Lu, H. Hua, F. Zhou, L. Lu and Y. Lin. 2012. Characterization and comparative analysis of small RNAs in three small RNA libraries of the Brown planthopper (*Nilaparvata lugens*). PLoS One 7: e32860.

Chi, S.W., G.J. Hannon and R.B. Darnell. 2012. An alternative mode of microRNA target recognition. Nat. Struct. Mol. Biol. 19: 321–327.

Chong, M.M., G. Zhang, S. Cheloufi, T.A. Neubert, G.J. Hannon and D.R. Litman. 2010. Canonical and alternate functions of the microRNA biogenesis machinery. Genes Dev. 24: 1951–1960.

Chung, W.J., P. Agius, J.O. Westholm, M. Chen, K. Okamura, N. Robine, C.S. Leslie and E.C. Lai. 2011. Computational and experimental identification of mirtrons in *Drosophila melanogaster* and *Caenorhabditis elegans*. Genome Res. 21: 286–300.

Chung, W.J., K. Okamura, R. Martin and E.C. Lai. 2008. Endogenous RNA interference provides a somatic defense against *Drosophila* transposons. Curr. Biol. 18: 795–802.

Conrad, K.D., F. Giering, C. Erfurth, A. Neumann, C. Fehr, G. Meister and M. Niepmann. 2013. MicroRNA-122 dependent binding of Ago2 protein to hepatitis C virus RNA is associated with enhanced RNA stability and translation stimulation. PLoS One 8: e56272.

Consortium, T.H.G. 2012. Butterfly genome reveals promiscuous exchange of mimicry adaptations among species. Nature 5: 94–98.

Corcoran, D.L., K.V. Pandit, B. Gordon, A. Bhattacharjee, N. Kaminski and P.V. Benos. 2009. Features of mammalian microRNA promoters emerge from polymerase II chromatin immunoprecipitation data. PLoS One 4: e5279.

Courteau, L.A., K.B. Storey and P.J. Morin. 2012. Differential expression of microRNA species in a freeze tolerant insect, *Eurosta solidaginis*. Cryobiology 65: 210–214.

Cristino, A.S., E.D. Tanaka, M. Rubio, P.M. D. and X. Bellés. 2011. Deep sequencing of organ- and stage-specific microRNAs in the evolutionarily basal insect *Blattella germanica* (L.) (Dictyoptera, Blattellidae). PLoS One 28: e19350.

Cruz, J., D. Martin, N. Pascual, J.L. Maestro, M.D. Piulachs and X. Bellés. 2003. Quantity does matter. Juvenile hormone and the onset of vitellogenesis in the German cockroach. Insect Biochem. Mol. Biol. 33: 1219–1225.

Czech, B. and G.J. Hannon. 2011. Small RNA sorting: matchmaking for Argonautes. Nat. Rev. Genet. 12: 19–31.

Czech, B., C.D. Malone, R. Zhou, A. Stark, C. Schlingeheyde, M. Dus, N. Perrimon, M. Kellis, J. Wohlschlegel, R. Sachidanandam, G.J. Hannon and J. Brennecke. 2008. An endogenous siRNA pathway in *Drosophila*. Nature 453: 798–802.

Daneshvar, K., S. Nath, A. Khan, W. Shover, C. Richardson and J.M. Goodliffe. 2013. MicroRNA miR-308 regulates dMyc through a negative feedback loop in *Drosophila*. Biol. Open 2: 1–9.

Davis, M.B., C.G.E., A.E. Robertson and M. Bender. 2005. Phenotypic analysis of EcR-A mutants suggests that EcR isoforms have unique functions during *Drosophila* development. Dev. Biol. 282: 385–396.

Delic, D., M. Dkhil, S. Al-Quraishy and F. Wunderlich. 2012. Hepatic miRNA expression reprogrammed by *Plasmodium chabaudi* malaria. Parasitol. Res. 1111–1121.

Dostert, C., E. Jouanguy, P. Irving, L. Troxler, D. Galiana-Arnoux, C. Hetru, J.A. Hoffmann and J.L. Imler. 2005. The Jak–STAT signaling pathway is required but not sufficient for the antiviral response of *Drosophila*. Nat. Immunol. 6: 946–953.

Ender, C., A. Krek, M.R. Friedlander, M. Beitzinger, L. Weinmann, W. Chen, S. Pfeffer, N. Rajewsky and G. Meister. 2008. A human snoRNA with microRNA-like functions. Mol. Cell 32: 519–528.

Etebari, K. and S. Asgari. 2013. Conserved microRNA miR-8 blocks activation of the Toll pathway by up-regulating Serpin 27 transcripts. RNA Biol. 10: 1356–1364.

Etebari, K., M. Hussain and S. Asgari. 2013. Identification of microRNAs from *Plutella xylostella* larvae associated with parasitization by *Diadegma semiclausum*. Insect Biochem. Mol. Biol. 43: 309–318.

Fabian, M.R., N. Sonenberg and W. Filipowicz. 2010. Regulation of mRNA translation and stability by microRNAs. Annu. Rev. Biochem. 79: 351–379.

Felippes, F.F., K. Schneeberger, T. Dezulian, D.H. Huson and D. Weigel. 2008. Evolution of *Arabidopsis thaliana* microRNAs from random sequences. RNA 14: 2455–2459.

Filipowicz, W. 2005. RNAi: the nuts and bolts of the RISC machine. Cell 122: 17–20.

Finnegan, E.F. and A.E. Pasquinelli. 2013. MicroRNA biogenesis: regulating the regulators. Crit. Rev. Biochem. Mol. Biol. 48: 51–68.

Flynt, A.S., W.J. Chung, J.C. Greimann, C.D. Lima and E.C. Lai. 2010. microRNA biogenesis via splicing and exosome-mediated trimming in *Drosophila*. Mol. Cell 38: 900–907.

Forstemann, K., Y. Tomari, T. Du, V.V. Vagin, A.M. Denli, D. Bratu, C. Klattenhoff, W.E. Theurkauf and P.D. Zamore. 2005. Normal microRNA maturation and germ-line stem cell maintenance requires Loquacious, a double-stranded RNA-binding domain protein. PLoS Biol. 3: e236.

Freitak, D., E. Knorr, H. Vogel and A. Vilcinskas. 2012. Gender- and stressor-specific microRNA expression in *Tribolium castaneum*. Biol. Lett. 8: 860–863.

Fukunaga, R., B.W. Han, J.H. Hung, J. Xu, Z. Weng and P.D. Zamore. 2012. Dicer partner proteins tune the length of mature miRNAs in flies and mammals. Cell 151: 533–546.

Fullaond, A. and S.Y. Lee. 2012. Identification of putative miRNA involved in *Drosophila melanogaster* immune response. Dev. Comp. Immunol. 36: 267–273.

Ge, W., Y.W. Chen, R. Weng, S.F. Lim, M. Buescher, R. Zhang and S.M. Cohen. 2012. Overlapping functions of microRNAs in control of apoptosis during *Drosophila* embryogenesis. Cell Death Differ. 19: 839–846.

Ghildiyal, M., J. Xu, H. Seitz, Z. Weng and P.D. Zamore. 2010. Sorting of *Drosophila* small silencing RNAs partitions microRNA* strands into the RNA interference pathway. RNA 16: 43–56.

Glazov, E.A., P.A. Cottee, W.C. Barris, R.J. Moore, B.P. Dalrymple and M.L. Tizard. 2008. A microRNA catalog of the developing chicken embryo identified by a deep sequencing approach. Genome Res. 18: 957–964.

Gomez-Orte, E. and X. Bellés. 2009. MicroRNA-dependent metamorphosis in hemimetabolan insects. Proc. Natl. Acad. Sci. USA 106: 21678–21682.

Grey, F., R. Tirabassi, H. Meyers, G. Wu, S. McWeeney, L. Hook and J.A. Nelson. 2010. A viral microRNA down-regulates multiple cell cycle genes through mRNA 5'UTRs. PLoS Pathog. 6: e1000967.

Grimson, A., K.K. Farh, W.K. Johnston, P. Garrett-Engele, L.P. Lim and D.P. Bartel. 2007. MicroRNA targeting specificity in mammals: determinants beyond seed pairing. Mol. Cell 27: 91–105.

Gundersen-Rindal, D.E. and M.J. Pedroni. 2010. Larval stage *Lymantria dispar* microRNAs differentially expressed in response to parasitization by *Glyptapanteles flavicoxis* parasitoid. Arch. Virol. 155: 787.

Guo, C., Y. Yan, H. Cui, X. Huang and Q. Qin. 2013a. miR-homoHSV of Singapore grouper iridovirus (SGIV) inhibits expression of the SGIV pro-apoptotic factor LITAF and attenuates cell death. PLoS One 8: e83027.

Guo, H., N.T. Ingolia, J.S. Weissman and D.P. Bartel. 2010. Mammalian microRNAs predominantly act to decrease target mRNA levels. Nature 466: 835–840.

Guo, Q., Y.-L. Tao and D. Chu. 2013b. Characterization and comparative profiling of miRNAs in invasive *Bemisia tabaci* (Gennadius) B and Q. PLoS One 8: e59884.

Hafner, M., M. Landthaler, L. Burger, M. Khorshid, J. Hausser, P. Berninger, A. Rothballer, M.J. Ascano, A.C. Jungkamp, M. Munschauer, A. Ulrich, G.S. Wardle, S. Dewell, M. Zavolan and T. Tuschl. 2010. Transcriptome-wide identification of RNA-binding protein and microRNA target sites by PAR-CLIP. Cell 141: 129–141.

Han, B.W., J.H. Hung, Z. Weng, P.D. Zamore and S.L. Ameres. 2011. The 3'-to-5' exoribonuclease Nibbler shapes the 3' ends of microRNAs bound to *Drosophila* Argonaute1. Curr. Biol. 21: 1878–1887.

Han, J., Y. Lee, K.H. Yeom, J.W. Nam, I. Heo, J.K. Rhee, S.Y. Sohn, Y. Cho, B.T. Zhang and V.N. Kim. 2006. Molecular basis for the recognition of primary microRNAs by the Drosha-DGCR8 complex. Cell 125: 887–901.

Hansen, I.A., G.M. Attardo, J.H. Park, Q. Peng and A.S. Raikhel. 2004. Target of rapamycin-mediated amino acid signaling in mosquito anautogeny. Proc. Natl. Acad. Sci. USA 101: 10626–10631.

He, P.A., Z. Nie, J. Chen, J. Chen, Z. Lv, Q. Sheng, S. Zhou, X. Gao, L. Kong, X. Wu, Y. Jin and Y. Zhang. 2008. Identification and characteristics of microRNAs from *Bombyx mori*. BMC Genomics 9: 248.

Helwak, A., G. Kudla, T. Dudnakova and D. Tollervey. 2013. Mapping the human miRNA interactome by CLASH reveals frequent noncanonical binding. Cell 153: 654–665.

Hilgers, V., N. Bushati and S.M. Cohen. 2010. *Drosophila* microRNAs 263a/b confer robustness during development by protecting nascent sense organs from apoptosis. PLoS Biol. 8: e1000396.

Huang, T., D. Xu and X. Zhang. 2012. Characterization of host microRNAs that respond to DNA virus infection in a crustacean. BMC Genomics 13: 159.

Huang, T. and X. Zhang. 2012. Functional analysis of a crustacean microRNA in host virus interactions. J. Virol. 86: 12997–13004.

Huang, X., J.T. Warren and L.I. Gilbert. 2008. New players in the regulation of ecdysone biosynthesis. J. Genet. Genomics 35: 1–10.

Hui, J.H., A. Marco, S. Hunt, J. Melling, S. Griffiths-Jones and M. Ronshaugen. 2013. Structure, evolution and function of the bi-directionally transcribed iab-4/iab-8 microRNA locus in arthropods. Nucleic Acid Res. 41: 3352–3361.

Hussain, M. and S. Asgari. 2010. Functional analysis of a cellular microRNA in insect host-ascovirus interaction. J. Virol. 84: 612–620.

Hussain, M. and S. Asgari. 2014. A microRNA-like viral small RNA from Dengue virus 2 autoregulates its replication in mosquito cells. Proc. Natl. Acad. Sci. USA 111: 2746–2751.

Hussain, M., F. Frentiu, L. Moreira, S. O'Neill and S. Asgari. 2011. *Wolbachia* utilizes host microRNAs to manipulate host gene expression and facilitate colonization of the dengue vector *Aedes aegypti*. Proc. Natl. Acad. Sci. USA 108: 9250–9255.

Hussain, M., R.J. Taft and S. Asgari. 2008. An insect virus-encoded microRNA regulates viral replication. J. Virol. 82: 9164–9170.

Hussain, M., S. Torres, E. Schnettler, A. Funk, A. Grundhoff, G.P. Pijlman, A.A. Khromykh and S. Asgari. 2012. West Nile virus encodes a microRNA-like small RNA in the 3' untranslated region which up-regulates GATA4 mRNA and facilitates virus replication in mosquito cells. Nucl Acid Res. 40: 2210–2223.

Hussain, M., S.L. O'Neill and S. Asgari. 2013a. *Wolbachia* interferes with the intracellular distribution of Argonaute 1 in the dengue vector *Aedes aegypti* by manipulating the host microRNAs. RNA Biol. 10: 1868–1875.

Hussain, M., T. Walker, S. O'Neill and S. Asgari. 2013b. Blood meal induced microRNA regulates development and immune associated genes in the Dengue mosquito vector, *Aedes aegypti*. Insect Biochem. Mol. Biol. 43: 146–152.

Hutvagner, G. and M. Simard. 2008. Argonaute proteins: key players in RNA silencing. Nat. Rev. Mol. Cell Biol. 9: 22–32.

Iovino, N., A. Pane and U. Gaul. 2009. miR-184 has multiple roles in *Drosophila* female germline development. Dev. Cell 17: 123–133.

Irvine, D.V., M. Zaratiegui, N.H. Tolia, D.B. Goto, D.H. Chitwood, M.W. Vaughn, L. Joshua-Tor and R.A. Martienssen. 2006. Argonaute slicing is required for heterochromatic silencing and spreading. Science 313: 1134.

Jagadeeswaran, G., Y. Zheng, N. Sumathipala, H. Jiang, E.L. Arrese, J.L. Soulages, W. Zhang and R. Sunkar. 2010. Deep sequencing of small RNA libraries reveals dynamic regulation of conserved and novel microRNAs and microRNA-stars during silkworm development. BMC Genomics 11: 52.

Jaubert-Possamai, S., C. Rispe, S. Tanguy, K. Gordon, T. Walsh, O. Edwards and D. Tagu. 2010. Expansion of the miRNA pathway in the hemipteran insect *Acyrthosiphon pisum*. Mol. Biol. Evol. 27: 979–987.

Jayachandran, B., M. Hussain and S. Asgari. 2013. Regulation of *Helicoverpa armigera* ecdysone receptor by miR-14 and its potential link to baculovirus infection. J. Invertebrate Pathol. 114: 151–157.

Jiang, J., X. Ge, Z. Li, Y. Wang, Q. Song, D.W. Stanely, A. Tan and Y. Huang. 2013. MicroRNA-281 regulates the expression of ecdysone receptor (EcR) isoform B in the silkworm, *Bombyx mori*. Insect Biochem. Mol. Biol. 43: 692–700.

Jin, H., V.N. Kim and S. Hyun. 2012. Conserved microRNA miR-8 controls body size in response to steroid signaling in *Drosophila*. Genes Dev. 26: 1427–1432.

Jinek, M. and J.A. Doudna. 2009. A three-dimensional view of the molecular machinery of RNA interference. Nature 457: 405–412.

Kadener, S., J. Rodriguez, K.C. Abruzzi, Y.L. Khodor, K. Sugino, M.T. Marr, S. Nelson and M. Rosbash. 2009. Genome-wide identification of targets of the drosha-pasha/DGCR8 complex. RNA 15: 537–545.

Kawahara, Y. and A. Mieda-Sato. 2012. TDP-43 promotes microRNA biogenesis as a component of the Drosha and Dicer complexes. Proc. Natl. Acad. Sci. USA 109: 3347–3352.

Kawai, S. and A. Amano. 2012. BRCA1 regulates microRNA biogenesis via the DROSHA microprocessor complex. J. Cell Biol. 197: 201–208.

Khajuria, C., C.E. Williams, M. El Bouhssini, R.J. Whitworth, S. Richards, J.J. Stuart and M.-S. Chen. 2013. Deep sequencing and genome-wide analysis reveals the expansion of microRNA genes in the gall midge *Mayetiola destructor*. BMC Genomics 14: 187.

Kim, V.N., J. Han and M.C. Siomi. 2009. Biogenesis of small RNAs in animals. Nature Rev. Mol. Cell Biol. 10: 126–139.

Koscianska, E., J. Starega-Roslan and W.J. Krzyzosiak. 2011. The role of Dicer protein partners in the processing of microRNA precursors. PLoS One 6: e28548.

Kudla, G., S. Granneman, D. Hahn, J. Beggs and D. Tollervey. 2011. Crosslinking, ligation, and sequencing of hybrids reveals RNA-RNA interactions in yeast. Proc. Natl. Acad. Sci. USA 108: 10010–10015.

Lane, D. and A. Levine. 2010. p53 Research: the past thirty years and the next thirty years. Cold Spring Harb. Perspect. Biol. 2: a000893.

Lazzaro, B.P. 2008. Natural selection on the *Drosophila* antimicrobial immune system. Curr. Opin. Microbiol. 11: 284–289.

Leaman, D., P.Y. Chen, J. Fak, A. Yalcin, M. Pearce, U. Unnerstall, D.S. Marks, C. Sander, T. Tuschl and U. Gaul. 2005. Antisense-mediated depletion reveals essential and specific functions of microRNAs in *Drosophila* development. Cell 121: 1097–1108.

Lee, Y., K. Jeon, J.T. Lee, S. Kim and V.N. Kim. 2002. MicroRNA maturation: stepwise processing and subcellular localization. EMBO J. 21: 4663–4670.

Lee, Y., M. Kim, J. Han, K.-H. Yeom, S. Lee, S.H. Baek and V.N. Kim. 2004. MicroRNA genes are transcribed by RNA polymerase II. EMBO J. 23: 4051–4060.

Legeai, F., G. Rizk, T. Walsh, O. Edwards, K. Gordon, D. Lavenier, N. Leterme, A. Mereau, J. Nicolas, D. Tagu and S. Jaubert-Possamai. 2010. Bioinformatic prediction, deep sequencing of microRNAs and expression analysis during phenotypic plasticity in the pea aphid, *Acyrthosiphon pisum*. BMC Genomics 11: 281.

Lemaitre, B. and J. Hoffmann. 2007. The host defense of *Drosophila melanogaster*. Annu. Rev. Immunol. 25: 697–743.

Liang, P., B. Feng, X. Zhou and X. Gao. 2013. Identification and developmental profiling of microRNAs in Diamondback Moth, *Plutella xylostella* (L.). PLoS One 8: e78787.

Lightfoot, H.L., A. Bugaut, J. Armisen, N.J. Lehrbach, E.A. Miska and S. Balasubramanian. 2011. A LIN28-dependent structural change in pre-let-7g directly inhibits dicer processing. Biochemistry 50: 7514–7521.

Lingel, A., B. Simon, E. Izaurralde and M. Sattler. 2003. Structure and nucleic-acid binding of the *Drosophila* Argonaute 2 PAZ domain. Nature 426: 465–469.

Lingel, A., B. Simon, E. Izaurralde and M. Sattler. 2004. Nucleic acid 3'-end recognition by the Argonaute2 PAZ domain. Nat. Struct. Mol. Biol. 11: 576–577.

Litman, G.W., J.P. Rast and S.D. Fugmann. 2010. The origins of vertebrate adaptive immunity. Nat. Rev. Immunol. 10: 543–553.

Liu, B., W. Dou, T.B. Ding, R. Zhong, C.Y. Liao, W.K. Xia and J.J. Wang. 2014. An analysis of the small RNA transcriptome of four developmental stages of the citrus red mite (*Panonychus citri*). Insect Mol. Biol. 23: 216–229.

Liu, S., S. Gao, D. Zhang, J. Yin, Z. Xiang and Q. Xia. 2010. MicroRNAs show diverse and dynamic expression patterns in multiple tissues of *Bombyx mori*. BMC Genomics 11: 85.

Lourenco, A.P., K.R. Guidugli-Lazzarini, F.C.P. Freitas, M.M.G. Bitondi and Z.L.P. Simoes. 2013. Bacterial infection activates the immune system response and dysregulates microRNA expression in honey bees. Insect Biochem. Mol. Biol. 43: 474–482.

Lucas, K. and A.S. Raikhel. 2013. Insect MicroRNAs: Biogenesis, expression profiling and biological functions. Insect Biochem. Mol. Biol. 43: 24–38.

Luciano, D.J., H. Mirsky, N.J. Vendetti and S. Maas. 2004. RNA editing of a miRNA precursor. RNA 10: 1174–1177.

Lund, E., S. Güttinger, A. Calado, J.E. Dahlberg and U. Kutay. 2004. Nuclear export of microRNA precursors. Science 303: 95–98.

Luo, Q., Q. Zhou, X. Yu, H. Lin, S. Hu and J. Yu. 2008. Genome-wide mapping of conserved microRNAs and their host transcripts in *Tribolium castaneum*. J. Genet. Genomics 35: 349–355.

Luplertlop, N., P. Surasombatpattana, S. Patramool, E. Dumas, L. Wasinpiyamongkol, L. Saune, R. Hamel, E. Bernard, D. Sereno, F. Thomas, D. Piquemal, H. Yssel, L. Briant and D. Misse. 2012. Induction of a peptide with activity against a broad spectrum of pathogens in the *Aedes aegypti* salivary gland, following infection with dengue virus. PLoS Pathog. 7: e1001252.

Lyons, P.J., J.J. Poitras, L.A. Courteau, K.B. Storey and P.J. Morin. 2013. Identification of differentially regulated micrornas in cold-hardy insects. Cryo Letters 34: 83–89.

Ma, E., I.J. MacRae, J.F. Kirsch and J.A. Doudna. 2008. Auto-inhibition of human Dicer by its internal helicase domain. J. Mol. Biol. 380: 237–243.

Ma, J.B., K. Ye and D.J. Patel. 2004. Structural basis for overhang-specific small interfering RNA recognition by the PAZ domain. Nature 429: 318–322.

Macias, S., M. Plass, A. Stajuda, G. Michlewski, E. Eyras and J.F. Cáceres. 2010. DGCR8 HITS-CLIP reveals novel functions for the Microprocessor. Nat. Struct. Mol. Biol. 19: 760–766.

Maines, J.Z., L.M. Stevens, X. Tong and D. Stein. 2004. Drosophila dMyc is required for ovary cell growth and endoreplication. Development 131: 775–786.

Marco, A., J.H. Hui, M. Ronshaugen and S. Griffiths-Jones. 2010. Functional shifts in insect microRNA evolution. Genome Biol. Evol. 2: 686–696.

Martinez, N.J., M.C. Ow, J.S. Reece-Hoyes, M.I. Barrasa, V.R. Ambros and A.J. Walhout. 2008. Genome-scale spatiotemporal analysis of *Caenorhabditis elegans* microRNA promoter activity. Genome Res. 18: 2005–2015.

Mehrabadi, M., M. Hussain and S. Asgari. 2013. MicroRNAome of *Spodoptera frugiperda* cells (Sf9) and its alteration following baculovirus infection. J. Gen. Virol. 94: 1385–1397.

Meng, Y., C. Shao, X. Ma and H. Wang. 2013. Introns targeted by plant microRNAs: a possible novel mechanism of gene regulation. Rice 6: 8.

Michlewski, G., S. Guil and J.F. Cáceres. 2010. Stimulation of pri-miR-18a Processing by hnRNP A1. Adv. Exp. Med. Biol. 700: 28–35.

Mohammed, J., A.S. Flynt, A. Siepel and E.C. Lai. 2014. The impact of age, biogenesis, and genomic clustering on *Drosophila* microRNA evolution. RNA 19: 1295–1308.

Monteys, A.M., R.M. Spengler, J. Wan, L. Tecedor, K.A. Lennox, Y. Xing and B.L. Davidson. 2010. Structure and activity of putative intronic miRNA promoters. RNA 16: 495–505.

Newman, M.A., J.M. Thomson and S.M. Hammond. 2008. Lin-28 interaction with the Let-7 precursor loop mediates regulated microRNA processing. RNA 14: 1539–1545.

Nozawa, M., S. Miura and M. Nei. 2010. Origins and evolution of microRNA genes in *Drosophila* species. Genome Biol. Evol. 2: 180–189.

Nykanen, A., B. Haley and P.D. Zamore. 2001. ATP requirements and small interfering RNA structure in the RNA interference pathway. Cell 107: 309–321.

O'Connell, R.M., D.S. Rao, A.A. Chaudhuri and D. Baltimore. 2010. Physiological and pathological roles for microRNAs in the immune system. Nat. Rev. Immunol. 10: 111–122.

Okamura, K., W.-J. Chung, J.G. Ruby, H. Guo, D.P. Bartel and E.C. Lai. 2008. The *Drosophila* hairpin RNA pathway generates endogenous short interfering RNAs. Nature 453: 803–806.

Okamura, K., J.W. Hagen, H. Duan, D.M. Tyler and E.C. Lai. 2007. The mirtron pathway generates microRNA-class regulatory RNAs in *Drosophila*. Cell 130: 89–100.

Okamura, K., A. Ishizuka, H. Siomi and M.C. Siomi. 2004. Distinct roles for Argonaute proteins in small RNA-directed RNA cleavage pathways. Genes Dev. 18: 1655–1666.

Orom, U.A., F.C. Nielsen and A.H. Lund. 2008. MicroRNA-10a binds the 5'UTR of ribosomal protein mRNAs and enhances their translation. Mol. Cell 30: 460–471.

Ortiz-Rivas, B., S. Jaubert-Possamai, S. Tanguy, J.-P. Gauthier, D. Tagu and R. Claude. 2012. Evolutionary study of duplications of the miRNA machinery in aphids associated with striking rate acceleration and changes in expression profiles. BMC Evol. Biol. 12: 216.

Osei-Amo, S., M. Hussain, S.L. O'Neill and S. Asgari. 2012. *Wolbachia*-induced aae-miR-12 miRNA negatively regulated the expression of MCT1 and MCM6 genes in *Wolbachia*-infected mosquito cell line. PLoS One 7: e50049.

Park, J.E., I. Heo, Y. Tian, D.K. Simanshu, D. Chang, D. Jee, D.J. Patel and V.N. Kim. 2011. Dicer recognizes the 59 end of RNA for efficient and accurate processing. Nature 475: 201–205.

Pek, J.W., A.K. Lim and T. Kai. 2009. *Drosophila maelstrom* ensures proper germline stem cell lineage differentiation by repressing microRNA-7. Dev. Cell 17: 417–424.

Piriyapongsa, J. and I.K. Jordan. 2007. A family of human microRNA genes from miniature inverted-repeat transposable elements. PLoS One 2: e203.

Place, R.F., L.C. Li, D. Pookot, E.J. Noonan and R. Dahiya. 2008. MicroRNA-373 induces expression of genes with complementary promoter sequences. Proc. Natl. Acad. Sci. USA 105: 1608–1613.

Raikhel, A.S., V.A. Kokoza, J. Zhu, D. Martin, S.-F. Wang, C. Li, G. Sun, A. Ahmed, N. Dittmer and G. Attardo. 2002. Molecular biology of mosquito vitellogenesis: from basic studies to genetic engineering of antipathogen immunity. Insect Biochem. Mol. Biol. 32: 1275–1286.

Reczko, M., M. Maragkakis, P. Alexiou, I. Grosse and A.G. Hatzigeorgiou. 2012. Functional microRNA targets in protein coding sequences. Bioinformatics 28: 771–776.

Rewitz, K.F., R. Rybczynski, J.T. Warren and L.I. Gilbert. 2006. The Halloween genes code for cytochrome P450 enzymes mediating synthesis of the insect moulting hormone. Biochem. Soc. Trans. 34: 1256–1260.

Reynolds, J.A., J. Clark, S.J. Diakoff and D.L. Denlinger. 2013. Transcriptional evidence for small RNA regulation of pupal diapause in the flesh fly, *Sarcophaga bullata*. Insect Biochem. Mol. Biol. 43: 982–989.

Riddiford, L.M. 2012. How does juvenile hormone control insect metamorphosis and reproduction. Gen. Comp. Endocrinol. 179: 477–484.

Rodrigueza, A., S. Griffiths-Jones, J.L. Ashurst and A. Bradley. 2004. Identification of mammalian microRNA host genes and transcription units. Genome Res. 14: 1902–1910.

Ronald, J.H., I.M.L. Billas, F. Bonneton, L.D. Graham and M.C. Lawrence. 2013. Ecdysone receptors: From the Ashburner model to structural biology. Annu. Rev. Entomol. 58: 251–271.

Rosewick, N., M. Momont, K. Durkin, H. Takeda, F. Caiment, Y. Cleuter, C. Vernin, F. Mortreux, E. Wattel, A. Burny, M. Georges and A. Van den Broeke. 2013. Deep sequencing reveals abundant noncanonical retroviral microRNAs in B-cell leukemia/lymphoma. Proc. Natl. Acad. Sci. USA 110: 2306–2311.

Rossi, J.J. 2011. A novel nuclear miRNA mediated modulation of a non-coding antisense RNA and its cognate sense coding mRNA. EMBO J. 30: 4340–4341.

Rubio, M. and X. Bellés. 2013. Subtle roles of microRNAs let-7, miR-100 and miR-125 on wing morphogenesis in hemimetabolan metamorphosis. J. Insect Physiol. 59: 1089–1094.

Rubio, M., A. de Horna and X. Bellés. 2012. MicroRNAs in metamorphic and non-metamorphic transitions in hemimetabolan insect metamorphosis. BMC Genomics 13: 386.

Rubio, M., R. Montañez, L. Perez, M. Milan and X. Bellés. 2013. Regulation of atrophin by both strands of the mir-8 precursor. Insect Biochem. Mol. Biol. 43: 1009–1014.

Ruby, J.G., C.H. Jan and D.P. Bartel. 2007. Intronic microRNA precursors that bypass Drosha processing. Nature 448: 83–86.

Saraiya, A.A. and C.C. Wang. 2008. snoRNA, a novel precursor of microRNA in *Giardia lamblia*. PLoS Pathog. 4: e1000224.

Sathyamurthy, G. and N.R. Swamy. 2010. Computational identification and characterization of putative miRNAs in *Nasonia* species. Int. J. Insect Sci. 2: 7–19.

Selbach, M., B. Schwanha¨usser, N. Thierfelder, Z. Fang, R. Khanin and N. Rajewsky. 2008. Widespread changes in protein synthesis induced by microRNAs. Nature 455: 58–63.

Sheng, Z., J. Xu, H. Bai, F. Zhu and S.R. Palli. 2011. Juvenile hormone regulates vitellogenin gene expression through insulin-like peptide signaling pathway in the red flour beetle, *Tribolium castaneum*. J. Biol. Chem. 286: 41924–41936.

Shin, C., J.W. Nam, K.K. Farh, H.R. Chiang, A. Shkumatava and D.P. Bartel. 2010. Expanding the microRNA targeting code: functional sites with centered pairing. Mol. Cell 38: 789–802.

Shin, S.W., Z. Zou, T.T. Saha and A.S. Raikhel. 2012. bHLH-PAS heterodimer of methoprene-toleratnt and cycle mediates cicradian expression of Juvenile hormone-induced mosquito genes. Proc. Natl. Acad. Sci. USA 109: 16576–16581.

Singh, C.P., J. Singh and J. Nagaraju. 2012. A baculovirus-encoded microRNA (miRNA) suppresses its host miRNA biogenesis by regulating the exportin-5 cofactor Ran. J. Virol. 86: 7867–7879.

Singh, J. and J. Nagaraju. 2008. *In silico* prediction and characterization of microRNAs from red flour beetle (*Tribolium castaneum*). Insect Mol. Biol. 17: 427–436.

Siomi, H. and M.C. Siomi. 2010. Posttranscriptional regulation of microRNA biogenesis in animals. Mol. Cell 38: 323–332.

Skalsky, R.L., D.L. Vanlandingham, F. Scholle, S. Higgs and B.R. Cullen. 2010. Identification of microRNAs expressed in two mosquito vectors, *Aedes albopictus* and *Culex quinquefasciatus*. BMC Genomics 11: 119.

Smibert, P., J.S. Yang, G. Azzam, J.L. Liu and E.C. Lai. 2013. Homeostatic control of Argonaute stability by microRNA availability. Nat. Rev. Mol. Cell. Biol. 20: 789–795.

Song, J., W. Guo, F. Jiang, L. Kang and S. Zhou. 2013. Argonaute 1 is indispensable for juvenile hormone mediated oogenesis in the migratory locust, *Locusta migratoria*. Insect Biochem. Mol. Biol. 43: 879–887.

Song, J.-J., S.K. Smith, G.J. Hannon and L. Joshua-Tor. 2004. Crystal structure of Argonaute and its implications for RISC slicer activity. Science 305: 1434–1437.

Soni, K., A. Choudhary, A. Patowary, A.R. Singh, S. Bhatia, S. Sivasubbu, S. Chandrasekaran and B. Pillai. 2013. miR-34 is maternally inherited in *Drosophila melanogaster* and *Danio rerio*. Nucleic Acid Res. 41: 4470–4480.

Stark, A., J. Brennecke, N. Bushati, R.B. Russell and S.M. Cohen. 2005. Animal microRNAs confer robustness to gene expression and have a significant impact on 3'UTR evolution. Cell 123: 1133–1146.

Stark, A., P. Kheradpour, L. Parts, J. Brennecke, E. Hodges, G.J. Hannon and M. Kellis. 2007. Systematic discovery and characterization of fly microRNAs using 12 *Drosophila* genomes. Genome Res. 17: 1865–1879.

Storey, K.B. and J.M. Storey. 2013. Molecular biology of freeze tolerance. Comp. Physiol. 3: 1283–1308.

Sun, B.S., Q.Z. Dong, Q.H. Ye, H.J. Sun, H.L. Jia, X.Q. Zhu, D.Y. Liu, J. Chen, Q. Xue, H.J. Zhou, N. Ren and L.X. Qin. 2008. Lentiviral-mediated miRNA against osteopontin suppresses tumor growth and metastasis of human hepatocellular carcinoma. Hepatology 48: 1834–1842.

Suzuki, H.I., M. Arase, H. Matsuyama, Y.L. Choi, T. Ueno, H. Mano, K. Sugimoto and K. Miyazono. 2011. MCPIP1 ribonuclease antagonizes dicer and terminates microRNA biogenesis through precursor microRNA degradation. Mol. Cell 44: 424–436.

Suzuki, H.I., K. Yamagata, K. Sugimoto, T. Iwamoto, S. Kato and K. Miyazono. 2009. Modulation of microRNA processing by p53. Nature 460: 529–533.

Talbot, W.S., E.A. Swyryd and D.S. Hogness. 1993. *Drosophila* tissues with different metamorphic responses to ecdysone express different ecdysone receptor isoforms. Cell 73: 1323–1337.

Tanaka, E.D. and M.-D. Piulachs. 2012. Dicer-1 is a key enzyme in the regulation of oogenesis in panoistic ovaries. Biol. Cell 104: 452–461.

Tanzer, A. and P.F. Stadler. 2004. Molecular evolution of a microRNA cluster. J. Mol. Biol. 339: 327–335.

Telfer, W.H. 2009. Egg formation in Lepidoptera. J. Insect Sci. 9: 50.

Tomari, Y., T. Du and P.D. Zamore. 2007. Sorting of *Drosophila* small silencing RNAs. Cell 130: 299–308.

Tomari, Y. and P.D. Zamore. 2005. Machines for RNAi. Genes Dev. 19: 517–529.

Van Wynsberghe, P.M., Z.S. Kai, K.B. Massirer, V.H. Burton, G.W. Yeo and A.E. Pasquinelli. 2011. LIN-28 co-transcriptionally binds primary let-7 to regulate miRNA maturation in *Caenorhabditis elegans*. Nat. Struct. Mol. Biol. 18: 302–308.

Varghese, J. and S.M. Cohen. 2007. microRNA miR-14 acts to modulate a positive autoregulatory loop controlling steroid hormone signaling in *Drosophila*. Genes Dev. 21: 2277–2282.

Vasudevan, S., Y. Tong and J.A. Steitz. 2007. Switching from repression to activation: microRNAs can up-regulate translation. Science 318: 1931–1934.

Voinnet, O. 2009. Origin, biogenesis, and activity of plant microRNAs. Cell 136: 669–687.

Wang, D., Z. Zhang, E. O'Loughlin, T. Lee, S. Houel, D. O'Carroll, A. Tarakhovsky, N.G. Ahn and R. Yi. 2012a. Quantitative functions of Argonaute proteins in mammalian development. Gene Dev. 26: 693–704.

Wang, G.H., L. Jiang, L. Zhu, T.C. Cheng, W.H. Niu, Y.F. Yan and Q.Y. Xia. 2012b. Characterization of Argonaute family members in the silkworm, *Bombyx mori*. Insect Sci. 20: 78–91.

Weaver, D.B., J.M. Anzola, J.D. Evans, J.G. Reid, J.T. Reese, K.L. Childs, E.M. Zdobnov, M.P. Samanta, J. Miller and C.G. Elsik. 2007. Computational and transcriptional evidence for microRNAs in the honey bee genome. Genome Biol. 8: R97.

Wei, Y., S. Chen, P. Yang, Z. Ma and L. Kang. 2009. Characterization and comparative profiling of the small RNA transcriptomes in two phases of locust. Genome Biol. 10: R6.

Weng, R., J.S.R. Chin, J.Y. Yew, N. Bushati and S.M. Cohen. 2013. miR-124 controls male reproductive success in *Drosophila*. eLIFE 2: e00640.

Winter, F., S. Edaye, A. Hüttenhofer and C. Brunel. 2007. *Anopheles gambiae* miRNAs as actors of defence reaction against *Plasmodium* invasion. Nucleic Acids Res. 35: 6953–6962.

Winter, J. and S. Diederichs. 2011. Argonaute proteins regulate microRNA stability: Increased microRNA abundance by Argonaute proteins is due to microRNA stabilization. RNA Biol. 8: 1149–1157.

Wu, P., S. Han, T. Chen, G. Qin, L. Li and X. Guo. 2013a. Involvement of microRNAs in infection of silkworm with *Bombyx mori* cytoplasmic polyhedrosis virus (BmCPV). PLoS One 8: e68209.

Wu, W., Q. Ren, C. Li, Y. Wang, M. Sang, Y. Zhang and B. Li. 2013b. Characterization and comparative profiling of microRNAs in a sexual dimorphism insect, *Eupolyphaga sinensis* Walker. PLoS One 8: e59016.

Wu, Y.L., C.P. Wu, C.Y. Liu, P.W. Hsu, E.C. Wu and Y.C. Chao. 2011. A non-coding RNA of insect HzNV-1 virus establishes latent viral infection through microRNA. Sci. Rep. 1: 60.

Wyatt, G.R., M.R. Kanost, B.C. Chin, K.E. Cook, B.M. Kawasoe and J.Z. Zhang. 1992. Juvenile hormone analog and injection effects on locust hemolymph protein synthesis. Arch. Insect Biochem. Physiol. 20: 167–180.

Xu, P., S.Y. Vernooy, M. Guo and B.A. Hay. 2003. The *Drosophila* microRNA miR-14 suppresses cell death and is required for normal fat metabolism. Curr. Biol. 13: 790–795.

Yamanaka, N., K.F. Rewitz and M.B. O'Connor. 2013. Ecdysone control of developmental transitions: lessons from *Drosophila* research. Annu. Rev. Entomol. 58: 497–516.

Yan, K.S., S. Yan, A. Farooq, A. Han, L. Zeng and M.M. Zhou. 2003. Structure and conserved RNA binding of the PAZ domain. Nature 426: 468–474.

Yang, G., L. Yang, Z. Zhao, J. Wang and X. Zhang. 2012. Signature miRNAs involved in the innate immunity of invertebrates. PLoS One 7: e39015.

Yang, J.-S., P. Simbert, J.O. Westholm, D. Jee, T. Maurine and E.C. Lai. 2014. Intertwined pathways for Argonaute-mediated microRNA biogenesis in *Drosophila*. Nucleic Acids Res. 42: 1987–2002.

Yang, J.S. and E.C. Lai. 2011. Alternative miRNA biogenesis pathways and the interpretation of core miRNA pathway mutants. Mol. Cell 43: 892–903.

Ye, X. and Q. Liu. 2008. Expression, purification, and analysis of recombinant *Drosophila* Dicer-1 and Dicer-2 enzymes. Meth. Mol. Biol. 442: 11–27.

Yi, R., G. Qin, I.G. Macara and B.R. Cullen. 2003. Exportin-5 mediates the nuclear export of pre-microRNAs and short hairpin RNAs. Genes Dev. 17: 3011–3016.

Yigit, E., P.J. Batista, Y. Bei, K.M. Pang, C.C. Chen, N.H. Tolia, L. Joshua-Tor, S. Mitani, M.J. Simard and C.C. Mello. 2006. Analysis of the *C. elegans* Argonaute family reveals that distinct Argonautes act sequentially during RNAi. Cell 127: 747–757.

Zhang, W. and S.M. Cohen. 2013. The Hippo pathway acts via p53 and microRNAs to control proliferation and proapoptotic gene expression during tissue growth. Biol. Open 2: 822–828.

Zhang, G., M. Hussain, S. O'Neill and S. Asgari. 2013a. *Wolbachia* uses a host microRNA to regulate transcripts of a methyltransferase, contributing to dengue virus inhibition in *Aedes aegypti*. Proc. Natl. Acad. Sci. USA 110: 10276–10281.

Zhang, X., K. Lu and Q. Zhou. 2013b. Dicer 1 is crucial for the oocyte maturation of telotrophic ovary in *Nilaparvata Lugens* (STÅL) (Hemiptera: Geometroidea). Arch. Insect Biochem. Physiol. 84: 194–208.

Zhou, J., Y. Zhou, J. Cao, H. Zhang and Y. Yu. 2013. Distinctive microRNA profiles in the salivary glands of *Haemaphysalis longicornis* related to tick blood-feeding. Exp. Appl. Acarol. 59: 339–349.

Zhu, L., Y. Masaki, T. Tatsuke, Z. Li, H. Mon, J. Xu, J.M. Lee and T. Kusakabe. 2013. A MC motif in silkworm Argonaute 1 is indispensible for translation repression. Insect Mol. Biol. 22: 320–330.

Zhuo, Y., G. Gao, J. an Shi, X. Zhou and X. Wang. 2013. miRNAs: Biogenesis, origin and evolution, functions on virus-host interaction. Cell. Physiol. Biochem. 32: 499–510.

5

Advances in Insect Physiology and Endocrinology Through Genomics and Peptidomics

Ian Orchard[a],* and Angela B. Lange[b]

Introduction

This chapter highlights advances in our understanding of insect physiology and endocrinology that have been realized in the post-genomic era. This is timely because of the rapid change brought about by genome projects, functional genomics and genetics (so called "omics" technologies) that include transcriptomics, proteomics, peptidomics, metabolomics, and reverse genetic tools (e.g., dsRNA) for gene silencing (see Boerjan et al. 2012, Hoffmann et al. this book, Burse and Boland this book). All of these help unravel complex regulatory processes. Thus, the field of insect physiology and endocrinology has advanced dramatically in recent times since the original sequencing of the *Drosophila* genome (Adams et al. 2000). Genomes from many insect species are now available and the importance of such information is illustrated by the i5k project (http://www.arthropodgenomes.org/wiki/i5K) whose ambition is to sequence genomes from 5,000 insects and related arthropods. This project will be

University of Toronto Mississauga, Department of Biology, 3359 Mississauga Rd., Mississauga, ON, Canada, L5L 1C6.
[a] Email: ian.orchard@utoronto.ca
[b] Email: angela.lange@utoronto.ca
* Corresponding author

transformative and consolidate the discipline in the 21st century. Its aims, as listed on the website above, are "to sequence the genomes of all insect species known to be important to worldwide agriculture, food safety, medicine, and energy production; all those used as models in biology; the most abundant in world ecosystems; and representatives in every branch of insect phylogeny so as to achieve a deep understanding of arthropod evolution and phylogeny". With such massive comparative data, researchers will be able to make relevant comparisons between molecular, cellular, tissue, organ, and organismal features. When coupled with another bold initiative, the Genome 10K project (which aims to assemble the genomes of 10,000 vertebrate species (http://www.genome10k.org/), the comparisons can truly be extended to evolution and phylogeny.

The value of genome sequences lies in the ability to exploit the information and to specifically interfere with gene expression. The resultant interference with phenotype expression provides evidence for functional roles for the genes. Researchers can truly capitalize on functional genomics and genetics. In the field of insect physiology and endocrinology, for example, there are now reasonable estimates of the total number of neuropeptides, peptide hormones and their G-protein-coupled receptors (GPCRs) in particular insect species. The genes and proteins at all levels of a signaling pathway are being defined. Temporal and spatial gene and peptide/protein expression can now be determined and quantified for individual cells and tissues, using techniques such as quantitative polymerase chain reaction (qPCR), *in situ* hybridization, and detailed mass spectrometry (MS) techniques; and the effects of experimental manipulation of expression levels can provide insight into the true physiological and endocrinological relevance of these signaling pathways.

The field of insect physiology and endocrinology has often been dominated by the use of "model" insects—insects that are easy to maintain under laboratory conditions; typically larger, and experimentally tractable and allowing detailed physiological and endocrinological analysis with experimental techniques that might still be difficult in *Drosophila* and other insects with sequenced genomes. Data from those insects for which the genomes have been sequenced, analyzed, and manipulated, are, however, becoming increasingly more important and useful for application to, and analysis of, similar physiological and endocrinological processes in insect models for which little genome data is available. In addition, whilst the size of the experimental insect used to be a determining factor, it is less so now, with more advanced techniques, especially of the "omics" variety. Typically, the advantages now afforded by working on *Drosophila* means that *Drosophila* gene technology and manipulations are ahead of those for research on other insects, so that in practice, basic processes are often first probed in *Drosophila*, and then this data is applied to the relevant insect model. A comparative approach is still meaningful and essential though, since genes have been lost in some species (including *Drosophila*), and may only be found in a limited number of insect species (see Hauser et al. 2008, Nygaard et al. 2011). Novel endocrine systems are still being discovered (e.g., Hansen et al. 2010, Jiang et al. 2013) and so research on a wide array of insect species must continue in order to advance the field—especially important in considering research on agriculturally- or medically-important insects.

In addition to the need for understanding the physiology and endocrinology of insects in themselves, due to their abundance (constituting over 80% of total animal species known) and to their human impact (beneficial and harmful, and the need for pest control strategies), research on insects has also played a major role in furthering our understanding of mammals, including humans. Unifying elements in regulatory processes can be identified at all levels through studies on insects. In particular, *Drosophila* is used as a genetic model organism for understanding molecular mechanisms of human disease (see Bilen and Bonini 2005, Botas 2007, Pandey and Nichols 2011). Fundamental processes are conserved between mammals and *Drosophila*, and it is now estimated that 75% of human disease-causing genes have functional homologs in *Drosophila*. Thus, *Drosophila* is aiding in our understanding of central nervous system (CNS) disorders, inflammatory disorders, cardiovascular disease, cancer, obesity, aging and diabetes (see Bilen and Bonini 2005, Botas 2007, Tweedie et al. 2009, Pandey and Nichols 2011). For instance, analysis of gene expression in Malpighian tubules (MTs) of *D. melanogaster* has not only helped determine the mechanisms of regulation of epithelial secretion and function in *D. melanogaster* and other insects, but has also illustrated that many genes are shared with renal and liver cells of mammals. Genes implicated in mammalian diseases have been identified in *Drosophila* MTs (see Dow 2009). Thus, *Drosophila* can serve as a model for aspects of human physiology, endocrinology and pathology (see FlyBase database at http://flybase.bio.indiana.edu). Further evidence of the importance of research into insect physiology and endocrinology can also be illustrated by original discoveries made in insects. The phosphatidylinositol signaling pathway and its role in mobilizing intracellular calcium was first described using blowfly salivary glands as a model (see Berridge 1993); the discovery of some neuropeptides, for example RFamides, was first made in molluscs, and then in insects, but these neuropeptides are now known to play critical roles in the control of reproduction, appetite and pain in mammals; the nuclear mechanisms of steroid action and involvement of transcription factors was first modelled using insect ecdysteroids (Ashburner et al. 1974,

King-Jones and Thummel 2005); and the involvement of Toll receptors in mediating the innate immune response was first shown in *D. melanogaster* (see Imler and Hoffmann 2001).

Insect physiology and endocrinology has advanced considerably in the last 10–15 years, and, although still focusing on many classical and fundamental questions, these can now be explored with a vast toolbox and with a greater comparative overview because of advances in comparative genomics, peptidomics, and related technologies (see also Rogers this book). We begin by discussing the fact that these "omics" technologies have been embraced and applied to insect physiology and endocrinology, and then subsequently illustrate this in practice, with regard to some case studies. The case studies are highly selective and only represent a small snapshot of the topics being examined. They do, however, set the scene and context for the exciting era we find ourselves in, and the depth of understanding that has come from this post-genomic revolution.

"Omics" and other Technologies

Gene sequences, but also including expressed sequence tags (ESTs), coupled with bioinformatics generates vast, searchable databases. These contain information that can be applied to the study of gene function and expression, and ultimately to our understanding of the physiological and endocrinological relevance of various genes and gene systems. In an aptly entitled review "Insect omics research coming of age" Boerjan et al. (2012) recently reviewed these "omics" technologies and how they have been integrated and applied to studies of insect biology. The disciplines discussed included functional genomics (transcriptomics, proteomics, and peptidomics), genetics (reverse genetics, such as RNA interference [see chapter by Hoffmann et al. this book], targeted gene knockdown and knockout and over expression; and forward genetics), and epigenomics. Additional technologies particularly relevant and essential to physiological and endocrinological research include reverse pharmacology for de-orphaning GPCRs, recently described and reviewed (Grimmelikhuijzen and Hauser 2012, Caers et al. 2014).

Peptidomics

Neuropeptides are now considered the most abundant, and by inference, therefore, possibly the most important chemical messengers of the nervous and endocrine systems. They are diverse in their structure (amino acid sequence), are versatile messengers, acting as neurotransmitters, neurohormones and neuromodulators (see Orchard et al. 2001 for definitions) and appear to influence almost all physiological processes;

they provide flexibility in the functioning of the nervous and endocrine system. Their presence in peripheral cells (see Orchard and Finlayson 1977, Žitňan et al. 1999) and endocrine-like cells of the midgut (see Žitňan et al. 1993) also points to their potential for bridging communication between the central and peripheral nervous systems, the digestive system and the endocrine system. It is of some historical importance (and illustrative of the advantages of newer technologies) that the first insect neuropeptide to be sequenced and synthesized was proctolin (Brown and Starratt 1975), extracted from 125,000 cockroaches (a monumental task), and shown by Edman degradation to have the sequence RYLPT. The second neuropeptide was locust adipokinetic hormone (Locmi-AKH), extracted and sequenced from 3000 glandular lobes of locust corpora cardiaca (Stone et al. 1976). No other insect peptides were further isolated until 1985 (Locmi-AKH II) and then an explosion in peptide sequences followed. Indeed, in a review from that era, Schoofs et al. (1997) reported the sequences of 56 locust peptides from an extract of 9000 brain–corpora cardiaca–corpora allata–suboesophageal ganglion complexes of locusts! Times have now changed, and MS is now a powerful tool for identifying the peptide content of individual cells, tissues and organs and thereby verifying the sequences predicted from bioinformatics (see Baggerman et al. 2005, Hummon et al. 2006, Predel et al. 2010). Thus, bioinformatics allows the prediction of neuropeptide and peptide sequences within the precursor, based upon dibasic cleavage sites, and MS studies can confirm that these peptides are actually processed from the precursor, i.e., that the peptide phenotype actually occurs. Post-translational modifications that are not predictable by bioinformatics have been identified through MS information, including, for example, sulfated tyrosine, N-terminal blocking of glutamine to pyro-glutamic acid, N-terminal acetylation of alanine, methylation of glutamic acid, oxidation of methionine, and phosphorylation of threonine. Early on in peptidomic studies, attempts were made to identify as many of the full complement of neuropeptides as possible from the CNS of an insect species. With improving sensitivities though, these neuropeptides have been mapped to single cells, clusters of cells, to single neurohemal sites, and to tissues. This neuropeptide profiling can provide insight into target tissues especially in a developmental and physiological context. Numerous studies have now detailed the peptide profile of many insect species (e.g., Baggerman et al. 2005, Predel et al. 2010, 2012, Huybrechts et al. 2010, Li et al. 2011, Ons et al. 2011) and within these species the neurons expressing the particular peptide are well described. Immunohistochemical analyses have also revealed target tissues for physiological investigation, or have revealed neuropeptides which are likely to act as neurohormones, controlling distant target tissues (e.g., TeBrugge et al. 1999, 2001, Patel et al. 2014). This powerful technique of peptidomics has provided the data for interesting comparisons between protostomes (most invertebrates) and deuterostomes (all vertebrates) and therefore the evolution of neuropeptide signaling. Though difficult, it is not impossible to find the protostome orthologs of the deuterostome neuropeptides (see Grimmelikhuijzen and Hauser 2012, Jékely 2013). Some solid examples include oxytocin/vasopressin, neuropeptide Y (NPY; neuropeptide F, NPF, in insects), sulfated gastrin/cholecystokinin (CCK)-like peptides (sulfakinins in insects; see Hoffmann et al. this book) and glycoprotein hormones related to follicle stimulating hormone (FSH), luteinizing hormone (LH) and thyroid stimulating hormone (TSH) (see Grimmelikhuijzen and Hauser 2012). Complete lists can be found in Liu et al. (2006) and in Jékely (2013).

Peptidomics has also revealed interesting facets of transcript expression and translation. For example, the transcript for the *capability* gene, encodes three predicted neuropeptides; CAPA-1, CAPA-2 and a pyrokinin (PK). In *Rhodnius prolixus* there are two *capability* genes, each encoding two CAPA neuropeptides and a PK. Only CAPA-2 shares the same amino acid sequence; the others are different from each other (see Table 1; Paluzzi et al. 2008, Paluzzi and Orchard 2010). Comparable amounts of each of the neuropeptides from both genes are found in all MS analyses from abdominal neurohemal sites, suggesting a common expression of both genes in the same neuroendocrine cells. Interestingly, though, CAPA-neurons of the fused abdominal ganglia and the suboesophageal ganglion (SOG) seem to process the transcripts differentially (Predel et al. 2010). In the fused abdominal ganglia, all predicted neuropeptides are detected whereas in the SOG only the PKs are found. Differential processing of the CAPA-precursor has also been found for CAPA-neurons of *D. melanogaster*, where only the PK but not the two CAPA peptides are expressed in two identified CAPA-neurons of the SOG (Predel et al. 2010). Differential neuropeptide precursor processing is considered rare in insects, but clearly cannot be discounted. The biochemical process is

Table 1. Sequences predicted from the Rhopr-CAPA gene and detected by mass spectrometry.

Gene	Peptide Name	Sequence
Rhopr-CAPA-α	Rhopr-CAPA-α1	SPISSVGLFPLRAamide
	Rhopr-CAPA-α2	EGGFISFPRVamide
	Rhopr-CAPA-αPK1	NGGGNGGGLWFGPRLamide
Rhopr-CAPA-β	Rhopr-CAPA-β1	SPITSIGLLPFLRAA
	Rhopr-CAPA-β2	EGGFISFPRVamide
	Rhopr-CAPA-βPK1	IGGGNGGGLWFGPRLamide

yet to be determined and so predictions are unavailable, but clearly this phenomenon points to another level of complexity in the complement of neuropeptides expressed. A further level of complexity has been highlighted by the demonstration of alternative splicing of a gene. For example, the *Manduca sexta* allatotropin gene (Manse-AT gene) is transcribed as at least three mRNAs, which differ from one another by alternative splicing (see Horodyski et al. 2001). Exons that are unique to a specific Manse-AT mRNA are located within the protein-coding region, so that three different propeptides can be derived from the mRNAs. The deduced translation product of each mRNA contains a sequence identical to that of Manse-AT. Additional peptides (allatotropin-like peptides) are predicted from regions of the precursor flanking Manse-AT. Some of these peptides are found in each deduced propeptide, whereas others are unique so that they are derived specifically from one Manse-AT mRNA. The multiple Manse-AT mRNAs obtained through alternative splicing are expressed in a tissue and stage specific manner. Since each mRNA codes for Manse-AT and a unique set of allatotropin-like peptides, the diversity of peptide products from a single gene can vary in a dynamic way during development and between tissues (see Horodyski 2013).

G-protein coupled receptors

With further regard to neuropeptides, GPCRs constitute a very large family of conserved proteins that includes olfactory and gustatory receptors, rhodopsin and receptors for classical neurotransmitters, neuropeptides, neuroamines and peptide hormones. Insect GPCRs have been identified and de-orphaned using reverse pharmacology (see Caers et al. 2012a,b). One of the fundamental techniques used to do this is the transfection of a cloned GPCR of interest into a cellular expression system, and then challenging the heterologous expressed receptor with candidate ligands (peptide library). In this functional receptor assay, receptor activation is then assessed, often by way of measuring changes in second messengers, such as calcium or cAMP (see Caers et al. 2014). This has now been done on almost the full complement of GPCRs from a number of insect species, and then homology helps de-orphan similar receptors in other insects (see Hauser et al. 2006, 2008, Fan et al. 2010). Another word of caution, however, is the possibility of alternative splicing of the GPCR gene (for examples see Paluzzi et al. 2010, Zandawala et al. 2013), resulting in variable GPCRs from the gene (as discussed for peptides). One prominent outcome of the genome sequencing projects has been the identification of the possibly full complement of GPCRs in particular insect species, and the recognition that many of the GPCRs are evolutionarily old and well conserved over time, even though their ligands might show more variability (see Grimmelikhuijzen and Hauser 2012, Jékely 2013). Thus, ancestrally-related GPCRs for neuropeptides have been identified throughout the animal kingdom and interestingly many neuropeptide-GPCR pairs are also conserved in terms of function. A signaling system that appears well conserved through evolution is that of vertebrate NPY and NPF of invertebrates, including insects. Another structurally and functionally conserved system is that of the insulin-like peptides (ILPs) and their tyrosine kinase type receptors, including downstream signaling components (see Wu and Brown 2006). Recent research on insect ILPs and NPFs and their roles in feeding, metabolism and longevity have again demonstrated the power of using insect models for basic physiological and endocrinological phenomena.

Gene silencing/overexpression

Reverse genetics (see Boerjan et al. 2012) refers to silencing of a candidate gene (once it has been identified) and observing the effects on the phenotype. This is particularly useful in insect physiology and endocrinology where a neuropeptide and/or its GPCR can be silenced and a phenotype therefore linked to a signaling pathway. Studies on *C. elegans* led to the discovery of RNA interference (RNAi, for detailed explanation see Scott et al. 2013) which has been used successfully in a number of insect species. Interestingly, and importantly, the efficiency of using double stranded RNA (dsRNA) for RNAi varies between insect species, the modes of delivery (injection or feeding) (Christiaens et al. 2014), the gene being targeted, and the tissue being targeted (see Wynant et al. 2012, Scott et al. 2013, Zhang et al. 2013). If successful though, dsRNA enables interference with the insect's physiology and endocrinology in which the target gene (e.g., neuropeptide or GPCR) plays a relevant role. Observation of the phenotype expressed in an insect in which a gene has been silenced can be used to assign a function to that gene. Thus, for example, knocking-down either the *Schistocerca gregaria* short NPF (Schgr-sNPF) precursor transcript or the Schgr-sNPF receptor transcript using RNAi results in an increase in the total food uptake in locusts compared to control insects indicating sNPF exerts an inhibitory effect on food uptake (Dillen et al. 2013, 2014). Similarly, in *Tribolium castaneum*, knocking-down the sulfakinin receptor 2 stimulates food intake indicating an involvement of sulfakinins in the regulation of food intake (Yu and Smagghe 2013). *Rhodnius prolixus* is also susceptible to RNAi and CCAP and the CCAP receptor transcripts can be knocked down by up to 92% with lethal consequences associated with the timing of ecdysis behavior (Lee et al. 2013). In addition to RNAi as a tool for understanding the physiology and endocrinology of insects, RNAi is being used for crop pest management, and also in the protection of beneficial insects (see Scott et al. 2013, Zhang et al. 2013). Thus, transgenic cotton plants expressing *cytochrome P450* dsRNA suffer less damage when exposed to cotton boll worm and the insects themselves have retarded growth (Mao et al. 2007). RNAi is also being used in the protection of insects against certain pathogens or parasites as exemplified for infections of bees (see Scott et al. 2013). As mentioned earlier, the genetic tool box is much more extensive for *Drosophila* and targeted knock-down or overexpression of a gene of interest has been frequently obtained using a variety of techniques, including Gal4-UAS driver-reporter gene mechanisms (see Boerjan et al. 2012). Of some encouragement with regard to other insects, is the discovery that these techniques can also be used in *Bombyx mori* and *T. castenaeum* (Dai et al. 2008, Schinko et al. 2010).

Case Studies

As mentioned earlier, many insect genomes have now been sequenced and annotated (see Boerjan et al. 2012) and omics technologies applied to furthering an understanding of insect physiology and endocrinology (see Fig. 1) in an ever increasing repertoire of cases. These include such areas as reproduction, growth and development, feeding and digestion, salt and water balance, immunity and ecdysis behavior. Highly selected examples, treated as case studies, are outlined here.

Neuroendocrine regulation of diuresis

There are a variety of insect models for studying the epithelial mechanisms of diuresis and their neuroendocrine control, and the reader is directed to a number of excellent reviews on this topic (Coast et al. 2002, Dow and Davies 2006, Coast 2009, Park 2012, Schooley et al. 2012). In these models a number of neuroendocrine factors have been identified, along with their modes of action at the molecular and cellular levels. Diuretic and antidiuretic hormones (DHs and ADHs) have been characterized, along with their receptors in a variety of insect species; the ligands include corticotropin-releasing-factor-related DH (CRF/DH), calcitonin-like DH (CT/DH), kinins, cardioacceleratory peptide 2b (CAP2b, now called CAPA), tachykinin, neuropeptide-like precursor (NPLP), arginine vasopressin-like (AVP-like) peptide, antidiuretic factors a and b (ADFa and b), ion transport peptide (ITP), neuroparsins, and the biogenic amines, serotonin and tyramine (for complete list see Park 2012). Whilst *Drosophila* and its genetic toolbox have provided particularly exciting data, including MTs as stress detectors (see Davies et al. 2014), we will focus on one of

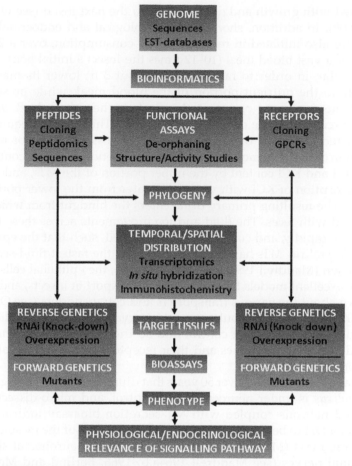

Figure 1. Flow diagram illustrating how advances in our understanding of neuropeptide signaling have benefitted from "omics" technologies.

the original model insects, *R. prolixus*, and its rapid post-feeding diuresis. The *R. prolixus* genome has only recently been sequenced, and therefore the advantages of a post-genomic era can be especially emphasized with regard to the recent progress in research in this insect.

Diuresis is a physiological process for maintaining water and ion homeostasis, and is also a mechanism for removing metabolic waste and environmental stressors. Tissues associated with diuresis can include the midgut, MTs, and hindgut. *Rhodnius prolixus* has been used as a model insect for physiology and endocrinology for almost 100 years because of the precise timing of events brought about by blood-gorging. Thus, unfed *R. prolixus* remain in a state of arrested development and gorging upon one blood meal triggers physiological and endocrinological events associated with growth and development to the next instar (see Orchard 2006, 2009). In addition, short term physiological and endocrinological events are also initiated in response to the consumption, over a 20 min period, of a vast blood meal (10–12 times the insect's initial body mass). In particular, in order to rapidly (over about 3 h) lower its mass and concentrate the nutrient portion of the blood meal, while preserving the volume, ionic and osmotic balance of the hemolymph, *R. prolixus* rapidly excretes a urine of high NaCl content. This process requires the coordinated transport of NaCl and water across the anterior midgut epithelium into the hemolymph, the rapid secretion of water containing high NaCl and KCl content by the upper portion of the MTs, and finally the reabsorption of KCl with very little water from the lower portion of the MTs. The resulting primary urine enters the hindgut from which it is eliminated with feces. The fluid and ion movements across these tissues occur very rapidly, and continue at a high speed, such that the epithelial cells of *R. prolixus* MTs have been described as the fastest fluid-secreting cells known (Maddrell 1991). For these reasons, the epithelial cells of the MTs are excellent models for studying ion transport in insects, and much is known about the various transporters and cotransporters (see Orchard 2006, 2009). Malpighian tubules are not innervated and the epithelial cells are under neuroendocrine control and so they are also useful models for studying neurohormones and their receptors that regulate epithelial transport.

It has been known for over 50 years that diuresis in response to gorging in *R. prolixus* is under neuroendocrine control, and micro-dissection of identified neurons coupled with MT secretion bioassay indicated the source of a DH to be posterior lateral neurons (PLNs) of the mesothoracic ganglionic mass (MTGM), and release from their neurohemal sites on abdominal nerves (see Maddrell 1964a,b, 1966, Berlind and Maddrell 1979). This DH was shown to be peptidergic in nature. Intriguingly though, serotonin was the first true DH to be identified and was shown to be released from dorsal unpaired median neurons (DUM neurons) of the MTGM (Lange et al. 1989, Orchard et al. 1989). But what of the peptide DH of the PLNs of the MTGM? The identification of the peptide had been elusive, until the post-genomic era (Fig. 2). The strategy to identify the peptide DHs was to BLAST search the genome database with queries based upon other insect neuropeptide DH genes, use gene specific primers to clone the transcript from cDNA, qPCR for transcript expression, MS of CNS tissue to confirm phenotypic expression of the peptides coupled with *in situ* hybridization and immunohistochemistry to locate the neurons (see Fig. 1), and finally use the synthetic peptides *in vitro* for physiological experimentation.

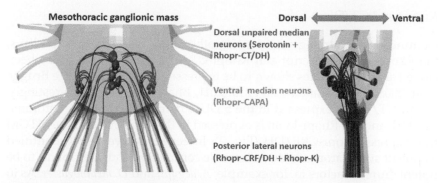

Figure 2. The location of neurons involved in the neuroendocrine control of diuresis in *Rhodnius prolixus*. Rhop-CT/DH is colocalized with serotonin in five dorsal unpaired median neurons; Rhopr-CAPA is localized to three pairs of ventral median neurons; and Rhopr-CRF/DH is co-localized with Rhopr-K(kinin) in posterior lateral neurons. All neurons are found within the mesothoracic ganglionic mass and project axons through abdominal nerves that result in neurohemal sites on abdominal nerves. Illustration created by Paul Hong.

CRF/DHs

The *R. prolixus* CRF-related peptide (Rhopr-CRF/DH) has been characterized and shown to be present in PLNs of the MTGM (the cells of Maddrell) and to be released from abdominal nerve neurohemal sites (see TeBrugge and Orchard 2002, Te Brugge et al. 2011a). In addition, Rhopr-CRF/DH is a potent DH, stimulating absorption across the anterior midgut and secretion by the upper MTs, and is synergistic with serotonin (see Paluzzi et al. 2012). It does not stimulate reabsorption of KCl from the lower MTs (this action is found to be restricted to serotonin). A CRF/DH was originally isolated from *M. sexta* (Kataoka et al. 1989) and found to be similar to its vertebrate counterpart. Homologous peptides are now known to be present in many insect species, and in some, multiple CRF/DHs are encoded by mutually alternative exons, such that in processing, closely related peptides are produced. Three such peptides are found in *B. mori*, two in *T. castaneum*; but only one in *D. melanogaster* and mosquitoes. The insect CRF/DH receptor is a member of the secretin family (family B) GPCR and related to the vertebrate CRF receptor. In some insects there are two CRF/DH receptors with different expression patterns (see Park 2012). The GPCRs for Rhopr-CRF/DH have recently been cloned and found to be homologous to other insect CRF GPCRs (unpublished data). Quantitative PCR confirms the physiological data with strong expression of one of the receptors in the anterior midgut and upper MTs and little or no expression in the lower MTs. The presence of the receptor in the CNS and other peripheral tissues indicates that Rhopr-CRF/DH has other physiological targets (see later under feeding).

Kinins and CT/DH

Continuing the strategy to identify other putative DHs in *R. prolixus*, the Rhopr-kinin transcript and Rhopr-CT/DH transcript have also been cloned and the peptides shown to be expressed in the CNS (see Te Brugge et al. 2008, 2011b, Zandawala et al. 2011, Bhatt et al. 2014). Interestingly, Rhopr-CT/DH is expressed in the 5 DUM neurons that release serotonin as a DH, and the Rhopr-kinin is expressed in the same PLNs of the MTGM that express Rhopr-CRF/DH (Fig. 2). Insect kinins were first identified as potent stimulators of visceral muscle contraction and later found to be potent diuretic factors in, for example, *A. aegypti* and *D. melanogaster*. As in *R. prolixus*, the kinin transcript codes for multiple kinin peptides. Unlike other insects, however, the Rhopr-kinins have no diuretic activity in *R. prolixus*.

Rhopr-CT/DH stimulates a very small (relative to Rhopr-CRF/DH or serotonin) increase in fluid secretion by the upper MTs and also stimulates a slight increase in cAMP content of the anterior midgut, and so may play some role in diuresis along with the other two very active DHs, serotonin and Rhopr-CRF/DH (see Zandawala 2012). The first CT/DH was called Dippu-DH$_{31}$ (Furiya et al. 2000), isolated from *Diploptera punctata*. Typically, single genes code for CT/DH; however, recently, the complete cDNA sequences of three Rhopr-CT/DH splice variants (Rhopr-CT/DH-A, Rhopr-CT/DH-B and Rhopr-CT/DH-C) were characterized in *R. prolixus* (Zandawala et al. 2011). This is the first example of three CT/DH splice variants in an arthropod species. The three splice variants are predominantly expressed in the CNS of unfed fifth-instars. Fluorescent *in situ* hybridization reveals that Rhopr-CT/DH is expressed in a variety of cells in the CNS, including the DUM neurons of the MTGM. Some of the target tissues include the dorsal vessel (heart), hindgut and MTs, which have been shown physiologically to respond to Rhopr-CT/DH. To identify other target tissues, two *R. prolixus* CT/DH receptor paralogs (Rhopr-CT/DH-R1 and Rhopr-CT/DH-R2) were isolated and functionally characterized using a novel heterologous assay of modified human embryonic kidney 293 (HEK293) cells (Zandawala et al. 2013). Both receptors are activated by Rhopr-CT/DH. CT/DH mediates its effects by binding to family B (secretin) GPCRs. Rhopr-CT/DH-R1 is orthologous to the previously characterized *D. melanogaster* CT/DH receptor (CG17415) while Rhopr-CT/DH-R2 is orthologous to the *D. melanogaster* receptor (CG4395). Receptor CG4395 was an orphan receptor although identified by Park (2012) as a paralogous gene in the CT/DH receptor group and likely an ancestral gene duplication. *Rhodnius prolixus* is the first example of an insect possessing at least two functionally different CT/DH receptors. This work also indirectly de-orphans the

D. *melanogaster* GPCR (CG4395) and its orthologs in other insects. Quantitative PCR demonstrates that Rhopr-CT/DH-R1 and Rhopr-CT/DH-R2 have distinct expression patterns, with both receptors expressed centrally and peripherally. At least one of the receptors is expressed in the dorsal vessel, foregut, salivary glands, hindgut and MTs; these tissues are associated with feeding and diuresis. Moreover, the expression analysis also identifies novel target tissues for this neuropeptide, including testes, ovaries and prothoracic glands, suggesting a possible role for Rhopr-CT/DH in reproductive physiology and development.

Access to the native *R. prolixus* neuropeptides and their receptors, and coupled to immunohistochemistry, qPCR, and bioassay has now shown that the Rhopr-DHs and their co-localized neuropeptides act on many tissues associated with feeding-related physiological events, and these signaling pathways co-ordinate many aspects of feeding, not just post-feeding diuresis (for reviews see Orchard 2006, 2009).

CAPA

Diuresis very rapidly ceases in *R. prolixus* when a sufficient amount of urine has been produced and expelled, thereby enabling the insect to be more mobile. Diuresis is terminated to avoid excessive loss of water and salts, and maintains homeostasis. An ADH in *R. prolixus* was postulated by Quinlan et al. (1997). These authors found that the cGMP content of the MTs is elevated *in vivo* at a time when diuresis is ceasing, and that the cardio acceleratory peptide, *M. sexta* CAP_{2b} (now referred to as Manse-CAPA-1) inhibits serotonin-induced secretion and elevates cGMP in MTs. The *R. prolixus* CAPA transcript has now been cloned and shown to be expressed in CAPA-like immunoreactive posterior ventral median neurons in the MTGM (Fig. 2), which have neurohemal sites on abdominal nerves (Paluzzi et al. 2008). It is now known that there are two CAPA genes in *R. prolixus* (Rhopr-CAPA-α and β). Each gene codes for three peptides, two of them being CAPA-related peptides and the third being a pyrokinin-related peptide (see Table 1). The second encoded peptide in each prepropeptide, Rhopr-CAPA-α2 (-β2), is identical in sequence in each paralog. Rhopr-CAPA-α2 inhibits serotonin-stimulated secretion by the MTs as well as absorption by the anterior midgut. Interestingly, Rhopr-CAPA-α2 does not inhibit Rhopr-CRF/DH-stimulated secretion. It appears that the antidiuretic activity of Rhopr-CAPA-α2 is due to its ability to eliminate the synergy between serotonin and Rhop-CRF/DH (see Paluzzi et al. 2012). The Rhopr-CAPA receptor transcript (the first receptor known to be involved in an antidiuretic strategy in insects) has also been cloned and belongs to the insect CAPA receptor family (i.e., activated by peptides encoded within the *capability* gene), and in addition, also shares similarity to the pyrokinin-1 receptor family in insects (Paluzzi et al. 2010). Expression of Rhopr-CAPA-R is localized to the digestive system, with highest transcript levels in MTs and anterior midgut (the known physiological targets of Rhopr-CAPA-α2). Two receptor transcripts are actually present in *R. prolixus*, CAPA-R1 and CAPA-R2; however the latter encodes an atypical GPCR lacking a region between the first and second transmembrane domains. The heterologous expression functional assay reveals the expressed CAPA-R1 is activated by Rhopr-CAPA-α2 (EC_{50} = 385 nM) but not by Rhopr-CAPA-α1. This receptor also has some minor sensitivity to the pyrokinin-related peptide, Rhopr-CAPA-αPK1, but with an efficacy approximately 40-fold less than Rhopr-CAPA-α2. Other peptides belonging to the PRXamide superfamily are inactive on the Rhop-rCAPA-R1. Manse-CAPA-1 was originally identified as stimulating increases in heart beat rate in *M. sexta*. Later, it was shown to be a potent diuretic in *D. melanogaster* (see Schooley et al. 2012) and it is therefore of some interest that it is antidiuretic in *R. prolixus*, especially considering the Rhopr-CAPA-R is orthologous to other insect CAPA receptors.

Neuroendocrine regulation of feeding

Diuresis is often a consequence of feeding, and advances in our understanding of neuroendocrine control of feeding (particularly food intake) has also benefited immensely from the post-genomic era and the use of "omics" technologies (see also Hoffmann et al. this book). The likely involvement of peptides in the regulation of food intake (and digestion, etc.) was suggested by the early discovery of a "brain-gut axis" whereby many CNS neuropeptides were also found to be synthesized and released from midgut endocrine cells. Indeed, the midgut may be considered the largest endocrine gland in insects. In addition, CNS neuropeptides are also directly delivered in the innervation to the digestive system where they can stimulate or inhibit contractions of the visceral muscles of the gut, and these, and/or neuropeptides from the midgut endocrine cells, can also modify digestive enzyme levels/activity. The peptidergic control of food intake and digestion in insects, in the context of the post-genomic era, has recently been reviewed (Audsley and Weaver 2009, Spit et al. 2012). Here we will focus on some specific examples of neuropeptides that regulate food intake as evidence of recent advances using "omics" (although many other neuropeptide families are also involved).

CRF/DH

As reviewed above, the CRF/DHs stimulate diuresis in a number of insects, and as such are released at the time of feeding. It was hypothesized that the CRF/DHs might be involved in other feeding—related activities and indeed, injection of Locmi-CRF/DH into the hemolymph reduces meal duration and increases the latency to feed in locusts (Goldsworthy et al. 2003). The Locmi-CRF/DH, released as a neurohormone to control diuresis, also regulates satiety and essentially acts as a message to terminate feeding. Similarly, in *S. gregaria* (Van Wielendaele et al. 2012) injection of Schgr-CRF/DH prior to the meal reduces food intake, whereas injection of dsRNA to down-regulate the precursor transcript increases food intake (an effect overcome by injecting Schgr-CRF/DH). Interestingly, injection of Schgr-CRF/DH in adult females retards oocyte growth and lowers ecdysteroid titers in hemolymph and ovaries. RNAi knockdown results in the opposite effect. The authors suggest that Schgr-CRF/DH is used to integrate several overlapping systems, including food intake and excretion, and oocyte growth and ecdysteroidogenesis following a meal.

sNPF

The family of short neuropeptide F (sNPF) has also been shown to regulate feeding (see Nässel and Wegener 2011, Spit et al. 2012). Earlier research in *D. melanogaster* indicated that sNPF promotes a positive influence on appetite whereby gain of function NPF mutants have higher food intake, and loss of function sNPF mutants have lower food intake (Lee et al. 2004). The sNPF receptor transcript is up-regulated during starvation, and the presence of sNPF in olfactory receptor neurons that signal odor-food searching indicates sNPF may be associated with starvation-dependent stimulation of food search behavior. In *B. mori*, the sNPF precursor codes for multiple sNPF and related peptides, and three of these, sNPF-1, 2 and 3 have been confirmed by MS studies on brain extracts (Nagata et al. 2012). The levels of all three peptides in the brain are lower in starved versus fed larvae. In addition, food deprivation decreases sNPF receptor transcript levels (determined by RT-PCR). Injections of sNPFs accelerate the onset of feeding in *B. mori* larvae. The authors conclude that sNPFs are involved in regulating feeding behavior and this may be linked to locomotory activity associated with foraging behavior. The decrease in receptor transcript may be connected to a strategy whereby larvae starved for longer periods may have decreased locomotory activity in order to conserve energy stores (Nagata et al. 2012). Interestingly, with this latter point in mind, sNPFs appear to be involved in a strategy to reduce feeding in some insect species. The hemolymph titer of sNPF increases after a blood meal in *A. aegypti*, coincident with an inhibition of host-seeking behavior (Brown et al. 1994), and injection of head peptide-1 (an sNPF), which is transferred from the male to female during copulation, inhibits host-seeking behavior (Brown et al. 1994, Naccarati et al. 2012). Here, sNPF appears to be acting as a satiety factor. In a similar manner, Dillen et al. (2013, 2014) recently identified sNPF as inhibiting food uptake in *S. gregaria*. Thus, the sNPF transcript was cloned confirming that two Schgr-sNPF previously identified by MS studies are coded on the same precursor. Starvation of locusts results in reduced transcript levels in the optic lobes, and down-regulation (90%) of the precursor transcript using RNAi significantly increases food intake. Injection of Schgr-sNPFs inhibits food intake. The *S. gregaria* sNPF GPCR has also been cloned and shown to be activated by both Schgr-sNPFs in the nanomolar range. Knocking-down the receptor transcript using RNAi results in an increase in total food uptake. Thus, in the locust, the sNPF signaling pathway is negatively correlated with feeding. In fire ants also, the expression of the sNPF receptor is lower in starved mated queens compared to non-starved congeners (Chen and Pietrantonio 2006). Furthermore, and adding complexity to sNPF signalling pathways, Castillo and Pietrantonio (2013) have shown sNPF is involved in mechanisms of worker division of labor and sensing or responding to colony nutritional requirements (e.g., protein requirements and availability).

Sulfakinin

Insect sulfakinins (SKs) are homologous to mammalian gastrin/cholecystokinin (which are known to induce satiety in mammals) and so are likely candidates for similar effects in insects. In early studies, in *Blattella germanica* (Maestro et al. 2001), injection of Peram-SK was found to inhibit food intake (the sulfate group was required for activity). In *T. castenaneum* both sulfated and non-sulfated peptides inhibit food intake by up to 70% (Yu et al. 2013a). This was further investigated (Yu et al. 2013b) with regard to the SK and SK-receptor1 transcripts. Both transcripts are expressed in all developmental stages and differentially expressed between tissues. Using RNAi, both transcripts could be knocked-down by 80–90% (SK transcript) and 30–50% (SK-receptor 1 transcript). The effect on the larvae was to stimulate food intake in SK-silenced larvae, with a similar but weaker effect (though not significant) in the SK- receptor 1-silenced larvae. Parallel experiments show that injection of SK analogs reduces food intake. Thus in *T. castaneum* the SK-signaling pathway appears to negatively regulate food intake. This was further investigated since there are two SK receptor genes in *T. castaneum* and silencing SK-receptor 1 transcript was less effective at stimulating food intake (Yu and Smagghe 2013). The SK-receptor 2 transcript is expressed throughout all developmental stages and highly expressed in the head of both larvae and adults, and in gut tissue. Knocking-down the SK-receptor 2 transcript (30–70%) in the first 4 days results in significantly more food being consumed than in controls, suggesting that the SK-receptor 2 is more important in the SK-signaling pathway for satiation than is the SK-receptor 1. In the Mediterranean field cricket, *Gryllus bimaculatus*, SK also appears to be involved in inhibiting food intake. Knocking down the SK transcript by injection or feeding dsRNA results in stimulation of food intake, suggesting that SKs are normally involved in inhibition (reducing) food intake (Meyering-Vos and Müller 2007, see Hoffmann et al. this book).

Pharmacology: Neuropeptides and their GPCRs

Neuropeptides constitute the largest and most diverse class of neuroactive chemical messengers in animals; and these neuropeptides along with their GPCRs regulate/coordinate a vast array of biological processes. Knowledge of the structure-activity relationships between the neuropeptide and its GPCR aids in our understanding of the physiological and endocrinological relevance of these messengers in any given insect species, but also furthers our knowledge of mammals, including evolution and phylogeny. In addition, neuropeptides and their GPCRs are candidate targets for the development of next generation pest-control agents. The natural neuropeptides themselves cannot be directly used for pest control because of certain limitations; they cannot pass through the cuticle, and are susceptible to degradation by peptidases in the insect gut, hemolymph and tissues. Structure-activity studies based upon bioassays (somewhat indirect measure of ligand binding/activation) or functional receptor assays (more direct measure of ligand binding/activation), have led to an understanding of the chemical structural and conformational requirements for ligand receptor interaction and thereby the design of analogs (mimetics) that overcome such limitations.

Fundamental knowledge of neuropeptide signaling and its evolution has been obtained from pharmacological analysis of many insect systems but two will be highlighted here. The first is that of *D. melanogaster* sex peptide (SP) and its receptor, and the second is that of adipokinetic hormone (AKH), corazonin, and the AKH/corazonin-related peptide (ACP) signaling system and co-evolution of peptides and receptors; we then highlight recent research examining analogs/mimetics of some other peptide families.

Sex Peptide

Sex peptide is one of several male accessory gland peptides/proteins that are transferred to the female during copulation in *Drosophila* species. Sex peptide induces long-term behavioral and physiological changes in the female, including male rejection behavior, and increased ovulation and oviposition (see Kubli 2003, Vandersmissen et al. 2013, Isaac et al. 2014). The SP consists of 36 amino acids and is only found in *Drosophila* species. It has a W(X₈)W sequence of residues at its core, and both W residues are required for full activation of the receptor. Analysis of the receptor reveals that it is in fact the receptor for another insect family of neuropeptides, the myoinhibiting peptide (MIP) family, also designated as allatostatin B or prothoracicotropic peptides (see Lange et al. 2012, Hoffmann et al. this book). The MIPs have a characteristic W(X₆)Wamide or W(X₇)Wamide C-terminal motif. Sex peptides may have evolved relatively recently as a ligand for the more ancestral MIP receptor. Both SP and MIPs activate the *D. melanogaster* receptor with EC_{50} values of about 15nM for SP and between 7 and 25 nM for different MIPs (Kim et al. 2010, Poels et al. 2010). The rejection behavior and increase in oviposition produced by SP following copulation is mediated via sensory neurons that respond to SP in the female reproductive tract. Sex peptide also crosses into the hemolymph, and circulating SP contributes to the post-mating behavioral changes (see Isaac et al. 2014). Injection of SP into the hemolymph induces female post-mating responses, but interestingly, injecting MIPs does not (Kim et al. 2010, Poels et al. 2010). In addition, MIPs cannot substitute for SP in SP-null males when expressed in male accessory glands followed by mating. One hypothesis for this concerns the stability of SP and instability of MIPs in the presence of peptidases, with MIPs being rendered inactive fairly quickly in the presence of peptidases. The functional receptor assay and behavioral assay have been used to examine structure-activity relationships of MIPs and SP in this context (Vandersmissen et al. 2013). MIPs with W(X₆)Wamide and W(X₇)Wamide motifs are potent activators of the SP receptor, as is SP with the core W(X₈)W motif (see Lange et al. 2012). The W residues are essential for receptor activation. Furthermore, a chimeric peptide, consisting of the N-terminus of SP fused with the Drome-MIP-4 (i.e., an N-terminally-extended MIP-4), elicits SP receptor activation comparable to the native Drome-MIP-4. This chimeric peptide is also capable of inducing egg-laying when injected into female *D. melanogaster*. This suggests that the N-terminal amino acid sequence of SP provides protection and stability against peptidases (Vandersmissen et al. 2013). This has now been confirmed by examining the degradome of SP and MIPs (Isaac et al. 2014). Sex peptide is stable whereas some MIPs tested are very susceptible to degradation by peptidases found within male accessory glands and hemolymph. The authors consider that metabolic stability is an important feature in the evolution of SP as an alternative recent ligand for the MIP receptor.

AKH/corazonin/ACP

As discussed earlier, it is possible, though difficult, to find mammalian neuropeptide orthologs in insects, but much easier to find orthologous GPCRs for the neuropeptides (Grimmelikhuijzen and Hauser 2012). An exemplary of this, and one that fully illustrates the advances that can be achieved by the post-genomic era is that of gonadotropin-releasing hormone (GnRH) and AKH, and also reveals receptor/ligand coevolution. The mammalian GnRH receptor has an ortholog in *D. melanogaster*, but the insect ligand is AKH; a metabolic hormone whose amino acid sequence has little in common with GnRH (Hansen et al. 2010, Grimmelikhuijzen and Hauser 2012). The GnRH receptor must be over 700 millon years old, and present before the split of the protostomes and deuterostomes. The ancestral ligand probably had a mixture of GnRH and AKH amino acid sequences, and has changed over evolution (Hansen et al. 2010, Grimmelikhuijzen and Hauser 2012). These authors question whether the ancestral GnRH signaling pathway may have controlled metabolism and/or reproduction. Furthermore though, insects have 3 closely-related receptors; one for AKH, one for corazonin (a cardioacceleratory peptide and also involved in ecdysis), and one for ACP (with unknown function). Although these peptides have closely-related sequences, none of the ligands cross-reacts with the receptor for the other ligands (as reveled by functional receptor assays). In biological assays also, in *R. prolixus*, Rhopr-AKH stimulates an elevation in hemolymph lipid, whereas corazonin does not; and corazonin increases heartbeat frequency whereas AKH does not (Patel et al. 2014). This system is an example of receptor/ligand coevolution, "where an ancestral receptor gene and its ligand gene have duplicated twice during evolution, yielding three independent neuropeptide/GPCR signaling systems" (Hansen et al. 2010, Grimmelikhuijzen and Hauser 2012).

Peptide analogs/mimetics

Numerous mimetic analogs have been designed to overcome limitations in the use of natural neuropeptides as pest control agents. A number of these have been generated against the kinin receptor, the diapause hormone receptor, and the pheromone biosynthesis activating neuropeptide (PBAN) receptor and will be discussed here as being illustrative of the principles.

Insect kinins are multifunctional neuropeptides, being stimulators of visceral muscle contraction, diuretic hormones, modulators of digestive enzyme release, and also capable of inhibiting weight gain in *H. zea* and *H. virescens* larvae (see Smagghe et al. 2010). These kinins have a C-terminal pentapeptide core which is the minimum sequence required for full biological activity when tested in bioassays or in functional receptor assays. The kinins have been extensively targeted for the development of biostable analogs and mimetics (see Nachman et al. 1997, Nachman and Pietrantonio 2010). Many such analogs and mimetics have been designed that are protected against peptidases whilst retaining activity in both biological and receptor assays. In addition, enhanced topical or oral delivery has been obtained with modifications to the kinin structure. For example, five biostable kinin analogs have been made which incorporate α,α-disubstituted amino acids, β-amino acids, or both to the C-terminal pentapeptide core. These analogs retain activity in bioassays or when tested against expressed receptors. These analogs have also been tested *in vivo* by feeding to the pea aphid, *Acyrthosiphon pisum*, and testing for antifeedant and aphicidal activity (Smagghe et al. 2010). Antifeedant activity can be monitored by measuring the amounts of honey dew (sugary liquid waste) produced following feeding. Three of the analogs are antifeedants, with reduced honey dew formation evident after one day, and mortality is high. The most active analog is a double Aib (alpha-amino-isobutyric acid), K-Aib-1 ([Aib]FF[Aib]WGamide). This analog is a potent aphicide with 0.063 nmol/µl capable of killing 50% of the aphids. The authors point out that this LC_{50} matches the potency of some commercially available aphicides. Although the precise mechanism that leads to death is not yet known, these kinin analogs appear to represent leads in the development of new generation pest control agents.

Another neuroendocrinological target involves that of the diapause hormone signaling pathway. Diapause is the dormant state that enables some insects to survive winter or other adverse seasons. Insects must exit diapause at the appropriate season and one of the hormones that terminates diapause in moths is diapause hormone. This 24 amino acid peptide is a pyrokinin and a member of the pyrokinin-PBAN superfamily. It initiates embryonic diapause in *B. mori* but breaks diapause in the pupae of the corn earworm *H. zea*. Recently the *H. zea* diapause receptor was cloned, and 68 diapause hormone analogs and mimetics tested using a functional receptor assay (Jiang et al. 2013). Many of these have also been tested in *in vivo* bioassays for their ability to break diapause (Zhang et al. 2011). These studies identified a number of analogs or mimetics with equivalent or greater potency to the native peptide. For example, 2Abf-Suc-FKPRLamide is a potent agonist when tested *in vivo*, with 50 fold higher potency than the diapause hormone (Zhang et al. 2011). Similarly, in the functional receptor assay, this mimetic has higher activity (125%) than the diapause hormone (Jiang et al. 2013). Another mimetic, 2Abf-Suc-F-[dA]-PRLamide, possesses the highest potency (144%) for the receptor although is yet to be tested for its ability to break pupal diapause. Interestingly, amphiphilic analogs of this pyrokinin-PBAN family of neuropeptides are capable of crossing the cuticle and foregut. Importantly, they retain potent biological activity in the *in vivo* pheromonotropic bioassay when topically applied or fed to *H. virescens* (Abernathy et al. 1996, Nachman et al. 2002).

Final Thoughts

The study of insect physiology and endocrinology has a rich history extending over 100 years, and this chapter is merely a snapshot of the current state of affairs. As a derivative scientific discipline, researchers in physiology and endocrinology embrace and apply new technologies taken from a wide array of disciplines, including engineering, physics, chemistry, mathematics, and computer science. This has been particularly dramatic since the sequencing of the first insect genome, *D. melanogaster* (Adams et al. 2000), and massive amounts of data have been obtained through advances in bioinformatics, gene sequencing technologies, imaging, MS technologies, and receptor identification and classification. Advances in functional genomic techniques can now be used to identify and then interfere with particular peptide-signaling pathways, providing an insight into the true physiological and endocrinological relevance of these pathways. A challenge for the future will be that these signaling pathways almost certainly overlap in functions. Indeed, it is likely that chemical substances (especially neuropeptides) bias, at many levels, neuronal, hormonal, and muscular events, leading them towards a new functional state in the animal (e.g., see Nässel 2012, Žitňan and Adams 2012). Thus, the same neuropeptides will likely co-ordinate many different facets of an insects' physiology and endocrinology. These complex scenarios in which many modulatory neuropeptides might alter several aspects of different behaviors, is what ultimately leads to a "successfully-behaving animal". We are only just appreciating these scenarios and therefore only just on the cusp of an understanding of such complexity. The next 100 years will be equally exciting!

Acknowledgements

Our research is supported by the Natural Sciences and Engineering Council of Canada.

Keywords: Neuroendocrinology, peptidomics, genomes, transcriptomics, mass spectrometry, diuresis, feeding, GPCR, RNAi, peptide mimetics, neuropeptides

References

Abernathy, R.L., P.E.A. Teal, J.A. Meredith and R.J. Nachman. 1996. Induction of pheromone production in a moth by topical application of a pseudopeptide mimic of a pheromonotropic neuropeptide. Proc. Natl. Acad. Sci. USA 93: 12621–12625.

Adams, M.D., S.E. Celniker, R.A. Holt, C.A. Evans, J.D. Gocayne, P.G. Amanatides, S.E. Scherer, P.W.Li, R.A. Hoskins, R.F. Galle, R.A. George, S.E. Lewis, S. Richards, M. Ashburner, S.N. Henderson, G.G. Sutton, J.R. Wortman, M.D. Yandell, Q. Zhang, L.X. Chen et al. 2000. The genome sequence of *Drosophila melanogaster*. Science 287: 2185–2195.

Ashburner, M., C. Chihara, P. Meltzer and G. Richard. 1974. Temporal control of puffing activity in polytene chromosomes. Cold Spring Harb. Symp. Quant. Biol. 38: 655–662.

Audsley, N. and R.J. Weaver. 2009. Neuropeptides associated with the regulation of feeding in insects. Gen. Comp. Endocrinol. 162: 93–104.

Baggerman, G., K. Boonen, P. Verleyen, A. De Loof and L. Schoofs. 2005. Peptidomic analysis of the larval *Drosophila melanogaster* central nervous system by two-dimensional capillary liquid chromatography quadrupole time-of-flight mass spectrometry. J. Mass Spectrom. 40: 250–260.

Berlind, A. and S.H.P. Maddrell. 1979. Changes in hormone activity of single neurosecretory cell bodies during a physiological secretion cycle. Brain Res. 161: 459–467.

Berridge, M.J. 1993. Inositol trisphosphate and calcium signaling. Nature 361: 315–325.

Bhatt, G., R. da Silva, R.J. Nachman and I. Orchard. 2014. The molecular characterization of the kinin transcript and the physiological effects of kinin in the blood-gorging insect, *Rhodnius prolixus*. Peptides 55: 148–158.

Bilen, J. and N.M. Bonini. 2005. *Drosophila* as a model for human neurodegenerative disease. Annu. Rev. Genet. 39: 153–171.

Boerjan, B., D. Cardoen, R. Verdonck, J. Caers and L. Schoofs. 2012. Insect omics research coming of age. Can. J. Zool. 90: 440–455.

Botas, J. 2007. *Drosophila* researchers focus on human disease. Nature Genetics 39: 589–591.

Brown, B.E. and A.N. Starratt. 1975. Isolation of proctolin, a myotropic peptide, from *Periplaneta americana*. J. Insect Physiol. 21: 1879–1881.

Brown, M.R., M.J. Klowden, J.W. Crim, L. Young, L.A. Shrouder and A.O. Lea. 1994. Endogenous regulation of mosquito host-seeking behavior by a neuropeptide. J. Insect Physiol. 40: 399–406.

Caers, J., L. Peeters, T. Janssen, W. De Haes, G. Gäde and L. Schoofs. 2012a. Structure-activity studies of *Drosophila* adipokinetic hormone (AKH) by a cellular expression system of dipteran AKH receptors. Gen. Comp. Endocrinol. 177: 332–337.

Caers, J., H. Verlinden, S. Zels, H.P. Vandersmissen, K. Vuerinckx and L. Schoofs. 2012b. More than two decades of research on insect neuropeptide GPCRs: an overview. Front. Endocrinol. 3: 151.

Caers, J., N. Suetens, L. Temmerman, T. Janssen, L. Schoofs and I. Beets. 2014. Characterization of G protein-coupled receptors by a fluorescence-based calcium mobilization assay. J. Vis. Exp. 89: e51516.

Castillo, P. and P.V. Pietrantonio. 2013. Difference in sNPF receptor-expressing neurons in brains of fire ant (*Solenopsis invicta* Buren) worker subcastes: Indicators for division of labor and nutritional status? PLoS One 8(12): e83966.

Chen, M.E. and P.V. Pietrantonio. 2006. The short neuropeptide F-like receptor from the red imported fire ant, *Solenopsis invicta* Buren (Hymenoptera: *Formicidae*). Arch. Insect Biochem. Physiol. 61: 195–208.

Christiaens, O., L. Swevers and G. Smagghe. 2014. DsRNA degradation in the pea aphid (*Acyrthosiphon pisum*) associated with lack of response in RNAi feeding and injection assay. Peptides 55: 307–314.

Coast, G.M. 2009. Neuroendocrine control of ionic homeostasis in blood-sucking insects. J. Exp. Biol. 212: 378–386.

Coast, G.M., I. Orchard, J.E. Phillips and D.A. Schooley. 2002. Insect diuretic and antidiuretic hormones. Adv. Insect Physiol. 29: 279–409.

Dai, H., L. Ma, J. Wang, R. Jiang, Z. Wang and J. Fei. 2008. Knockdown of ecdysis-triggering hormone gene with a binary UAS/GAL4 RNA interference system leads to lethal ecdysis deficiency in silkworm. Acta Biochim. Biophys. Sin. 40: 790–795.

Davies, S.A., P. Cabrero, G. Overend, L. Aitchison, S. Sebasian, S. Terhzaz and J.A.T. Dow. 2014. Cell signalling mechanisms for insect stress tolerance. J. Exp. Biol. 217: 119–128.

Dillen, S., S. Zels, H. Verlinden, J. Spit, P. Van Wielendaele and J. Vanden Broeck. 2013. Functional characterization of the short neuropeptide F receptor in the desert locust, *Schistocerca gregaria*. PLoS One 8: e53604.

Dillen, S., R. Verdonck, S. Zels, P. Van Wielendaele and J. Vanden Broeck. 2014. Identification of short neuropeptide F precursor in the desert locust: evidence for an inhibitory role of sNPF in the control of feeding. Peptides 55: 134–139.

Dow, J.A.T. 2009. Insights into the Malpighian tubule from functional genomics. J. Exp. Biol. 212: 435–445.

Dow, J.A.T. and S.A. Davies. 2006. The Malpighian tubule: Rapid insights from post-genomic biology. J. Insect Physiol. 52: 365–378.

Fan, Y., P. Sun, Y. Wang, X. He, X. Deng, X. Chen, G. Zhang, X. Chen and N. Zhou. 2010. The G protein-coupled receptors in the silkworm, *Bombyx mori*. Insect Biochem. Molec. Biol. 40: 581–591.

Furuya, K., R.J. Milchak, K.M, Schegg, J. Zhang, S.S. Tobe, G.M. Coast and D.A. Schooley. 2000. Cockroach diuretic hormones: characterization of a calcitonin-like peptide in insects. Proc. Natl. Acad. Sci. USA 97: 6469–6474.

Goldsworthy, G.J., J.S. Chung, M.S.J. Simmonds, M. Tatarı, S. Varouni and C.P. Poulos. 2003. The synthesis of an analogue of the locust CRF-like diuretic peptide, and the biological activities of this and some C-terminal fragments. Peptides 24: 1607–1613.

Grimmelikhuijzen, C.J.P. and F. Hauser. 2012. Mini-review: The evolution of neuropeptide signaling. Reg. Pept. 177: S6–S9.

Hansen, K.K., E. Stafflinger, M. Schneider, F. Hauser, G. Cazzamali, M. Williamson, M. Kollman, J. Schachtner and C.J.P. Grimmelikhuijzen. 2010. Discovery of a novel insect neuropeptide signaling system closely related to the insect adipokinetic hormone and corazonin hormonal systems. J. Biol. Chem. 285: 10736–10747.

Hauser, F., G. Cazzamali, M. Williamson, W. Blenau and C.J.P. Grimmelikhuijzen. 2006. A review of neurohormone GPCRs present in the fruitfly *Drosophila melanogaster* and the honey bee *Apis mellifera*. Prog. Neurobiol. 80: 1–19.

Hauser, F., G. Cazzamali, M. Williamsoon, Y. Park, B. Li, Y. Tanaka, R. Predel, S. Neupert, J. Schachtner, P. Verleyen and C.J.P. Grimmelikhuijzen. 2008. A genome-wide inventory of neurohormone GPCRs in the red flour beetle *Tribolium castaneum*. Front. Neuroendocrinol. 29: 142–165.

Horodyski, F.M. 2013. Allatotropin. pp. 197–202. In: A. Kastin (ed.). Handbook of Biologically Active Peptides. Academic Press, San Diego.

Horodyski, F.M., S.R. Bhatt and K.-Y. Lee. 2001. Alternative splicing of transcripts expressed by the *Manduca sexta* allatotropin (Mas-AT) gene is regulated in a tissue-specific manner. Peptides 22: 263–269.

Hummon, A.B., A. Amnare and J.V. Sweedler. 2006. Discovering new invertebrate neuropeptides using mass spectrometry. Mass Spectrom. Reviews 25: 77–98.

Huybrechts, J., J. Bonhomme, S. Minoli, N. Prunier-Leterme, A. Dombrovsky, M. Abdel-Latief, A. Robichon, J.A. Veenstra and D. Ragu. 2010. Neuropeptide and neurohormone precursors in the pea aphid, *Acyrthosiphon pisum*. Insect Molec. Biol. 19: 87–95.

Imler, J.-L. and J.A. Hoffmann. 2001. Toll receptors in innate immunity. Trends Cell Biol. 11: 304–311.

Isaac, R.E., Y.-J. Kim and N. Audsley. 2014. The degradome and the evolution of *Drosophila* sex peptide as a ligand for the MIP receptor. Peptides 55: 258–264.

Jékely, G. 2013. Global view of the evolution and diversity of metazoan neuropeptide signaling. Proc. Natl. Acad. Sci. USA 110: 8702–8707.

Jiang, H., A. Lkhagva, I. Daubnerová, H.-S. Chae, L. Šimo, S.-H. Jung, Y.-K. Yoon, N.-R. Lee, J.Y. Seong, D. Žitňan, Y. Park and Y.-J. Kim. 2013. Natalisin, a tachykinin-like signaling system, regulates sexual activity and fecundity in insects. Proc. Natl. Acad. Sci. USA 110: E3526–E3534.

Kataoka, H., R.G. Troetschler, J.P. Li, S.J. Dramer, R.L. Carney and D.A. Schooley. 1989. Isolation and identification of a diuretic hormone from the tobacco hornworm, *Mandua sexta*. Proc. Natl. Acad. Sci. USA 86: 2976–2980.

Kim, Y.J., K. Bartalska, N. Audsley, N. Yamanaka, N. Yapici, J.Y. Lee, Y.C. Kim, M. Markovic, E. Isaac, Y. Tanaka and B.J. Dickson. 2010. MIPs are ancestral ligands for the sex peptide receptor. Proc. Natl. Acad. Sci. USA 107: 6520–6525.

King-Jones, K. and C.S. Thummel. 2005. Nuclear receptors—a perspective from *Drosophila*. Nat. Rev. Genet. 6: 311–323.

Kubli, E. 2003. Sex-peptide: seminal peptide of the *Drosophila* male. Cell Mol. Life Sci. 60: 1689–1704.

Lange, A.B., I. Orchard and F.M. Barrett. 1989. Changes in haemolymph serotonin levels associated with feeding in the blood-sucking bug, *Rhodnius prolixus*. J. Insect Physiol. 35: 393–399.

Lange, A.B., U. Alim, H.P. Vandersmissen, A. Mizoguchi, J. Vanden Broeck and I. Orchard. 2012. The distribution and physiological effects of the myoinhibiting peptides in the kissing bug, *Rhodnius prolixus*. Front. Neurosci. 6: 98.

Lee, D.H., I. Orchard and A.B. Lange. 2013. Evidence for a conserved CCAP-signaling pathway controlling ecdysis in a hemimetabolous insect, *Rhodnius prolixus*. Front. Neurosci. 7: 207.

Lee, K.S., K.H. You, J.K. Choo, Y.M. Han and K. Yu. 2004. *Drosophila* short neuropeptide F regulates food intake and body size. J. Biol. Chem. 279: 50781–50789.

Li, B., R. Predel, S. Neupert, F. Hauser, Y. Tanaka, G. Cazzamali, M. Williamson, Y. Arakane, P. Verleyen, L. Schoofs, J. Schachtner, C.J.P. Grimmelikhuijzen and Y. Park. 2011. Genomics, transcriptomics, and peptidomics of neuropeptides and protein hormones in the red flour beetle *Tribolium castaneum*. Genome Res. 18: 113–122.

Liu, F., G. Baggerman, L. Schoofs and G. Wets. 2006. Uncovering conserved patterns in bioactive peptides in Metazoa. Peptides 27: 3137–3153.

Maddrell, S.H.P. 1964a. Excretion in the blood-sucking bug, *Rhodnius prolixus* Stal. II. The normal course of diuresis and the effect of temperature. J. Exp. Biol. 41: 163–170.

Maddrell, S.H.P. 1964b. Excretion in the blood-sucking bug, *Rhodnius prolixus* Stal. III. The control of the release of the diuretic hormone. J. Exp. Biol. 41: 459–472.

Maddrell, S.H.P. 1966. The site of release of the diuretic hormone in *Rhodnius*—a new neurohaemal system in insects. J. Exp. Biol. 45: 499–508.

Maddrell, S.H.P. 1991. The fastest fluid-secreting cell known: The upper Malpighian tubule cell of *Rhodnius*. Bioessays 13: 357–362.

Maestro, J.L., R. Aguilar, N. Pascual, M.-L. Valero, M.-D. Piulachs, D.Andreu, I. Navarro and X. Bellés. 2001. Are arthropod sulfakinins homologous to vertebrate gastrins-cholecystokinins? Eur. J Biochem. 268: 5824–5830.

Mao, Y.-B., W.-J. Cai, J.-W. Wang, G.-J. Hong, X.-Y. Tao, L.-J. Wang, Y.-P. Huang and X.-Y. Chen. 2007. Silencing a cotton bollworm P450 monooxygenase gene by plant-mediated RNAi impairs larval tolerance of gossypol. Nature Biotech. 25: 1307–1313.

Meyering-Vos, M. and A. Müller. 2007. RNA interference suggests sulfakinins as satiety effectors in the cricket *Gryllus bimaculatus*. J. Insect Physiol. 53: 840–848.

Naccarati, C., N. Audsley, J.N. Keen, J.-H. Kim, G.J. Howell, Y.-J. Kim and R.E. Isaac. 2012. The host-seeking inhibitory peptide, Aea-HP-1, is made in the male accessory gland and transferred to the female during copulation. Peptides 34: 150–157.

Nachman, R.J. and P.V. Pietrantonio. 2010. Interaction of mimetic analogs of insect kinin neuropeptides with arthropod receptors. pp. 27–48. *In*: T.G. Geary and A.G. Maule (eds.). Neuropeptide Systems as Targets for Parasite and Pest Control. Springer Science + Business Media, LLC, New York, USA.

Nachman, R.J., P.E.A. Teal and A. Strey. 2002. Enhanced oral availability/pheromonotropic activity of peptide-resistant topical amphiphilic analogs of pyrokinin/PBAN insect neuropeptides. Peptides 23: 2035–2043.

Nachman, R.J., R.E. Isaac, G.M. Coast and G.M. Holman. 1997. Aib-containing analogues of the insect kinin neuropeptide family demonstrate resistance to an insect angiotensin-converting enzyme and potent diuretic activity. Peptides 18: 53–57.

Nagata, S., S. Matsumoto, T. Nakane, A. Ohara, N. Morooka, T. Konuma, C. Nagai and H. Nagasawa. 2012. Effects of starvation on brain short neuropeptide F-1, -2, and -3 levels and short neuropeptide F receptor expression levels of the silkworm, *Bombyx mori*. Front. Endocrin. 3: 3.

Nässel, D.R. 2012. Insulin-producing cells and their regulation in physiological and behavior of *Drosophila*. Can. J. Zool. 90: 476–488.

Nässel, D.R. and C. Wegener. 2011. A comparative review of short and long neuropeptide F signaling in invertebrates: any similarities to vertebrate neuropeptide Y signaling? Peptides 32: 1335–1355.

Neupert, S., W.K. Russell, D.H. Russell and R. Predel. 2010. Two *capa*-genes are expressed in the neuroendocrine system of *Rhodnius prolixus*. Peptides 31: 408–411.

Nygaard, S., G. Zhang, M. Schiott, C. Li, Y. Wurm, H. Hu, J. Zhou, L. Ji, F. Qiu, M. Rasmussen, H. Pan, F. Hauser, A. Krogh, C.J.P. Grimmelikhuijzen, J. Wang and J.J. Boomsma. 2011. The genome of the leaf-cutting ant *Acromyrmex echinatior* suggests key adaptations to advanced social life and fungus farming. Genome Res. 21: 1339–1348.

Ons, S., M. Sterkel, L. Diambra, H. Urlaub and R. Rivera-Pomar. 2011. Neuropeptide precursor gene discovery in the Chagas disease vector *Rhodnius prolixus*. Insect Molec. Biol. 20: 29–44.

Orchard, I. 2006. Serotonin: A coordinator of feeding-related activities in *Rhodnius prolixus*. Comp. Biochem. Physiol. 144: 316–324.

Orchard, I. 2009. Peptides and serotonin control feeding-related events in *Rhodnius prolixus*. Front. Biosci. E1: 250–262.

Orchard, I. and L.H. Finlayson. 1997. Electrically excitable neurosecretory cell bodies in the periphery of the stick insect, *Carausius morosus*. Experientia 33: 226–228.

Orchard, I., A.B. Lange and W.G. Bendena. 2001. FMRFamide-related peptides: A multifunctional family of structurally related neuropeptides in insects. Adv. Insect Physiol. 28: 267–329.

Orchard, I., A.B. Lange, H. Cook and J.-M. Ramirez. 1989. A subpopulation of dorsal unpaired median neurons in the blood-feeding insect, *Rhodnius prolixus*, displays serotonin-like immunoreactivity. J. Comp. Neurol. 289: 118–128.

Paluzzi, J.-P. and I. Orchard. 2010. A second gene encodes the anti-diuretic hormone in the insect, *Rhodnius prolixus*. Molec. Cell. Endocrinol. 317: 53–63.

Paluzzi, J.-P., W. Naikkhwah and M.J. O'Donnell. 2012. Natriuresis and diuretic hormone synergism in *R. prolixus* upper Malpighian tubules is inhibited by the anti-diuretic hormone, RhoprCAPA-α2. J. Insect Physiol. 58: 534–542.

Paluzzi, J.-P., W.K. Russell, R.J. Nachman and I. Orchard. 2008. Isolation, cloning and expression mapping of a gene encoding an anti-diuretic hormone and other CAPA-related peptides in the disease vector, *Rhodnius prolixus*. Endocrinol. 149: 4638–4646.

Paluzzi, J.P., Y. Park, R.J. Nachman and I. Orchard. 2010. Isolation, expression analysis and functional characterization of the first anti-diuretic hormone receptor in insects. Proc. Natl. Acad. USA 107: 10290–10295.

Pandey, U.B. and C.D. Nichols. 2011. Human disease models in *Drosophila melanogaster* and the role of the fly in therapeutic drug discovery. Pharmacol. Rev. 63: 411–436.

Park, Y. 2012. Endocrine regulation of insect diuresis in the early postgenomic era. Can. J. Zool. 90: 507–520.

Patel, H., I. Orchard, J.A. Veenstra and A.B. Lange. 2014. The distribution and physiological effects of three evolutionarily and sequence-related neuropeptides in *Rhodnius prolixus*: Adipokinetic hormone, corazonin and adipokinetic hormone/corazonin-related peptide. Gen. Comp. Endocrinol. 195: 1–8.

Poels, J., T. Van Loy, H.P. Vandersmissen, B. Van Hiel, S. Van Soest, R.J. Nachman and J. Vanden Broeck. 2010. Myoinhibiting peptides are the ancestral ligands of the promiscuous *Drosophila* sex peptide receptor. Cell. Mol. Life Sci. 67: 3511–3522.

Predel, R., S. Neupert, S.F. Garczynski, J.W. Crim, M.R. Brown, W.K. Russell, J. Kahnt, D.H. Russell and R.J. Nachman. 2010. Neuropeptidomics of the mosquito *Aedes aegypti*. J. Proteome Res. 9: 2006–2015.

Predel, R., W.K. Russell, D.H. Russell, C.P.-C. Suh and R.J. Nachman. 2012. Neuropeptides of the cotton fleahopper, *Pseudatomoscelis seriatus* (Reuter). Peptides 34: 39–43.

Quinlan, M.C., N.J. Tublitz and M.J. O'Donnell. 1997. Anti-diuresis in the blood-feeding insect *Rhodnius prolixus* stal: The peptide CAP-2b and cyclic GMP inhibit Malpighian tubule fluid secretion. J. Exp. Biol. 200: 2363–2367.

Schoofs, L., D. Veelaert, J. Vanden Broeck and A. De Loof. 1997. Peptides in the locusts, *Locusta migratoria* and *Schistocerca gregaria*. Peptides 18: 145–156.

Schinko, J.B., M. Weber, I. Viktorinova, A. Kiupakis, M. Averof, M. Klingler, E.A. Wimmer and G. Bucher. 2010. Functionality of the GAL4/UAS system in *Tribolium* requires the use of endogenous core promoters. BMC Develop. Biol. 10: 53–64.

Schooley, D.A., F.M. Horodyski and G.M Coast. 2012. Hormones controlling homeostasis in insects. pp. 366–429. *In*: L.I. Gilbert (ed.). Insect Endocrinology. Academic Press, San Diego.

Scott, J.G., K. Michel, L. Bartholomay, B.D. Siegfried, W.B. Hunter, G. Smagghe, K.Y. Zhu and A.E. Douglas. 2013. Towards the elements of successful insect RNAi. J. Insect Physiol. 59: 1212–1221.

Smagghe, G., K. Mahdian, P. Zubrzak and R.J. Nachman. 2010. Antifeedant activity and high mortality in the pea aphid *Acrythosiphon pisum* (Hemipetera: Aphidae) induced by biostable insect kinin analogs. Peptides 31: 498–505.

Spit, J., L. Badisco, H. Verlinden, P. Van Wielendaele, S. Zels, S. Dillen and J. Vanden Broeck. 2012. Peptidergic control of food intake and digestion in insects. Can. J. Zool. 90: 489–506.

Stone, J.V., W. Mordue, L.E. Batley and H.R. Morris. 1976. Structure of locust adipokinetic hormone, a neurohormone that regulates lipid utilization during flight. Nature 263: 207–211.

Te Brugge, V.A. and I. Orchard. 2002. Evidence for CRF-like and kinin-like peptides as diuretic neurohormones in the blood-feeding bug, *Rhodnius prolixus*. Peptides 23: 1967–1980.

Te Brugge, V.A., S.M. Miksys, G.M. Coast, D.A. Schooley and I. Orchard. 1999. The distribution of a CRF-like diuretic peptide in the blood-feeding bug *Rhodnius prolixus*. J. Exp. Biol. 202: 2017–2027.

Te Brugge, V.A., D.R. Nässell, G.M. Coast, D.A. Schooley and I. Orchard. 2001. The distribution of a kinin-like peptide and its co-localization with a CRF-like peptide in the blood-feeding bug, *Rhodnius prolixus*. Peptides 22: 161–173.

Te Brugge, V.A., V.A. Lombardi, D.A. Schooley and I. Orchard. 2008. Amino acid sequence and biological activity of a calcitonin-like diuretic hormone (DH$_{31}$) from *Rhodnius prolixus*. J. Exp. Biol. 211: 382–390.

Te Brugge, V.A., J.-P. Palluzzi, D.A. Schooley and I. Orchard. 2011a. Identification of the elusive peptidergic diuretic hormone in the blood-feeding bug, *Rhodnius prolixus*: a CRF-related peptide. J. Exp. Biol. 214: 371–381.

Te Brugge, V.A., J.-P. Paluzzi, S. Neupert, R.J. Nachman and I. Orchard. 2011b. Identification of kinin-related peptides in the disease vector, *Rhodnius prolixus*. Peptides 32: 469–474.

Tweedie, S., M. Ashburner, K. Falls, P. Leyland, P. McQuilton, S. Marygold, G. Millburn, D. Osumi-Sutherland, A. Schroeder, R. Seal, H. Zhang and The FlyBase Consortium. 2009. FlyBase: enhancing *Drosophila* gene ontology annotations. Nucl. Acids Res. 37: D555–D559.

Vandermissen, H.P., R.J. Nachman and J. Vanden Broeck. 2013. Sex peptides and MIPs can activate the same G protein-coupled receptor. Gen. Comp. Endocrinol. 188: 137–143.

Van Wielendaele, P., S. Dillen, E. Marchal, L. Badisco and J. Vanden Broeck. 2012. CRF-like diuretic hormone negatively affects both feeding and reproduction in the desert locust, *Schistocerca gregaria*. PLoS One 7:e31425.

Wynant, N., H. Verlinden, B. Breugelmans, G. Simonet and J. Vanden Broeck. 2012. Tissue-dependence and sensitivity of the systemic RNA interference response in the desert locust, *Schistocerca gregaria*. Insect Biochem. Molec. Biol. 42: 911–917.

Wu, Q. and M.R. Brown. 2006. Signaling and function of insulin-like peptides in insects. Annu. Rev. Entomol. 51: 1–24.

Yu, N. and G. Smagghe. 2014. Characterization of sulfakinin receptor 2 and its role in food intake in the red flour beetle, *Tribolium castaneum*. Peptides 55: 232–237.

Yu, N., R.J. Nachman and G. Smagghe. 2013a. Characterization of sulfakinin and sulfakinin receptor and their roles in food intake in the red flour beetle *Tribolium castaneum*. Gen. Comp. Endocrinol. 188: 196–203.

Yu, N., V. Benzi, M.J. Zotti, D. Staljanssens, K. Kaczmarek, J. Zabrocki, R.J. Nachman and G. Smagghe. 2013b. Analogs of sulfakinin-related peptides demonstrate reduction in food intake in the red flour beetle, *Tribolium castaneum*, while putative antagonists increase consumption. Peptides 41: 107–112.

Zandawala, M. 2012. Calcitonin-like diuretic hormones in insects. Insect Biochem. Molec. Biol. 42: 816–825.

Zandawala, M., J.-P. Paluzzi and I. Orchard. 2011. Isolation and characterization of the cDNA encoding DH_{31} in the kissing bug: *Rhodnius prolixus*. Molec. Cell. Endocrinol. 331: 79–88.

Zandawala, M., S. Li, F. Hauser, C.J.P. Grimmelikhuijzen and I. Orchard. 2013. Isolation and functional characterization of calcitonin-like diuretic hormone receptors in *Rhodnius prolixus*. PLoS One 8: e82466.

Zhang, H., H.-C. Li and X.-X. Miao. 2013. Feasibility, limitation and possible solutions of RNAi-based technology for insect pest control. Insect Sci. 20: 15–30.

Zhang, Q., R.J. Nachman, K. Kaczmarek, J. Zabrocki and D.L. Denlinger. 2011. Disruption of insect diapause using agonists and an antagonist of diapause hormone. Proc. Natl. Acad. Sci. USA 108: 16922–16926.

Žitňan, D. and M.E. Adams. 2012. Neuroendocrine regulation of ecdysis. pp. 253–309. *In*: L.I. Gilbert (ed.). Insect Endocrinology. Academic Press, San Diego, California, USA.

Žitňan, D., I. Šauman and F. Sehnal. 1993. Peptidergic innervation and endocrine cells of insect midgut. Arch. Insect Biochem. Physiol. 22: 113–132.

Žitňan, D., L.S. Ross, I. Žitňanova, J.L. Hermesman, S.S. Gill and M.E. Adams. 1999. Steroid induction of a peptide hormone gene leads to orchestration of a defined behavioral sequence. Neuron 23: 523–535.

6

Neuropeptide Signaling and RNA Interference

Klaus H. Hoffmann,[a],* Sandy Weidlich[b] and Franziska Wende[c]

Insect Neuropeptides

Neuropeptides represent an important class of signal molecules involved in diverse aspects of metazoan development and homeostasis (Yamanaka et al. 2008). Neuropeptides are generated in neurons of the brain and peripheral nerve system by enzymatic prohormone cleavage. Present in both cnidarians and bilaterians, they represent an ancient and widespread mode of neurohumoral communication (Jekely 2013). Neuropeptides are essential brain peptides and modulate animal physiology by affecting the activity of almost every neural circuit (Hummon et al. 2006). As signaling molecules they commonly act through G-protein coupled receptors (GPCRs) and affect the levels of cyclic nucleotides (e.g., c-AMP and c-GMP) and Ca^{2+} in target tissues (Mykles et al. 2010). The G-protein coupled receptor family is characterized by seven transmembrane domain proteins and together with the respective peptide ligands they play an important role in nerve transmission and sensory perception, locomotion, development and reproduction, metabolism, and neuromodulation (Bendena et al. 2012). Clustering of neuropeptide preprohormones and

Animal Ecology I, BayCEER, University of Bayreuth, Universitätsstraße 30, 95440 Bayreuth, Germany.
[a] Email: klaus-hoffmann@uni-bayreuth.de
[b] Email: sandy.weidlich@uni-bayreuth.de
[c] Email: franziska.wende@web.de
* Corresponding author

neuropeptide GPCRs allowed the reconstruction of the phylogenetic distribution of hormone and receptor families and revealed extensive conservation and long-term coevolution of the receptor-ligand pairs (Jekely 2013).

Within insects, neuropeptides represent the largest and most diverse single class of signal compounds and are involved in the regulation of most major physiological and behavioral processes. The first neuropeptide structures were identified through peptide purification and amino acid sequencing, but in recent years, a tremendous increase in our knowledge on neuropeptide precursor genes, processed bioactive peptides, and on their cognate receptors became known from insect genome projects (Nässel and Winther 2010, Orchard and Lange 2012). The receptor-ligand conservation also allowed ligand predictions for many uncharacterized GPCRs from non-model species (Jekely 2013). Three examples for recently discovered insect neuropeptide families and their receptors are the orcokinin B in the hemipteran, *Rhodnius prolixus* (Sterkel et al. 2012), the insect RYamide neuropeptides (Collin et al. 2011), and the ACP (AKH/ corazonin-related peptide) signaling system (Hansen et al. 2010). In *Drosophila melanogaster*, whose genome was first sequenced in 2000, about 200 GPCRs have been annotated and about 50 of these receptors were predicted to have neuropeptides as their ligands (Hauser et al. 2006). Within the *Apis mellifera* genome, more than 200 neuropeptides were inferred and the sequences of 100 peptides were confirmed (Hummon et al. 2006).

Most insect neuropeptides are pleiotropic in function and defining neuropeptide action in different insect species and systems has become a challenge. The small peptides and their receptors are promising targets for a novel generation of non-neurotoxic insecticides. Due to their short half-life, pharmacokinetics and poor bioavailability, the peptides themselves cannot be used as insect control agents, but research designed to find stable neuropeptide mimetics has increased (Scherkenbeck and Zdobinsky 2009, Xie et al. 2011). The application of molecular techniques to transform insect neuropeptide or neuropeptide receptor genes, or the knockout of specific genes, also offer promising pest control targets (Bendena 2010).

Neuropeptides Associated with Development, Reproduction and Feeding

The regulation of development, life-phase transition, and reproduction by neuropeptides is a mechanism common to multicellular animals and has been studied in great detail in insects (Gäde and Hoffmann 2005, Van Wielendaele et al. 2013). Various functions are attributed to neuropeptides that regulate the production of ecdysteroids or molting hormones, juvenile hormones, oocyte growth and yolk deposition, ecdysis and courtship behavior (see also Rogers this book).

Neuropeptides that regulate the functioning of the prothoracic glands or other ecdysteroid-producing cells and tissues either activate (ecdysiotropin or prothoracicotropic hormone, PTTH) or inhibit (ecdysiostatin or prothoracicostatic hormone) the rate of ecdysteroid biosynthesis (Gäde et al. 1997, Gäde and Hoffmann 2005). The structure of PTTH is known only from a few insect species, mainly silkworms. PTTH occurs in multiple forms, both large (11–30 kDa) and small PTTHs (4–6 kDa). A prothoracicostatic hormone was isolated from larval brains of *Bombyx mori* (Bommo-PTSP) and it turned out to be identical to a type B allatostatin (see below) isolated from *Manduca sexta* (Hua et al. 1999).

Juvenile hormone (JH) biosynthesis in the corpora allata (CA) may be controlled by stimulatory (allatotropin, AT) and inhibitory (allatostatin, AS) neuropeptides. To date only three allatotropins, compared with several allatostatins have been identified from various insect species (Hoffmann et al. 1999, Stay 2000, Gäde 2002, Weaver and Audsley 2009, Bendena and Tobe 2012). Manse-AT was first isolated from the head of pharate adults of the tobacco hornworm, *M. sexta* (Kataoka et al. 1989). Manse-AT has also been identified from other lepidopteran species (Audsley et al. 2000, Oeh et al. 2000) and is generally termed as lepidopteran AT. A structurally similar peptide was isolated from the mosquito, *Aedes aegypti* (Veenstra and Costes 1999). Besides its allatotropic activity, Manse-AT appears to be involved in the control of gut contractility, acts as a cardioactive peptide, and controls the functioning of the reproductive system (Nässel 2002). The recently characterized *Rhodnius prolixus* AT (Rhopr-AT) failed to show any myotropic activity on the gut, but stimulated contractions of muscles surrounding the salivary glands (Masood and Orchard 2014). Another peptide that is related structurally to Manse-AS has been detected in the noctuid moth, *Spodoptera frugiperda*, by Abdel-latief et al. (2004a) through gene cloning. The decapeptide RVRGNPISCF-OH strongly stimulated the synthesis of JH by the CA of adult moths, *in vitro*, and was therefore code-named Spofr-AT2.

Allatostatins are structurally diverse peptides and were originally shown to inhibit JH biosynthesis *in vitro* in a variety of insects (review in Hoffmann et al. 1999, Bendena and Tobe 2012). There are three families of peptides referred to as allatostatins, the AS type A (FGLamides) or cockroach type, the AS type B [W(X)$_6$Wamide] or cricket type, and the AS type C (Manse-AS) or lepidopteran type. In *D. melanogaster*, three genes have been identified that encode the three types of AS (Lenz et al. 2000, Williamson et al. 2001a,b) and it is suggested that members of the three peptide families are present in most, but not all, insect species. In addition to their allatoregulatory function, members of the three peptide families seem to inhibit visceral muscles and may act as neuromodulators in the central nervous system. $W(X)_{6-8}$Wamides were first identified in *Locusta migratoria*, where they inhibited muscle contraction *in vitro* and were named MIP (myoinhibiting neuropeptide) (Schoofs et al. 1991). In insect in general, MIP-containing neurons are responsible for the initiation and execution of ecdysis behavior, a form of life-transition (Schoofs and Beets 2013). Gene expression of the three types of AS preprohormones in the digestive tract and in reproductive tissues suggest further functions of the peptides. The FGLamides exert their action through galanin-like receptors, the $W(X)_6$Wamides through sex peptide-binding receptor, and the PISCF-peptides through somatostatin-like receptor (Bendena and Tobe 2012).

Other neuropeptides may act on the gonads directly, and are thus designated as gonadotropins and antigonadotropins. In contrast to vertebrates, insect gonadotropins are quite diverse in structure. The first peptidic gonadotropin isolated from insects was the ovary maturating parsin from *L. migratoria* (Locmi-OMP) (Girardie et al. 1991). Another peptide with sequence homology to Locmi-OMP was found in *Ae. aegypti* (Aedae-OEH I) (Brown et al. 1998). The only brain gonadotropin known to target the testes is the so-called testes ecdysiotropin from the gypsy moth, *Lymantria dispar* (Wagner et al. 1997, Loeb et al. 2001).

Some neuropeptides are either directly or indirectly involved in the processes of edysis and eclosion [eclosion hormone (EH), pre-ecdysis triggering hormone (PETH), ecdysis-triggering hormone (ETH), bursicon, corazonin, and crustacean cardioactive peptide (CCAP)], whereas others control courtship behavior or mating [e.g., pheromone biosynthesis activating peptide (PBAN)] (Gäde and Hoffmann 2005).

Some of the peptides associated with insect development and reproduction also regulate feeding behavior and might therefore influence the nutritional status of the insect (Van Wielendaele et al. 2013). Peptides known to control feeding and digestion in general are found within the allatoregulatory peptides (Alzugaray et al. 2013, see above), diuretic hormones, FMRFamide-related peptides (short), neuropeptide F (see Orchard and Lange 2012 this book), proctolin, saliva production stimulating peptide, kinins, and tachikinins (Spit et al. 2012). Several of these peptides have been shown to stimulate or inhibit contraction of the foregut (Audsley and Weaver 2009) while others show stimulatory or inhibitory effects on enzyme activity levels in the gut. Besides their presence in the nervous system, myoactive neuropepides have been localized in endocrine cells of the midgut.

The superfamily of FMRFamide-like peptides consists of at least three families, the FMRFamides, the HMRFamides, and the FLRFamides (Orchard et al. 2001). Different members of these peptide families strongly act on the activity of visceral (gut) muscles. They also modulate the release of digestive enzymes and influence feeding behavior (Spit et al. 2012). The HMRFamides in insects are also termed sulfakinins (SK) because of their sulfated tyrosine residue, which is essential for their biological activity (Schoofs and Nachman 2006). Sulfakinins bear structural homology to the vertebrate gastrin and cholecystokinin peptides that are involved in the control of digestion also in humans.

In summary, the regulation of reproduction and feeding in insects are very complex processes, which include positive and negative sensory feedback, the release of hormones and peptides, well coordination of the relevant processes, and their adjustment to external factors of abiotic and biotic nature (Van Wielendaele et al. 2013, Perez-Hedo et al. 2014). Because of the long evolutionary history of many insect orders, it is not always clear to what extent functional data obtained for a given species can be extrapolated to other taxa. Some of the neuropeptide effects seem to be species-dependent and further studies on their functions in other taxa will be needed before we can fully understand such physiological complex systems as reproduction and feeding (Spit et al. 2012). Peptide functions can be further explored by employing genetic tools such as knockout or overexpression of the respective genes (Boerjan et al. 2012). Such tools include RNA interference (RNAi) and are currently being investigated in diverse insect species (Bellès 2010, Scott et al. 2013, Orchard and Lange this book).

RNA Interference in Entomology

RNA interference (RNAi) is a natural cellular process by which a mRNA is targeted for degradation by a small interfering RNA that contains a strand complementary to a fragment of the target mRNA. This results in a sequence-specific inhibition of gene expression (Mito et al. 2011). RNAi is highly conserved in many organisms. Plants and insects have evolved to use RNAi as protection against viruses (Katoch and Thakur 2013). RNAi offers great opportunities for entomologists, especially to analyze gene functions in non-model insects, manage pest populations, or reduce disease pathogens (Scott et al. 2013).

The discovery of RNAi enabled the use of loss-of-function analyses in many non-model insects with still unknown genome. When a double-stranded RNA (dsRNA) with a strand complementary to a fragment of the target mRNA is either injected or orally administered to an insect, phenotypes appear, depending on the concentration of dsRNA, the type of target gene and other still unknown factors (Mito et al. 2011). RNAi can be applied in all developmental stages of an insect. When dsRNA for a target gene is injected into developing eggs, the RNAi phenotype can appear in the embryo, in the larvae and nymphs, or even in the adults. Injection into larvae can be used to create pupal and adult loss-of-function phenotypes and dsRNA injection into adults may be used to study reproductive strategies.

Bellés (2010) reviewed RNAi studies on insects covering some 30 species representing nine orders. Although, many positive results have been published, not all reported species show the same degree of sensitivity towards RNAi. In general, less derived species seem to be more sensitive to systemic RNAi. The best known of the poorly sensitive species is *D. melanogaster*, whereas *Tribolium* shows a robust systemic RNAi response. Within the Lepidoptera, RNAi was particularly successful in the family Saturniidae and genes involved in immunity (Terenius et al. 2011). Within a given species, efficiency of RNAi may also depend on the developmental stage of the insect. Other parameters affecting the success of RNAi experiments are the choice of sequence for the dsRNA preparation and the RNAi dosage administered. Even under optimal conditions, RNAi usually leads to an only partial silencing of the target gene (Boerjan et al. 2012). A full targeted inactivation of a gene can be achieved by mutagenic and homologous recombination approaches. The mechanisms underlying species or tissue-dependent differences in the sensitivity to dsRNA, are not fully understood (Mito et al. 2011). At least in some cases, differences in the RNAi efficiency may reflect differences in the dsRNA uptake mechanisms.

Functional Analysis of Neuropeptides by RNA Interference

In insects, neuropeptides can be released as neuromodulators within the central nervous system (CNS) or as neurohormones from neurohemal tissues into the hemolymph (Boerjan et al. 2012). In the initial experiments of peptidomic research as many as possible neuropeptides from entire CNS of various insect species were extracted and identified (Schoofs and Baggerman 2003). With increasing sensitivity of mass-spectrometric techniques, the presence of these neuropeptides could be associated with different neurohemal organs (Predel et al. 2004, Clynen and Schoofs 2009). Such neuropeptide profiling is well suited to study peptide functions in a developmental context. Today, bioinformatic and molecular tools allow us to predict neuropeptide sequences within their precursors and to study differential neuropeptide expression in defined parts of the nervous system and other tissues. When neuropeptide precursors are identified by transcriptomic, proteomic or peptidomic profiling, the physiological roles of the peptides can be unraveled by a reverse genetic approach such as the RNA interference. However, only a few examples exist for RNAi inhibition of neuropeptide genes (Bendena 2010).

In *Drosophila*, dsRNA for CCAP (crustacean cardioactive peptide) was targeted to CCAP neurons and reduced the expression of the peptide, but the reduction was not sufficient to compromise the CCAP effects on ecdysis or heartbeat (Dulcis et al. 2005). Huang et al. (2007) injected double-stranded bursicon RNA into *B. mori* pupae to test RNA interference. Bursicon is a developmental hormone responsible for cuticula tanning and wing expansion. RNAi significantly reduced the bursicon mRNA in pupae, and a defect in wing expansion was observed in adults. To understand the roles of bursicon, CCAP, eclosion hormone (EH), and ecdysis triggering hormone (ETH) and their receptors in the red flour beetle, *T. castaneum*, Arakane et al. (2008) performed systemic RNAi experiments utilizing post-embryonic injections of dsRNA targeted to the respective neuropeptide or its receptor gene. RNAi of *eh* or *eth* disrupted preecdysis behavior and prevented subsequent ecdysis behavior, *ccap* RNAi interrupted ecdysis behavior, and *bursicon* knockdown resulted in incomplete wing expansion. RNAi of genes encoding the receptors for the peptides produced phenotypes comparable to those of their respective cognate neuropeptides. Another neuropeptide linked to mating behavior and reproductive function is natalisin, which had recently been identified in three to four neurons of three insect species, *D. melanogaster, T. castaneum,* and *B. mori* (Jiang et al. 2013). Natalisin RNAi induced significant defects in the mating behavior of both females and males and led to reduced fecundity. Knockdown of neuroparsins in the locust *Schistocerca gregaria* affected vitellogenin transcript levels and oocyte growth (Badisco et al. 2011).

A wide range of behavior including olfactory learning is controlled by peptides from the mushroom bodies (MB) of the brain. Several potential neuromodulators of MB neurons, the so-called Kenyon cells, have been anatomically identified, but their function remained unclear. Knapek et al. (2013) showed by RNAi experiments that a neuropeptide precursor gene encoding four types of short neuropeptide F (SNPF) are required in the Kenyon cells for appetite olfactory memory.

A broad range of physiological functions for another pleiotropic neuropeptide, the pheromone biosynthesis activating neuropeptide (PBAN), was evaluated by RNA interference (Choi et al. 2012). In various moth species, PBAN acts on sex pheromone production in the pheromone gland, whereas in other species PBAN knockdown affected pupal development, adult mortality, and embryonic diapause. Pheromone biosynthesis in moths can be inhibited by RNA knockdown of the *PBAN precursor* gene as well as the *PBAN receptor* gene (Lee et al. 2011).

In summary, RNAi technology applied to neuropeptide hormones presents a promising strategy for finding unknown physiological function(s) of the peptides, but also for the development of alternative insect pest control tools (Scherkenbeck and Zdobinsky 2009, Bendena 2010, Choi et al. 2012). In the following sections, we present examples of our recent research on the evaluation of neuropeptide functions (allatostatins, allatotropins, sulfakinins) associated with development, reproduction, and feeding by using RNA interference.

Case Studies

The Regulation of Juvenile Hormone Biosynthesis by Neuropeptides during Larval Development and Reproduction in the Fall Armyworm, Spodoptera frugiperda

The fall armyworm, *S. frugiperda*, is an important agricultural pest species. In the polyandric moth, oogenesis is completed only after adult emergence and stimulation of egg production as well as oviposition is strictly dependent on mating proceeding with an elevated JH titer (Ramaswamy et al. 1997). Lepidopterans contain three homologs of JH in their hemolymph, JH I, JH II, and JH III, and the synthesis of JH in the CA is regulated mainly by the "classical" lepidopteran allatoregulatory peptides, Manse-AT and Manse-AS (C-type allatostatin) (Stay 2000). The existence of Manse/Spofr-AT and Manse/Spofr-AS in *S. frugiperda* was verified by cloning the cDNA, which encodes the precursors of the two peptides (Abdel-latief et al. 2003). Besides Spofr-AT and Spofr-AS, a gene encoding *Spofr-AS A-type* was isolated from *S. frugiperda*, which is expressed in the brain and midgut of larvae and adults in a time- and tissue-specific manner, but also in the reproductive tissues of both sexes (Abdel-latief et al. 2004b). Another cDNA encoding a hitherto unknown *S. frugiperda* allatoregulating peptide was isolated and cloned by Abdel-latief et al. (2004a) and code-named Spofr-AT2. Spofr-AT2 is expressed in the brain, midgut, and ovary in a tissue- and developmental-specific manner. Functional analysis of the *Spofr-AS C-type* gene product by RNAi in *S. frugiperda* larvae revealed clear allatostatic activity for this peptide (Griebler et al. 2008). Injection of *Spofr-AS C-type* dsRNA into one day old penultimate instar larvae (L5/1) resulted in an elevated JH titer (mainly JH III) and the duration of the last larval stage (L6) was prolonged by 24 hours. Injection of *Spofr-AT* dsRNA into young penultimate instar larvae did not affect the JH titer of the caterpillars (Hassanien et al. 2014a). However, a single injection of *Spofr-AT* dsRNA into freshly molted last instar larvae (L6/1) resulted in an acceleration of the prepupal commitment and the transformation into the pupa by about 24 hours. This earlier prepupal commitment was accompanied by a significant increase in the amount of free ecdysteroids (ecdysone, 20-hydroxyecdysone) in the hemolymph of last instar larvae shortly before pupation. Here, Spofr-AT seems to act as an ecdysiostatin, but not as an allatotropin. Injection of *Spofr-AS A-type* dsRNA into freshly molted L5 larvae resulted in a drastic increase in the JH titer 48 hours later, mainly due to an increase of JH I and JH III (Meyering-Vos et al. 2006). This indicates that Spofr-AS A-type peptides act as true allatostatins in *S. frugiperda* L5 larvae. However, in spite of the high JH titer in the hemolymph of the larvae following *Spofr-AS A-type* RNAi, no additional larval molting or "superlarvae" were observed. Injection of *Spofr-AT2* dsRNA into penultimate instar larvae also resulted in an increased JH titer in last instar caterpillars and the L6 stage was prolonged. This suggests an allatostatic activity of that peptide in the larva (Griebler et al. 2008). Concentrations of JH homologs and free ecdysteroids in the hemolymph and tissues (see below) of animals were quantified by the LC-MS method as described earlier (Westerlund and Hoffmann 2004).

In adult females of *S. frugiperda*, we observed that mated animals show much higher JH titers in their hemolymph (average of 90 pg/µL and maxima of 130 pg/µL total JH) than virgins (average of 40 pg/µL and maxima of 75 pg/µL total JH) and that they lay almost twice the number of eggs than unmated animals (1200 eggs per female in mated animals compared to 600 eggs per female in virgins) (Hassanien et al. 2014a). Moreover, in contrast to virgins the mated females contained considerable amounts of JH I and JH II in their hemolymph, which are thought to be received from the males during copulation. To confirm this transfer of JH from the male accessory reproductive gland (AG) to the female bursa copulatrix (BC) during mating, we measured the amount of JH homologs in the male AG before mating and in the BC and the hemolymph of the females after mating. Then we used the RNAi technique to evaluate the role of allatoregulating neuropeptides for JH biosynthesis in the AG and its transfer to the female (Hassanien et al. 2014b).

The AG of virgin males contained high amounts of JH I and JH II, whereas JH III remained low (Fig. 1). This confirms earlier suggestions in the literature that male AG of some Lepidoptera (Park et al. 1998), Diptera (Borovsky et al. 1994), and Coleoptera (Tian et al. 2010) synthesize and store JH. Mating resulted in a drastic decrease of JH I and JH II in the male AG. The BC of newly emerged females contained only traces of JH, but an increase in JH I and JH II was observed a few hours after mating (Fig. 2). One day after mating, JH I and JH II disappeared from the BC and the hormones could then be found in the hemolymph of the females (Hassanien et al. 2014a).

Injection of dsRNA targeted against the leptidopteran AS (Spofr-AS C-type) into freshly emerged males led to a further accumulation of JH, especially JH I and JH II, in the AG compared to Ringer-injected controls (Table 1), indicating an allatostatic activity of Spofr-AS C-type on the biosynthesis of JH in the male AG. Accordingly, mating of such males

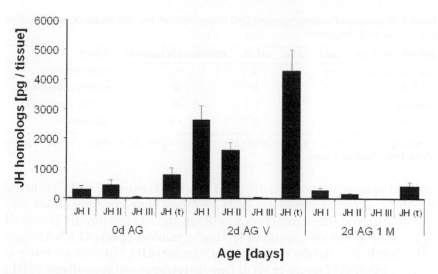

Figure 1. Amounts of JH homologs [JH I, JH II, JH III, total (t) JH] in the male accessory reproductive glands (AG) of unmated (virgin, V) and mated (M) males at days 0 and 2 of adult life. 2d AG 1 M = AG from 2 day old males mated once at day 1 after emergence. Means ± SEM; n = 12. After Hassanien (2013).

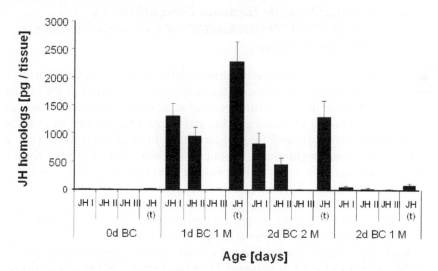

Figure 2. Amount of JH homologs [JH I, JH II, JH III, total (t) JH] in the bursa copulatrix (BC) of virgin (day 0) and mated (M) females. 1d BC 1 M = BC from 1 day old mated female; 2d BC 2 M = BC from 2 day old female mated twice on day 1 and day 2 of adult life; 2d BC 1 M = BC from 2 day old female mated once on day 1 of adult life. Means ± SEM; n = 12. After Hassanien (2013).

Table 1. Regulation of juvenile hormone (JH) biosynthesis in the male accessory reproductive glands (AG) of *S. frugiperda*.

dsRNA Injection	JH I	JH II	JH III	Male/female transfer	Effect
Spofr-AS C-type	++	+	0	++	allatostatic
Spofr-AS A-type	++	0	0	0	slightly allatostatic
Spofr-AT	–	0	0	0	slightly allatotropic
Spofr-AT2	– –	–	0	++	allatotropic

++ strong positive effect, + weak positive effect, – – strong negative effect, – weak negative effect, 0 no effect observed.

resulted in an enhanced transfer of JH I and JH II from the male AG to the female BC. Knockdown of the *Spofr-AS A-type* gene in young males only slightly affected the accumulation of JH in the AG (mainly JH I) as well as the male to female transfer of JH during mating. Spofr-AT RNAi again only slightly affected the synthesis and transfer of JH, but in an allatotropic manner. Spofr-AT2 seems to act as true allatotropin on the synthesis of JH I and JH II in the AG of young males as well as its transfer to the female BC during copulation (Hassanien et al. 2014b).

The Regulation of Juvenile Hormone Biosynthesis by Neuropeptides in Caste Differentiation of a Primitive Termite, Mastotermes darwiniensis

Termites are eusocial insects with differentiated castes. They exhibit two developmental systems, the linear pathway system as found in the family Kalotermitidae and in some Rhinotermitidae, and the bifurcated system as found in the families Mastotermitidae and Termitidae as well as in some Rhinotermitidae (Korb and Hartfelder 2008). Juvenile hormones play a central role in caste regulation (Nijhout 1994). Several studies have shown that JH synthesis is elevated during soldier development (Elliott and Stay 2008). The synthesis of JH in the CA is regulated by allatostatins of the FGLamide family (A-type), as in cockroaches and crickets (Yagi et al. 2005, 2008). The gene for the *AS A-type* preprohormone has been identified in one termite species, *Reticulitermes flavipes* (Elliott et al. 2009).

Recently, we have identified the gene for the prepro-AS peptides in the only genus of the basal termite family Mastotermitidae, *M. darwiniensis* (HQ110056.1, Wende 2012). *M. darwiniensis* shows a bifurcated development with an apterous line (sterile line), in which soldiers develop from first instar larvae through five larval stages, and only very rarely neotenics occur, and the nymphal line (fertile line), in which the alates develop through eleven nymphal stages. The *AS A-type* gene of *M. darwiniensis* encodes a prohormone precursor with 14 putative peptides.

The precursor is similar in structure to those in *R. flavipes*, cockroaches and crickets, but shows highest homology with the preprohormone sequence from a primitive, wood-feeding cockroach species, *Cryptocercus darwini*, which had been identified in the same study (Wende 2012). The *AS A-type* gene is predominantly expressed in the brain of *M. darwiniensis* and to less extent in the midgut of the termites. The expression rate in the brain, as measured by quantitative real-time PCR (qPCR), differed in different developmental stages and was generally higher in "adult" stages than in juvenile stages (Fig. 3).

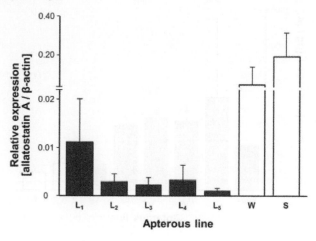

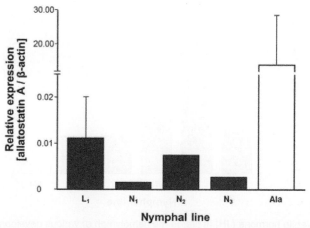

Figure 3. Relative expression of *AS A-type* gene in the brain of various developmental stages of the apterous (above) and the nymphal line (below) of *M. darwiniensis*. Expression was measured by qPCR and results were normalized to β-actin mRNA. Means + SD; n = 1–6 with 11 to 60 individuals per sample. No SD is shown when only one sample was available. Ala = alates, L = larval stage, N = nymphal stage, S = soldier, W = workers. After Wende (2012).

In most developmental stages, we were able to measure the JH III concentration in the hemolymph by LC-MS (Fig. 4). Termites contain only JH III in the hemolymph. JH III titers were highest in young larvae (L_1). Larval stages (L_2 to L_4) had generally higher JH titers than nymphs, which indicates their potential to develop into soldiers. During the development of reproductives, JH III content was low. This is true for the development of primary reproductives from nymphs (alates) as well as for the development of secondary reproductives from workers (neotenics).

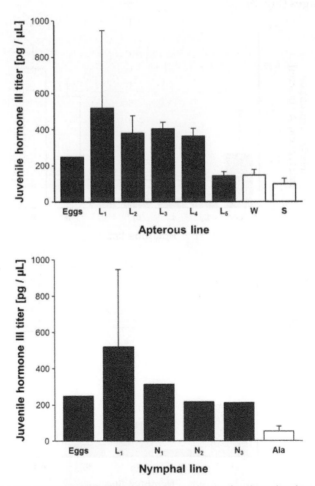

Figure 4. Juvenile hormone (JH) III titer in the hemolymph of various developmental stages of the apterous (above) and the nymphal line (below) of *M. darwiniensis* as measured by LC-MS. For eggs and first instar larvae whole body extracts were used. Means + SD; n = 1–8. No SD is shown when only one sample was available. Abbreviations are as in Fig. 3. Neotenics, which appeared twice in our colonies, had only low JH III titers (about 50 pg/µL) in the hemolymph. After Wende (2012).

The two parameters, JH III titer in the hemolymph of the animals and allatostatin A-type expression in the brain, showed a strong negative correlation, which means that development stages with high JH III titer had a low AS A-type expression and vice versa (Fig. 5). This indicates a clear allatostatic activity for the A-type allatostatins in *M. darwiniensis*.

RNA interference was used to further analyze the function of the *AS A-type* gene in the largest caste by number of individuals, the workers. Injection of 1 μg dsRNA targeted against A-type AS into workers of both sexes resulted in a minor reduction of expression in the brain (21%), but a drastic reduction of gene expression in the midgut (87%) three days after the treatment (Fig. 6).

At that time, the JH III titer in the hemolymph of such treated workers had changed only slightly, whereas the concentration of 20-hydroxyecdysone showed a significant increase (Fig. 7). These results indicate that in workers of *M. darwiniensis* A-type AS clearly acts as a brain-gut peptide and it may show an ecdysiostatic instead of an allatostatic activity.

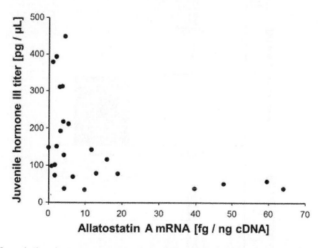

Figure 5. Correlation between expression of the *AS A-type* gene in the brain and the JH III titer in the hemolymph of the same animals of various developmental stages of *M. darwiniensis*. Spearman rang correlation; r_s = −0.539, P = 0.005; n = 25. After Wende (2012).

Functional Activity of Allatoregulating Neuropeptides in the Pupal Stage of the Red Flour Beetle, Tribolium castaneum

In holometabolous insects larvae differ in form and life style from the adults and an intermediate pupal stage occurs. JH reappears during the larval-pupal transformation and prevents precocious differentiation

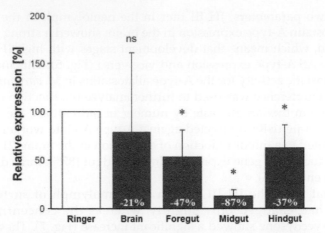

Figure 6. Relative expression of the *AS A-type* gene in various tissues of *M. darwiniensis* workers three days after injection of 1 μg *AS A-type* dsRNA. Ringer controls were injected with 2 μL termite Ringer. Means + SD; n = 3 with 12 to 16 individuals per sample; ns = not significantly different from the Ringer control, * = P < 0.05. After Wende (2012).

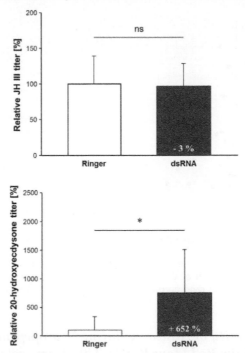

Figure 7. Effects of AS A-type RNAi on the concentrations of juvenile hormone (JH) III (above) and 20-hydroxyecdysone (20E) (below) in the hemolymph of the dsRNA treated workers. Means + SD; n = 8–14; ns = not significantly different from the Ringer control (students t-test: T = 0.241, df = 24, P = 0.81), * = P < 0.05 (Mann-Whitney-U-test: U = 16, Z = –2.341, P = 0.019). After Wende (2012).

of adult structures (Jindra et al. 2013). Adult development is caused by ecdysteroids in the absence of JH during the pupal stage (Hiruma and Kaneko 2013). The recent proteomic and genomic studies on the red flour beetle, *T. castaneum*, have identified a single AT-like peptide, an AS type C peptide, an AS CC (Veenstra 2009), and six AS type B peptides and their respective precursors (Amare and Sweedler 2007, Li et al. 2008, Tribolium Genome Sequencing Consortium 2008), but no AS type A preprohormone. Functional studies have shown that allatotropin and allatostatins act as brain-gut peptides (allatoregulatory and myoregulatory properties) in larvae and adults of *T. castaneum* (Abdel-latief and Hoffmann 2010, Vuerinckx et al. 2011, Audsley et al. 2013). Whether the allatoregulatory neuropeptides function during the pupal stage, where JH is presumably absent and the animals do not feed, has not yet been studied. To provide further insight into the presence and activity of allatoregulatory peptides in the pupa of *T. castaneum*, we cloned the cDNAs of the suggested allatotropin (Trica-AT) and allatostatins (Trica-AS B- and Trica-AS C-type) and analyzed the developmental expression patterns during the pupal stage. Furthermore, we knocked down the peptide precursors by RNAi at the beginning of the pupal stage and followed the adult development, the phenotypes of the emerging adults, and the fecundity of the adult females (Abdel-latief and Hoffmann 2014).

Cloning of the cDNA and gene structure analyzes of the *Trica-AT* gene confirmed that the gene is expressed in three mRNA isoforms, AT1 to AT3 (Fig. 8).

Real time PCR data demonstrated that the three *Trica-AT* isoforms and the *AS* genes, *Trica-AS C-type* and *Trica-AS B-type*, are expressed in discerning developmental and tissue-specific patterns with significant expression rates also in brain and gut of the pupa (Abdel-latief and Hoffmann 2014). Single injections of 500 ng dsRNA targeted against the respective preprohormones into the body cavity of young pupae (4 h old) silenced the respective transcripts specifically by more than 90% and resulted in abnormal adult phenotypes, whereby about half of the animals looked relatively normal but the females laid only a low number of eggs (P1 phenotype) (Figs. 9, 10). The other half (P2 phenotypes) exhibited strong developmental defects with abnormal long duration of the pupal stage, abnormal head and body sizes, short elytra, and incomplete sclerotization. These females laid no eggs and died within one week after emergence. *Trica-AT* dsRNA treated animals had a significantly shorter pupal stage than the controls (4.8 vs. 6.9 days) and the emerging adults heads were smaller (0.17 vs. 0.26 mm³; LxWxH) as well as the body (1.8 vs. 2.6 mm³; LxWxH). Injection of *Trica-AS C-type* dsRNA resulted in a slightly prolonged pupal stage (7.3 vs. 6.9 days) and the heads and bodies of the emerging adults were also smaller compared with the controls. In contrast,

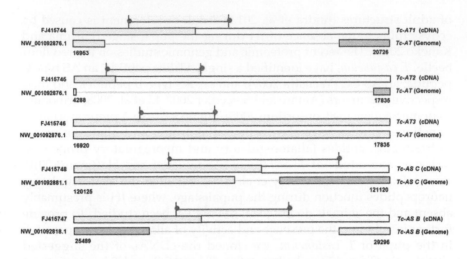

Figure 8. Gene structure analysis of the *T. castaneum* allatoregulatory neuropeptides cDNAs (Trica-AT1, -AT2, and -AT3, Trica-AS C-type and Trica-AS B-type). The analyses were performed using Splign server http://www.ncbi.nlm.nih.gov/sutils/splign/splign.cgi. In the graphs, each compartment is plotted separately with the cDNA sequence at the top, and the genome sequence at the bottom. Spaces between the segments in the genomic sequence indicate consensus-spliced introns.

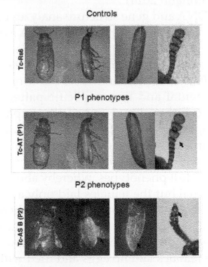

Figure 9. Representative adult phenotypes obtained from 4 h old *T. castaneum* pupae injected with 500 ng of dsRNA *for Trica-AT* and *Trica-AS B-type*. Controls were injected with dsRNA for *Trica-Rs6* (*T. castaneum* ribosomal protein 6). Ventral and lateral view onto the whole body of the adult beetles (left). Elytra and antenna views (right). Arrows indicate the sites of phenotype characters in knockdown animals which are different from the controls. For more data see Abdel-latief and Hoffmann (2014). With permission from Elsevier.

Color image of this figure appears in the color plate section at the end of the book.

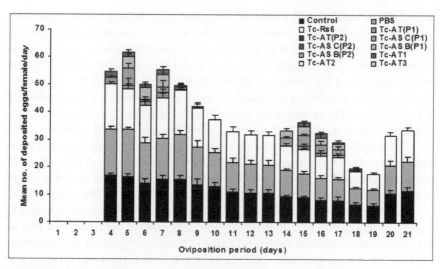

Figure 10. Effects of dsRNA injections for *Trica-AT, Trica-AS C-type*, and *Trica-AS B-type* as well as the *Trica-AT1, -AT2*, and *-AT3* mRNA isoforms on the egg deposition rates of *T. castaneum* adult females. The dsRNAs (0.5 µg/insect) were injected into 4 h old pupae (n = 20; two biological pools). As controls the pupae were either untreated (control) or injected with PBS buffer or dsRNA for *Tc-Rs 6* (n = 20 each). The data are presented as means + SE. After Abdel-latief and Hoffmann (2014).

Trica-AS B-type treated animals had a significantly longer pupal stage (7.8 vs. 6.9 days) than controls and the adult beetles showed a much smaller body (1.5 vs. 2.6 mm³; LxWxH) but somewhat larger head size (2.8 vs. 2.6 mm³; LxWxH).

Our results clearly indicate a significant role of allatotropin and allatostatins in the pupal development. Injection of the respective dsRNAs into freshly molted pupae caused specific adult loss-of-function. A normal metamorphosis seems to require the presence of allatoregulating neuropeptides within the pupa. Moreover, they are *sine qua non* for proper ovarian maturation and egg laying in adults. The distinct mechanisms of action, however, remain to be determined. We were unable to measure JH titers in the pupa, which would be necessary to confirm allatoregulatory properties of the peptides.

The Regulation of Food Intake and Digestion by Neuropeptides in the Cricket, Gryllus bimaculatus

The function of the insect digestive system is largely coordinated by neurohormones (Audsley and Weaver 2009, Spit et al. 2012) acting through the stomatogastric nervous system. Many regulatory peptides are also produced by endocrine cells located in the gut epithelium (brain-gut peptides). Several of the insect gut regulating peptides have vertebrate homologs, but less is known about the specific functions and modes of action of the insect gut peptides, compared to those in vertebrates. Unraveling the peptidergic control of insect feeding and digestion was greatly advanced by genomic approaches during the last two decades. Many peptides exhibited myoactivity on the visceral muscles, especially of the foregut. These peptides include proctolin, the insect kinins (e.g., myokinins, leucokinins, tachykinins), FMRFamide-related peptides, and the allatoregulating peptides. Since one of the first consequences of food intake is to stretch the foregut followed by contraction and relaxation of the gut muscles, it is suggested that the control of the gut motility plays an important role in the regulation of feeding (Wei et al. 2000). Some of the peptides also regulate feeding behavior and stimulate or inhibit enzyme activity levels in the gut (Fusé et al. 1999, Aguilar et al. 2003). Studies in our laboratory have demonstrated that in the fall armyworm, *S. frugiperda*, a Spofr-AS A-type peptide inhibits amylase and trypsin activity in the caeca, whereas Spofr-AT stimulates amylase and trypsin release *in vitro*. The Spofr-AS A-type peptide also inhibited ileum myoactivity and Spofr-AT stimulated myoactivity (Lwalaba et al. 2009). In the Mediterranean field cricket, *G. bimaculatus*, Grybi-AS A peptide treatment resulted in an elevated amylase activity/release in the caeca of 2 day old fed females (Woodring et al. 2009), but the effects were weaker than for trypsin, lipase, cellulase, and aminopeptidase (Fig. 11). Sulfakinins, which bear structural and functional homology to the vertebrate gastrin/cholecystokinin

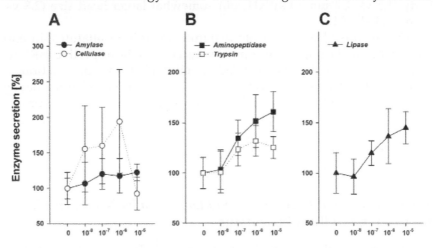

Figure 11. *In vitro* effects of type-A allatostatin 5 (Grybi-AS 5) on carbohydrase (A), protease (B) and lipase (C) secretion from caecal tissue of 2 day old adult *G. bimaculatus* females. Means ± SE; n = 10. After Weidlich (2013).

peptides, regulate digestion, but also influence feeding behavior and food intake (Baldwin 2010). Some of the peptides exert their action at multiple levels, possibly having biological functions depending on their site of delivery (Spit et al. 2012).

The aim of our studies using RNA interference was to gain more insight into the putative role of A-type allatostatins controlling the synthesis of digestive enzymes in the midgut of *G. bimaculatus* (Weidlich 2013). Two µg *Grybi-AS A* dsRNA in 10 µL cricket Ringer were injected into either freshly molted last instar larvae or newly emerged adult females and the activity of digestive enzymes was measured 48 hours after treatment in caecal tissue homogenates. The enzyme activity in the tissue homogenate represents the amount of enzymes stored in the tissue cells and, therefore, represents an indication of enzyme synthesis rate. Controls were injected with 10 µL cricket Ringer. Gene silencing of *Grybi-AS A-type* reduced the synthesis of trypsin, lipase, and amylase in the caecal tissue of female and male last instar larvae and of adult crickets (Fig. 12).

The results confirm an activating effect of A-type allatostatins on the synthesis/activity of digestive enzymes, also *in vivo*. Crickets, however, seem to compensate the reduced rate of enzyme synthesis in the absence of an allatostatin by an increase in the amount of enzyme released from the caecal tissue into the gut lumen, especially in female last instar larvae (Fig. 13).

Sulfakinins (SK) are known to inhibit food intake in various insects (Wei et al. 2000, Maestro et al. 2001, Downer et al. 2007, Yu et al. 2013a). Moreover, SK stimulated the release of amylase in the beetle *Rhynchophorus ferrugineus* (Nachman et al. 1997) and the moth *Opisina arenosella* (Harshini et al. 2002). Recently, we have identified the structure of the SK cDNA sequence of *G. bimaculatus* (Meyering-Vos and Müller 2007a) and could then use RNAi to study the effects of SK gene silencing on food intake and food transport through the gut of the cricket (Meyering-Vos and Müller 2007b). In contrast to other insects, where SK expression was found in various tissues (Yu et al. 2013), SK expression in *G. bimaculatus* was restricted to the brain (Meyering-Vos and Müller 2007a). Injection of *SK* dsRNA (2 to 10 µg) into the abdomen of female adults and last instar larvae led to a systemic, specific, and dose-dependent silencing of the *SK* gene, as was shown by RT-PCR (Meyering-Vos and Müller 2007b). Treatment of young female adults by *SK* dsRNA injection (2 µg) led to a significant stimulation of the food intake up to 110% of controls within a feeding period of 80 min. Feeding of the *SK* dsRNA solution to freshly ecdysed adults also induced a significant higher food intake by the animals (Fig. 14).

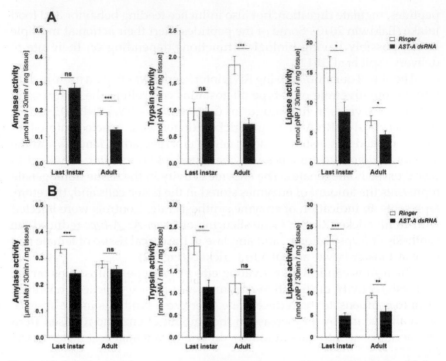

Figure 12. Activity of amylase, trypsin and lipase in the caecal tissue homogenate of 2 day old female (A) and male (B) *G. bimaculatus* last instar larvae and adults injected with either 2 μg *Grybi-AS A-type* dsRNA in 10 μL Ringer (black) or 10 μL Ringer (white) (control) at the preceding molt. Means ± SE; n = 9–10; * P<0.05, ** P<0.01, *** P<0.001, ns = not significantly different. After Weidlich (2013).

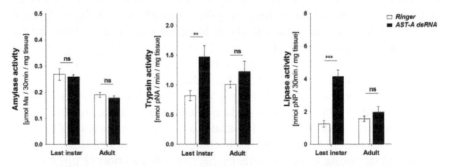

Figure 13. Activity of amylase, trypsin and lipase in the caecal tissue incubation medium of 2 day old female *G. bimaculatus* last instar larvae and adults injected with either 2 μg *Grybi-AS A-type* dsRNA in 10 μL Ringer or 10 μL Ringer (control) at the preceding molt. Means ± SE; n = 9–10; ** P<0.01, *** P<0.001, ns = not significantly different. After Weidlich (2013).

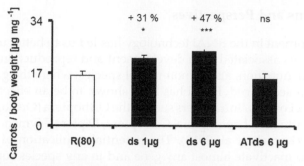

Figure 14. Effect of feeding *SK* dsRNA to adult crickets on their food intake. 1 day old adult females that were starved directly after emergence were fed either with 10 μL cricket Ringer (R) or 1 and 6 μg *SK* dsRNA (ds) or 6 μg control *allatotropin* dsRNA (ATds) from *S. frugiperda*. Animals were then starved for 22 h and afterwards fed with carrot powder for 80 min. Means + SEM; n = 12–16. Percentages of stimulation are indicated. * P<0.05, *** P<0.005, ns = not significantly different from control. From Meyering-Vos and Müller (2007b).

Assuming that gene silencing is followed by a depletion of the SK peptide in tissues and/or hemolymph, our results confirm an inhibitory role of the native SK peptide on food intake of crickets. Similar results were recently found by Yu et al. (2013a), where injection of *SK* dsRNA stimulated the food intake of *T. castaneum* larvae, and, consequently, injection of a synthetic SK analog strongly reduced food intake.

Our further studies on effects of *SK* gene silencing on the activity of digestive enzymes in the gut of *G. bimaculatus* did not exhibit a stimulation of amylase release, as it was found in *R. ferrugineus*, but indicate that SK peptides inhibit carbohydrate digestion, at least in females (Fig. 15).

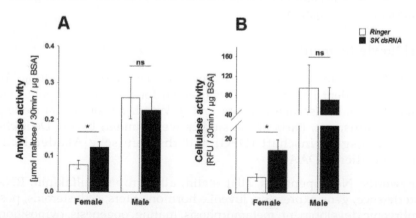

Figure 15. Effect of Ringer and *sulfakinin* (*SK*) dsRNA injection on the secretion of amylase (A) and cellulase (B) into the caeca of 2 day old adult females and males of *G. bimaculatus*. 2 μg *Grybi-SK* dsRNA in 10 μL Ringer or 10 μL Ringer (control) were injected into crickets at the day of imaginal molt. Means ± SE; n = 9–10; * P<0.05, ns = not significantly different. After Weidlich (2013).

Conclusions and Perspectives

The development in the RNAi technology has led us to better understand gene functions associated with development and reproduction of insects, or with their nutrition, also in non-model species, where the genome has not yet been sequenced. RNAi has been shown to be an effective tool in various insect orders/infraorders such as the Orthoptera (Orthopteroidea), Isoptera (Blattodea), Lepidoptera, Coleoptera, Diptera, Hymenoptera, and Hemiptera (Katoch et al. 2013). The potential application of the RNAi technique to inactivate almost any gene and in any species could lead to comparative studies of gene functions in many insect groups providing insight into evolutionary processes that have modeled these functions (Bellés 2010).

RNAi also has enormous potential for applied entomology (Price and Gatehouse 2008, Xue et al. 2012, Katoch et al. 2013). RNAi can be used for insect pest control, for example, by suppressing essential genes leading to reduced fertility, fitness, and mortality (Scott 2013). Neuropeptides represent such a group of biomolecules essential for life. In contrast to chemical insecticides, which also kill non-targeted animals, RNAi offers the possibility to kill only the target species by down regulating specific and essential genes (Mito et al. 2011). From the different methods of dsRNA application used so far, feeding seems to be most promising because it represents the natural route of delivery (Yu et al. 2013b). Recently, a direct delivery of dsRNA through spray has been demonstrated for the Asian corn borer, *Ostrinia furnacalis* (Wang et al. 2011). The results show that at least in a lepidopteran insect, dsRNA can penetrate the integument without loss of function.

Acknowledgements

We would like to thank our former coworker Dr. Martina Meyering-Vos and PhD student Dr. Intisar T.E. Hassanien for providing unpublished data. We would also like to thank Prof. J. Woodring for critical reading of the manuscript. Parts of the studies were funded by the Deutsche Forschungsgemeinschaft (DFG) and the Deutsche Akademische Austauschdienst (DAAD).

Keywords: Neuropeptides, allatostatin, allatotropin, sulfakinin, RNA interference, gene expression, juvenile hormone titers, ecdysteroids, post embryonic development, metamorphosis, mating, oogenesis, oviposition, food uptake, digestion, caste differentiation

References

Abdel-latief, M. and K.H. Hoffmann. 2010. Neuropeptide regulators of the juvenile hormone biosynthesis (*in vitro*) in the beetle, *Tenebrio molitor* (Coleoptera: Tenebrionidae). Arch. Insect Biochem. Physiol. 74: 135–146.

Abdel-latief, M. and K.H. Hoffmann. 2014. Functional activity of allatotropin and allatostatin in the pupal stage of a holometabolous insect, *Tribolium castaneum* (Coleoptera, Tenebrionidae). Peptides 53: 172–184.

Abdel-latief, M., M. Meyering-Vos and K.H. Hoffmann. 2003. Molecular characterization of cDNAs from the fall armyworm *Spodoptera frugiperda* encoding *Manduca sexta* allatotropin and allatostatin preprohormone peptides. Insect Biochem. Mol. Biol. 33: 467–476.

Abdel-latief, M., M. Meyering-Vos and K.H. Hofffmann. 2004a. Characterization of a novel peptide with allatotropic activity in the fall armyworm *Spodoptera frugiperda*. Reg. Pept. 122: 69–78.

Abdel-latief, M., M. Meyering-Vos and K.H. Hoffmann. 2004b. Type A allatostatins from the fall armyworm, *Spodoptera frugiperda*: Molecular cloning, expression and tissue-specific localization. Arch. Insect Biochem. Physiol. 55: 120–132.

Aguilar, R., J.L. Maestro, L. Vilaplana, N. Pascual, M.D. Piulachs and X. Bellés. 2003. Allatostatin gene expression in brain and midgut and activity of synthetic allatostatins on feeding-related processes in the cockroach *Blattella germanica*. Reg. Pept. 115: 171–177.

Alzugaray, M.E., M.L. Adami, L.A. Diambra, S. Hernandez-Martinez, C. Damborenea, F.G. Noriega and J.R. Ronderos. 2013. Allatotropin: an ancestral myotropic neuropeptide involved in feeding. PloS One 8: e77520.

Amare, A. and J.V. Sweedler. 2007. Neuropeptide precursors in *Tribolium castaneum*. Peptides 28: 1282–1291.

Arakane, Y., B. Li, S. Muthukrishnan, R.W. Beeman. J.J. Kramer and Y. Park. 2008. Functional analysis of four neuropeptides, EH, ETH, CCAP and bursicon, and their receptors in adult ecdysis behavior of the red flour beetle, *Tribolium castaneum*. Mech. Develop. 125: 984–995.

Audsley, N., H. Duve, A. Thorpe and R.J. Weaver. 2000. Morphological and physiological comparisons of two types of allatostatin in the brain and retrocerebral complex of the tomato moth, *Lacanobia oleracea* (Lepidoptera: Noctuidae). J. Comp. Neurol. 424: 37–46.

Audsley, N. and R.J. Weaver. 2009. Neuropeptides associated with the regulation of feeding in insects. Gen. Comp. Endocrinol. 162: 93–104.

Audsley, N., H.P. Vandermissen, R. Weaver, P. Dani, J. Matthews, R. Down, K. Vuerinckx, Y.-J. Kim and J. Vanden Broeck. 2013. Characterization and tissue distribution of the PISCF allatostatin receptor in the red flour beetle, *Tribolium castaneum*. Insect Biochem. Mol. Biol. 43: 65–74.

Badisco, L., E. Marchal, P. Van Wielendaele, H. Verlinden, R. Vleugels and J. Vanden Broeck. 2011. RNA interference of insulin-related peptide and neuroparsins affects vitellogenesis in the desert locust *Schistocerca gregaria*. Peptides 32: 573–580.

Baldwin, G.S., O. Patel and A. Shulkes. 2010. Evolution of gastrointestinal hormones: the cholecystokinin/gastrin family. Curr. Opin. Endocrinol. Diabetes Obes. 17: 77–88.

Bellès, X. 2010. Beyond *Drosophila*: RNAi *in vivo* and functional genomics in insects. Annu. Rev. Entomol. 55: 111–128.

Bendena, W.G. 2010. Neuropeptide physiology in insects. pp. 166–191. *In*: T.G. Geary and A.G. Maule (eds.). Neuropeptide Systems as Targets for Parasite and Pest Control. Landes Bioscience and Springer Science+ Business Media, LLC, New York.

Bendena, W.G. and S.S. Tobe. 2012. Families of allatoregulator sequences: a 2011 perspective. Can. J. Zool. 90: 521–544.

Bendena, W.G., J. Campell, L. Zara, S.S. Tobe and I.D. Chin-sang. 2012. Select neuropeptides and their G-protein coupled receptors in *Caenorhabditis elegans* and *Drosophila melanogaster*. Front. Endocrinol. 3: article 93.

Boerjan, B., D. Cardoen, R. Verdonck, J. Caers and L. Schoofs. 2012. Insect omics research coming of age. Can. J. Zool. 90: 440–455.

Borovsky, D., D.A. Carlson, R. Hancock, G.H. Rembold and E. van Handel. 1994. *De novo* biosynthesis of juvenile hormone III and I by the accessory glands of the male mosquito. Insect Biochem. Mol. Biol. 24: 437–444.

Brown, M.R., R. Graf, K.M. Swiderek, D. Fendley, T.H. Stracker, D.E. Champagne and A.O. Lea. 1998. Identification of a steroidogenic neurohormone in female mosquitoes. J. Biol. Chem. 273: 3967–3971.

Choi, M.-Y., R.K. Vander Meer, M. Coy and M.E. Scharf. 2012. Phenotypic impacts of PBAN RNA interference in an ant, *Solenopsis invicta*, and a moth, *Helicoverpa zea*. J. Insect Physiol. 58: 1159–1165.

Clynen, E. and L. Schoofs. 2009. Peptidomic survey of the locust neuroendocrine system. Insect Biochem. Mol. Biol. 39: 491–507.

Collin, C., F. Hauser, P. Krogh-Meyer, K.K. Hansen, E.G. de Valdivia, E. Gonzalez, M. Williamson and C.J.P. Grimmelikhuijzen. 2011. Identification of the *Drosophila* and *Tribolium* receptors for the recently discovered insect RYamide neuropeptides. Biochem. Biophys. Res. Commun. 412: 578–583.

Downer, K.E., A.T. Haselton, R.J. Nachman and J.G. Stoffolano. 2007. Insect satiety: sulfakinin localization and the effect of drosulfakinin on protein and carbohydrate ingestion in the blow fly, *Phormia regina* (Diptera: Calliphoridae). J. Insect Physiol. 53: 106–112.

Dulcis, D., R.B. Levine and J. Ewer. 2005. Role of the neuropeptide CCAP in *Drosophila* cardiac function. J. Neurobiol. 64: 259–272.

Elliott, K.L. and B. Stay. 2008. Changes in juvenile hormone biosynthesis in the termite *Reticulitermes flavipes* during development of soldiers and neotenic reproductive from groups of isolated workers. J. Insect Physiol. 54: 492–500.

Elliott, K.L., G.L. Hehman and B. Stay. 2009. Isolation of the gene for the precursor of Phe-Gly-Leu-amide allatostatins in the termite *Reticulitermes flavipes*. Peptides 30: 855–860.

Fusé, M., J.R. Zhang, E. Partridge, R.J. Nachman, I. Orchard, W.G. Bendena and S.S. Tobe. 1999. Effects of an allatostatin and a myosuppressin on midgut carbohydrate enzyme activity in the cockroach *Diploptera punctata*. Peptides 20: 1285–1293.

Gäde, G. 2002. Allatoregulatory peptides—molecules with multiple functions. Invertebr. Reprod. Develop. 41: 127–135.

Gäde, G. and K.H. Hoffmann. 2005. Neuropeptides regulating development and reproduction in insects. Physiol. Entomol. 30: 103–121.

Gäde, G., K.H. Hoffmann and J.H. Spring. 1997. Hormonal regulations in insects: Facts, gaps and future directions. Physiol. Reviews 77: 693–1032.

Girardie, J., O. Richard, J.C. Huet, C. Nespoulous, A. Van Dorsselaer and J.C. Pernollet. 1991. Physical characterization and sequence identification of the ovary maturating parsin. A new neurohormone purified from the nervous corpora cardiaca of the African locust (*Locusta migratoria migratorioides*). Eur. J. Biochem. 202: 1121–1126.

Griebler, M., S.A. Westerlund, K.H. Hoffmann and M. Meyering-Vos. 2008. RNA interference with the allatoregulating neuropeptide genes from the fall armyworm *Spodoptera frugiperda* and its effects on the JH titer in the hemolymph. J. Insect Physiol. 54: 997–1007.

Hansen, K.K., E. Stafflinger, M. Schneider, F. Hauser, G. Cazzamali, M. Williamson, M. Kollmann, J. Schachtner and C.J.P. Grimmelikhuijzen. 2010. Discovery of a novel insect neuropeptide signaling system closely related to the insect adipokinetic hormone and corazonin hormonal systems. J. Biol. Chem. 285: 10736–10747.

Harshini, S., R.J. Nachman and S. Sreekumar. 2002. Inhibition of digestive enzyme release by neuropeptides in larvae of *Opisina arenosella*. Comp. Biochem. Physiol. 132B: 353–358.

Hassanien, I.T.E. 2013. RNA interference with allatoregulating neuropeptide genes affecting circadian rhythm, development, mating and reproduction of *Spodoptera frugiperda* (Lepidoptera: Noctuidae). PhD Thesis, University of Bayreuth, Bayreuth, Germany.

Hassanien, I.T.E., M. Meyering-Vos and K.H. Hoffmann. 2014a. RNA interference reveals allatotropin functioning in larvae and adults of *Spodoptera frugiperda* (Lepidoptera, Noctuidae). Entomologia 2: 169.

Hassanien, I.T.E., M. Grötzner, M. Meyering-Vos and K.H. Hoffmann. 2014b. Neuropeptides affecting the transfer of juvenile hormones from males to females during mating in *Spodoptera frugiperda*. J. Insect Physiol. 66: 45–52.

Hauser, F., M. Williamson, G. Cazzamali and C.J.P. Grimmelikhuijzen. 2006. Identifying neuropeptide and protein hormone receptors in *Drosophila melanogaster* by exploiting genomic data. Brief. Funct. Genom. Proteom. 4: 321–330.

Hiruma, K. and Y. Kaneko. 2013. Hormonal regulation of insect metamorphosis with special reference to juvenile hormone biosynthesis. Curr. Top. Dev. Biol. 103: 73–100.

Hoffmann, K.H., M. Meyering-Vos and M.W. Lorenz. 1999. Allatostatins and allatotropins: is regulation of corpora allata activity their primary function? Europ. J. Entomol. 96: 255–266.

Hua, Y.-J., Y. Tanaka, K.I. Nakamura, M. Sakakibara, S. Nagata and H. Kataoka. 1999. Identification of a prothoracicostatic peptide in the larval brain of the silkworm, *Bombyx mori*. J. Biol. Chem. 274: 31169–31173.

Huang, J., Y. Zhang, M. Li, S. Wang, W. Liu, P. Couble, G. Zhao and Y. Huang. 2007. RNA interference-mediated silencing of the bursicon gene induces defects in wing expansion of silkworm. FEBS Lett. 581: 697–701.

Hummon, A.B., T.A. Richmond, P. Verleyen, G. Baggerman, J. Huybrechts, M.A. Ewing, E. Vierstraete, S.L. Rodriguez-Zas, L. Schoofs, G.E. Robinson and J.V. Sweedler. 2006. From the genome to the proteome: uncovering peptides in the *Apis* brain. Science 314: 647–649.

Jekely, G. 2013. Global view of the evolution and diversity of metazoan neuropeptide signaling. Proc. Natl. Acad. Sci. USA 110: 8702–8707.

Jiang, H., A. Lkhagva, I. Daubnerová, H.S. Chae, L. Šimo, S.H. Jung, Y.K. Yoon, N.R. Lee, J.Y. Seong, D. Žitnan, Y. Park and K.J. Kim. 2013. Natalisin, a tachikinin-like signaling system, regulates sexual activity and fecundity in insect. Proc. Natl. Acad. Sci. USA 110: E3526–3534.

Jindra, M., S.R. Palli and L.M. Riddiford. 2013. The juvenile hormone signalling pathway in insect development. Annu. Rev. Entomol. 58: 181–204.

Kataoka, H., A. Toschi, J.P. Li, R.L. Carney, D.A. Schooley and S.J. Kramer. 1989. Identification of an allatotropin from adult *Manduca sexta*. Science 243: 1481–1483.

Katoch, R. and N. Thakur. 2013. Advances in RNA interference technology and its impact on nutritional improvement, disease and insect control in plants. App. Biochem. Biotechnol. 169: 1579–1605.

Katoch, R., A. Sethi, N. Thakur and L.L. Murdock. 2013. RNAi for insect control: current perspective and future challenges. Appl. Biochem. Biotechnol. 171: 847–873.

Knapek, S., L. Kahsai, A.M.E. Winther, H. Tanimoto and D.R. Nässel. 2013. Short neuropeptide F acts as a functional neuromodulator for olfactory memory in Kenyon cells of *Drosophila* mushroom bodies. J. Neurosci. 33: 5340–5345.

Korb, J. and K. Hartfelder. 2008. Life history and development—a framework for understanding developmental plasticity in lower termites. Biol. Rev. 83: 295–313.

Lee, D.-W., S. Shrestha, A.Y. Kim, S.J. Park, C.Y. Yang, Y. Kim and Y.H. Koh. 2011. RNA interference of pheromone biosynthesis-activating neuropeptide receptor suppresses mating behavior by inhibiting sex pheromone production in *Plutella xylostella* (L.). Insect Biochem. Mol. Biol. 41: 236–243.

Lenz, C., M. Williamson and C.J.P. Grimmelikhuijzen. 2000. Molecular cloning and genomic organization of an allatostatin preprohormone from *Drosophila melanogaster*. Biochem. Biophys. Res. Commun. 273: 1126–1131.

Li, B., R. Predel, S. Neupert, F. Hauser, Y. Tanaka, G. Cazzamali, M. Williamson, Y. Arakane, P. Verleyen, L. Schoofs, J. Schachtner, C.J.P. Grimmelikhuijzen and Y. Park. 2008. Genomics, transcriptomics, and peptidomics of neuropeptides and protein hormones in the red flour beetle *Tribolium castaneum*. Genome Res. 18: 113–122.

Loeb, M.J., A. De Loof, D.B. Gelman, R.S. Hakim, J. Howard, J.P. Kochansky, S.M. Meola, L. Schoofs, C. Steel, X. Vafopoulou, R.M. Wagner and C.W. Woods. 2001. Testis ecdysiotropin, an insect gonadotropin that induces synthesis of ecdysteroids. Arch. Insect Biochem. Physiol. 47: 181–188.

Lwalaba, D., K.H. Hoffmann and J. Woodring. 2009. Control of the release of digestive enzymes in the larvae of the fall armyworm, *Spodoptera frugiperda*. Arch. Insect Biochem. Physiol. 73: 14–29.

Maestro, J.L., R. Aguilar, N. Pascual, M.L. Valero, M.D. Piulachs, D. Andreu, I. Navarro and X. Bellés. 2001. Screening of antifeedant activity in brain extracts led to the identification of sulfakinin as a satiety promoter in the German cockroach. Eur. J. Biochem. 268: 5824–5830.

Masood, M. and I. Orchard. 2014. Molecular characterization and possible roles of allatotropin in *Rhodnius prolixus*. Peptides 53: 159–171.

Meyering-Vos, M. and A. Müller. 2007a. Structure of the sulfakinin cDNA and gene expression from the Mediterranean field cricket *Gryllus bimaculatus*. Insect Biochem. Mol. Biol. 16: 445–454.

Meyering-Vos, M. and A. Müller. 2007b. RNA interference suggests sulfakinins as satiety effectors in the cricket *Gryllus bimaculatus*. J. Insect Physiol. 53: 840–848.

Meyering-Vos, M., S. Merz, M. Sertkol and K.H. Hoffmann. 2006. Functional analysis of the allatostatin A gene in the cricket *Gryllus bimaculatus* and the fall armyworm, *Spodoptera frugiperda*. Insect Biochem. Mol. Biol. 36: 492–504.

Mito, T., T. Nakamura, T. Bando, H. Ohucho and S. Noji. 2011. The advent of RNA interference in entomology. Entomol. Sci. 14: 1–8.

Mykles, D.L., M.E. Adams, G. Gäde, A.B. Lange, H.G. Marco and I. Orchard. 2010. Neuropeptide action in insects and crustaceans. Physiol. Biochem. Zool. 83: 836–846.

Nachman, R.J., W. Giard, P. Favrel, T. Suresh, S. Sreekumar and G.M. Holman. 1997. Insect myosuppressins and sulfakinins stimulate release of the digestive enzyme α-amylase in two invertebrates. Ann. New York Acad. Sci. 814: 335–338.

Nässel, D.R. 2002. Neuropeptides in the nervous system of *Drosophila* and other insects: multiple roles as neuromodulators and neurohormones. Progr. Neurobiol. 68: 1–84.

Nässel, D.R. and A.M.E. Winther. 2010. *Drosophila* neuropeptides in regulation of physiology and behavior. Progr. Neurobiol. 92: 42–104.

Nijhout, H.F. 1994. Insect Hormones. Princeton University Press, Princeton.

Oeh, U., H. Dyker, P. Lösel and K.H. Hoffmann. 2000. *In vivo* effects of *Manduca sexta* allatotropin and allatostatin on development and reproduction in the fall armyworm, *Spodoptera frugiperda* (Lepidoptera: Noctuidae). Invertebr. Reprod. Dev. 39: 239–247.

Orchard, I., A.B. Lange and W.G. Bendena. 2001. FMRFamide-related peptides: a multifunctional family of structurally related neuropeptides in insects. Adv. Insect Physiol. 28: 267–329.

Orchard, I. and A.B. Lange. 2012. Advances in insect physiology and endocrinology through genomics, peptidomics, and related technologies. Can. J. Zool. 90: 435–439.

Park, Y.I., S. Shu, S.B. Ramaswamy and A. Srinivasan. 1998. Mating in *Heliothis virescens*: Transfer of juvenile hormone during copulation by male to female and stimulation of biosynthesis of endogenous juvenile hormone. Arch. Insect Biochem. Physiol. 38: 100–107.

Perez-Hedo, M., C. Rivera-Perez and F.G. Noriega. 2014. Starvation increases insulin sensitivity and reduces juvenile hormone synthesis in mosquitoes. PloS One 9: e86183.

Predel, R., C. Wegener, W.K. Russell, S.E. Tichy, D.H. Russell and R.J. Nachman. 2004. Peptidomics of CNS-associated neurohemal systems of adult *Drosophila melanogaster*: a mass spectrometric survey of peptides from individual flies. J. Comp. Neurol. 474: 379–392.

Price, D.R.G. and J.A. Gatehouse. 2008. RNAi-mediated crop production against insects. Trends Biotechnol. 26: 393–400.

Ramaswamy, S.B., S. Shu, Y. Park and F. Zeng. 1997. Dynamics of juvenile hormone-mediated gonadotropism in the Lepidoptera. Arch. Insect Biochem. Physiol. 35: 539–558.

Scherkenbeck, J. and T. Zdobinsky. 2009. Insect neuropeptides: structures, chemical modifications and potential for insect control. Bioorg. Med. Chem. 17: 4071–4084.

Schoofs, L. and G. Baggerman. 2003. Peptidomics in *Drosophila melanogaster*. Brief Funct. Genom. Proteom. 114–120.

Schoofs, L. and R.J. Nachman. 2006. Sulfakinins. pp. 183–187. *In*: A.J. Kasti (ed.). Handbook of Biologically Active Peptides. Academic Press, London.

Schoofs, L. and I. Beets. 2013. Neuropeptides control life-phase transitions. Proc. Natl. Acad. Sci. USA 110: 7973–7974.

Schoofs, L., G.M. Holman, T.K. Hayes, R.J. Nachman and A. DeLoof. 1991. Isolation, identification and synthesis of locusta myoinhibiting peptide (Loc-MIP), a novel biologically active neuropeptide from *Locusta migratoria*. Regul. Pept. 36: 111–119.

Scott, J.G., K. Michel, L.C. Bartholomay, B.D. Siegfried, W.B. Hunter, G. Smagghe, K.Y. Zhu and A.E. Douglas. 2013. Towards the elements of successful insect RNAi. J. Insect Physiol. 59: 1212–1221.

Spit, J., L. Badisco, H. Verlinden, P. Van Wielendaele, S. Zels, S. Dillen and J. Vanden Broeck. 2012. Peptidergic control of food intake and digestion in insects. Can. J. Zool. 90: 489–506.

Stay, B. 2000. A review of the role of neurosecretion in the control of juvenile hormone synthesis: a tribute to Berta Scharrer. Insect Biochem. Mol. Biol. 30: 653–662.

Sterkel, M., P.L. Oliveira, H. Urlaub, S. Hernandez-Martinez, R. Rivera-Pomar and S. Ons. 2012. OKB, a novel family of brain-gut neuropeptides from insects. Insect Biochem. Mol. Biol. 42: 466–473.

Terenius, O., A. Papanicolaou, J.S. Garbutt, I. Eleftherianos, H. Huvenne, S. Kanginakudru, M. Albrechtsen, C.J. An, J.L. Aymeric, A. Barthel, P. Bebas, K. Bitra, A. Bravo, F. Chevalieri, D.P. Collinge, C.M. Crava, R.A. de Maagd, B. Duvic, M. Erlandson, I. Faye, G. Felfoldi, H. Fujiwara, R. Futahashi, A.S. Gandhe, H.S. Gatehouse, L.N. Gatehouse, J.M. Giebultowicz, I. Gomez, C.J.P. Grimmelikhuijzen, A.T. Groot, F. Hauser, D.G. Heckel, D.D. Hegedus, S. Hrycaj, L.H. Huang, J.J. Hull, K. Iatrou, M. Iga, M.R. Kanost, J. Kotwica, C.Y. Li, J.H. Li, J.S. Liu, M. Lundmark, S. Matsumoto, M. Meyering-Vos, P.J. Millichap, A. Monteiro, N. Mrinal, T. Niimi, D. Nowara, A. Ohnishi, V. Oostra, K. Ozaki, M. Papakonstantinou, A. Popadic, M.V. Rajam, S. Saenko, R.M. Simpson, M. Soberon, M.R. Strand, S. Tomita, U. Toprak, P. Wang, C.W. Wee, S. Whyard, W.Q. Zhang, J. Nagaraju, R.H. Ffrench-Constant, S. Herrero, K. Gordon, L. Swelters and G. Smagghe. 2011. RNA interference in Lepidoptera: An overview of successful and unsuccessful studies and implications for experimental design. J. Insect Physiol. 51: 1105–1116.

Tian, L., B.Z. Ji, S.W. Liu, F. Jin, J. Gao and L. Sheng. 2010. Juvenile hormone III produced in male accessory glands of the longhorned beetle, *Apriona germari*, is transferred to female ovaries during copulation. Arch. Insect Biochem. Physiol. 75: 57–67.

Tribolium Genome Sequencing Consortium. 2008. The genome of the model beetle and pest *Tribolium castaneum*. Nature 452: 949–955.

Van Wielendaele, P., L. Badisco and J. Vanden Broeck. 2013. Neuropeptidergic regulation of reproduction in insects. Gen. Comp. Endocrinol. 188: 23–34.

Veenstra, J.A. 2009. Allatostatin C and its paralog allatostatin double C: The arthropod somatostatins. Insect Biochem. Mol. Biol. 39: 161–170.

Veenstra, J.A. and L. Costes. 1999. Isolation and identification of a peptide and its cDNA from the mosquito *Aedes aegypti* related to *Manduca sexta* allatotropin. Peptides 20: 1145–1151.

Vuerinckx, K., H. Verlinden, H. Lindemans and J. Vanden Broeck. 2011. Characterization of an allatotropin-like peptide receptor in the red flour beetle, *Tribolium castaneum*. Insect Biochem. Mol. Biol. 41: 815–822.

Wagner, R.M., M.J. Loeb, J.P. Kochansky, D.B. Gelman, W.R. Lusby and R.A. Bell. 1997. Identification and characterization of an ecdysiotropic peptide from brain extracts of the gypsy moth, *Lymantria dispar*. Arch. Insect Biochem. Physiol. 34: 175–189.

Wang, Y., H. Zhang, H. Li and X. Miao. 2011. Second-generation sequencing supply an effective way to screen RNAi targets in large scale for potential application in pest insect control. PloS One 6: e18644.

Weaver, R.J. and N. Audsley. 2009. Neuropeptide regulators of juvenile hormone synthesis: structures, functions, distribution, and unanswered questions. Ann. New York Acad. Sci. 1163: 316–329.

Wie, Z., G. Baggerman, R.J. Nachman, G. Goldworthy, P. Verhaert, A. De Loof and L. Schoofs. 2000. Sulfakinins reduce food intake in the desert locust, *Schistocerca gregaria*. J. Insect Physiol. 46: 1259–1265.

Weidlich, S. 2013. The regulation of digestive enzyme release in the two-spotted field cricket *Gryllus bimaculatus* (de Geer): effects of endogenous and environmental factors. PhD Thesis, University of Bayreuth, Bayreuth, Germany.

Wende, F. 2012. Endokrine Signale in der Kastendetermination der ursprünglichen Termite *Mastotermes darwiniensis* Froggatt (Isoptera: Mastotermitidae). PhD Thesis, University of Bayreuth, Bayreuth, Germany.

Westerlund, S.A. and K.H. Hoffmann. 2004. Rapid quantification of juvenile hormones and their metabolites in insect haemolymph by liquid chromatography-mass spectrometry (LC-MS). Analyt. Bioanalyt. Chem. 379: 540–543.

Williamson, M., C. Lenz, A.M.E. Winther, D.R. Nässel and C.J.P. Grimmelikhuijzen. 2001a. Molecular cloning, genomic organization, and expression of a B-type (cricket-type) allatostatin preprohormone from *Drosophila melanogaster*. Biochem. Biophys. Res. Commun. 281: 544–550.

Williamson, M., C. Lenz, A.M.E. Winther, D.R. Nässel and C.J.P. Grimmelikhuijzen. 2001b. Molecular cloning, genomic organization, and expression of a C-type (*Manduca sexta* type) allatostatin preprohormone from *Drosophila melanogaster*. Biochem. Biophys. Res. Commun. 282: 124–130.

Woodring, J., S. Diersch, D. Lwalaba, K.H. Hoffmann and Martina Meyering-Vos. 2009. Control of the release of digestive enzymes in the caeca of the cricket *Gryllus bimaculatus*. Physiol. Entomol. 34: 144–151.

Xie, Y., Z.P. Kai, S.S. Tobe, X.L. Deng, Y. Ling, X.Q. Wu, J. Huang, L. Zhang and X.L. Yang. 2011. Design, synthesis and biological activity of peptidomimetic analogs of insect allatostatins. Peptides 32: 581–586.

Xue, X.Y., Y.B. Mao, X.Y. Tao, Y.P. Huang and X.Y. Chen. 2012. New approaches to agricultural insect pest control based on RNA interference. Adv. Insect Physiol. 42: 73–117.

Yagi, K.J., R. Kwok, K.K. Chan, R.R. Setter, T.G. Myles, S.S. Tobe and B. Stay. 2005. Phe-Gly-Leu-amide allatostatin in the termite *Reticulitermes flavipes*: Content in brain and corpus allatum and effect of juvenile hormone synthesis. J. Insect Physiol. 51: 357–365.

Yagi, K.J., K.L. Elliott, L. Teesch, S.S. Tobe and B. Stay. 2008. Isolation of cockroach Phe-Gly-Leu-amide allatostatins from the termite *Reticulitermes flavipes* and their effect on juvenile hormone synthesis. J. Insect Physiol 54: 930–948.

Yamanaka, N., S. Yamamoto, D. Žitňan, K. Watanabe, T. Kawada, H. Satake, Y. Kaneko, K. Hiruma, Y. Tanaka, T. Shinoda and H. Kataoka. 2008. Neuropeptide receptor transcriptome reveals unidentified neuroendocrine pathways. PloS One 3: e3048.

Yu, N., R. Nachman and G. Smagghe. 2013a. Characterization of sulfakinin and sulfakinin receptor and their roles in food intake in the red flour beetle *Tribolium castaneum*. Gen. Comp. Endocrinol. 188: 196–203.

Yu, N., O. Christiaens, J. Liu, J. Niu, K. Cappelle, S. Caccia, H. Huvenne and G. Smagghe. 2013b. Delivery of dsRNA for RNAi in insects: an overview and future directions. Insect Sci. 20: 4–14.

7

Insect Photoperiodism

Shin G. Goto[1],* and *Hideharu Numata*[2]

Seasonal Schedules and the Life History in Insects

The Earth's rotation on its own axis gives rise to daily changes in light and temperature, while the tilt of the Earth relative to its plane of rotation about the sun causes annual changes in light and temperature and, consequently, the changing of seasons. The length of day and its rate of change are functions of latitude; at higher latitudes, days are longer and day length increases more rapidly than in areas at lower latitudes after the spring equinox. Conversely, at higher latitudes, days are shorter and day length decreases more rapidly as compared to lower latitudes after the autumn equinox. Colder climates are typically associated with higher latitudes, which restrict the duration of seasons that are ideal for growth and reproduction even with longer days. Insects have responded to these geographic variations in climate by evolving appropriate modifications to their photoperiodism.

Photoperiodism is an adaptive, seasonal timing system that enables organisms to coordinate their development and physiology to annual changes in the environment using day length (photoperiod) as a cue (Nelson et al. 2010). Insects must concentrate their reproductive efforts into seasons that favor the development and survival of their offspring,

[1] Graduate School of Science, Osaka City University, Sumiyoshi, Osaka 558-8585, Japan.
  Email: shingoto@sci.osaka-cu.ac.jp
[2] Graduate School of Science, Kyoto University, Sakyo, Kyoto 606-8502, Japan.
  Email: numata@ethol.zool.kyoto-u.ac.jp
* Corresponding author

and other activities, such as growth, migration, and diapause, must also be timed relative to seasonal abiotic and biotic environmental changes. Photoperiodism was first demonstrated in plants nearly a century ago (Garner and Allard 1920). Shortly thereafter, the first published report of photoperiodism in animals described the switch to oviparity in response to short days in the female strawberry root aphid *Aphis forbesi* Weed (Marcovitch 1923). Since this pioneering study, many examples of photoperiodic response in animals have been described; amongst insects, induction and termination of diapause and the control of seasonal morphs are the most widespread, and have been extensively investigated (Saunders 2002).

Temperature is also a useful cue for timing periodic behaviors that affect reproductive and developmental success. However, temperature can be unreliable, particularly in a terrestrial environment. A survey of daily average temperature and day length in Kyoto, Japan (35.0° N, 135.5° E) in 2011–2012 and 2012–2013 revealed dramatic day to day fluctuations and interannual variation (Fig. 1). In contrast, photoperiod is perfectly correlated with the time of year without interannual variation. Moreover, compared with temperature, which can undergo abrupt fluctuations, changes in the photoperiod are gradual, which potentially facilitates monitoring of the adaptive responses by organisms. Photoperiodic changes also occur prior to seasonal transitions. The longest day length (i.e., summer solstice) occurs approximately 1.5 month earlier than the period of highest summer temperatures, while the shortest day length (i.e., winter solstice) proceeds the period of lowest winter temperatures by a comparable length of time. As such, organisms can anticipate the seasons much earlier by using photoperiod as a cue.

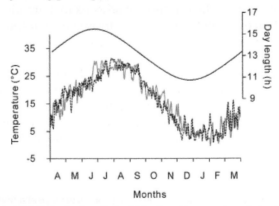

Figure 1. Annual changes in day length (solid line) and temperature during 2011–2012 (grey line) and 2012–2013 (broken line) in Kyoto, Japan (35.0° N, 135.5° E). Day length includes the length of civil twilight. Data are from the Japan Meteorological Agency (2014). There is no interannual difference in day length.

The significance of photoperiodism in the life history of an organism is illustrated by the example of the bean bug *Riptortus pedestris* (F.) from Kyoto (Fig. 2). This insect has a photoperiodic reproductive cycle, laying eggs during longer days, while suppressing the development of their reproductive organs and replenishing energy reserves during shorter days (i.e., reproductive diapause). The critical day length (CDL) for the induction of diapause in the Kyoto strain of *R. pedestris* is between 13 and 14 h; that is, insects avert or enter diapause when the day length is longer than 14 h or shorter than 13 h, respectively (Kobayashi and Numata 1993). The first and second generations of their offspring reproduce during the spring and summer when day lengths are longer than the CDL. However, third generation progeny grow during the autumn, when the day length is shorter than the CDL, enabling adults to enter reproductive diapause and overwinter while deferring oviposition until the following spring (Fig. 2). Thus, the life history of *R. pedestris*, which is characterized by three generations per year, is regulated by photoperiodism.

Photoperiodic responses have been recorded in many insect species (Danilevskii 1965, Tauber et al. 1986, Danks 1987, Saunders 2002). The physiological mechanisms underlying photoperiodic responses have been discovered by observing responses to natural or unnatural photoperiods, and in some species, the biochemical and molecular components have been identified.

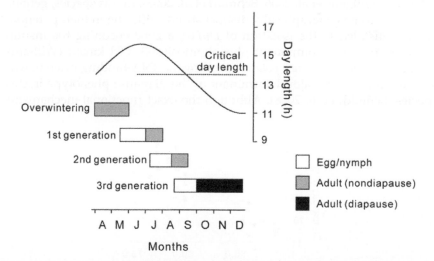

Figure 2. Day length and the life cycle of the bean bug *Riptortus pedestris* from Kyoto, Japan. Photoperiodism plays a significant role in establishing the life cycle of this species: adults in the first and second generations reproduce in response to long days; however, nymphs and adults in the third generation experience a day length that is shorter than the critical day length, inducing adults to enter diapause and overwinter.

Geographic Variations in Photoperiodism and Their Genetic Bases

Geographic variations or clines in photoperiodic responses are present in a large number of insect species (Danilevskii 1965, Saunders 2002). In an exhaustive survey, 41 geographical strains of *Drosophila littoralis* (Meigen) were collected from localities ranging from the Black Sea coast (41.6° N) to northern Finland (69.0° N), revealing geographic variations in CDLs for the induction of adult diapause (Fig. 3, Lankinen 1986). The CDL ranged from 11.6 to 20.3 h (in the south and north, respectively)and was highly correlated with latitude.

Cross matings between strains from different geographical areas have also been performed to investigate the genetic basis of photoperiodism (Saunders 2002). For example, the St. Petersburg (formerly Leningrad, 59.9° N) and Sukhumi (43.0° N) strains of *Acronycta rumicis* (L.) had CDLs of 19 and 15 h, respectively; their F_1 and F_2 progeny had an intermediate CDL of approximately 17 h (Danilevskii 1965). This was similar to the CDL of a population of *A. rumicis* from latitude of 50° N. These results indicate that a continuous latitudinal gradient exists for CDL-determining gene frequencies with continuous hybridization.

Genes that determine differences in photoperiodism have been identified from studies in the model organism *Drosophila melanogaster* (Meigen) (Williams et al. 2006, Schmidt et al. 2008). In this species, genetic crosses between geographically distinct strains differing in their potential for diapause led to the isolation of *Dp110*, a gene encoding the insulin signaling pathway component phosphoinositol-3- OH kinase (Williams et al. 2006). The gene *couch potato*, encoding an RNA-binding protein, was also identified as a major determinant of the diapause phenotype in this species (Schmidt et al. 2008). Although the exact functional mechanisms

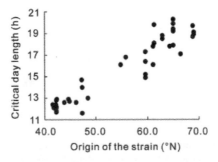

Figure 3. Critical day length (CDL) for the induction of adult diapause in *Drosophila littoralis*. Each point on the graph represents a strain collected from a different locality ranging from 41.6 N to 69.0° N (data from Lankinen 1986). A high correlation is apparent between CDL and strain origin.

of these genes are still unknown, they are thought to directly regulate diapause and not photoperiodism. Quantitative trait loci (QTL) analysis is a powerful tool for identifying specific loci or genomic regions in non-model organisms, and in the first QTL map of photoperiodism for the pitcher plant mosquito *Wyeomyia smithii* (Coquillett), a geographic variation in the photoperiodic control of diapause was revealed: nine QTL for CDL and four for diapause stage were found, although the genes at these loci have yet to be identified (Mathias et al. 2007).

Physiological Cascades and the Molecular Basis of Photoperiodism

The photoperiodic response in an organism comprises a sequence of several events (Fig. 4, Saunders 2002): (i) photoreception; (ii) assessment of day or night length by a photoperiodic time measurement system;

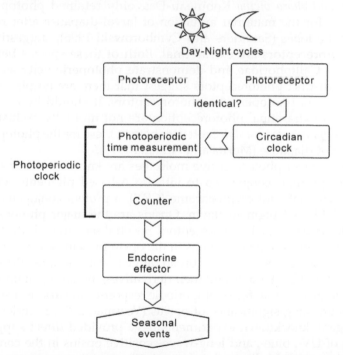

Figure 4. Establishment of photoperiodism through various modules. Light/dark signals are received by photoreceptors for photoperiodism and for the circadian clock, which may or may not be identical depending on the species. The photoperiodic time measurement system measures the length of day or night and involves the circadian clock, while the counter system counts the number of photoperiodic cycles; together, these constitute the photoperiodic clock. When the number of cycles exceeds an internal threshold, the release/ restraint of endocrine effectors is triggered, inducing seasonal events.

(iii) simultaneous evaluation of the number of photoperiodic cycles by a counter system; and (iv) activation of endocrine effectors that initiate the seasonal event. In this section, each physiological system and the associated molecular mechanism will be discussed in turn.

Photoreceptors

Visual retinal or nonvisual extraretinal photoreceptors allow organisms to detect photoperiodic information from the environment. Photoperiodic photoreceptors have been described in more than a dozen species from five different orders (Goto et al. 2010). For example, in the blowfly *Protophormia terraenovae* (Robineau-Desvoidy), photoperiodic induction of adult diapause was lost after the surgical removal of their compound eyes (Shiga and Numata 1997), underscoring the importance of retinal photoreceptors for initiating this physiological process. In contrast, the blowfly *Calliphora vicina* Robineau-Desvoidy retained photoperiodic sensitivity for the maternal induction of larval diapause after removal of the optic lobes (Saunders and Cymborowski 1996), suggesting that their photoreceptors are extraretinal. Both of these species belong to the family Calliphoridae and demonstrate photoperiodicity as adults, but their distinct photoreceptors suggest that there are no phylogenetic constraints on photoperiodic photoreceptors. It should be noted that retinal and extraretinal photoreception are not mutually exclusive: the stink bug *Plautia crossota stali* Scott uses both of these for the photoperiodic induction of diapause (Morita and Numata 1999).

Two types of photoreceptive molecules are known in insects: opsin, a class of proteins conjugated to vitamin A-based pigments retinal or 3-hydroxyretinal, and cryptochrome (CRY), a protein conjugated to the vitamin B2-based pigment flavin. Opsins are the major photoreceptive molecules in the retinal photoreceptors. Spectral sensitivity is determined by specific amino acid side chains in the opsin protein, and most species have multiple opsins covering a broad range of wavelengths (Briscoe and Chittka 2001). Carotenoids and their derivatives, including vitamin A, are verified to be essential for photoperiodic responses in various insects and mites, suggesting significance of opsins (Veerman 2001, Saunders 2012). Recent gene knockdown experiments have provided direct support for the role of UV-, blue-, and long-wave-sensitive opsins in the compound eyes of the cricket *Modicogryllus siamensis* Chopard that are responsible for the photoperiodism of nymphal diapause (Sakamoto and Tomioka 2007, Tamaki et al. 2013). There have been no similar experiments performed in species that use extraretinal photoreceptors; however, in the aphid *Megoura viciae* (Buckton), in which photoperiodic photoreceptors are restricted to a small area of the protocerebrum (Lees 1964; Steel and Lees

1977). Gao et al. (1999) detected a crescent-shaped opsin-immunoreactive region in the anterior ventral part of the brain. The opsin-immunoreactive region is not exactly the same as the putative photoreceptor site (Steel and Lees 1977), but they are very close. The long wavelength-sensitive opsin boceropsin is expressed in the cerebral ganglion of *Bombyx mori* (L.), in which the cerebral ganglion is the photoperiodic photoreceptor, implying that boceropsin is their photoperiodic photoreceptor (Shimizu et al. 2001); however, this observation still awaits functional validation.

CRY absorbs short wavelength light in the range from UV to blue, with little or no sensitivity to wavelengths greater than 500 nm (Berndt et al. 2007). CRY resets the circadian clock by phase delay or advancement (Stanewsky et al. 1998, see below). In one study, a light pulse delivered during either early or late scotophase prevented pupal diapause in the flesh fly *Sarcophaga similis* Meade; nonetheless, certain physiological differences were apparent in the responses for each phase, suggesting that distinct mechanisms are involved. The late scotophase was sensitive to a broad range of wavelengths from 395 to 660 nm, indicating that light is perceived by more than one type of opsin. In contrast, the early scotophase detected light at wavelengths of 470 nm or less, and was insensitive to wavelengths of 583 nm or greater (Fig. 5); moreover, a greater responsiveness to low light intensity was observed in the early phase. These results imply that different photoreceptors operate in early and late scotophases. Although

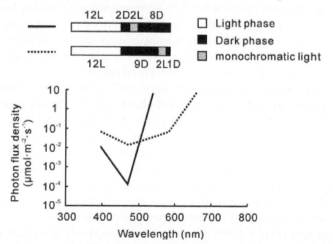

Figure 5. Spectral sensitivity in the flesh fly *Sarcophaga similis* during night interruption photoperiods with various photon flux densities of monochromatic light. Diapause-destined larvae, which were reared under 12:12 h light/dark (12L:12D) conditions, were exposed to night interruption photoperiods of 12L:2D:2L [monochromatic light]: 8D (solid line) or 12L:9D:2L [monochromatic light]:1D (broken line). Lines in the graph indicate estimated values of photon flux density at which a half of individuals entered diapause [from Goto and Numata (2009b)].

the identity of these photoreceptors is unknown, CRY is a candidate photoreceptor for the early scotophase that could act as the photoperiodic time measurement system based on the external coincidence model (see below, Goto and Numata 2009b).

Photoperiodic Time Measurement

Organisms measure the length of day or night based on information on photoperiod acquired through photoreceptors. Bünning (1936) first proposed the involvement of a circadian clock in photoperiodism. Bünning's hypothesis posited that the 24-h circadian clock consisted of two 12-h half-cycles: the photophil and scotophil (light- and dark-loving phases, respectively) (Fig. 6A), and that short-day effects are observed when light is restricted to the photophil, while long day effects are produced when light penetrates the scotophil. Although this idea is too simple to explain the range of photoperiodic responses in organisms, the basic concept of a circadian clock in photoperiodic time measurement is now widely accepted, not only in insects (Saunders and Bertossa 2011) but also in other organisms from fungi to mammals (Nelson et al. 2010).

Experiments designed to reveal the role of the circadian system in photoperiodic time measurement are based on the known effects of environmental light pulses on the phase shifting and entrainment of circadian oscillations (Saunders 2002). For instance, a short-day light phase of 10–12 h can be coupled with periods of darkness varying from 4 to 72 h (Nanda and Hamner 1958); alternatively, insects can be exposed to 48- or 72-h cycles consisting of a 12-h photophase with a light pulse systematically interrupting an extended period of perceived night (Bünsow 1953). In both types of experiment, these aberrant light cycles are repeated throughout the photoperiod-sensitive period, and short-day effects are assessed for each condition at the end of the experiment. A circadian involvement is suspected when short-day effects are observed as occurring in alternating peaks and troughs with 24-h periodicity in the extended night; conversely, the absence of this pattern is evidence for an hour glass-like timer, of which *M. viciae* (Lees 1973) is an example. However, this can be considered as a heavily dampened circadian oscillator (Saunders 2009), which has been shown to be important for photoperiodic timing even in *M. viciae* (Vaz Nunes and Hardie 1993). Thus, the functional role of a circadian clock in photoperiodic time measurement is now widely accepted, although some details are still disputed (Bradshaw and Holzapfel 2007b), as discussed below.

A variety of photoperiodic time measurement models have been established by incorporating data accrued under different experimental conditions (Vaz Nunes and Saunders 1999); among these, the external

A: Bünning's hypothesis

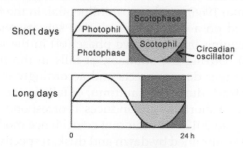

B: External coincidence model

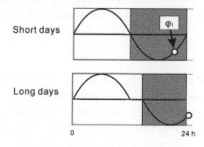

C: Internal coincidence model

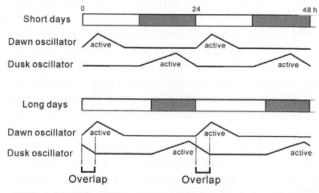

Figure 6. Conceptual diagrams of Bünning's hypothesis and the external and internal coincidence models. (A) In Bünning's hypothesis, the 24-h circadian clock comprises two 12-h half-cycles (photophil and scotophil). Short-day effects are seen when light is restricted to the photophil, while long-day effects are produced when light penetrates the scotophil. (B) Time measurement in the external coincidence model is based on a circadian clock, which sets its phase at dusk and positions the defining light-sensitive phase (φi) in the latter half of scotophase. Under short days, φi is not exposed to light, eliciting a short-day response; under long days, φi is exposed to light and a long-day response is induced. (C) The internal coincidence model proposes two oscillators entrained by dawn and dusk, respectively, whose internal phase relationship changes with photophase length. Specific phases for each oscillator are active phases, and long-day responses occur when these overlap.

and internal coincidence models (Pittendrigh 1960, Pittendrigh and Minis 1964, Tyshchenko 1966) are the most influential. In the former, time measurement is based on a circadian clock that sets its phase at dusk and positions the defining light-sensitive phase (φi) in the latter half of scotophase (Fig. 6B). During the summer, φi falls in the light period, which represents a longer day to insects and accordingly elicits a long-day response. In contrast, during the autumn, φi falls in the dark phase, which is interpreted as a short day and induces a corresponding response (see Saunders 2002, Goto 2013). The internal coincidence model proposes two oscillators that are entrained by dawn and dusk, respectively, with an internal phase that changes with the photophase length (Tyshchenko 1966, Fig. 6C). Certain phases of these oscillators are active phases, and long-day responses are induced when these overlap (see Danilevsky et al. 1970).

Photoperiodic Counter

Following the measurement of day or night length by the photoperiodic time measurement system, a photoperiodic counter registers successive cycles during the sensitive period until an internal threshold is reached, which triggers a physiological response mediated by endocrine effectors (Saunders 2002). The required day number (RDN) is defined as the number of calendar days or photoperiodic cycles needed to produce a specific seasonal event. The RDN has a capacity for temperature compensation. When females of the parasitic wasp *Nasonia vitripennis* (Walker) were maintained at various temperatures under a 12:12 h light/dark (12L:12D) cycle, they produced a physiological response associated with short day length and initiated the production of diapause larvae. The uniformity in the response over a range of temperatures suggested that a high degree of temperature compensation had occurred (Saunders 1966). A similar finding was reported for the photoperiodic induction of pupal diapause in the flesh fly (Saunders 1971). In one model of photoperiodic summation, insects accumulate or reduce (under short or long day conditions, respectively) a hypothetical diapause titer in the counter system after processing photoperiodic information in the time measurement system (Gibbs 1975). Thus, a short-day response is elicited upon exceeding the internal threshold, whereas a long day response is induced at sub-threshold values (Gibbs 1975). This putative substance is quantitatively accumulated in a photoperiod-dependent manner in some species. For example, *S. similis* enters pupal diapause in response to short days, but this can be reversed upon long-day exposure: diapause was averted when larvae progressing toward diapause were exposed to 6 d of 15L:9D or 16L:8D, indicating that both 15 and 16 h of light were both interpreted as long days. However, after 4 d of exposure, some larvae failed to avert diapause under 15L:9D conditions, although all averted diapause under 16L:8D (Goto and Numata 2009a). This type of quantitative discrimination of photoperiod can be incorporated into Gibbs's model (Fig. 7, Tagaya et al. 2010).

The molecular components of the photoperiodic counter have been investigated in the cabbage moth *Mamestra brassicae* (L.) and the giant oak silkmoth *Antheraea pernyi* (Guérin-Méneville). *M. brassicae* enters pupal diapause in response to short days, during which dopamine levels are higher in prepupal and early pupal stages. When larvae were fed the dopamine precursor L-dihydroxyphenylalanine during the final instar, diapause was induced even during long days (Noguchi and Hayakawa 1997). These results indicate that dopamine in the hemolymph and nervous system acts as a putative diapause-promoting substance. The

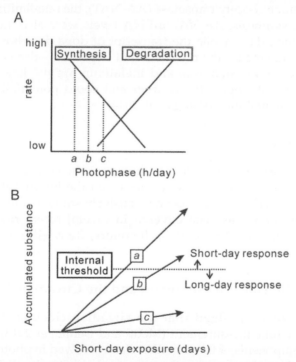

Figure 7. Conceptual diagrams of quantitative photoperiodic time measurement and counter systems. (A) A hypothetical substance is synthesized or degraded according to photoperiod in a quantitative manner (quantitative photoperiodic time measurement). The substance accumulates at a higher rate under shorter (photoperiod *a*) compared to longer (photoperiod *c*) days. (B) A short-day response is elicited when the amount of the substance exceeds an internal threshold, but a long-day response is induced for sub-threshold amounts [based on Tagaya et al. (2010)].

Receptor for activated protein kinase C1 (*Rack 1*) gene was upregulated in response to short days as well as dopamine treatment in *M. brassicae* (Uryu et al. 2002). The response of the *Rack 1* gene to short day cycles was quantitative; that is, expression was low for a 1-d and high for a 3-d exposure, while an intermediate response was observed upon exposure to 2 short days. This suggests that *Rack 1* is involved in the photoperiodic counter system. Rack 1 protein binds to and stimulates the nuclear translocation of protein kinase C, transducing signals that regulate a variety of cellular functions. In *A. pernyi*, the photoperiodic termination of pupal diapause was induced by injection of melatonin and flupentixol, a dopamine receptor antagonist, even under diapause-promoting short day conditions. Conversely, the injection of dopamine and luzindole, a melatonin receptor antagonist, delayed adult emergence even under diapause-terminating long day conditions. The transition from short to long day conditions was accompanied by the transcriptional upregulation of arylalkylamine N-acetyltransefase (AA-NAT), the rate-limiting enzyme in melatonin synthesis. *AA-NAT* mRNA levels were also increased by exposure to long days, while the transcript of dopa decarboxylase—the rate-limiting enzyme for the production of dopamine—decreased. These results suggest that dopamine and melatonin are the key molecules involved in the photoperiodic counter and could potentially function through mutual inhibition (Wang et al. 2014).

Endocrine Effectors

The hypothetical substance in the photoperiodic counter triggers the release/restraint of endocrine effectors when the internal threshold is exceeded. These effectors have been extensively studied at the molecular level, and the topic has been covered in several recent reviews (e.g., Denlinger et al. 2012, Nylin 2013); therefore, they will not be further addressed here.

Photoperiodic Time Measurement and the Circadian Clock

It is widely acknowledged that the circadian clock is involved in photoperiodic time measurement (Meuti and Denlinger 2013). Although molecular components of the circadian clock involved in photoperiodism is far from clear, those of the clock governing behavioral rhythmicity have been extensively studied. Such circadian clocks comprise several genes that self-regulate through negative feedback loops (Tomioka and Matsumoto 2010). The core molecular components are well-established, and no alternative, putative clocks have been proposed in insects; thus, it is hypothesized that the molecules of the circadian clock involved in photoperiodic time measurement and those regulating circadian behavior are the same. This section provides an overview of the current knowledge of the circadian clock that governs behavioral rhythmicity, before discussing the roles of clock genes in photoperiodism.

Molecular Machinery of the Circadian Clock

In a highly influential report that paved the way for the molecular dissection of circadian clocks, Konopka and Benzer (1971) reported on three mutants of *D. melanogaster* with abnormal rhythms of circadian-based adult eclosion and locomotor activity. The mutations were mapped to a single gene known as *period* (*per*). Subsequent forward and reverse genetics approaches have led to the discovery of several additional circadian clock genes (Sandrelli et al. 2008, Tomioka and Matsumoto 2010).

The *Drosophila* circadian clock consists of interlocked transcriptional and translational negative feedback loops (Fig. 8). Positive regulation is provided by the Per-Arnt-Sim domain-containing proteins CLOCK (CLK) and CYCLE (CYC), which heterodimerize and induce the transcription of *per*, *timeless* (*tim*) and other clock-controlled genes. PER and TIM then act as negative regulators by forming a heterodimer that suppresses CYC–CLK activity. This feedback regulation results in an oscillation in the expression levels of the core clock components *per* and *tim*. The expression of *per* and *tim* mRNA is low during photophase but is elevated during scotophase; protein levels have a similar pattern, but with peaks that are delayed by a few hours compared to those of the transcript. The expression of *Clk* mRNA and CLK protein are in antiphase with respect to *per* and *tim* mRNA levels; however, there is no oscillation in the expression of *cyc* transcript or protein.

In insects, two types of CRY are known: a *Drosophila*-type CRY (CRY-d, also known as CRY1) and a mammalian-type CRY (CRY-m, also known as CRY2). CRY-d is a flavin-based UV- and blue light-sensitive photopigment that induces the degradation of TIM in a light-dependent manner in the central clock. The *Drosophila* genome does not contain the *cry-m* gene, but it has been identified in other insect species. CRY-m does not act as a photoreceptor but as a transcriptional suppressor, possibly working with PER and TIM (Fig. 8) in the monarch butterfly *Danaus plexippus* (Yuan et al. 2007, Zhu et al. 2008). *tim* and *cry-d* are not found in the genomes of the honey bee *Apis mellifera* or *N. vitripennis*. Instead, PER is thought to function without TIM in the feedback loop, and opsin-mediated resetting of the clock is expected to play a significant role in these species. Although the diversity of circadian clocks components in insects is recognized, the essential features—positive regulation by CYC and CLK, and negative

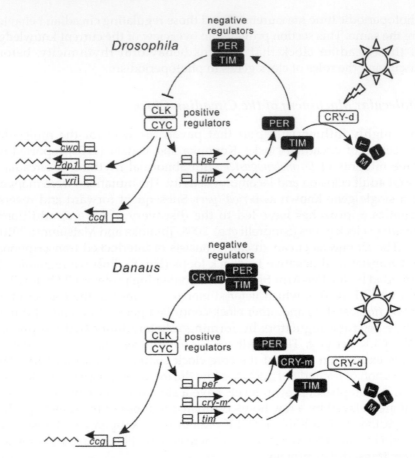

Figure 8. Circadian clock models for *Drosophila melanogaster* and the monarch butterfly *Danaus plexippus*. In *Drosophila*, CLOCK (CLK) and CYCLE (CYC) act as positive regulators to induce transcription of *period* (*per*), *timeless* (*tim*), and other circadian clock genes participating in the interlocked feedback loops of the clock such as *clockwork orange* (*cwo*), *Par-domain protein 1* (*Pdp1*), and *vrille* (*vri*), as well as *clock-controlled genes* (*ccg*) that mediate downstream signaling cascades. PER and TIM act as negative regulators that inhibit the transcriptional activity of the CYC–CLK heterodimer. *Drosophila* CRYPTOCHROME (CRY-d) induces the degradation of TIM in a light-dependent manner. *Danaus* has a near-identical clock, except that mammalian CRY (CRY-m), which is not found in the *Drosophila* genome, also acts as a negative regulator by forming a complex with PER and TIM to inhibit the transcriptional activity of CYC–CLK [modified from Yuan et al. (2007), Zhu et al. (2008), and Tomioka et al. (2012)].

feedback provided by PER—are largely conserved among species (Tomioka and Matsumoto 2010).

Role of Circadian Clock Genes in Photoperiodism

In an early study of the role of circadian clock genes in photoperiodic time measurement, *per null* (*per⁰*) and deletion mutants of *D. melanogaster* were able to discriminate photoperiods for the induction of diapause (Saunders et al. 1989). This finding suggested that *per* is not necessary for sensing photoperiods in this species. However, it is premature to conclude that a *per*-containing circadian clock is dispensable for photoperiodic time measurement, as several lines of evidence suggest otherwise (Koštál 2011); indeed, *per⁰* mutants had a CDL that was 2 h shorter than in wild-type (Saunders et al. 1989).

It is now known that *tim* is involved in the photoperiodic induction of diapause in *D. melanogaster*. The incidence of diapause in northern European populations of this species is higher than that in the south. Two alleles of *tim*, *s-tim* and *ls-tim*, have been identified, the former producing truncated TIM (S-TIM) and the latter generating both S-TIM and full-length TIM (L-TIM) proteins. Interestingly, the incidence of diapause was altered by introducing natural or artificial *tim* alleles into the different genetic backgrounds (Tauber et al. 2007): a higher frequency was observed among females carrying *ls-tim* than in those with *s-tim* for any given photoperiod. In addition, diapause was also induced in the null mutant *tim⁰¹*. There were no significant interactions found between photoperiod and the *tim* alleles with respect to the incidence of diapause. These results indicate that while *tim* is not a component of the photoperiodic time measurement system *per se*, it can confer a predisposition for the regulation of diapause (Bradshaw and Holzapfel 2007b). It is also worth noting that *tim⁰¹* flies exhibit an ambiguous photoperiodic response curve, and enter diapause irrespective of the photoperiod (Tauber et al. 2007). This raises the possibility that *tim* is directly involved in the photoperiodic time measurement system.

L-TIM was shown to interact with CRY more weakly than S-TIM; as such, flies carrying the *ls-tim* allele had significantly smaller phase responses in locomotor activity compared to *s-tim* allele carriers (Sandrelli et al. 2007), which promoted higher levels and amplitude of oscillation of TIM in the *ls-tim* flies. It is still unclear how the strength of the L-TIM/CRY interaction contributes to the photoperiodic response or diapause itself. However, it is likely that phase resetting of the clock is involved in seasonal timing.

As a well-established model insect, *D. melanogaster* provides a variety of genetic tools for the study of photoperiodism; nonetheless, it has several limitations. Diapause in this species is shallow, the incidence varies widely (see Saunders et al. 1989), and it is induced at temperatures that are close to the lower developmental threshold. It was recently determined that in two natural populations of *D. melanogaster* representing latitudinal extremes in eastern North America, temperature was the main determinant of ovarian dormancy, while there was no significant effect of photoperiod (Emerson et al. 2009).

An association between clock genes in the photoperiodic induction of diapause has been demonstrated in several other insect species. A genetic variant of the drosophilid fly *Chymomyza costata* had abnormal photoperiodic response, and failed to enter diapause even under short-day conditions (Riihimaa and Kimura 1988); an arrhythmic pattern of adult eclosion was also observed (Lankinen and Riihimaa 1992), suggesting a causal link between the circadian clock and photoperiodic response. Daily oscillations in *per* and *tim* expression were seen in wild-type flies, but *per* was expressed arrhythmically and at low levels in the variant. In addition, *tim* mRNA was completely absent (Koštál and Shimada 2001, Pavelka et al. 2003) due to a large deletion in a crucial *cis*-regulatory element and minimal promoter (Kobelkova et al. 2010). A genetic linkage analysis mapped the gene responsible for the non-diapause phenotype to the locus containing *tim* (Pavelka et al. 2003), providing evidence for the role of *tim* in the photoperiodic induction of diapause in *C. costata*.

RNA interference (RNAi) is a powerful tool that allows the function of a gene of interest to be studied in an organism, and is particularly useful for non-model organisms (Mito et al. 2011) (see also Orchard and Lange this book, Hoffmann et al. this book). This technique was used to examine the role of *tim* in the photoperiodic response of *C. costata*; however, due to quite small effect on diapause phenotype, a clear interpretation of the results was not possible (Pavelka et al. 2003). In the cricket *M. siamensis*, the induction of nymphal diapause is photoperiodic; under short-day conditions, nymphs take more time to reach adults. In this species, *per* RNAi caused arrhythmic locomotor activity under light-dark conditions and constant darkness, confirming the requirement for *per* in circadian rhythmicity. In addition, irrespective of the photoperiod, adult emergence patterns upon *per* knockdown were similar to those of crickets maintained under constant darkness (Sakamoto et al. 2009), indicating that the circadian clock is indispensable for photoperiodic discrimination.

In a series of studies on *R. pedestris*, the expression of *per, cry-m, cyc*, and *Clk* was knocked down by RNAi (Ikeno et al. 2010, 2011a,b,c, 2013). Because of the ambiguity of behavioral rhythmicity in this species in the laboratory, cuticle deposition rhythm was used as an output of clock function. The insect endocuticle thickens by the alternating deposition of chitin microfibrils in two different orientations (lamellate and non-lamellate layers) (Fig. 9A), and in *R. pedestris* and in several other insect species, the rhythm of cuticle deposition is regulated by a circadian clock (Neville 1975, Ito et al. 2008, Ikeno et al. 2010). Disruption of

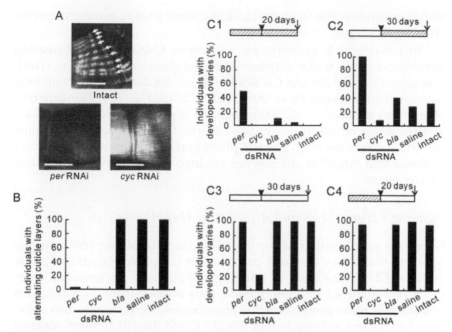

Figure 9. Endocuticle and ovarian development in the bean bug *Riptortus pedestris*. Gene knockdown via RNA interference (RNAi) was performed by injecting double-stranded RNA (dsRNA) on the day of adult emergence. (A) Cross-sections of the tibia in the hind leg of intact, *per* RNAi, and *cyc* RNAi insects 20 d after emergence. Alternating lamellate (arrows) and non-lamellate layers of the endocuticle are observed in intact insects. Knockdown of *per* or *cyc* produced a single, thickened non-lamellate or lamellate layer, respectively. Scale bar, 25 μm. (B) Fraction of insects with normal, alternating cuticle layers 20 d after the injection of *per*, *cyc*, or *β-lactamase* (*bla*; negative control) dsRNA or saline. (C) Effects of *per* and *cyc* RNAi on ovarian development. The experimental schedules are shown as horizontal hatched and open bars (short- and long-day conditions, respectively). Arrowheads indicate the day of adult emergence and of dsRNA injection, and arrows indicate the day of dissection. Insects were maintained under short-day conditions (C1), transferred from long- to short-day conditions (C2), maintained under long-day conditions (C3), or transferred from short- to long-day conditions (C4) [from Ikeno et al. (2010)].

either negative or positive regulators arrests the clock and results in the production of single layers with distinct phenotypes: when *per* or *cry-m*, negative regulators, are knocked down, only a thick non-lamellate layer is produced, whereas the silencing of *cyc* and *Clk*, positive regulators, gives rise to only a lamellate layer (Fig. 9A,B). RNAi-mediated knockdown of the negative regulators failed to suppress CYC–CLK activity, causing consistent activation of the downstream cascade; this was in turn suppressed by inhibiting the positive regulators, as evidenced by the expression of other clock genes (Ikeno et al. 2010, 2011a,b). Thus, the removal of either the positive or negative regulators caused the clock to stop its oscillation, but stuck the clock at distinct phases, causing a specific phenotypic output.

In females of *R. pedestris*, *per* and *cry-m* RNAi promoted ovarian development even under diapause-inducing short-day conditions, which was suppressed by *cyc* and *Clk* RNAi even under diapause-averting long day conditions (Ikeno et al. 2010, 2011b, 2013, Fig. 9C). The different phenotypes induced by silencing negative or positive regulators indicate that the arrest of the circadian clock at specific phases activates distinct signaling cascades that govern photoperiodic response, providing evidence that circadian clock genes are involved in photoperiodic time measurement.

Bünning's Hypothesis and Alternative Hypothesis

Given the participation of the circadian clock in photoperiodism, the question that arises is which processes in the photoperiodic response are affected. On the basis of Bünning's hypothesis, the circadian clock is actively involved in photoperiodic time measurement; however, it is possible that other processes in the photoperiodic response are also under circadian control independent of this clock, and directly induces seasonal events. Support for this idea comes from the observation that the circadian clock regulates a wide array of physiological processes in an organism, such as behavior, learning, feeding, metabolism, chemosensation, and immunity (Allada and Chung 2010). It is also known that in addition to a central circadian oscillator in the brain, peripheral clocks exist in a variety of organs, including the compound eyes, antennae, wings, legs, Malpighian tubules, and epidermis (Ito et al. 2008). These oscillators have an inherent rhythmicity that is independent of the central clock (Tomioka et al. 2012). In experiments where RNAi is introduced into an organism by feeding or injection, gene knockdown is not tissue- or cell type-specific, and it is therefore not possible to identify whether the central or peripheral clocks are being affected (Bradshaw and Holzapfel 2010). Recent studies examining the function of clock genes in peripheral tissues in relation to photoperiodism found that there they played significant "non-circadian" role (Bajgar et al. 2013a,b).

Like *R. pedestris*, the linden bug *Pyrrhocoris apterus* belongs to the infraorder Pentatomomorpha and enters reproductive diapause in response to short days. RNAi of *per* induced oviposition in about one-third of females maintained under conditions promoting diapause, but silencing *cyc* or *cry-m* had no effect on reproduction (Bajgar et al. 2013a). *cry-m* expression was highly upregulated in the gut of *P. apterus* in diapause, while that of the circadian clock gene *Par domain protein 1* (*Pdp1*) was concomitantly reduced. The juvenile hormone (JH) receptor

Methoprene-tolerant (*Met*), *Clk* and *cyc* are all required in the gut to activate the *Pdp1* gene during reproduction and to simultaneously suppress *cry-m* gene that promotes the diapause program (Bajgar et al. 2013b). Reproductive diapause is induced in a wide variety of insect species in the absence of JH. In *P. apterus* maintained under long-day conditions, ablation of the corpus allatum—the site of JH synthesis and secretion—induced diapause-specific expression patterns of the circadian clock genes in the gut (Bajgar et al. 2013a), implying that their expression is regulated by JH. Thus, while CYC, CLK, PDP1, and CRY-m play no role in regulating the peripheral circadian clock in the gut, PDP1 and CRY-m are nonetheless under the control of a hormonal signal that dictates the diapause- or nondiapause-specific physiological state of the organ and CYC and CLK works with MET to regulate the JH signaling (Bajgar et al. 2013b).

Although these results emphasize the role of circadian clock genes in the peripheral tissues as the output system of photoperiodism, they do not exclude their involvement in a photoperiodic time measurement system within the brain, which is discussed in the following section.

Location of the Circadian Clock for Photoperiodism

Microcautery experiments have determined that neurosecretory cells (NSCs) located at the anterior extremity of the protocerebrum (Group I NSCs) secrete effectors that control photoperiodic regulation of virginopara production in *M. viciae* (Steel and Lees 1977). Lesioning a region lateral to Group I NSCs also effectively disrupted the normal photoperiodic response. On the basis of these observations, this region has been assumed to harbor the photoperiodic clock; however, clear functional evidence is still lacking.

Immunocytochemical approaches were used to map neurons expressing PER, TIM, and the neuropeptide prothoracicotropic hormone (PTTH) in *A. pernyi*, which enters pupal diapause in response to short days (Sauman and Reppert 1996a,b). PTTH triggers the synthesis and secretion of ecdysteroids, and in most cases the arrest of PTTH secretion is the critical determinant of larval and pupal diapause (Denlinger et al. 2012). In *A. pernyi*, PER and TIM were coexpressed in the cytoplasm and axons of eight cells in the dorsal lateral protocerebrum of pupae and adults. A pair of PER-positive cells was located adjacent to the PTTH-expressing NSCs; their physical proximity and the extensive dendritic arborization of the NSCs in this region suggest routes of communication between these two cell populations that are important for the circadian control of PTTH release (Sauman and Reppert 1996a,b). However, there is no report on a link between these cells and photoperiodic responses in this species. In the tobacco hornworm *Manduca sexta* (L.), four NSCs in the pars lateralis

(PL) of each brain hemisphere are immunoreactive for PER. These cells were identified as type-Ia1 NSCs that produce the neuropeptide hormone corazonin (Wise et al. 2002). When these cells were removed under diapause-inducing short-day conditions, the incidence of nondiapause pupae increased relative to controls (Shiga et al. 2003). The dendrites of type-Ia1 and PTTH cells coexist in the same region of neuropil (Shiga et al. 2003), and intracellular staining of lateral NSCs has revealed that their dendritic fields overlap (Carrow et al. 1984). This provides a basis for possible paracrine or synaptic inhibition of the PTTH NSCs by type-Ia1 cells, which would implicate the latter as an element of the photoperiodic time measurement system.

In *P. terraenovae*, neurons with cell bodies in the PL projecting to the retrocerebral complex (designated as PL neurons) are necessary for the induction of reproductive diapause under short-day and low temperature conditions (Shiga and Numata 2000). Synaptic connections exist between PL neurons and pigment-dispersing factor (PDF)-immunoreactive neurons (Hamanaka et al. 2005). PDF is considered to function as an output signal from the circadian clock to regulate locomotor activity (Renn et al. 1999). Significant role of PDF in the circadian rhythm has also been reported in other insect species (Petri and Stengl 1997, Lee et al. 2009, Hassaneen et al. 2011). In *P. terraenovae*, PDF-immunoreactive fibers with synaptic connections to PL neurons probably originate from PDF-expressing neurons whose cell bodies are located at the base of the medulla in the optic lobe (Hamanaka et al. 2005). These neurons, called s-LNvs, were also immunoreactive for PER; when the brain region containing s-LNvs was removed, flies showed arrhythmic activity patterns under constant darkness, and failed to discriminate photoperiods, suggesting that s-LNvs, which function as circadian clock neurons in the brain, are also responsible for photoperiodic time measurement (Shiga and Numata 2009).

A similar neural circuit governing photoperiodism also exists in *R. pedestris* (Fig. 10, Ikeno et al. 2014). PDF-immunoreactive neurons in this insect have cell bodies located in the optic lobe, with fibers extending to the protocerebrum. Surgical removal of the region containing PDF-positive cell bodies disrupted the photoperiodic regulation of diapause (Fig. 10), although RNAi-mediated silencing of *pdf* expression had no effect on photoperiodic response. This suggests that PDF-expressing neurons, which are involved in circadian regulation, mediate photoperiodic responses in *R. pedestris* through a factor other than PDF. Although the identification of these neurons as definitive circadian clock components in this species awaits confirmation, their morphological and biochemical similarities to PDF-secreting neurons in other insects provide strong support for this possibility (Stengl and Homberg 1994, Vafopoulou et al. 2010, Vafopoulou and Steel 2012, see also the references in Ikeno et al. 2014).

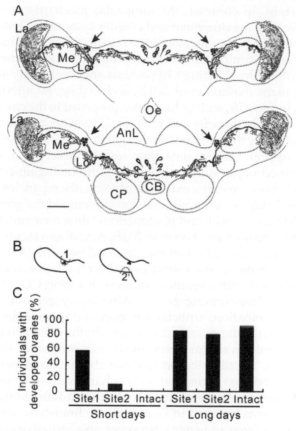

Figure 10. Pigment-dispersing factor (PDF)-immunoreactive neurons in the brain of the bean bug *Riptortus pedestris* and effect of their removal on photoperiodism. (A) Tracings of PDF-expressing neurons in the brain (top, anterior view; bottom, dorsal view). Arrows indicate cell bodies. AnL, antennal lobe; CB, central body; CP, corpus pedunculatus; La, lamina; Lo, lobulla; Me, medulla; Oe, oesophagus. Scale bar, 100 μm. (B) Schematic illustrations of ablated optic lobe regions. Anterior and posterior bases of the medulla are designated as Sites 1 and 2, respectively. PDF-positive neurons are shown as closed circles. (C) Effect of eliminating Sites 1 or 2 on photoperiodic induction of ovarian development. The brain regions were surgically removed in insects reared under diapause-inducing short-day conditions, after which they were maintained under the same conditions or transferred to diapause-averting long-day conditions [from Ikeno et al. (2014)].

Future Prospects

Endocrine effectors that direct seasonal events in an organism have been extensively studied at the molecular level (Denlinger et al. 2012). Recently, evidence for the causal involvement of opsins in photoperiodic photoreception was obtained (Tamaki et al. 2013), while the role of CRY is still uncertain. In contrast, the molecular mechanisms underlying photoperiodic time measurement and counter systems are still obscure.

A remaining point of contention in the photoperiodic time measurement system is whether components of the circadian clock are involved. The principles outlined by Bünning's hypothesis are now widely accepted for various organisms (Nelson et al. 2012), and to date, no alternatives to a model involving *per*, *Clk*, and *cyc* have been proposed in the circadian clock. Because circadian rhythmicity and photoperiodism are both dependent on photoperiod, linking the mechanisms that govern daily and seasonal timing has intrinsic appeal. Indeed, data from several insect species support a functional connection, although others have cautioned against drawing too many parallels because of the existence of some discrepancies (Bradshaw and Holzapfel 2012). For example, one study examined the genetic linkage between the circadian clock and photoperiodic time measurement system in *D. littoralis* (Lankinen and Forsman 2006). A northern strain with a long CDL for diapause, early phase of entrained eclosion rhythm in response to extremely short days, and a short period for the free-running eclosion rhythm was crossed with a southern strain with a short CDL, late eclosion phase, and long free-running period. After many generations, during which free recombination, artificial selection, and genetic drift occurred, a novel strain with diapause and eclosion rhythm characteristic of the southern and northern strains, respectively, was produced. The complete separation of the circadian-based eclosion rhythm from photoperiodic behavior revealed that independent mechanisms underlie these two processes. A similar incongruence between circadian rhythmicity and photoperiodism was reported in *W. smithii* (see Bradshaw and Holzapfel 2007a, 2012 and references herein). However, an alternative explanation is that a single oscillator is responsible for both processes, and that circadian or photoperiodic responses are affected through distinct downstream pathways (Lankinen and Forsman 2006, Goto 2013).

On the basis of the existing data, the most parsimonious approach to elucidating the photoperiodic time measurement system is to assume the involvement of the circadian clock genes, particularly as there is still no clear evidence to the contrary. It is also useful to approach the problem using forward genetics in order to identify non-circadian genes that are involved in photoperiodic time measurement. As photoperiodic responses have evolved multiple times in insects and the core circadian clock differs among insect taxa, it is likely that circadian clocks play species-specific roles in photoperiodic responses. Comparative studies involving diverse insect species are therefore welcome.

Apart from the question of whether circadian clock genes participate in photoperiodic time measurement, there is still a lack of basic understanding about process of photoperiodic time measurement itself. For instance, the physiological entity represented by φi in the external coincidence model has yet to be identified; in the internal coincidence model, the components that measure the overlap between the two oscillators are still unclear. Similarly, in the counter system, the accumulation of hypothetical substances and existence of internal thresholds have been hypothesized, but these concepts still require substantiation with physiological evidence. Dopamine, melatonin, and the *Rack 1* gene have all been proposed to be involved (Noguchi and Hayakawa 1997, Uryu et al. 2003, Wang et al. 2014) but functional experiments are required in order to confirm their roles.

In conclusion, there is still much to be discerned in the cascade of photoperiodism, and a molecular dissection of this process represents a promising direction in the field. The integration of experimental approaches such as classic physiological experiments, fine-scale mapping of genes by QTL analyses, genome-editing and high-throughput sequencing technologies, which would be particularly powerful for non-model organisms, and loss of function experiments using RNAi will greatly enhance future studies on photoperiodism.

Acknowledgements

We thank Dr. Sakiko Shiga at Osaka City University for critical reading of this manuscript. This work was supported in part by Grants-in-Aid for Scientific Research from the Japan Society for the Promotion of Science to H.N. (20380038) and to S.G.G. (25450488).

Keywords: Photoperiodism, photoreceptor, photoperiodic time measurement, circadian clock, counter, endocrine effectors, diapause, Bünning's hypothesis, brain, day length, circadian clock genes, opsin, cryptochrome, pigment-dispersing factor

References

Allada, R. and B.Y. Chung. 2010. Circadian organization of behavior and physiology in *Drosophila*. Annu. Rev. Physiol. 72: 605–624.

Bajgar, A., D. Dolezel and M. Hodkova. 2013a. Endocrine regulation of non-circadian behavior of circadian genes in insect gut. J. Insect Physiol. 59: 881–886.

Bajgar, A., M. Jindra and D. Dolezel. 2013b. Autonomous regulation of the insect gut by circadian genes acting downstream of juvenile hormone signaling. Proc. Natl. Acad. Sci. USA 110: 4416–4421.

Berndt, A., T. Kottke, H. Breitkreuz, R. Dvorsky, S. Hennig, M. Alexander and E. Wolf. 2007. A novel photoreaction mechanism for the circadian blue light photoreceptor *Drosophila* cryptochrome. J. Biol. Chem. 282: 13011–13021.

Bradshaw, W.E. and C.M. Holzapfel. 2007a. Evolution of animal photoperiodism. Annu. Rev. Ecol. Evol. Sys. 38: 1–25.

Bradshaw, W.E. and C.M. Holzapfel. 2007b. Tantalizing *timeless*. Science 316: 1851–1852.

Bradshaw, W.E. and C.M. Holzapfel. 2010. Circadian clock genes, ovarian development and diapause. BMC Biol. 8: 115.

Bradshaw, W.E., K.J. Emerson and C.M. Holzapfel. 2012. Genetic correlations and the evolution of photoperiodic time measurement within a local population of the pitcher-plant mosquito, *Wyeomyia smithii*. Heredity 108: 473–479.

Briscoe, A.D. and L. Chittka. 2001. The evolution of color vision in insects. Annu. Rev. Entomol. 46: 471–510.

Bünning, E. 1936. Die endogene Tagesrhythmik als Grundlage der photoperiodischen Reaktion. Ber. Dtsch. Bot. Ges. 54: 590–607.

Bünsow, R.C. 1953. Uber tages- und jahresrhythmische Änderungen der photoperiodischen Lichtempfindlichkeit bei *Kalanchoe blossfeldiana* und ihre Beziehungen zur endogenen Tagesrhythmik. Z. Bot. 41: 257–276.

Carrow, G.M., R.L. Calabrese and C.M. Williams. 1984. Architecture and physiology of insect cerebral neurosecretory cells. J. Neurosci. 4: 1034–1044.

Danilevskii, A.S. 1965. Photoperiodism and Seasonal Development of Insects. Oliver & Boyd, Edinburgh.

Danilevsky, A.S., N.I. Goryshin and V.P. Tyshchenko. 1970. Biological rhythms in terrestrial arthropods. Annu. Rev. Entomol. 15: 201–244.

Danks, H.V. 1987. Insect Dormancy: An Ecological Perspective. Biological Survey of Canada, Ottawa.

Denlinger, D.L., G.D. Yocum and J.P. Rinehart. 2012. Hormonal control of diapause. pp. 430–463. In: L.I. Gilbert (ed.). Insect Endocrinology. Academic Press, London.

Emerson, K.J., A.M. Uyemura, K.L. McDaniel, P.S. Schmidt, W.E. Bradshaw and C.M. Holzapfel. 2009. Environmental control of ovarian dormancy in natural populations of *Drosophila melanogaster*. J. Comp. Physiol. A 195: 825–829.

Gao, N., M. von Schantz, R.G. Foster and J. Hardie. 1999. The putative brain photoperiodic photoreceptors in the vetch aphid, *Megoura viciae*. J. Insect Physiol. 45: 1011–1019.

Garner, W.W. and H.A. Allard. 1920. Effect of the relative length of day and night and other factors of the environment on growth and reproduction in plants. J. Agric. Res. 18: 553–606.

Gibbs, D. 1975. Reversal of pupal diapause in *Sarcophaga argyrostoma* by temperature shifts after puparium formation. J. Insect Physiol. 21: 1179–1186.

Goto, S.G. 2013. Roles of circadian clock genes in insect photoperiodism. Entomol. Sci. 16: 1–16.

Goto, S.G., S. Shiga and H. Numata. 2010. Perception of light and the role of clock genes. pp. 258–286. In: R.J. Nelson, D.L. Denlinger and D.E. Somers (eds.). Photoperiodism: The Biological Calendar. Oxford University Press, Oxford.

Goto, S.G. and H. Numata. 2009a. Alteration of the pupal diapause program and regulation of larval duration by photoperiod in the flesh fly *Sarcophaga similis* Meade (Diptera: Sarcophagidae). Appl. Entomol. Zool. 44: 603–609.

Goto, S.G. and H. Numata. 2009b. Possible involvement of distinct photoreceptors in the photoperiodic induction of diapause in the flesh fly *Sarcophaga similis*. J. Insect Physiol. 55: 401–407.

Hamanaka, Y., K. Yasuyama, H. Numata and S. Shiga. 2005. Synaptic connections between pigment-dispersing factor-immunoreactive neurons and neurons in the pars lateralis of the blow fly, *Protophormia terraenovae*. J. Comp. Neurol. 491: 390–399.

Hassaneen, E., A. El-Din Sallam, A. Abo-Ghalia, Y. Moriyama, S.G. Karpova, S. Abdelsalam, A. Matsushima, Y. Shimohigashi and K. Tomioka 2011. Pigment-dispersing factor affects nocturnal activity rhythms, photic entrainment, and the free-running period of the circadian clock in the cricket *Gryllus bimaculatus*. J. Biol. Rhythms 26: 3–13.

Ikeno, T., H. Numata and S.G. Goto. 2008. Molecular characterization of the circadian clock genes in the bean bug, *Riptortus pedestris*, and their expression patterns under long- and short-day conditions. Gene 419: 56–61.

Ikeno, T., S.I. Tanaka, H. Numata and S.G. Goto. 2010. Photoperiodic diapause under the control of circadian clock genes in an insect. BMC Biol. 8: 116.

Ikeno, T., C. Katagiri, H. Numata and S.G. Goto. 2011a. Causal involvement of *mammalian-type cryptochrome* in the circadian cuticle deposition rhythm in the bean bug *Riptortus pedestris*. Insect Mol. Biol. 20: 409–415.

Ikeno, T., H. Numata and S.G. Goto. 2011b. Photoperiodic response requires *mammalian-type cryptochrome* in the bean bug *Riptortus pedestris*. Biochem. Biophys. Res. Commun. 410: 394–397.

Ikeno, T., H. Numata and S.G. Goto. 2011c. Circadian clock genes *period* and *cycle* regulate photoperiodic diapause in the bean bug *Riptortus pedestris* males. J. Insect Physiol. 57: 935–938.

Ikeno, T., H. Numata, S.G. Goto and S. Shiga. 2014. The involvement of the brain region containing pigment-dispersing factor-immunoreactive neurons in the photoperiodic response of the bean bug *Riptortus pedestris*. J. Exp. Biol. 217: 453–462.

Ito, C., S.G. Goto, S. Shiga, K. Tomioka and H. Numata. 2008. Peripheral circadian clock for the cuticle deposition rhythm in *Drosophila melanogaster*. Proc. Natl. Acad. Sci. USA 105: 8446–8451.

Japan Meteorological Agency. 2014. URL http://www.jma.go.jp/jma/indexe.html.

Kobayashi, S. and H. Numata. 1993. Photoperiodic responses controlling the induction of adult diapause and the determination of seasonal form in the bean bug, *Riptortus clavatus*. Zool. Sci. 10: 983–990.

Konopka, R.J. and S. Benzer. 1971. Clock mutants of *Drosophila melanogaster*. Proc. Natl. Acad. Sci. USA 68: 2112–2116.

Koštál, V. 2011. Insect photoperiodic calendar and circadian clock: Independence, cooperation, or unity? J. Insect Physiol. 57: 538–556.

Koštál, V. and K. Shimada. 2001. Malfunction of circadian clock in the non-photoperiodic-diapause mutants of the drosophilid fly, *Chymomyza costata*. J. Insect Physiol. 11: 1269–1274.

Kobelková, A., A. Bajgar and D. Dolezel. 2010. Functional molecular analysis of a circadian clock gene *timeless* promoter from the drosophilid fly, *Chymomyza costata*. J. Biol. Rhythms 25: 399–409.

Lankinen, P. 1986. Geographical variation in circadian eclosion rhythm and photoperiodic adult diapause in *Drosophila littoralis*. J. Comp. Physiol. A 159: 123–142.

Lankinen, P. and P. Forsman. 2006. Independence of genetic geographical variation between photoperiodic diapause, circadian eclosion rhythm and Thr-Gly repeat region of the *period* gene in *Drosophila littoralis*. J. Biol. Rhythms 21: 3–12.

Lankinen, P. and A.J. Riihimaa. 1992. Weak circadian eclosion rhythmicity in *Chymomyza costata* (Diptera: Drosophilidae), and its independence of diapause type. J. Insect Physiol. 38: 803–811.

Lee, C.M., M.T. Su and H.J. Lee. 2009. Pigment dispersing factor: an output regulator of the circadian clock in the German cockroach. J. Biol. Rhythms 24: 35–43.

Lees, A.D. 1964. The location of the photoperiodic receptors in the aphid *Megoura viciae* Buckton. J. Exp. Biol. 41: 119–133.

Lees, A.D. 1973. Photoperiodic time measurement in the aphid *Megoura viciae*. J. Insect Physiol. 19: 2279–2316.

Marcovitch, S. 1923. Plant lice and light exposure. Science 58: 537–538.

Mathias, D., L. Jacky, W.E. Bradshaw and C.M. Holzapfel. 2007. Quantitative trait loci associated with photoperiodic response and stage of diapause in the pitcher-plant mosquito, *Wyeomyia smithii*. Genetics 176: 391–402.

Meuti, M.E. and D.L. Denlinger. 2013. Evolutionary links between circadian clocks and photoperiodic diapause in insects. Integr. Comp. Biol. 53: 131–143.

Mito, T., T. Nakamura, T. Bando, H. Ohuchi and S. Noji. 2011. The advent of RNA interference in Entomology. Entomol. Sci. 14: 1–8.

Morita, A. and H. Numata. 1999. Localization of the photoreceptor for photoperiodism in the stink bug, *Plautia crossota stali*. Physiol. Entomol. 24: 189–195.

Nanda, K.K. and K.C. Hamner. 1958. Studies on the nature of the endogenous rhythm affecting photoperiodic response of *Biloxi* soybean. Bot. Gaz. 120: 14–25.

Nelson, R.J., D.L. Denlinger and D.E. Somers (eds.). 2010. Photoperiodism: Biological Calendar. Oxford University Press, New York.

Neville, A.C. 1975. Biology of the Arthropod Cuticle. Springer, Berlin.

Noguchi, H. and Y. Hayakawa. 1997. Role of dopamine at the onset of pupal diapause in the cabbage armyworm *Mamestra brassicae*. FEBS Lett. 413: 157–161.

Nylin, S. 2013. Induction of diapause and seasonal morphs in butterflies and other insects: knowns, unknowns and the challenge of integration. Physiol. Entomol. 38: 96–104.

Pavelka, J., K. Shimada and V. Koštál. 2003. TIMELESS: a link between fly's circadian and photoperiodic clocks? Eur. J. Entomol. 100: 255–265.

Petri, B. and M. Stengl. 1997. Pigment-dispersing hormone shifts the phase of the circadian pacemaker of the cockroach *Leucophaea maderae*. J. Neurosci. 17: 4087–4093.

Pittendrigh, C.S. 1960. Circadian rhythms and the circadian organization of living systems. Cold Spring Harb. Symp. Quant. Biol. 25: 159–184.

Pittendrigh, C.S. and D.H. Minis. 1964. The entrainment of circadian oscillations by light and their role as photoperiodic clocks. Am. Nat. 98: 261–294.

Renn, S.C., J.H. Park, M. Rosbash, J.C. Hall and P.H. Taghert. 1999. A *pdf* neuropeptide gene mutation and ablation of PDF neurons each cause severe abnormalities of behavioral circadian rhythms in *Drosophila*. Cell 99: 791–802.

Riihimaa, A.J. and M.T. Kimura. 1988. A mutant strain of *Chymomyza costata* (Diptera: Drosophilidae) insensitive to diapause-inducing action of photoperiod. Physiol. Entomol. 13: 441–445.

Sakamoto, T. and K. Tomioka. 2007. Effects of unilateral compound-eye removal on the photoperiodic responses of nymphal development in the cricket *Modicogryllus siamensis*. Zool. Sci. 24: 604–610.

Sakamoto, T., O. Uryu and K. Tomioka. 2009. The clock gene *period* plays an essential role in photoperiodic control of nymphal development in the cricket *Modicogryllus siamensis*. J. Biol. Rhythms 24: 379–390.

Sandrelli, F., R. Costa, C.P. Kyriacou and E. Rosato. 2008. Comparative analysis of circadian clock genes in insects. Insect Mol. Biol. 17: 447–463.

Sandrelli, F., E. Tauber, M. Pegoraro, G. Mazzotta, P. Cisotto, J. Landskron, R. Stanewsky, A. Piccin, E. Rosato, M. Zordan, R. Costa and C.P. Kyriacou. 2007. A molecular basis for natural selection at the *timeless* locus in *Drosophila melanogaster*. Science 316: 1898–1900.

Sauman, I. and S.M. Reppert. 1996a. Circadian clock neurons in the silkmoth *Antheraea pernyi*: novel mechanisms of period protein regulation. Neuron 17: 889–900.

Sauman, I. and S.M. Reppert. 1996b. Molecular characterization of prothoracicotropic hormone (PTTH) from the giant silkmoth *Antheraea pernyi*: Developmental appearance of PTTH-expressing cells and relationship to circadian clock cells in central brain. Dev. Biol. 178: 418–429.

Saunders, D.S. 1966. Larval diapause of maternal origin—II. The effect of photoperiod and temperature on *Nasonia vitripennis*. J. Insect Physiol. 12: 569–581.

Saunders, D.S. 1971. The temperature-compensated photoperiodic clock 'programming' development and pupal diapause in the flesh-fly, *Sarcophaga argyrostoma*. J. Insect Physiol. 17: 801–812.

Saunders, D.S. 2002. Insect Clocks. 3rd edn. Elsevier Science, Amsterdam.

Saunders, D.S. 2009. Circadian rhythms and the evolution of photoperiodic timing in insects. Physiol. Entomol. 34: 301–308.

Saunders, D.S. 2012. Insect photoperiodism: seeing the light. Physiol. Entomol. 37: 207–218.

Saunders, D.S. and R.C. Bertossa. 2011. Deciphering time measurement: the role of circadian 'clock' genes and formal experimentation in insect photoperiodism. J. Insect Physiol. 57: 557–566.

Saunders, D.S. and B. Cymborowski. 1996. Removal of optic lobes of adult blow flies *Calliphora vicina* leaves photoperiodic induction of larval diapause intact. J. Insect Physiol. 42: 807–811.

Saunders, D.S., V.C. Henrich and L.I. Gilbert. 1989. Induction of diapause in *Drosophila melanogaster*: Photoperiodic regulation and the impact of arrhythmic clock mutations on time measurement. Proc. Natl. Acad. Sci. USA 86: 3748–3752.

Schmidt, P.S., C.-T. Zhu, J. Das, M. Batavia, L. Yang and W.F. Eanes. 2008. An amino acid polymorphism in the *couch potato* gene forms the basis for climatic adaptation in *Drosophila melanogaster*. Proc. Natl. Acad. Sci. USA 105: 16207–16211.

Shiga, S. and H. Numata. 1997. Induction of reproductive diapause via perception of photoperiod through the compound eyes in the adult blow fly, *Protophormia terraenovae*. J. Comp. Physiol. A 181: 35–40.

Shiga, S. and H. Numata. 2000. The roles of neurosecretory neurons in the pars intercerebralis and pars lateralis in reproductive diapause of the blow fly, *Protophormia terraenovae*. Naturwissenschaften 87: 125–128.

Shiga, S. and H. Numata. 2009. Roles of PER-immunoreactive neurons in the circadian rhythm and photoperiodism in the blow fly, *Protophormia terraenovae*. J. Exp. Biol. 212: 867–877.

Shiga, S., N.T. Davis and J.G. Hildebrand. 2003. Role of neurosecretory cells in the photoperiodic induction of pupal diapause of the tobacco hornworm *Manduca sexta*. J. Comp. Neurol. 462: 275–285.

Shimizu, I., Y. Yamanaka, Y. Shimazaki and T. Iwasa. 2001. Molecular cloning of *Bombyx* cerebral opsin (boceropsin) and cellular localization of the expression in the silkworm brain. Biochem. Biophys. Res. Commun. 287: 37–34.

Stanewsky, R., M. Kaneko, P. Emery, B. Beretta, K. Wager-Smith, S.A. Kay, M. Rosbash and J.C. Hall. 1998. The *cry*[b] mutation identifies cryptochrome as a circadian photoreceptor in *Drosophila*. Cell 95: 681–692.

Steel, C.G.H. and A.D. Lees. 1977. The role of neurosecretion in the photoperiodic control of polymorphism in the aphid *Megoura viciae*. J. Exp. Biol. 67: 117–135.

Stengl, M. and U. Homberg. 1994. Pigment-dispersing hormone-immunoreactive neurons in the cockroach *Leucophaea maderae* share properties with circadian pacemaker neurons. J. Comp. Physiol. A 175: 203–213.

Tagaya, J., H. Numata and S.G. Goto. 2010. Sexual difference in the photoperiodic induction of pupal diapause in the flesh fly *Sarcophaga similis*. Entomol. Sci. 13: 311–319.

Tamaki, S., S. Takemoto, O. Uryu, Y. Kamae and K. Tomioka. 2013. Opsins are involved in nymphal photoperiodic responses in the cricket *Modicogryllus siamensis*. Physiol. Entomol. 38: 163–172.

Tauber, E., M. Zordan, F. Sandrelli, M. Pegoraro, N. Osterwalder, C. Breda, A. Daga, A. Selmin, K. Monger, C. Benna, E. Rosato, C.P. Kyriocou and R. Costa. 2007. Natural selection favors a newly derived *timeless* allele in *Drosophila melanogaster*. Science 316: 1895–1898.

Tauber, M.J., C.A. Tauber and S. Masaki. 1986. Seasonal Adaptations of Insects. Oxford University Press, New York.

Tomioka, K. and A. Matsumoto. 2010. A comparative view of insect circadian clock systems. Cell. Mol. Life Sci. 67: 1397–1406.

Tomioka, K., O. Uryu, Y. Kamae, Y. Umezaki and T. Yoshii. 2012. Peripheral circadian rhythms and their regulatory mechanism in insects and some other arthropods: a review. J. Comp. Physiol. B 182: 729–740.

Tyshchenko, V.P. 1966. Two-oscillatory model of the physiological mechanism of insect photoperiodic reaction. Zh. Obshch. Biol. 27: 209–222 (In Russian).

Uryu, M., Y. Ninomiya, T. Yokoi, S. Tsuzuki and Y. Hayakawa. 2003. Enhanced expression of genes in the brains of larvae of *Mamestra brassicae* (Lepidoptera: Noctuidae) exposed to short day length or fed Dopa. Eur. J. Entomol. 100: 245–250.

Vafopoulou, X. and C.G.H. Steel. 2012. Metamorphosis of a clock: remodeling of the circadian timing system in the brain of *Rhodnius prolixus* (Hemiptera) during larval-adult development. J. Comp. Neurol. 520: 1146–1164.

Vafopoulou, X., K.L. Terry and C.G.H. Steel. 2010. The circadian timing system in the brain of the fifth larval instar of *Rhodnius prolixus* (Hemiptera). J. Comp. Neurol. 518: 1264–1282.

Vaz Nunes, M. and J. Hardie. 1993. Circadian rhythmicity is involved in photoperiodic time measurement in the aphid *Megoura viciae*. Experientia 49: 711–713.

Vaz Nunes, M. and D.S. Saunders. 1999. Photoperiodic time measurement in insects: a review of clock models. J. Biol. Rhythms 14: 84–104.

Veerman, A. 2001. Photoperiodic time measurement in insects and mites: a critical evaluation of the oscillator-clock hypothesis. J. Insect Physiol. 47: 1097–1109.

Wang, Q., Y. Egi, M. Takeda, K. Oishi and K. Sakamoto. 2014. Melatonin pathway transmits information to terminate pupal diapause in the Chinese oak silkmoth, *Antheraea pernyi*, and through reciprocated inhibition of dopamine pathway functions as a photoperiodic counter. Entomol. Sci. DOI: 10.1111/ens.12083.

Wise, S., N.T. Davis, E. Tyndale, J. Noveral, M.G. Folwell, V. Bedian and K.K. Siwicki. 2002. Neuroanatomical studies of *period* gene expression in the hawkmoth, *Manduca sexta*. J. Comp. Neurol. 447: 366–380.

Williams, K.D., M. Busto, M.L. Suster, A.K.C. So, Y. Ben-Shahar, S.J. Leevers and M.B. Sokolowski. 2006. Natural variation in *Drosophila melanogaster* diapause due to the insulin-regulated PI3-kinase. Proc. Natl. Acad. Sci. USA 103: 15911–15915.

Yuan, Q., D. Metterville, A.D. Briscoe and S.M. Reppert. 2007. Insect cryptochromes: Gene duplication and loss define diverse ways to construct insect circadian clocks. Mol. Biol. Evol. 24: 948–955.

Zhu, H., I. Sauman, Q. Yuan, A. Casselman, M. Emery-Le, P. Emery and S.M. Reppert. 2008. Cryptochromes define a novel circadian clock mechanism in monarch butterflies that may underlie sun compass navigation. PLoS Biol. 6: E4.

8

Insects in Winter
Metabolism and Regulation of Cold Hardiness

Kenneth B. Storey[a],* and *Janet M. Storey*[b]

Introduction

Insects are found in most of coldest places on Earth—the Arctic, the Antarctic, high on mountains. Many temperate and polar species display outstanding tolerances of cold temperatures and have amazing survival stories. For example, the Arctic Woolly Bear moth (*Gynaephora groenlandica*) is a caterpillar for seven years, feeds for only one month each summer, spends most of its life frozen, and endures temperatures as low as –70°C (Morewood et al. 1998). Insect cold hardiness is widely studied not only to understand the diversity and mechanisms of adaptations that contribute to survival below 0°C but also for applied purposes in agriculture and forestry to analyze (and possibly manipulate) the cold hardiness of beneficial or pest species. Not surprisingly, several main models for insect cold hardiness research are economically important pests including the rice stem borer (*Chilo suppressalis*), onion maggot (*Delia antiqua*), and spruce budworm (*Choristoneura fumiferana*).

Institute of Biochemistry, Carleton University, 1125 Colonel By Drive, Ottawa, Ontario, Canada K1S5B6.
[a] Email: kenneth.storey@carleton.ca
[b] Email: jan.storey@carleton.ca
* Corresponding author

Insect cold hardiness has been reviewed many times including fundamental early reviews (Salt 1961, Somme 1982) and multiple reviews from the perspectives of ecology, physiology, metabolism, ice-active proteins, and genome/proteome screening (among others: Storey and Storey 1988, 2012, 2013, Zachariassen and Kristiansen 2000, Duman 2001, Bale 2002, Block 2003, Sinclair et al. 2003, Danks 2006, Clark and Worland 2008, Doucet et al. 2009, Lee and Denlinger 2010, Bale and Hayward 2010, Wharton 2011, MacMillan and Sinclair 2011). Two multi-author books are also devoted to the subject (Lee and Denlinger 1991, Denlinger and Lee 2010).

Four strategies for insect cold hardiness have been defined. Freeze avoidance and freeze tolerance are by far the most widely used and studied whereas cryoprotective dehydration and vitrification are known mostly for insects in extreme polar environments.

Freeze avoidance involves deep supercooling of body fluids stabilized by the seasonal production of antifreeze proteins (AFPs) and low molecular mass carbohydrate cryoprotectants (Fig. 1A) as well as physical methods to avoid inoculation by environmental ice (e.g., water-proofed cuticle or cocoon, moderate body water loss, voiding the gut of non specific ice-nucleators). This widely used strategy is likely the least stressful and the least specialized; insects can maintain normal aerobic metabolism and extreme dehydration is avoided but animals run the risk of instantaneous lethal freezing if they chill below their supercooling point (SCP) or are nucleated via contact with environmental ice. Freeze avoidance may have developed out of two common characteristics of insects: (a) an exoskeleton that provides good resistance against inoculation by external ice, and (b) use of polyhydric alcohols (generally glycerol) as a means of desiccation resistance during diapause (Danks 1987).

Freeze tolerance involves acceptance of ice formation in extracellular fluid spaces (typical tolerance is ~65% of total body water frozen) while defending the liquid state of the cytoplasm with high levels of cryoprotectants (Fig. 1B). Ice nucleators, either non-specific agents or specific ice-nucleating proteins, are often used to trigger ice formation at relatively high subzero temperatures allowing a slow rate of freezing that allows time for metabolic adjustments that optimize survival. Tolerance of substantial cell volume reduction and of anoxia is required while frozen.

Cryoprotective dehydration combines extreme dehydration of the body (loss of almost all freezable water) with high cryoprotectant levels (mainly trehalose) that stabilizes membranes and other macromolecules. The strategy is common among polar soil invertebrates including nematodes and springtails (Wharton 2003, Sørensen and Holmstrup 2011) and serves the Antarctic chironomid *Belgica antarctica* (Elnitsky et al. 2008). The mechanisms used mirror those of highly desiccation tolerant insects

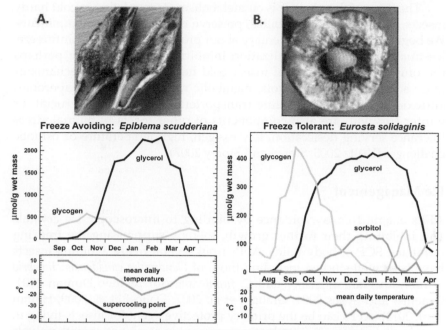

Figure 1. (A) Caterpillars of the goldenrod gall moth (*Epiblema scudderiana*) live inside elliptical galls on the stems of goldenrod and use the freeze avoidance strategy of winter survival. During autumn cold hardening, the larvae convert their stores of glycogen to glycerol, which rises to over 2 molar or ~19% of larval body mass. High glycerol plus antifreeze proteins suppress larval supercooling point (SCP) from −14°C in September to −38°C by December, values well below environmental temperature extremes. (B) Larvae of the goldenrod gall fly (*Eurosta solidaginis*) live in ball galls and are freeze tolerant. They produce glycerol early in the autumn and sorbitol once ambient temperatures get colder. Figure is modified from Storey and Storey (2004) with permission. Photographs by J.M. Storey.

around the world such as the African sleeping chironomid *Polypedilum vanderplanki* (Cornette and Kikawada 2011).

Vitrification allows deeply supercooled small insects to undergo a glass transition, thereby avoiding damage due to instantaneous freezing of body fluids at very low temperatures. Recently confirmed for two Arctic insects (Sformo et al. 2010, Kostál et al. 2011), vitrification is favored by extensive dehydration (to ~20% of summer values), accumulation of AFPs, and high levels of polyols and proline. An interesting case is that of the Arctic bark beetle *Cucujus clavipes puniceus* where two groups of larvae were identified, one group supercooled to around −40°C using freeze avoidance and the other survived to about −100°C by undergoing a glass transition between −58 to −70°C (Sformo et al. 2010, Carrasco et al. 2012). Perhaps this is a form of bet-hedging for winter survival. It is not yet known how widely vitrification is used in nature but since glass transition temperatures are typically very low, it is likely limited to insects in extreme polar or alpine climates.

The focus of this chapter is on metabolism, its regulation in cold hardy insects, and methods to stabilize/preserve macromolecular components. We begin with a brief commentary about proteins that provide antifreeze, ice-nucleating, or recrystallization inhibition since these are perhaps the only unique features of insect cold hardiness. Other mechanisms (e.g., accumulation of polyols, metabolic rate depression, chaperones, antioxidants, water and solute transporters, etc.), although crucial for winter survival, are also components of other forms of animal stress tolerance serving desiccation, low oxygen, high temperature or osmotic challenges (Kültz 2005, Storey and Storey 2007).

Ice Management

AFPs enable freeze avoidance by binding to microscopic ice crystals and inhibiting their further growth, thereby suppressing the freezing point and SCP of body fluids. AFPs from several freeze avoiding insects have been well-studied including those of *Dendroides canadensis, Tenebrio molitor, Rhagium inquisitor* and *C. fumiferana* (for review: Duman 2001, Graether and Sykes 2004, Graham et al. 2007, Doucet et al. 2009, Hakim et al. 2013). AFPs can be the primary protective agent for insects living in well-insulated sites (e.g., under the snow) but in combination with polyols they can push SCP to –40°C or lower to protect insects wintering in more exposed sites (Zachariassen and Kristiansen 2000). Insect AFPs are beta-helical proteins with a high cysteine content and the ice-binding surface is a threonine repeat (TxT) within a larger sequence repeat; e.g., the *T. molitor* AFP has a 12-amino acid repeat (TCTxSxxCxxAx) in a right-handed beta-helix whereas the *C. fumiferana* AFP is a left-handed helix with a 15 residue repeat (Graether and Sykes 2004). The regularly placed cysteines form disulfide bridges creating a stable barrel conformation that allows threonine hydroxyl groups to bind to oxygens in the ice lattice. Insects can have a suite of different-sized but related AFP peptides that have tissue-specific distributions; e.g., *D. canadensis* AFPs are found in hemolymph, midgut, hindgut and urine fluid compartments and in tissues where there can be as few as 4 (epidermis) and as many as 24 (Malpighian tubules) variants (Nickell et al. 2013). The variants may be needed to deal with diverse non-specific ice-nucleators in different areas of the body. Seasonal regulation of insect AFPs and the influences of photoperiod, temperature and hormones on their production have been well characterized (Duman 2001). AFPs are also found in freeze tolerant insects where they have a different crucial action—they inhibit recrystallization, the process of restructuring of small ice crystals into larger ones (Yu et al. 2010). A new class of biological antifreeze agents, xylomannan glycolipids, was also recently discovered in both freeze avoiding and freeze tolerant insects (Walters et al. 2009). The glycolipid is associated with the plasma membrane and exhibits both antifreeze action and ice recrystallization inhibition. For freeze tolerant species, the glycolipid might potentially block ice propagation from extracellular into liquid intracellular spaces.

Freeze tolerant insects also typically exhibit INAs or INPs (ice nucleating agents or proteins) that trigger ice formation at high subzero temperatures (above −10°C) and/or manage ice growth through extracellular compartments (Duman 2001). Both protein and lipoprotein INPs are known and Trautsch et al. (2011) suggest that lipoprotein INPs may also have an important second role as metal binding proteins that prevent heavy metals from reaching toxic levels during freeze concentration of body fluids.

Genomics and Proteomics: New Insights into Cold Hardiness

Methodological advances in genome, transcriptome and proteome sequencing are giving researchers unprecedented insights into the gene/ protein adaptations that support insect cold hardiness and diapause (Michaud and Denlinger 2010, Storey and Storey 2011, 2013). Extensive resources are available for researchers through the Arthropod Genomic Consortium (http://arthropodgenomes.org/wiki) including information and links to transcriptome (and some genome) projects for various cold hardy species including *B. antarctica, C. fumiferana, Sarcophaga crassipalpis, Ostrinia nubilalis, Rhagoletis pomonella* and *Bombyx mori*. Studies with flesh flies *S. crassipalpis* and *S. bullata* have examined gene responses to both diapause and rapid cold hardening (Li and Denlinger 2008, Ragland et al. 2010, Rinehart et al. 2010, Teets et al. 2012a). Responses to cold have also been profiled in *Drosophila melanogaster* (Sinclair et al. 2007, Telonis-Scott et al. 2009, Zhang et al. 2011). Recent work by Hao et al. (2012) analyzed gene expression differences between summer and winter diapause in *D. antiqua* pupae identifying up-regulation of stress-responsive genes in both conditions including Hsp70, alcohol dehydrogenase, delta9-desaturase, trehalose-6-P synthase (TPS) and immune-related genes (among others) and showing strong sustained TPS expression across the whole diapause period in winter. Gene responses associated with the cryoprotective dehydration strategy have been profiled in *B. antarctica* (Teets and Denlinger 2013, 2014, Teets et al. 2012b, 2013) as well as in the springtail *Megaphorura arctica* (Clark et al. 2009). Up-regulated genes in *M. arctica* included those encoding antioxidants (GST), Hsps, trehalose metabolism, membrane transporters (including aquaporins) and genes associated with skeletal reconstruction, very similar to the responses described for desiccation resistance in *P. vanderplanki* (Cornette et al. 2010). Multiple genes associated with autophagy were also up-regulated by dehydration (40% water loss) in *B. antarctica* suggesting that long term winter survival is enhanced by the limited catabolism of cell contents to recycle endogenous nutrients, probably necessitated by an inability to circulate nutrients between tissues in the dehydrated state (Teets and Denlinger 2013).

Proteomics approaches have also been applied to identify proteins that supported cryoprotective dehydration in *M. arctica*; these included proteins of carbohydrate metabolism (e.g., UDP-glucose pyrophosphorylase of trehalose biosynthesis), several chaperones, and proteins involved in cytoskeletal organization (Thorne et al. 2011). A study of mountain pine beetle larvae, *Dendroctonus ponderosae*, revealed proteins associated with winter hardiness (e.g., trehalose and glycerol metabolism, antioxidants, adenylate metabolism and ATP synthase, among others) and spring arousal (Bonnett et al. 2012). Protein responses to rapid cold hardening (2 h to 0°C) were evaluated in brain of *S. crassipalpis* pharate adults (Li and Denlinger 2008) showing enhanced levels of ATP synthase alpha subunit, Hsp23 and tropomyosin-1 isoforms 33/34 (another tropomyosin-1 isoform was suppressed). Cold exposure of a parasitic wasp, *Aphidius colemani*, similarly led to enhanced expression of the F0F1 ATP synthase alpha subunit as well as proteins of energy metabolism (glycolysis, TCA cycle), chaperones (Hsp70/Hsp 90), and the proteasome (Colinet et al. 2007). This suggests a need to maintain/readjust mitochondrial ATP synthesis in the cold.

Studies of the proteomics of cold hardiness by *C. c. puniceus* larvae identified both summer vs. winter changes (multiple cytoskeletal and cuticle proteins) (Carrasco et al. 2011a) and differences between the non-deep cooling and deep cooling groups of larvae; for the deep cooling group this included enhanced levels of cytoskeleton and cuticle proteins, Hsp70, ATP synthase subunits, and some immune proteins (Carrasco et al. 2012). Overall, then, genomic and proteomic studies of diapause and/or cold tolerance have revealed common themes including the importance of trehalose biosynthesis, lipid desaturation, membrane transporters, antioxidants, cuticular and cytoskeletal rearrangements, heat shock proteins, and ATP synthase restructuring.

Metabolism and Cold Hardiness

Cryoprotectant Synthesis

A central feature of insect cold hardiness is the accumulation of low molecular weight polyhydric alcohols as cryoprotectants (Fig. 1); these provide colligative suppression of freezing and supercooling points, promote vitrification, defend against excessive cell volume reduction in freeze tolerant species, and stabilize/protect macromolecules including lipids and proteins at low water contents (Anchordoguy et al. 1987, Storey and Storey 1988). Glycerol is by far the most common (often the sole polyol in freeze-avoiding insects) but other polyols occur in different species including sorbitol, mannitol, myo-inositol, ribitol, erythritol, threitol and ethylene glycol. For example, pupae of *Arimania comaroffi* accumulate a mixture of sorbitol, myo-inositol and trehalose for cryoprotection (Bemani et al. 2012). Trehalose and proline are typically present in substantial levels in freeze avoiding or freeze tolerant insects where they are effective in membrane stabilization (Anchordoguy et al. 1987). However, trehalose accumulates in high levels in species using cryoprotective dehydration, this disaccharide being particularly suited for stabilizing dry biological systems (Crowe et al. 2001). The prominence of glycerol as a cryoprotectant likely derives from multiple factors: (a) its ease of synthesis from glycogen, (b) the optimal output of osmotically active molecules (1 C6 unit from glycogen makes 2 C3 glycerol versus only1 C6 polyol), (c) a minimum net carbon loss in making glycerol compared with CO_2 losses to make C2, C4 or C5 polyols, (d) an easy metabolic pathway to catabolise glycerol as a spring fuel source, and (e) the normal presence in animal tissues of an aquaporin isoform that can transport glycerol across cell membranes (Storey and Storey 1988, 1991, 2013). Trehalose synthesis has an significant energy cost (1 UTP needed per trehalose made) and, as a disaccharide, is less efficient for colligative action but, because it is the blood sugar of insects, the biosynthetic pathway for trehalose synthesis is constitutive as are trehalose transporters. Furthermore, the sugar can be used as a metabolic fuel both during the winter and for spring resumption of insect activity. Although TPS expression is widely up-regulated in cold hardy insects, the specific enzymatic controls that mediate cold-induced trehalose accumulation are not yet known.

Cryoprotectant production in insects is a massive undertaking; e.g., glycerol levels rise to over 2 molar or 18% of the total body mass of larvae of the goldenrod gallmoth, *Epiblema scudderiana*, by midwinter (Fig. 1). Almost all of the glycogen built up during summer/autumn feeding is converted to polyols with calculated efficiencies of carbon conversion to glycerol or sorbitol of 84% and 92%, respectively (Storey and Storey 1991). Carbon losses are mainly due to the need to channel glucose-6-P (G6P) (derived from glycogen breakdown) through the pentose phosphate pathway (PPP) to produce the NADPH needed to reduce sugars to polyols (Fig. 2). For every 6 G6P cycled through the PPP, a net of 1 G6P is lost as $6CO_2$ and 12 NADPH are produced; the NADPH supports 6 C6 units converted to 12 glycerol or 12 C6 units converted to sorbitol. Production of polyols can be cued to changes in photoperiod, thermoperiod, desiccation, and/or entry into diapause so that cryoprotectant buildup is largely in place well

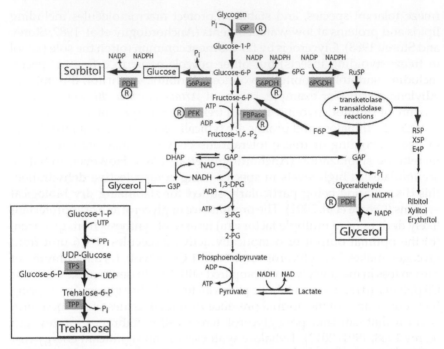

Figure 2. Pathways of cryoprotectant synthesis in cold hardy insects. Glycogen is the carbon source and a high percentage of total carbon is cycled through the pentose phosphate cycle (PPP) to produce NADPH-reducing power and sugar phosphates. After removal of the phosphate, sugars are reduced by polyol dehydrogenase (PDH) to produce the corresponding polyol. Glycerol is the polyol used by most insect species and is made mainly from the glyceraldehyde-3-phosphate (GAP) output of the PPP, although an alternative output via glycerol-3-phosphate (G3P) may also be used. Inset shows trehalose biosynthesis on the lower left. Polyol synthesis is regulated (®) by reversible protein phosphorylation at multiple sites to achieve directed flux of carbon. Regulation of trehalose synthesis undoubtedly occurs but enzymatic controls that apply to trehalose accumulation in cold hardy insects are not yet known. Selected enzymes are highlighted: GP, glycogen phosphorylase; G6Pase, glucose-6-phosphatase; G6PDH, glucose-6-phosphate dehydrogenase; 6PGDH,6-phosphgluconate dehydrogenase; PFK, 6-phosphofructo-6-kinase; FBPase; fructose-1,6-bisphosphatase; TPS, trehalose-6-phosphate synthase; TPP, trehalose-6-phosphate phosphatase.The figure is modified from Storey and Storey (2010) with permission.

in advance of its actual need (Storey and Storey 1988, 1991, Furusawa et al. 1999, Levis et al. 2012). In *Eurosta solidaginis* (and perhaps other species using dual cryoprotectants), both the seasonal timing and the ambient trigger temperatures for glycerol versus sorbitol production are distinctly different (Fig. 1B) (Storey and Storey 2004a). Seasonal increases in the necessary biosynthetic enzymes occur including glycogen phosphorylase, polyol dehydrogenase (PDH), sugar phosphate phosphatases, and TPS as well as PPP enzymes that supply NADPH (Fig. 2).

The need to generate large amounts of NADPH reducing power and route much carbon through the PPP raised questions about how this was regulated. Important insights were gained from experiments in which autumn-collected *E. solidaginis* and *E. scudderiana* larvae were challenged with anoxia exposure (Churchill and Storey 1989, Storey and Storey 1990). As expected, both species produced lactate and alanine as end products of anaerobic glycolysis, but in both cases this was accompanied by large accumulations of polyols. Polyol synthesis makes no contribution to anaerobic ATP production and seems to be a waste of carbon. Therefore, the only logical conclusion that could be drawn from this anomalous result was that in autumn larvae the G6P formed from glycogen breakdown is blocked from entering the upper portion of glycolysis and directed almost exclusively into the PPP (Storey and Storey 1991). Carbon then exits the PPP as glyceraldehyde-3-phosphate (GAP) and continues down the triose-phosphate portion of glycolysis and into the ATP-producing reactions. Hence, large amounts of NADPH accumulate and anaerobic metabolism can only remain in redox balance if additional carbon is mobilized and converted to polyols.

This example provided strong evidence that polyol synthesis is tightly regulated and involves seasonal changes that control carbon flux through competing pathways of glycolysis versus the PPP. How is this done? It has long been known that insect glycogen phosphorylase (GP) is cold-activated and that low temperature triggers the phosphorylation of inactive GPb to convert it to the active GPa form (Ziegler et al. 1979) as a result of differential temperature effects on the activities of the phosphorylase kinase (enhanced) and phosphorylase phosphatase (suppressed) that regulate GP (Storey and Storey 1988). However, GP activation only provides the substrate for polyol synthesis and cannot influence the final product so controls on other enzymes are needed. Initial studies focused on known regulatory enzymes of glycolysis and gluconeogenesis and the probability that they were also regulated by reversible protein phosphorylation (RPP). To promote one-way synthesis of polyols in the autumn, backflow of carbon into glycogen needs to be prevented and this was linked to dephosphorylation-mediated inhibition of the gluconeogenic enzyme fructose-1,6-bisphosphatase (FBPase) plus phosphorylation inhibition of glycogen synthase to prevent GAP and fructose-6-phosphate (F6P) outputs of the PPP from recycling back into glycogen (Fig. 1) (Muise and Storey 1997, 1999). Differential temperature effects on phosphofructokinase (PFK) that contribute to streaming carbon into glycerol versus sorbitol production in *E. solidaginis* are also likely due to RPP control, as defined for PFK in *B. mori* (Furusawa et al. 1999).

The above controls create the context for directing carbon flow during autumn cold hardening but evidence of direct regulation of the enzymes that determine the ultimate polyol product was still missing until recent research discovered that PDH, that catalyzes the NADPH-dependent reduction of a sugar to its corresponding polyol, is also a target of RPP (Holden and Storey 2011). Studies showed that although the maximum activity of PDH did not change with acclimation of autumn-collected *E. scudderiana* larvae to different temperatures, enzyme substrate affinity for glyceraldehyde was greatest for larvae held at 5°C (the temperature where natural glycerol synthesis is high) and 2-fold lower than for the enzyme from –15°C larvae. K_m values proved to be interconvertible when extracts from +5 and –15°C larvae were treated to promote the actions of protein kinases (raised K_m) or protein phosphatases (lowered K_m). Furthermore, both phosphoprotein staining and immunoblot detection of phosphoserine residues on purified PDH showed a much higher phosphate content on the enzyme from –15°C larvae (Holden and Storey 2011). Hence, the PDH reaction is promoted by a low phosphate form of the enzyme that has a low K_m for glyceraldehyde and dominates at temperatures that are permissive for the rapid accumulation of glycerol pools. New research also shows that the first enzyme of the PPP, glucose-6-phosphate dehydrogenase (G6PDH), is also regulated by RPP in *E. scudderiana* with high and low phosphate forms that have different affinities for G6P and are interconverted by treatments with protein kinases vs. phosphatases (K. Abnous and K. Storey unpublished data). Overall, then, we have strong evidence that glycerol synthesis is a highly regulated event in cold-hardy insects.

Transmembrane Water and Cryoprotectant Carriers

Insect cold hardiness in all forms requires movements of water and/or small molecular weight cryoprotectants across membranes separating intra-and extracellular spaces. Polyols/sugars must be transported from their tissue of origin (fat body holds the main glycogen reserves) to destination tissues and major water fluxes occur during freeze/thaw and cryoprotective dehydration. Water can move passively across membranes but when flow needs to be high or regulated, animals used integral membrane water channel proteins called aquaporins (AQPs). Multiple AQP isoforms exist and, notably, the AQP3 subgroup also transports glycerol (Campbell et al. 2008).

Recent studies provide evidence to support a crucial role of AQPs in insect freeze tolerance. Working with the *C. suppressalis*, Izumi et al. (2006) showed that glycerol uptake and water loss from fat body cells during freezing was suppressed in the presence of an aquaporin inhibitor, mercuric chloride. Furthermore, cold exposure and diapause interacted to up-regulate aquaporins (Izumi et al. 2007). Studies with *E. solidaginis* larvae identified both water channels (AQP2, AQP4) and a glycerol channel (AQP3) and AQP3 levels rose ~50% in response to desiccation or –20°C freezing (compared to 4°C controls) (Philip et al. 2008). Larval tissues frozen *in vitro* also showed reduced viability after thawing when incubated with glycerol+ mercuric chloride, compared with glycerol alone. Aquaporin and aquaglyceroporin levels in *E. solidaginis* also rose between summer to midwinter and in October, when larvae are naturally poised for high rates of glycerol synthesis, both acute cold and desiccation exposures rapidly up-regulated AQP3 (Philip and Lee 2010). Aquaporins also responded to dehydration, rehydration and freezing in the freeze and desiccation tolerant *B. antarctica* and blocking AQPs reduced the freeze tolerance of midgut and Malpighian tubules (Yi et al. 2011).

Information on membrane transporters that can move sorbitol or other polyols is currently lacking but it is possible that sorbitol might make use of a modified glucose transporter. Although impermeable to mammalian cells, sorbitol challenge to rat adipocytes triggered the translocation of GLUT4 transporters to the plasma membrane using a signaling pathway similar to that triggered by insulin (Sajan et al. 2002). Hence, it would be interesting to examine the substrate specificity of insect GLUT isoforms to determine if those from sorbitol-producing insects can transport the polyol. As the circulating sugar in insects, it was just a matter of time until a trehalose transporter was identified. Recent work discovered TRET1 in *P. vanderplanki* and other insects (Kikawada et al. 2007, Kanamori et al. 2010) and a proton-dependent form (H-TRET1) was subsequently found in Malpighian tubules that probably functions to reclaim trehalose from the tubule lumen (Kituta et al. 2012). When Chinese hamster ovary cells were transfected with TRET1 they showed a 7-fold increase in trehalose uptake and a marked increase in viability after desiccation, compared with non-transfected cells (Chakraborty et al. 2012).

Hypometabolism

Metabolic rate depression and entry into a hypometabolic state is a wide spread organismal response to environmental stress and an integral part of survival strategies including diapause, estivation, hibernation, anaerobiosis, and others (Storey and Storey 2004b, 2007). It might be argued that low winter temperatures themselves would effectively impose metabolic depression on ectotherms and allow them to survive the long winter season based on a slow catabolism of internal energy reserves. However, this is doubtful for several reasons:

1) Temperature change has differential effects on the rates of individual metabolic reactions and these can disrupt the integration of metabolic functions. The classic example is the differential effect of hypothermia on oppositely directed transmembrane ion channels (facilitated diffusion) versus ion pumps (ATP-driven) that leads to collapse of membrane potential difference (Hochachka 1986); e.g., human heart function arrests below ~25°C. This is likely the basis of chill coma in insects which has recently been linked with perturbations of Na^+ and water balance (MacMillan et al. 2012). Chill coma occurs when insects are cooled to their critical thermal minimum (CT_{min}), a reversible state where neuromuscular transmission and movement are halted; the CT_{min} is often used in ecophysiology as a measure of evolved or acquired low temperature tolerance (MacMillan and Sinclair 2011). Hence, this argues for the need to regulate metabolism in overwintering insects to maintain homeostasis in the face of wide temperature variations on both diurnal and seasonal time frames.

2) Temperature change affects the rates of all metabolic processes but does not provide a way to differentially suppress cell functions that are not needed or not feasible over the winter months (e.g., digestion and nutrient processing, macromolecular biosynthesis, growth, development, reproduction, etc.) while sustaining crucial vital functions.

3) Despite cold winter temperatures, insects are still vulnerable to abiotic stress (e.g., dehydration, oxygen limitation when frozen or encased in ice, lack of food, UV radiation, etc.) (Teets and Denlinger 2014) and entry into a hypometabolic state is a virtually universal survival response to environmental stress (Storey and Storey 2004b, 2007). Hence, a regulated entrance into hypometabolism, whether developmentally regulated (diapause) or a quiescent state, would be an important defense against abiotic stress over the winter.

Entry into hypometabolism supports insect survival by coordinately suppressing metabolic activity and energy use over the winter months, thereby extending the time that animals can survive in a non-feeding state while also typically maintaining sufficient internal fuel reserves to support activities such as metamorphosis, breeding and reproduction when spring arrives. In *E. solidaginis* for example, the metabolic rate of diapause larvae in midwinter was just 35–40% of the early autumn value at the same temperature (Irwin et al. 2001, Levin et al. 2003). Coordinated reversible metabolic suppression also allows the cellular infrastructure to be maintained and available immediately when animals rewarm, thaw or rehydrate. Hence, the known mechanisms of metabolic rate depression are mainly reversible ones and include mechanisms such as post translational modifications of proteins, regulation and storage of mRNA transcripts via the actions of microRNA (see Asgari and Hussain, this book), and gene silencing by epigenetic mechanisms. This latter is a newly emerging idea in metabolic control. Whereas epigenetic mechanisms that alter DNA (e.g., cytosine methylation) or histones (phosphorylation, acetylation, methylation) have long been regarded as a way of permanently silencing genes during development and differentiation, new studies are showing that these can also be used for reversible control of gene expression and chromatin packing during stress-responsive metabolic rate depression (Storey and Wu 2013). Although not yet studied among cold hardy insects, reversible changes in histone H3 modification that increase chromatin packing and provide global transcriptional suppression in hypometabolic states are now known for both mammalian hibernation and turtle anaerobiosis (Morin and Storey 2009, Krivoruchko and Storey 2010).

Higher level signaling in the triggering and control of hypometabolism in cold hardy insects has been associated with two sources to date. In the freeze tolerant larvae of *E. solidaginis* the hypoxia-inducible transcription factor (HIF-1) has been implicated. The larvae respond to anoxia, cold and freezing exposures with strong increases in transcript levels of the *hif-1* alpha subunit and HIF-1α protein levels (Morin et al. 2005). Seasonal patterns of HIF-1α protein expression also occur with a progressive increase over the winter months to peak in February. Given that freezing imposes an ischemic/hypoxic state on the larvae, the up-regulation of the oxygen-sensitive HIF-1α subunit would predictably lead to gene expression changes that help the larvae to endure prolonged oxygen restriction during freezing. The AMP-activated kinase (AMPK) is known as the energy-sensor of the cell and recent studies show that AMPK activity was strongly elevated by 70–90% in February larvae (midwinter fully cold-hardened) as compared with September (prior to substantial cold-hardening) in both goldenrod gall formers (Rider et al. 2011). In *E. scudderiana* a 3-fold increase in Thr172 phosphorylation on the catalytic α-subunit also occurred, a modification that is known to activate AMPK. This correlated with enhanced phosphorylation and inhibition of acetyl-CoA carboxylase (ACC) on Ser79 in winter larvae, a well-known mechanism for inhibition of lipid synthesis, and fits with the observed winter suppression of the activities of lipogenic enzymes (e.g., ATP-citratelyase, malic enzyme) (Joanisse and Storey 1996a). Furthermore, inhibition of ACC is well-known to reduce malonyl-CoA levels and this, in turn, stimulates mitochondrial long chain fatty acid oxidation. Thus, AMPK regulation of ACC, along with elevated activities of fatty acid oxidizing enzymes, shows that lipid catabolism is favored over the winter in the freeze avoiding species. By contrast, freeze tolerant *E. solidaginis* generally showed reduced levels of lipolytic enzymes during the winter in line with the inability to metabolize lipids under anoxic conditions when frozen.

Reversible Protein Phosphorylation

Studies with multiple systems (e.g., mammalian hibernation, anaerobiosis, estivation) have documented the consistent involvement of reversible protein phosphorylation (RPP) in regulating entry/arousal from hypometabolism and adjusting/suppressing key metabolic functions in a coordinated manner including intermediary metabolism, biosynthesis, ion transport, protein translation, and transcription factor action, to name a few (Storey and Storey 2004b, 2007). In cold hardy insects, the role of RPP in contributing to metabolic rate depression is best illustrated to date in studies of ion motive ATPases in the larvae of goldenrod gall formers. The sodium-potassium ion pump is often the single greatest consumer of ATP in cells and so energy savings in hypometabolism typically include a net suppression of ATP turnover by Na^+K^+-ATPase balanced by oppositely-directed control of membrane ion channels to maintain membrane potential difference. Another major ion pump is the sarco(endo)plasmic reticulum Ca^{2+}-ATPase (SERCA); Ca^{2+} release from the sarco(endo)plasmic reticulum provides crucial cytoplasmic signaling and/or triggering of muscle contraction and then SERCA pumps the ions back into storage. Seasonal changes in these ion pumps were prominent in both freeze tolerant *E. solidaginis* and freeze avoiding *E. scudderiana* larvae; midwinter activities of Na^+K^+-ATPase were strongly reduced by 75–85% in the two species whereas SERCA activity fell by 65–80% (McMullen and Storey 2008a, 2010). Notably, the changes in SERCA activities over the winter months occurred without a change in SERCA protein content in *E. solidaginis* larvae over the winter and despite a six-to eight fold increase in SERCA protein in *E. scudderiana* between September to April, as assessed by immunoblotting (McMullen and Storey 2010). *In vitro* studies confirmed that the mechanism involved was RPP; incubation of cell free extracts from 15°C acclimated larvae under conditions that stimulated the action of protein kinases A, G or C reduced Na^+K^+-ATPase activity by >90% and SERCA activity by 25–50% whereas alkaline phosphatase treatment had no effect (McMullen and Storey 2008a, 2010). One functional effect of RPP on SERCA proved to be a decrease in enzyme affinity for ATP, the enzyme from –20°C acclimated larvae having a 3.2-fold higher K_m ATP than the 15°C control enzyme (McMullen et al. 2010). The probability that protein kinase A (PKA) was the mediator of ion channel suppression was supported by other studies that showed that total PKA activity increased over autumn/winter in both species whereas protein phosphatase 1 activity decreased

(Pfister and Storey 2006a,b). Furthermore, given that PKA activates multiple enzymes involved in polyol metabolism (as discussed above), this signal transduction enzyme provides a way to integrate both the implementation of cryoprotection and the suppression of ATP-expensive metabolic activities as the larvae enter into the winter season.

MicroRNA

The recent discovery and exploration of small non-coding RNA types has led to a revolution in our understanding of the posttranscriptional regulation of gene transcripts. In particular, microRNAs (18–23 nt in length) are now known to be crucial to the regulation of virtually every facet of metabolism. In insects, microRNAs are involved in controlling development, metamorphosis, aging, social behavior and disease, among others (Lucas and Raikhel 2013, Asgari 2013, also see the chapter by Asgari and Hussain in this book). Animal genomes encode hundreds of miRNAs and more than 60% of protein-coding genes are computationally predicted to be targets for miRNA binding (Ebert and Sharp 2012). Mature microRNAs anneal to the 3′-untranslated region of target mRNA transcripts within an RNA-induced silencing complex (miRISC). When base-pairing between mRNA and miRNA is high or perfect, mRNA transcripts are directed to degradation by the argonaute endonuclease in the RISC but imperfect pairing results in reversible translational repression of transcripts via storage into P-bodies and stress granules (Bartel 2009). The latter fate is crucial to preserving mRNA transcripts during hypometabolism, particularly transcripts that need to be rapidly translated during arousal.

MicroRNAs are now known as players in animal responses to abiotic stress including hibernation, freeze tolerance (frogs, snails) and anaerobiosis and are consistently associated with metabolic rate depression (Biggar and Storey 2011, Biggar et al. 2012, Storey and Wu 2013). Recent work also documented changes in microRNA expression in cold hardy insects. Courteau et al. (2012) reported increased levels of miR-284, miR-3791-5p and miR-92c-3p in *E. solidaginis* larvae frozen at −15°C, compared with 5°C controls, and a subsequent study found that miR-1 levels were unchanged by −5°C cold exposure but increased by 1.9-fold when larvae froze at −15°C (Lyons et al. 2013). By contrast, neither −5°C nor −15°C exposures affected miR-1 expression in freeze avoiding *E. scudderiana*. Interestingly, miR-1 has been linked with translational repression of genes including histone deacetylase 4 and Hsp60. Hence, this could be a mechanism behind the strong suppression of Hsp60 levels over the winter in *E. solidaginis* but not in *E. scudderiana*.

Mitochondrial Metabolism

Metabolic rate depression must include attention to mitochondrial metabolism to lower substrate input and oxygen consumption in coordination with ATP demand and controls at other loci of metabolism. Hence, one aspect of cold hardiness or diapause may be a reduction in mitochondrial load, particularly for species that use freeze tolerance or cryoprotective dehydration, strategies that are largely incompatible with continued aerobic ATP generation by mitochondria over most of the winter. By reducing mitochondrial load under conditions where metabolic rate will remain low for many months, insects could reduce the investment that would otherwise be needed to cryoprotect these organelles and divest themselves of a major endogenous source of free radical production (complex I of the electron transport chain).

Evidence from two freeze tolerant species, *G. groenlandica* and *E. solidaginis*, indicate a strong reduction in mitochondrial numbers over the winter. Levin et al. (2003) reported that mitochondrial DNA content was reduced by ~50% in winter versus summer larvae and cytochrome c oxidase (COX) activity also decreased by 30–50% over the winter in *E. solidaginis* (McMullen and Storey 2008b). COX activity also decreased in freeze avoiding *E. scudderiana* larvae but markers of mitochondrial numbers did not change including 12S rRNA content and both the DNA and transcript levels of the COX 1 subunit that is encoded on the mitochondrial genome (McMullen and Storey 2008b). However, both freeze tolerant *E. solidaginis* and freeze avoiding *E. scudderiana* showed reduced activities (by ~50%) of mitochondrial enzymes over the winter (e.g., pyruvate dehydrogenase, citrate synthase, glutamate dehydrogenase, NAD-isocitrate dehydrogenase) (Joanisse and Storey 1994, Rider et al. 2011) in line with overall reduced metabolic requirements in the cold, non-feeding, diapause state. However, two key exceptions were noted: (a) fatty acid oxidizing enzymes rose in *E. scudderiana* indicating that aerobic lipid catabolism fuels winter ATP needs of freeze avoiding insects, and (b) SOD activity, a crucial antioxidant, rose over the winter in *E. solidaginis* suggesting the need to deal with reactive oxygen species (ROS) produced as a result of freeze/thaw events in the freeze tolerant insect. Hence, freeze avoiding species show selective reorganization of mitochondrial metabolism during the winter whereas freeze tolerant species appear to undergo substantial mitochondrial degradation.

Protection of Macromolecules

Although typically inactive throughout the winter, insects are still vulnerable to abiotic stress (e.g., temperature extremes, UV radiation, oxygen availability, desiccation, etc.) that can compromise macromolecular structure/function. Coupled with reduced capacities for degrading damaged macromolecules or resynthesizing new ones (especially under frozen, dehydrated or diapause conditions), this means that preservation mechanisms that stabilize and protect macromolecules are needed to ensure long term viability. Cryoprotectants contribute to this (e.g., polyols are excellent at stabilizing native protein structure) but two other preservation strategies are also crucial: (1) chaperones act to fold or refold proteins, and (2) antioxidants deal with challenges from ROS generation. Both are components of the cellular stress response (Kültz 2005) as well as integral to hypometabolism (Storey and Storey 2007) and recent studies confirm their role in insect cold tolerance (Storey and Storey 2010a, 2011, 2012). For example, gene screening studies on small polar invertebrates that use cryoprotective dehydration showed up-regulation of a core group of genes in response to desiccation or freezing by both nematode and collembolan species (Adhikari et al. 2009, Clark et al. 2009). These included genes encoding antioxidant enzymes, heat shock proteins, late embryogenesis abundant proteins, enzymes of trehalose metabolism and transporters for water, trehalose, and other solutes. Notably, these are the same responses that support anhydrobiosis in the African sleeping chironomid, *P. vanderplanki* (Cornette et al. 2010).

Chaperone Proteins

Chaperones were first identified as responders to acute heat stress in *Drosophila* and named heat shock proteins (HSPs); subsequently the list of chaperone proteins has grown long and includes other types of proteins including some crystallins and the glucose-regulated proteins (GRPs) that work in the endoplasmic reticulum. Chaperones are crucial to the integrity of the proteome, acting to facilitate folding of naïve proteins or re-folding of malfolded proteins, prevent aggregation of unfolded proteins, and direct protein trafficking and assembly (Feder and Hofmann 1999). Increased chaperone expression is well known to accompany insect diapause (Storey and Storey 2011, 2012) but the need for chaperones by cold-hardy species was under-appreciated for a long time partly because accumulation is slow over a long period of cold-hardening and partly because expression is often coincident with diapause entry. However, several studies have now documented chaperone involvement in insect cold tolerance of (for review: Clark and Worland 2008, Storey and Storey 2011, 2012). For example, cold acclimation triggered enhanced expression of *hsp90* in nondiapausing *C. suppressalis* larvae but not in diapausing larvae that already had high transcript levels (Sonoda et al. 2006). Similarly, cold hardy pupae of *D. antiqua* showed up-regulation of *hsp70*, *hsp60* and *tcp1* (t-complex polypeptide-1) genes (Kayukawa et al. 2005, Chen et al. 2006, Kayukawa and Ishikawa 2009) and HSP up-regulation occurred during both summer and winter diapause in this species (Hao et al. 2012). Continuous expression of *hsp70, hsp90* and small *hsps* also occurs in *B. antarctica* larvae (that overwinter) but expression is reduced in the short-lived summer adult stage (Rinehart et al. 2006). Increased expression of HSPs is also a feature of rapid cold hardening or the recovery after cold exposure (Lee and Denlinger 2010, Wang et al. 2011).

Chaperones are also elevated in larvae of the both goldenrod gall insects over the winter. Protein levels of Hsp110, Hsp70, Hsp60 and Hsp40 all increased by 2–2.5 fold in *E. scudderiana* by midwinter and GRP78, Hsp110, Hsp70 and Hsp40 increased in *E. solidaginis* (Storey and Storey 2011, Zhang et al. 2011). Hsp110, Hsp70 and Hsp40 actually work in partnership so it is not surprising that they are coordinately up-regulated (Dragovic et al. 2006). The discordant response of Hsp60—up in freeze avoiding *E. scudderiana* but down ~50% in freeze tolerant *E. solidaginis*—is intriguing but likely explained by the winter degradation of mitochondria in *E. solidaginis* and with it a reduction in the levels of this mitochondria-specific HSP.

In *E. solidaginis*, levels of the heat shock transcription factor (HSF1) that controls *hsp* gene expression were high in early autumn when HSP levels were rising but dropped to low values through out the winter suggesting that HSPs, once synthesized, are likely retained (Zhang et al. 2011). Sustained chaperone levels would preserve the proteome against multiple challenges (e.g., freeze/thaw, water/osmotic stress, oxygen restriction, etc.) in a situation that is not conducive to high energy expenditures to degrade damaged proteins or resynthesize new ones. Hence, stabilization of the proteome is imperative. Kayukawa and Ishikawa (2009) provided an example of this; non-hardy pupae of *D. antiqua* experienced cold-induced depolymerization of actin but cold-hardy pupae did not due high levels of a chaperone called CCT (chaperonin containing t-complex polypeptide-1).

Antioxidant Defenses

Antioxidant defenses come in multiple forms. Various low molecular weight metabolites are antioxidants (e.g., glutathione, ascorbic acid, tocopherols). Antioxidant enzymes include superoxide dismutase (SOD), catalase and glutathione peroxidase (GPox) that directly destroy ROS such as superoxide radicals, hydrogen peroxide and peroxyl radicals on macromolecules (Hermes-Lima 2004) whereas glutathione S-transferases (GSTs) neutralize various xenobiotics and aldehydic products of lipid peroxidation. Proteins that bind iron and copper also have antioxidant actions because, by sequestering these ions, Fe^{2+} or Cu^+ catalyzed formation hydroxyl and lipid radicals (the most highly reactive ROS) by the Fenton reaction is reduced (Hermes-Lima 2004).

$$H_2O_2 + Fe^{2+} \text{ (or } Cu^+) \rightarrow Fe3^+ \text{ (or } Cu^{2+}) + OH^- + {}^{\bullet}OH \qquad \text{Fenton reaction}$$

The importance of antioxidant defenses to insect cold hardiness is indicated by several studies. For example, in *B. antarctica*, the presence of high activities of SOD and catalase has been suggested to provide protection against ROS generated from intense ultraviolet radiation in the Antarctic (Lopez-Martinez et al. 2008). Proteomic profiling of the mountain pine beetle, *D. ponderosae* showed strong autumn increases in SOD and GPox, the latter decreasing again in the spring (Bonnett et al. 2012). In freeze tolerant *E. solidaginis*, activities of catalase, GPox and GST were high in September (while the larvae were still feeding) but decreased by ~50% during midwinter, before rising again in the spring prior to pupation (Joanisse and Storey 1996b). This pattern paralleled changes in mitochondria numbers and activities of mitochondrial enzymes suggesting that antioxidant capacity is linked with rates of oxygen consumption and ROS generation by these organelles and/or with entry into diapause (Joanisse and Storey 1994, 1996a,b, Levin et al. 2003, McMullen and Storey 2008a). However, SOD activity stayed high over the winter. A possible reason for this is that SOD can be effective in limiting peroxidation of polyunsaturated fatty acids (PUFAs) (Hermes-Lima 2004, McCord 2008).

Cold acclimation typically includes changes in lipid composition to increase levels of mono-and poly-unsaturated fatty acids in triglycerides and phospholipids thereby enhancing lipid fluidity. In turn, this sustains membrane functions and allows lipid mobilization from fuel depots at low temperatures. Lipid modification is carried out by desaturases; e.g., in *D. antiqua*, delta-9 desaturase mRNA levels increased 2–10 fold in cold-acclimated and diapause pupae correlated with increases of palmitoleic and oleic acids in phospholipids (Kayukawa et al. 2007). PUFAs are particularly sensitive to oxidation because ROS attack forms lipid peroxyl radicals that then propagate through lipids in a chain reaction of peroxidation. Hence, antioxidant defenses that protect lipids from peroxidation are important, not just for winter survival but to preserve large lipid depots (e.g., ~18% of larval wet mass in *E. solidaginis*) (Joanisse 1996a) that are needed for activities such as spring metamorphosis and reproduction. In addition to the enzymatic action of SOD, tocopherols (the best studied being α-tocopherol) are well known to neutralize peroxyl PUFAs, forming a tocopheryl radical and 'LOOH; the radical is then recycled nonenzymatically by other antioxidants (Hermes-Lima 2004). New work by Koštál et al. (2013) showed seasonal patterns of γ- and δ-tocopherol levels in the heteropteran *Pyrrhocoris apterus* that were minimal in summer and maximum in diapausing, cold-acclimated bugs in winter, supporting the idea that inhibition of lipid peroxidation is necessary for winter antioxidant defense.

Metal catalyzed ROS generation is kept in check by minimizing free Fe^{2+} and Cu^+ by binding them to blood/hemolymph proteins during transport (transferrin, ceruloplasmin), intracellular storage proteins (ferritin, metallothionein), or locking them in functional proteins (e.g., hemoglobin, hemocyanin, cytochromes, etc.). Links between heavy metal binding proteins and insect cold hardiness have been discovered. Screening of cDNA libraries revealed that ferritin heavy chain was cold up-regulated by 4–5 fold in *E. solidaginis* (Storey and Storey 2010b) and ferritin protein levels increased 4-fold in the autumn in *D. ponderosae* and decreased strongly (–17.5 fold) in the spring (Bonnett et al. 2012). Beetles, *Apriona germari*, also showed 2–4 fold increases in transferrin transcripts in response to cold (4°C), heat (37°C) and other stresses (Lee et al. 2006). In freeze or desiccation tolerant species, enhanced metal binding might be especially important to lower metal-catalyzed ROS formation because freezing or desiccation greatly concentrates remaining body fluids. Furthermore, iron sequestering is an important part of innate immunity in animals; by denying a source of iron, microbial growth is inhibited (Cherayil 2011). Hence, up-regulation of iron binding proteins would lower the risk of bacterial infection at times when insects are poorly able to respond to infection (i.e., frozen or in diapause). Apart from copper binding, Viarengo et al. (1999) suggested that metallothionein may also act directly as an antioxidant due to its high content of cysteine residues (~30% of total amino acids). The high cysteine AFPs of many freeze avoiding insects may also have secondary actions as antioxidants (Zachariassen et al. 2004) and the lipoprotein INPs of some insects are also potent metal binding proteins (Trautsch et al. 2011).

Acknowledgements

Thanks to the many past and present members of the Storey lab who have contributed to unraveling the molecular secrets of insect cold hardiness. Research in the Storey lab is supported by a discovery grant from the NSERC Canada and the Canada Research Chair in Molecular Physiology to KBS.

Keywords: Metabolic rate depression, freeze tolerance, freeze avoidance, cryoprotective dehydration, cryoprotectants, antioxidant defense, heat shock proteins, antifreeze proteins, microRNA, proteomics

References

Anchordoguy, T.J., A.S. Rudolph, J.F. Carpenter and J.H. Crowe. 1987. Modes of interaction of cryoprotectants with membrane phospholipids during freezing. Cryobiology 24: 324–331.

Asgari, S. 2013. MicroRNA functions in insects. Insect Biochem. Mol. Biol. 43: 388–397.

Bale, J.S. 2002. Insects and low temperatures: from molecular biology to distributions and abundance. Philos. Trans. R. Soc. Lond. B Biol. Sci. 357: 849–862.

Bale, J.S. and S.A. Hayward. 2010. Insect overwintering in a changing climate. J. Exp. Biol. 213: 980–994.

Bartel, D.P. 2009. MicroRNAs: target recognition and regulatory functions. Cell 136: 215–233.

Bemani, M., H. Izadi, K. Mahdian, A. Khani and M. Amin Samih. 2012. Study on the physiology of diapause, cold hardiness and supercooling point of overwintering pupae of the pistachio fruit hull borer, *Arimania comaroffi*. J. Insect Physiol. 58: 897–902.

Biggar, K.K. and K.B. Storey. 2011. The emerging roles of microRNAs in the molecular responses of metabolic rate depression. J. Mol. Cell Biol. 3: 167–175.

Biggar, K.K., S.F. Kornfeld, Y. Maistrovski and K.B. Storey. 2012. MicroRNA regulation in extreme environments: differential expression of microRNAs in the intertidal snail *Littorina littorea* during extended periods of freezing and anoxia. Genom. Proteom. Bioinform. 10: 302–309.

Block, W. 2003. Water or ice? The challenge for invertebrate cold survival. Sci. Prog. 86: 77–101.

Bonnett, T.R., J.A. Robert, C. Pitt, J.D. Fraser, C.I. Keeling, J. Bohlmann and D.P. Huber. 2012. Global and comparative proteomic profiling of overwintering and developing mountain pine beetle, *Dendroctonus ponderosae* (Coleoptera: Curculionidae) larvae. Insect Biochem. Mol. Biol. 42: 890–901.

Campbell, E.M., A. Ball, S. Hoppler and A.S. Bowman. 2008. Invertebrate aquaporins: a review. J. Comp. Physiol. B 178: 935–955.

Carrasco, M.A., S.A. Buechler, R.J. Arnold, T. Sformo, B.M. Barnes and J.G. Duman. 2012. Investigating the deep supercooling ability of an Alaskan beetle, *Cucujus clavipes puniceus*, via high throughput proteomics. J. Proteomics 75: 1220–1234.

Chakraborty, N., M.A. Menze, H. Elmoazzen, H. Vu, M.L. Yarmush, S.C. Hand and M. Toner. 2012. Trehalose transporter from African chironomid larvae improves desiccation tolerance of Chinese hamster ovary cells. Cryobiology 64: 91–96.

Chen, B., T. Kayukawa, A. Monteiro and Y. Ishikawa. 2006. Cloning and characterization of the HSP70 gene, and its expression in response to diapauses and thermal stress in the onion maggot, *Delia antiqua*. J. Biochem. Mol. Biol. 39: 749–758.

Cherayil, B.J. 2011. The role of iron in the immune response to bacterial infection. Immunol. Res. 50: 1–9.

Churchill, T.A. and K.B. Storey. 1989. Metabolic correlates to glycerol biosynthesis in a freeze avoiding insect, *Epiblema scudderiana*. J. Comp. Physiol. B 159: 461–472.

Clark, M.S. and M.R. Worland. 2008. How insects survive the cold: molecular mechanisms—a review. J. Comp. Physiol. B 178: 917–933.

Clark, M.S., M.A. Thorne, J. Purać, G. Burns, G. Hillyard, Z.D. Popović, G. Grubor-Lajsić and M.R. Worland. 2009. Surviving the cold: molecular analyses of insect cryoprotective dehydration in the Arctic springtail *Megaphorura arctica* (Tullberg). BMC Genomics 10: 328.

Cornette, R. and T. Kikawada. 2011. The induction of anhydrobiosis in the sleeping chironomid: current status of our knowledge. IUBMB Life 63: 419–429.

Courteau, L., K.B. Storey and P.J. Morin. 2012. Differential expression of microRNA species in a freeze tolerant insect, *Eurosta solidaginis*. Cryobiology 65: 210–221.

Crowe, J.H., L.M. Crowe, A.E. Oliver, N. Tsvetkova, W. Wolkers and F. Tablin. 2001. The trehalose myth revisited: introduction to a symposium on stabilization of cells in the dry state. Cryobiology 43: 89–105.

Danks, H.V. 1987. Insect Dormancy: an Ecological Perspective. Biological Survey of Canada (Terrestrial Arthropods), Ottawa.

Danks, H.V. 2006. Insect adaptations to cold and changing environments. Can. Entomol. 138: 1–23.

Denlinger, D.L. and R.E. Lee. 2010. Low Temperature Biology of Insects. Cambridge University Press, Cambridge.

Doucet, D., V.K. Walker and W. Qin. 2009. The bugs that came in from the cold: molecular adaptations to low temperatures in insects. Cell Mol. Life Sci. 66: 1404–1418.

Dragovic, Z., S.A. Broadley, Y. Shomura, A. Bracher and F.U. Hartl. 2006. Molecular chaperones of the Hsp110 family act as nucleotide exchange factors of Hsp70s. EMBO J. 25: 2519–2528.

Duman, J.G. 2001. Antifreeze and ice nucleator proteins in terrestrial arthropods. Annu. Rev. Physiol. 63: 327–357.

Ebert, M.S. and P.A. Sharp. 2012. Roles for microRNAs in conferring robustness to biological processes. Cell 149: 515–524.

Elnitsky, M.A., S.A.L. Haywood, J.P. Rinehart, D.A. Denlinger and R.E. Lee. 2008. Cryoprotective dehydration and the resistance to inoculative freezing in the Antarctic midge, *Belgica antarctica*. J. Exp. Biol. 211: 524–530.

Feder, M.E. and G.E. Hofmann. 1999. Heat-shock proteins, molecular chaperones, and the stress response: evolutionary and ecological physiology. Annu. Rev. Physiol. 6: 243–282.

Furusawa, T., A. Konishi, D. Sakano, E. Kotani, Y. Sugimura, J.M. Storey and K.B. Storey. 1999. Regulatory properties of phosphofructokinase in the eggs of the silkworm, *Bombyx mori*. J. Sericulture Sci. Japan 68: 181–194.

Graether, S.P. and B.D. Sykes. 2004. Cold survival in freeze-intolerant insects: the structure and function of beta-helical antifreeze proteins. Eur. J. Biochem. 271: 3285–3296.

Hao, Y.J., W.S. Li, Z.B. He, F.L. Si, Y. Ishikawa and B. Chen. 2012. Differential gene expression between summer and winter diapause pupae of the onion maggot *Delia antiqua*, detected by suppressive subtractive hybridization. J. Insect Physiol. 58: 1444–1449.

Hakim, A., J.B. Nguyen, K. Basu, D.F. Zhu, D. Thakral, P.L. Davies, F.J. Isaacs, Y. Modis and W. Wuyi Meng. 2013. Crystal structure of an insect antifreeze protein and its implications for ice binding. J. Biol. Chem. 288: 12295–12304.

Hermes-Lima, M. 2004. Oxygen in biology and biochemistry: role of free radicals. pp. 319–368. *In*: K.B. Storey (ed.). Functional Metabolism: Regulation and Adaptation. Wiley-Liss, Hoboken.

Hochachka, P.W. 1986. Defense strategies against hypoxia and hypothermia. Science 231: 234–241.

Holden, H.A. and K.B. Storey. 2011. Reversible phosphorylation regulation of NADPH-linked polyol dehydrogenase in the freeze-avoiding gall moth, *Epiblema scudderiana*: role in glycerol metabolism. Arch. Insect Biochem. Physiol. 77: 32–44.

Irwin, J.T., V.A. Bennett and R.E. Lee. 2001. Diapause development in frozen larvae of the goldenrod gall fly, *Eurosta solidaginis* Fitch (Diptera: Tephritidae). J. Comp. Physiol. B 171: 181–188.

Izumi, Y., S. Sonoda and H. Tsumuki. 2007. Effects of diapause and cold-acclimation on the avoidance of freezing injury in fat body tissue of the rice stem borer, *Chilo suppressalis* Walker. J. Insect Physiol. 53: 685–690.

Izumi, Y., S. Sonoda, H. Yoshida, H.V. Danks and H. Tsumuki. 2006. Role of membrane transport of water and glycerol in the freeze tolerance of the rice stem borer, *Chilo suppressalis* Walker (Lepidoptera: Pyralidae). J. Insect Physiol. 52: 215–220.

Joanisse, D.R. and K.B. Storey. 1994. Mitochondrial enzymes during overwintering in two species of cold-hardy gall insects. Insect Biochem. Molec. Biol. 24: 145–150.

Joanisse, D.R. and K.B. Storey. 1996a. Fatty acid content and enzymes of fatty acid metabolism in overwintering cold-hardy gall insects. Physiol. Zool. 69: 1079–1095.

Joanisse, D.R. and K.B. Storey. 1996b. Oxidative stress and antioxidants during overwintering in larvae of cold-hardy goldenrod gall insects. J. Exp. Biol. 199: 1483–1491.

Kanamori, Y., A. Saito, Y. Hagiwara-Komoda, D. Tanaka, K. Mitsumasu, S. Kikuta, M. Watanabe, R. Cornette, T. Kikawada and T. Okuda. 2010. The trehalose transporter 1 gene sequence is conserved in insects and encodes proteins with different kinetic properties involved in trehalose import into peripheral tissues. Insect Biochem. Mol. Biol. 40: 30–37.

Kayukawa, T. and Y. Ishikawa. 2009. Chaperonin contributes to cold hardiness of the onion maggot *Delia antiqua* through repression of depolymerization of actin at low temperatures. PLoS One 4(12): e8277.

Kayukawa, T., B. Chen, S. Hoshizaki and Y. Ishikawa. 2007. Upregulation of a desaturase is associated with the enhancement of cold hardiness in the onion maggot, *Delia antiqua*. Insect Biochem. Mol. Biol. 37: 1160–1167.

Kayukawa, T., B. Chen, S. Miyazaki, K. Itoyama, T. Shinoda and Y. Ishikawa. 2005. Expression of the mRNA for the t-complex polypeptide-1, a subunit of chaperonin CCT, is upregulated in association with increased cold hardiness in *Delia antiqua*. Cell Stress Chaperones 10: 204–210.

Kikawada, T., A. Saito, Y. Kanamori, Y. Nakahara, K. Iwata, D. Tanaka, M. Watanabe and T. Okuda. 2007. Trehalose transporter 1, a facilitated and high-capacity trehalose transporter, allows exogenous trehalose uptake into cells. Proc. Natl. Acad. Sci. USA 104: 11585–11590.

Kikuta, S., Y. Hagiwara-Komoda, H. Noda and T.A. Kikawada. 2012. A novel member of the trehalose transporter family functions as an H^+-dependent trehalose transporter in the reabsorption of trehalose in Malpighian tubules. Front. Physiol. 3: 290.

Koštál, V., T. Urban, L. Rimnáčová, P. Berková and P. Simek. 2013. Seasonal changes in minor membrane phospholipid classes, sterols and tocopherols in overwintering insect, *Pyrrhocoris apterus*. J. Insect Physiol. 59: 934–941.

Koštál, V., H. Zahradníčková and P. Šimek. 2011. Hyperprolinemic larvae of the drosophilid fly, *Chymomyza costata*, survive cryopreservation in liquid nitrogen. Proc. Natl. Acad. Sci. USA 108: 13041–13046.

Krivoruchko, A. and K.B. Storey. 2010. Forever young: mechanisms of anoxia tolerance in turtles and possible links to longevity. Oxid. Med. Cell. Longevity 3: 186–198.

Morin, P. and K.B. Storey. 2009. Mammalian hibernation: differential gene expression and novel application of epigenetic controls. Int. J. Devel. Biol. 53: 433–442.

Kültz, D. 2005. Molecular and evolutionary basis of the cellular stress response. Annu. Rev. Physiol. 67: 225–257.

Lee, K.S., B.Y. Kim, H.J. Kim, S.J. Seo, H.J. Yoon, Y.S. Choi, I. Kim, Y.S. Han, Y.H. Je, S.M. Lee, D.H. Kim, H.D. Sohn and B.R. Jin. 2006. Transferrin inhibits stress-induced apoptosis in a beetle. Free Rad. Biol. Med. 41: 1151–1161.

Lee, R.E. and D.L. Denlinger. 1991. Insects at Low Temperature. Chapman and Hall, New York.

Lee, R.E. and D.L. Denlinger. 2010. Rapid cold-hardening: ecological significance and underpinning mechanisms. pp. 35–58. *In*: D.L. Denlinger and R.E. Lee (eds.). Low Temperature Biology of Insects. Cambridge University Press, Cambridge.

Levis, N.A., S.X. Yi and R.E. Lee. 2012. Mild desiccation rapidly increases freeze tolerance of the goldenrod gall fly, *Eurosta solidaginis*: evidence for drought-induced rapid cold-hardening. J. Exp. Biol. 215: 3768–3773.

Li, A. and D.L. Denlinger. 2008. Rapid cold hardening elicits changes in brain protein profiles of the flesh fly, *Sarcophaga crassipalpis*. Insect Mol. Biol. 17: 565–572.

Lopez-Martinez, G., M.A. Elnitsky, J.B. Benoit, R.E. Lee and D.L. Denlinger. 2008. High resistance to oxidative damage in the Antarctic midge *Belgica antarctica*, and developmentally linked expression of genes encoding superoxide dismutase, catalase and heat shock proteins. Insect Biochem. Mol. Biol. 38: 796–804.

Lucas, K. and A.S. Raikhel. 2013. Insect microRNAs: biogenesis, expression profiling and biological functions. Insect Biochem. Mol. Biol. 43: 24–38.

Lyons, P.J., J.J. Poitras, L.A. Courteau, K.B. Storey and P.J. Morin. 2013. Identification of differentially regulated microRNAs in cold-hardy insects. Cryo-Lett. 34: 83–89.

MacMillan, H.A. and B.J. Sinclair. 2011. Mechanisms underlying insect chill-coma. J. Insect Physiol. 57: 12–20.

MacMillan, H.A., C.M. Williams, J.F. Staples and B.J. Sinclair. 2012. Reestablishment of ion homeostasis during chill-coma recovery in the cricket *Gryllus pennsylvanicus*. Proc. Natl. Acad. Sci. USA 109: 20750–20755.

McCord, J.M. 2008. Superoxide dismutase, lipid peroxidation, and bell-shaped dose response curves. Dose Response 6: 223–238.

McMullen, D.C. and K.B. Storey. 2008a. Suppression of Na^+K^+-ATPase activity by reversible phosphorylation over the winter in a freeze-tolerant insect. J. Insect Physiol. 54: 1023–1027.

McMullen, D.C. and K.B. Storey. 2008b. Mitochondria of cold hardy insects: responses to cold and hypoxia assessed at enzymatic, mRNA and DNA levels. Insect Biochem. Mol. Biol. 38: 367–373.

McMullen, D.C., C.J. Ramnanan and K.B. Storey. 2010. In cold-hardy insects, seasonal, temperature, and reversible phosphorylation controls regulate sarco/endoplasmic reticulum Ca^{2+}-ATPase (SERCA). Physiol. Biochem. Zool. 8: 677–686.

Michaud, M.R. and D.L. Denlinger. 2010. Genomics, proteomics and metabolomics: finding the other players in insect cold-tolerance. pp. 91–115. *In*: D.L. Denlinger and R.E. Lee (eds.). Low Temperature Biology of Insects. Cambridge University Press, Cambridge.

Morewood, W.D. and R.A. Ring. 1998. Revision of the life history of the high Arctic moth *Gynaephora groenlandica* (Wocke) (Lepidoptera: Lymantriidae). Can. J. Zool. 76: 1371–1381.

Muise, A.M. and K.B. Storey. 1997. Reversible phosphorylation of fructose-1,6-bisphosphatase mediates enzyme role in glycerol metabolism in the freeze avoiding gall moth *Epiblema scudderiana*. Insect Biochem. Mol. Biol. 27: 617–623.

Muise, A.M. and K.B. Storey. 1999. Regulation of glycogen synthetase in a freeze-avoiding insect: role in cryoprotectant glycerol metabolism. Cryo-Lett. 20: 223–228.

Nickell, P.K., S. Sass, D. Verleye, E.M. Blumenthal and J.G. Duman. 2013. Antifreeze proteins in the primary urine of larvae of the beetle *Dendroides canadensis*. J. Exp. Biol. 216: 1695–1703.

Pfister, T.D. and K.B. Storey. 2006a. Insect freeze tolerance: roles of protein phosphatases and protein kinase A. Insect Biochem. Mol. Biol. 36: 18–24.

Pfister, T.D. and K.B. Storey. 2006b. Responses of protein phosphatases and cAMP-dependent protein kinase in a freeze-avoiding insect, *Epiblema scudderiana*. Arch. Insect Biochem. Physiol. 62: 43–54.

Philip, B.N. and R.E. Lee. 2010. Changes in abundance of aquaporin-like proteins occur concomitantly with seasonal acquisition of freeze tolerance in the goldenrod gall fly, *Eurosta solidaginis*. J. Insect Physiol. 56: 679–685.

Philip, B.N., S.X. Yi, M.A. Elnitsky and R.E. Lee. 2008. Aquaporins play a role in desiccation and freeze tolerance in larvae of the goldenrod gall fly, *Eurosta solidaginis*. J. Exp. Biol. 211: 1114–1119.

Ragland, G.J., D.L. Denlinger and D.A. Hahn. 2010. Mechanisms of suspended animation are revealed by transcript profiling of diapause in the flesh fly. Proc. Natl. Acad. Sci. USA 107: 14909–14914.

Rider, M.H., N. Hussain, S.M. Dilworth, J.M. Storey and K.B. Storey. 2011. AMP-activated protein kinase and metabolic regulation in cold-hardy insects. J. Insect Physiol. 57: 1453–1462.

Rinehart, J.P., S.A. Hayward, M.A. Elnitsky, L.H. Sandro, R.E. Lee and D.L. Denlinger. 2006. Continuous up-regulation of heat shock proteins in larvae, but not adults, of a polar insect. Proc. Natl. Acad. Sci. USA 103: 14223–14227.

Rinehart, J.P., A. Li, G.D. Yocum, R.M. Robich, S.A. Hayward and D.L. Denlinger. 2007. Up-regulation of heat shock proteins is essential for cold survival during insect diapause. Proc. Natl. Acad. Sci. USA 104: 11130–11137.

Rinehart, J.P., R.M. Robich and D.L. Denlinger. 2010. Isolation of diapause-regulated genes from the flesh fly, *Sarcophaga crassipalpis*, by suppressive subtractive hybridization. J. Insect Physiol. 56: 603–609.

Sajan, M.P., G. Bandyopadhyay, Y. Kanoh, M.L. Standaert, M.J. Quon, B.C. Reed, I. Dikic and R.V. Farese. 2002. Sorbitol activates atypical protein kinase C and GLUT4 glucose transporter translocation/glucose transport through proline-rich tyrosine kinase-2, the extracellular signal-regulated kinase pathway and phospholipase D. Biochem. J. 362: 665–674.

Salt, R.W. 1961. Principles of insect cold hardiness. Annu. Rev. Entomol. 6: 55–74.

Sformo, T., K. Walters, K. Jeannet, B. Wowk, G.M. Fahy, B.M. Barnes and J.G. Duman. 2010. Deep supercooling, vitrification and limited survival to 100°C in the Alaskan beetle *Cucujus clavipes puniceus* (Coleoptera: Cucujidae) larvae. J. Exp. Biol. 213: 502–509.

Sinclair, B.J., A. Addo-Bediako and S.L. Chown. 2003. Climatic variability and the evolution of insect freeze tolerance. Biol. Rev. Cambr. Philos. Soc. 78: 181–195.

Sinclair, B.J., A.G. Gibbs and S.P. Roberts. 2007. Gene transcription during exposure to, and recovery from, cold and desiccation stress in *Drosophila melanogaster*. Insect Mol. Biol. 16: 435–443.

Somme, L. 1982. Supercooling and winter survival in terrestrial arthropods. Comp. Biochem. Physiol. A 73: 519–543.

Sonoda, S., K. Fukumoto, Y. Izumi, H. Yoshida and H. Tsumuki. 2006. Cloning of heat shock protein genes (hsp90 and hsc70) and their expression during larval diapause and cold tolerance acquisition in the rice stem borer, *Chilo suppressalis* Walker. Arch. Insect Biochem. Physiol. 63: 36–47.

Sørensen, J.G. and M. Holmstrup. 2011. Cryoprotective dehydration is widespread in Arctic springtails. J. Insect Physiol. 57: 1147–1153.

Storey, J.M. and K.B. Storey. 1990. Carbon balance and energetics of cryoprotectant synthesis in a freeze tolerant insect: responses to perturbation by anoxia. J. Comp. Physiol. B 160: 77–84.

Storey, J.M. and K.B. Storey. 2004a. Cold hardiness and freeze tolerance. pp. 473–503. *In*: K.B. Storey (ed.). Functional Metabolism: Regulation and Adaptation. Wiley-Liss, Hoboken.

Storey, K.B. and J.M. Storey. 1988. Freeze tolerance in animals. Physiol. Rev. 68: 27–84.

Storey, K.B. and J.M. Storey. 1991. Biochemistry of cryoprotectants. pp. 64–93. *In*: R.E. Lee and D.L. Denlinger (eds.). Insects at Low Temperature. Chapman and Hall, New York.

Storey, K.B. and J.M. Storey. 2007. Tribute to P.L. Lutz: Putting life on 'pause'—molecular regulation of hypometabolism. J. Exp. Biol. 210: 1700–1714.

Storey, K.B. and J.M. Storey. 2010. Oxygen: stress and adaptation in cold hardy insects. pp. 141–165. *In*: D.L. Denlinger and R.E. Lee (eds.). Low Temperature Biology of Insects. Cambridge University Press, Cambridge.

Storey, K.B. and J.M. Storey. 2011. Heat shock proteins and hypometabolism: adaptive strategy for proteome preservation. Res. Rep. Biol. 2: 57–68.

Storey, K.B. and J.M. Storey. 2012. Insect cold hardiness: recent advances in metabolic, gene and protein adaptation. Can. J. Zool. 90: 456–475.

Storey, K.B. and J.M. Storey. 2013. Molecular biology of freeze tolerance in animals. Comprehensive Physiol. 3: 1283–1308.

Storey, K.B. and C.-W. Wu. 2013. Stress response and adaptation: a new molecular toolkit for the 21st century. Comp. Biochem. Physiol. A 165: 417–428.

Teets, N.M. and D.L. Denlinger. 2013. Autophagy in Antarctica: combating dehydration stress in the world's southernmost insect. Autophagy 9: 629–631.

Teets, N.M. and D.L. Denlinger. 2014. Surviving in a frozen desert: environmental stress physiology of terrestrial Antarctic arthropods. J. Exp. Biol. 217: 84–93.

Teets, N.M., Y. Kawarasaki, R.E. Lee and D.L. Denlinger. 2013. Expression of genes involved in energy mobilization and osmoprotectant synthesis during thermal and dehydration stress in the Antarctic midge, *Belgica antarctica*. J. Comp. Physiol. B. 183: 189–201.

Teets, N.M., J.T. Peyton, G.J. Ragland, H. Colinet, D. Renault, D.A. Hahn and D.L. Denlinger. 2012a. Combined transcriptomic and metabolomic approach uncovers molecular mechanisms of cold tolerance in a temperate flesh fly. Physiol. Genomics 44: 764–777.

Teets, N.M., J.T. Peyton, H. Colinet, D. Renault, J.L. Kelley, Y. Kawarasaki, R.E. Lee and D.L. Denlinger. 2012b. Gene expression changes governing extreme dehydration tolerance in an Antarctic insect. Proc. Natl. Acad. Sci. USA 109: 20744–20749.

Telonis-Scott, M., R. Hallas, S.W. McKechnie, C.W. Wee and A.A. Hoffmann. 2009. Selection for cold resistance alters gene transcript levels in *Drosophila melanogaster*. Insect Physiol. 55: 549–555.

Thorne, M.A., M.R. Worland, R. Feret, M.J. Deery, K.S. Lilley and M.S. Clark. 2011. Proteomics of cryoprotective dehydration in *Megaphorura arctica* Tullberg 1876 (Onychiuridae: Collembola). Insect Mol. Biol. 20: 303–310.

Trautsch, J., B.O. Rosseland, S.A. Pedersen, E. Kristiansen and K.E. Zachariassen. 2011. Do ice nucleating lipoproteins protect frozen insects against toxic chemical agents? J. Insect Physiol. 57: 1123–1126.

Viarengo, A., B. Burlando, M. Cavaletto, B. Marchi, E. Ponzano and J. Blasco. 1999. Role of metallothionein against oxidative stress in the mussel *Mytilus galloprovincialis*. Am. J. Physiol. 277: R1612–R1619.

Walters, K.R., A.S. Serianni, Y. Voituron, T. Sformo, B.M. Barnes and J.G. Duman. 2011. A thermal hysteresis-producing xylomannan glycolipid antifreeze associated with cold tolerance is found in diverse taxa. J. Comp. Physiol. B 181: 631–640.

Wang, H., Z. Lei, X. Li and R.D. Oetting. 2011. Rapid cold hardening and expression of heat shock protein genes in the B-biotype *Bemisia tabaci*. Environ. Entomol. 40: 132–139.

Wharton, D.A. 2003. The environmental physiology of Antarctic terrestrial nematodes: a review. J. Comp. Physiol. B 173: 621–628.

Wharton, D.A. 2011. Cold tolerance of New Zealand alpine insects. J. Insect Physiol. 57: 1090–1095.

Yu, S.O., A. Brown, A.J. Middleton, M.M. Tomczak, V.K. Walker and P.L. Davies. 2010. Ice restructuring inhibition activities in antifreeze proteins with distinct differences in thermal hysteresis. Cryobiology 61: 327–334.

Zachariassen, K.E. and E. Kristiansen. 2000. Ice nucleation and antinucleation in nature. Cryobiology 41: 257–279.

Zachariassen, K.E., E. Kristiansen and S.A. Pedersen. 2004. Inorganic ions in cold-hardiness. Cryobiology 48: 126–133.

Zhang, G., J.M. Storey and K.B. Storey. 2011. Chaperone proteins and winter survival by a freeze tolerant insect. J. Insect Physiol. 57: 1115–1122.

Zhang, J., K.E. Marshall, J.T. Westwood, M.S. Clark and B.J. Sinclair. 2011. Divergent transcriptomic responses to repeated and single cold exposures in *Drosophila melanogaster*. J. Exp. Biol. 214: 4021–4029.

9

Evolutionary Ecology of Insect Immunity

Gerrit Joop[1] and *Andreas Vilcinskas*[1,2,*]

Introduction

The evolutionary ecology of insect immunity is the integrated study of insect immune defenses. This not only takes into account the evolutionary background or history of the focal organism and ecological factors such as the species ecology or species interactions (including host–parasite interactions), but also the underlying molecular mechanisms at the genetic and epigenetic levels.

Insects are the most successful taxonomic group in terms of species diversity, thanks to their evolutionary ability to adapt to different environments, including pressure from a diverse range of parasites and pathogens (Schmid-Hempel 2011). Parasites and pathogens on the other hand need to avoid, overcome or suppress host defenses. The comparatively short generation times and often small genomes of parasites and pathogens allow them to evolve rapidly (Berenos et al. 2009, Schulte et al. 2013) forcing the insect host to co-evolve and genetically adapt (Schmid-Hempel 2011), as shown in a number of experimental co-evolution approaches (Wegner et al. 2008, Schulte et al. 2010, Masri et al.

[1] Institute for Phytopathology and Applied Zoology University of Giessen, Heinrich-Buff-Ring 26–32, 35392 Gießen, Germany.
   Email: gerrit.joop@agrar.uni-giessen.de
[2] Fraunhofer IME, Projectgroup Bioressources, Winchesterstr.2, 35394 Gießen, Germany.
   Email: Andreas.Vilcinskas@agrar.uni-giessen.de
* Corresponding author

2013, reviewed in Kerstes and Martin 2013). This is often described as an arms race (Van Valen 1973) and requires a dynamic host immune system and genetic plasticity comparable to that of parasites and pathogens (Sackton et al. 2007). This also explains why host immunity genes evolve much faster than other parts of the genome (Lazzaro and Little 2009).

Insects also interact with beneficial microbes such as symbionts (see also Dettner, this book). Here, recent studies suggest that the insect immune system also plays a role in the maintenance and protection of endosymbionts (Login et al. 2011). Again, this requires a dynamic host immune system and genetic plasticity, because the balance between maintaining and protecting symbionts on the one hand and the ability to recognize and fend off parasites and pathogens on the other hand need to be traded off against each other carefully (Haine et al. 2008).

The insect defense tool box includes the innate immune system, consisting of a complex array of molecules and layered cellular mechanisms that recognize and combat intruders. It also includes the recently-discovered phenomena such as (trans-generational) immune priming (Moret and Schmid-Hempel 2000, Roth et al. 2009), symbionts (Haine 2008) and the environmental impact on immunity, including dietary effects (Freitak et al. 2007) and microbiota (Dong et al. 2009). Recent research shows the importance of extended or integrated immune defenses (*sensu* Otti et al., pers. comm.), e.g., by the means of chemical defense (Gross et al. 2008, Joop et al. 2014) or potential parasites carried along as predation-defenses (Vilcinskas et al. 2013a). The innate immune system and the extended immune defense mechanisms appear to be carefully balanced with each other (Schmidberg et al. 2013, Joop et al. 2014).

The availability of genome sequences and immune gene expression profiles for several insect species allows us to understand the evolution and ecology of insect immunity by combining experimental approaches and comparative studies between species or within species exposed to different parasites and pathogens (Vilcinskas 2013). Such studies often emphasize genes encoding antimicrobial peptides and small proteins (AMPs), i.e., naturally-occurring chemical defense molecules used by eukaryotic cells against a broad spectrum of pathogens and parasites, including viruses, bacteria, fungi and even protozoans (Zasloff 2002). Although some AMPs are evolutionarily conserved across taxa, including nematodes, ancient insects and humans (Kim and Ausubel 2005, Altincicek and Vilcinskas 2007, Wiesner and Vilcinskas 2010) others are more restricted. For example, gloverins are only found in the Lepidoptera, and coleptericins appear to be restricted to the Coleoptera (Vilcinskas 2013). Vertebrate AMPs are not only involved in killing pathogens, but also promote immunomodulation, wound repair, and apoptosis. Besides killing pathogens, also invertebrate AMPs cover additional functions, such as immunomodulation (Easton et al. 2009) and the control of endosymbionts (Login et al. 2011).

In this article, we discuss the immense evolutionary plasticity of immunity in insects, starting with an overview of insect immune defenses. Based on this, we introduce two levels of extended immune defense (external and internal), and consider how these may be traded-off with each other or with further aspects of insect immune defense. We conclude by discussing the consequences of such trade-offs, focusing on experimental evolution and co-evolutionary approaches, as well as making use of comparative genomics, proteomics and latest technologies in this evolutionary framework.

Insect Innate Immune Defense

Classical insect immune defense comprises three levels: (i) the physicochemical barrier, i.e., the cuticle including the periotrophic membrane that need to be crossed; (ii) the cellular defense, consisting of hemocytes, which are involved in wound healing, phagocytosis and multicellular encapsulation; and (iii) the humoral immune defense. The latter includes AMPs as well as the phenoloxidase (PO) cascade with melanin as the endpoint, which again is involved in wound healing.

The cellular and humoral defenses, and particularly the encapsulation response and the activity of PO, involve a number of potential trade-offs with other life-history parameters. The underlying general hypothesis is that immunity is costly and that in addition to resource allocation costs there might also be physiological costs (Sheldon and Verhulst 1996, Lochmiller and Deerenberg 2000). Because reproduction is a critical fitness trait, this is the major focus for research into the costs of immune defense. The first investigation of fitness costs associated with insect immunity was conducted by König and Schmid-Hempel (1995), who found that increased foraging effort by bumblebees reduced their encapsulation response. Similarly in calopterygid damselflies, reproductive effort in both sexes reduced the encapsulation response (Siva-Jothy et al. 1998). In the mealworm beetle *Tenebrio molitor*, the trade-off between reproduction and immune defense (measured as PO activity) was shown to be mediated by juvenile hormone (Rolff and Siva-Jothy 2002). In crickets, the trade-off between reproduction and immune defense differs between the sexes (Adamo et al. 2001). Similarly, female red flour beetles generally invest more in PO activity and survival than males (Hangartner et al. 2013), whereas female coenagrionid damselflies are more resistant towards fungal infection than males (Joop et al. 2006). These findings agree with Bateman's principle (Bateman 1948, Rolff 2002), which states that males gain fitness through increased mating whereas females gain fitness through increased longevity and therefore sexes should follow different immune investment strategies.

Like the trade-offs with the encapsulation response and PO activity, it has been shown that the induction of antimicrobial activity imposes reproductive fitness costs (e.g., Ahmed et al. 2002) and reduces survival (Moret and Schmid-Hempel 2000). Antimicrobial defense has also been considered by one of the rare studies investigating underlying genetic correlations. Cotter and colleagues (2004) found that *Spodoptera littoralis* hemocyte density was genetically correlated with the immunity-related traits cuticular melanization, PO activity and antibacterial activity. Although no genetic correlations were found in a similar study involving *Tribolium castaneum* (Hangartner et al. 2013), the genetic correlations among immune traits in the closely related species *T. molitor* are highly dependent on environmental conditions (Prokkola et al. 2013), highlighting the often underestimated impact of environmental factors.

The studies of antimicrobial activity discussed above mainly used inhibition assays as a proxy for AMPs, which do not distinguish between the various AMPs described in insects even though the number and presence of AMPs appears to be diverse among different species. At least 34 AMPs were identified in *Drosophila melanogaster* (Vilcinskas 2013), more than 50 in *Harmonia axyridis* (Vilcinskas et al. 2013b), but only 15 in *Anopheles gambiae* (Christophides et al. 2002) and 20 in *T. castaneum* (Zou et al. 2007, Altincicek et al. 2008). These differences probably reflect the varying selection pressures which correspond to the lifestyle of each species. For example, *D. melanogaster* feeds on rotten fruits and molds and may therefore be exposed to more pathogens compared to *A. gambiae*, which has a highly specialized blood diet normally containing fewer pathogens, but also blood cells that need to remain unharmed. *T. castaneum* feeds on grain and flour, which are of relatively poor nutritional quality and appear not to be highly exposed to microorganisms. In contrast, invasive species such as *H. axyridis* (Roy and Wajnberg 2008) require strong immune defenses to facilitate the colonization of new habitats and the unknown pathogens living there (Lee and Klasing 2004), potentially explaining the large number of AMPs in this species and the invasive success coming with it. Furthermore, whereas PO activity is a constitutive defense (Schmid-Hempel and Ebert 2003), AMPs are induced, with different AMPs expressed depending on the pathogen, thus providing a certain level of specificity (Le Maitre and Hoffmann 2007). Recent studies suggest that even greater specificity is possible, potentially even differentiating between bacterial strains. This defense capability (immune priming) is gained on first encounter with a pathogen, and protects following a second encounter with the same challenge (Sadd and Schmid-Hempel

2006, Pham et al. 2007, Roth et al. 2009). The underlying mechanisms remain unknown, but current studies are focusing on DSCAM alternative splicing (reviewed by Armitage et al. 2012), hemocytes (Pham et al. 2007) and epigenetic effects (Mukherjee 2012).

Another aspect of insect immune defense which thus far has only been described phenomenologically is trans-generational immune priming. This has been shown in a number of different insect species and occurs when the offspring of the primed individuals gain a protection benefit when encountering the same bacterial (Moret and Schmid-Hempel 2000, Sadd et al. 2005, Moret 2006, Roth et al. 2010, Rosengaus et al. 2013) or viral (Tidbury et al. 2011) pathogen as their parents. Again, the mechanism remains to be determined, but, similar to birds (Grindstaff et al. 2006), transfer might involve components of the egg (Sadd et al. 2005, Freitak et al. 2014) or epigenetic mechanisms may be involved (Jokela 2010, Roth et al. 2010, Chambers and Schneider 2012).

However these specific phenotypes may be gained mechanistically, the application of different AMPs in concert may induce a more specific host immune defense (Park et al. 2011). Comparative analysis of gene expression data in the same host species infected with different parasites or pathogens might provide helpful insight here. Comparative genomics also revealed another surprise in AMP research, namely the low number of genes contributing to the recognition and deterrence of parasites and pathogens in the so-called social insects (e.g., Evans et al. 2006 in *Apis mellifera*, Suen et al. 2011 in *Atta cephalotes*) as well as in pea aphids (Gerardo et al. 2010). The close coevolution of pea aphids with their symbionts has been discussed (Vilcinskas 2013), but in social insects it may be the presence of social defenses that allow for the reduced number of AMPs. Both the presence of symbionts or the reliance on alternative defense mechanisms allow the term 'immune defense' to be extended to include aspects such as behavior and chemical defenses.

Extended Immune Defenses—External

Insect immune defenses are generally considered to take place inside the body, but the first lines of defense are found well beyond the organism (Boughton et al. 2011, Fig. 1). In vertebrates, the inclusion of behavioral defenses is widely accepted, e.g., avoidance (e.g., Rohner et al. 2000) and grooming (e.g., Nunn and Altizer 2006) which occur prior to infection and reduce risk. After infection, further protective behaviors include a switch in dietary intake (Horak et al. 2003), the application of glandular secretions (Martin-Vivaldi et al. 2010), self-medication (Huffman 2001) and behavioral fever (Seebacher and Franklin 2005). Invertebrates use the same strategies, i.e., diet (König and Schmid-Hempel 1996), secretion

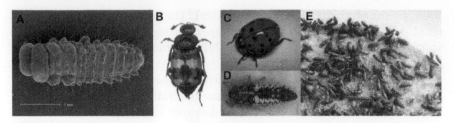

Figure 1. Model beetles in extended external and internal defenses. (A) Larva of the chrysomelid beetle *Phratora vitellinae* with nine pairs of dorsally located exocrine glands, where secretion has been shown to protect larvae from entomopathogenic fungi (Gross et al. 2008). (B) The burying beetle *Nicrophorus vespilloides*. In this biparental beetle, 34 compounds have been identified in head space, anal and oral secretions, the majority showing antimicrobial activity, supporting the hypothesis on burying beetles actively regulating the microbial community of carrion (Degenkolb et al. 2011). Adult (C) and larva (D) of the harlequin ladybird *Harmonia axyridis*. In both life stages, high concentrations of the antimicrobial substance harmonine are found, which can also inhibit parasites such as *Plasmodium falciparum* (Röhrich et al. 2012). (E) In the red flour beetle *Tribolium castaneum* adult secretions condition their environment, potentially also to the benefit of their offspring.

Color image of this figure appears in the color plate section at the end of the book.

(Fernandez-Martin et al. 2009), self-medication (Lee et al. 2006) and behavioral fever (Adamo 1998) although these are often overlooked. None of these defenses is without cost, and one may also expect trade-offs between the different levels of defense including resource-allocation trade-offs, true physiological trade-offs or the recently discussed phenomenon of simple energetic depletion (Bashir-Tanoli and Tinsley 2014).

One long neglected example is the external immune defense of the red flour beetle *T. castaneum* (Fig. 2). Although this species is an emerging model organism for immune defense (e.g., Roth et al. 2009, Roth et al. 2010), the ecological importance of its external immune defense strategy is rarely considered (but see Li et al. 2013, Joop et al. 2014). The beetles secrete antimicrobial substances into their environment (Loconti and Roth 1953) and by doing so condition their environment, which is also their food source. This secretion depends on the availability of tyrosine, which is also essential for one major aspect of the internal immune defense, the PO cascade (Cerenius and Söderhäll 2004), so a trade-off between these strategies might be expected. In a previous study (Joop et al. 2014) we tested experimentally the relationship between external and internal immune defenses, making use of the fact that internal and external immune defenses can be measured separately in *T. castaneum*. First, we maintained recently collected beetle strains under a selection regime for either high or low external defense for several generations and were able to obtain strains which showed stable differences. Then we measured several external and

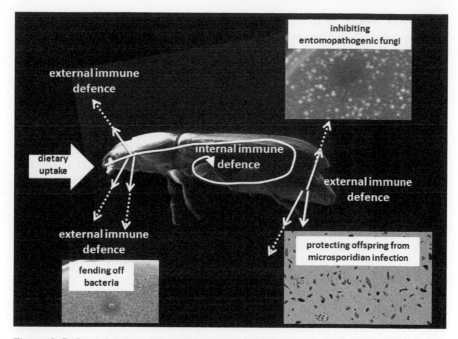

Figure 2. Defense levels in the red flour beetle, *Tribolium castaneum*. Whereas larvae are capable of the standard insect innate immune response, adults also possess an extended external defense involving two prothoracal and two post-abdominal glands. This external immune defense with methyl-para-benzoquinone and ethyl-para-benzoquinone as major active substances (Villaverde et al. 2007) covers the beetle's cuticle (solid arrows), spreads into the environment (dashed arrows) and conditions it. The quinone secretion is effective against various microorganisms (compare Unruh et al. 1997, Yezerski et al. 2007). Here, bacterial inhibition is shown using *Staphylococcus epidermis*, a common member of the human skin flora, which can cause severe problems on implants and in immunocompromised patients. Inhibition of entomopathogenic fungi is shown using *Beauveria bassiana*, an entomopathogenic fungus also used in pest control. The substances tested were methyl-para-benzoquinone (bacteria, Mitschke unpublished), a main component of the beetle secretion (Villaverde et al. 2007), and the complete beetle secretion (fungi, Rafaluk unpublished). Both the external secretion and the internal innate immune defense rely on the uptake of tyrosine (block arrow), so a trade-off between these two levels of defense is anticipated.

internal defense parameters in these lines in addition to several life history parameters and compared them with non-selected controls. Interestingly, we found that the beetles might initially have been in an optimum state (Viney et al. 2005) because both selection regimes (high and low external defense) led to reductions in some parameters. Finally, we found that the conditioning of the flour by the beetles influenced the survival of the offspring but that this depended on the presence of parasites. Essentially, environmental conditioning reduced the survival of the offspring in the absence of parasites but enhanced survival in their presence, confirming a trade-off related to external immune defense and supporting the idea of such antimicrobial secretions keeping the environment clean (Sokoloff 1974, Fig. 1E , Fig. 2).

The example discussed above illustrates the importance of extended external immune defenses. The external secretions in chrysomelid beetles have been studied extensively (e.g., Gross et al. 2008, Kirsch et al. 2011; also compare Burse and Boland this book, Fig. 1A), and other examples include the secretion of chemical defenses by earwigs (Gasch et al. 2013), and the secretions produced by burying beetles (Fig. 1B) that are used to preserve cadavers as food (Cotter and Kilner 2010a, Degenkolb et al. 2011). Extended immune defenses in social insects are more widely accepted (reviewed by Cremer et al. 2007). Such defenses include behavior, e.g., in carpenter ants, where it has been shown that trophallaxis supports colony growth even under high parasite burdens (Hamilton et al. 2010). Similarly, honeybees actively remove *Varroa*-infected brood from their nests, and transcriptomic analysis suggests that pleiotropic behavioral genes could be involved in social immunity and further behavioral phenotypes, potentially reducing the costs of defense (Le Conte et al. 2011). In ants and honeybees it has even been shown that social defenses reduce the investment in individual immune defenses (Castella et al. 2008, Simone et al. 2009, and compare above). This may be interpreted as another example of a trade-off between different levels of defense, here the external social strategy traded against an internal one. Individuals might pay the price if the social level fails, e.g., if new parasites are introduced, but under standard conditions the costs of immune defense and its maintenance might be reduced (compare Cotter and Kilner 2010b).

Social insects are also well known for their parental care, but they are not the only insects that provide such intense support for their offspring. A well-known example of parental care is the burying beetle *Nicrophorus vespilloides*, where both parents care for the larvae and provide them with pre-digested meat (Pukowski 1933). Burying beetles feed on carcasses, which they prevent from decay by applying antimicrobial substances (Cotter and Kilner 2010a, Degenkolb et al. 2011). The preserved meat provides food for adults and their offspring, as well as the environment for larval development. Therefore, such anal exudates may be seen as extended external immune defenses or even as a social immune response (*sensu* Cotter and Kilner 2010b). This secretion is phenotypically plastic, produced only during reproduction (Cotter and Kilner 2010a) and likely to be traded-off with other immune functions. Although no trade-off between individual internal defenses, measured as the encapsulation rate, and extended external immune defenses, measured as anal exudation, was found in *N. orbicollis* (Steiger et al. 2011), this was not the case in the closely related species *N. vespilloides*. Here, following wounding to induce individual internal immune defenses, the antibacterial activity of the anal exudates declined compared to non-wounded controls (Cotter et al. 2013), representing a trade-off between external and internal immune defenses. This supports the hypothesis, developed for social insects, that reduced individual internal immune defenses can be afforded if counterbalanced by external or social immunity. More such examples, ideally over a gradient of non-social to social insects, need to be tested before this hypothesis can be formally proven.

Extended Immune Defenses—Internal

An extended immune defense does not necessarily need to be outside the host, but may also be found in the gut, and may involve microbiota and symbionts. Microbiota refers to the microbial gut community of the host. As has been shown in a variety of organisms (Azambuja et al. 2005, Engel and Moran 2013), the composition of the microbiota is important for defense against pathogens and parasites. If it becomes unbalanced, e.g., due to stress (Feldhaar 2011), the door is left open for infections. However, it has also been shown that the microbiota composition can change after infection (Jakubowska et al. 2013) and may even be actively mediated by the dietary choice of the host (Ridley et al. 2012), back-linking to dietary intake and self-medication of the extended external immune defense.

Symbionts may be classified in a continuum with two endpoints, i.e., the primary symbionts, which are essential for the host and are usually transmitted vertically, and the secondary symbionts, which are not necessarily essential and tend to be transmitted horizontally. Symbionts help insects to adapt or survive in different environments or feed in a poor diet by substituting missing nutrients (Feldhaar 2011). This level of defense also has costs, e.g., aphids feeding on plant phloem have a much reduced repertoire of immunity-related genes, apparently paying the price for harboring symbionts (Gerado et al. 2010). In other species, part of the internal immune defense may be needed to keep symbionts under control (Login et al. 2011). Even chemical defenses may be produced by symbionts rather than by the insects themselves, as described in *Paederus* beetles (Piel 2002). Here, more research is needed to understand how such substances are synthesized and how insects apply them without poisoning themselves, their offspring or the environment.

A third class of extended internal immune defenses, so called biological weapons, has recently been described in the harlequin ladybird beetle (*Harmonia axyridis*), also known as the Asian ladybird or multicolored ladybird (Vilcinskas et al. 2013b, Fig. 1C and D). This ladybird originated from central Asia but is a successful invasive species in many parts of the world, where it was introduced to control aphids and other insect pests. Today, it causes severe problems by outcompeting indigenous ladybird species such as *Coccinella septempunctata* in Central Europe (Roy and Wanjnberg 2008). The tremendous success of *H. axyridis* over *C. septempunctata* in part reflects the lethality of *H. axyridis* eggs and larvae when they are eaten by the native ladybird, whereas the reciprocal form of intraguild predation is non-lethal to *H. axyridis* (Kajita et al. 2010). Intraguild predation is the major selective force among competing ladybird species (Gardiner et al. 2012). We have shown recently that this story of invasion and competition may be even more exciting than anticipated, because *H. axyridis* is not simply toxic to other ladybird species upon ingestion, but harbors microorganisms that are tolerated by the host but pathogenic to native predators (Vilcinskas et al. 2013a). *H. axyridis* carries large numbers of microsporidia closely related to *Nosema thomsoni* in its hemolymph (Fig. 3A). Microsporidia are obligate killing parasites, which need to enter eukaryotic cells for reproduction. In the ladybird system, the microsporidia appear to remain inactive in *H. axyridis*, as indicated by the absence of microsporidian gene expression in the eggs (Vilcinskas et al. 2013b), but become actively pathogenic when consumed by *C. septempunctata*, thus reducing the fitness of the competitor (Vilcinskas et al. 2013a) and hence protecting the original host (compare Haine 2008) and supporting its invasive success, as proposed in bumblebees (Colla et al. 2006) and honeybees (Chen and Huang 2010). The success of the biological weapon strategy depends on the original host (*H. axyrids*) tolerating and surviving the presence of millions of microsporidia in the hemolymph, effectively treating them as symbionts rather than pathogens. This could potentially be achieved by AMPs as part of the innate immune defense system, and it has been shown that *H. axyridis* produces more than 50 different AMPs that can be induced by bacterial injection (Vilcinskas et al. 2013b). However, also in *C. septempunctata* AMPs are upregulated upon infection (Schmidberg et al. 2013, Fig. 3C), making them an unlikely control mechanism against the microsporidia. Furthermore, AMPs are part of the induced innate immune defense system in insects and only become active following infection. However, the microsporidia associated with *H. axyridis* appear to remain inactive in the hemolymph throughout all life stages, from the egg to the adult beetle (Vilcinskas et al. 2013a). Therefore, AMPs seem unlikely to represent the control mechanism. Some form of constitutive control is more likely, and the most likely candidate is (17R,9Z)-1,17-diaminooctadec-9-ene more commonly known as harmonine, an alkaloid compound present at high concentrations in the hemolymph of *H. axyridis* but not any of the other ladybird species investigated thus far (Röhrich et al. 2012). This secondary metabolite has already been shown to have potent

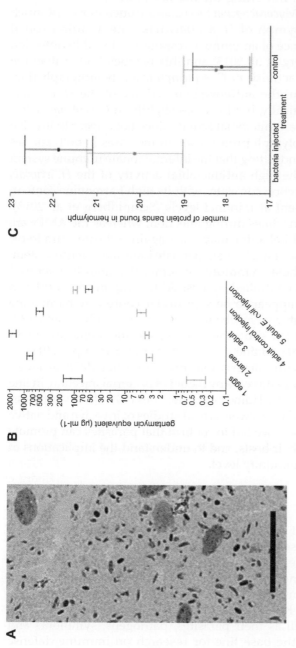

Figure 3. Defense trade-off in the harlequin ladybird *Harmonia axyridis*. Its microsporidia appear to be controlled constitutively, with harmonine as the most likely candidate substance (Schmidtberg et al. 2013). AMPs are unlikely candidates because these are induced by infection, and only reach the levels present in *C. septempunctata* hemolymph when infections are established, whereas harmonine levels decline following infection (Schmidtberg et al. 2013). This provides a classic scenario for a trade-off between constitutive (chemical) and induced (AMP) immune defense systems. (A) Light microscopy image on *H. axyridis* hemolymph, showing the high ratio of microsporidia to the relatively small number of hemocytes. Scale bar = 20 µm. (B) Inhibition of *Escherichia coli* by eggs, larval and adult hemolymph of two ladybird species (black: *H. axyridis*, gray: *C. septempunctata*), using agar diffusion tests. In addition, adults were injected with either buffer only (control) or with a bacterial suspension to identify differences following immune challenge (values are means ± SE). Figure re-drawn from Schmidtberg et al. (2013) using the supplementary raw data. (C) Number of SDS-PAGE protein bands following the fractionation of hemolymph from untreated and immune-challenged adults of two ladybird species (black: *H. axyridis*, gray: *C. septempunctata*). Bands analyzed using Quantity One 1-D analysis software, v4.6.9. Gel pictures kindly provided by H. Schmidtberg (Schmidtberg et al. 2013, Schmidtberg unpublished). Values are means ± SE.

biological activity, even inhibiting the human pathogens *Plasmodium falciparum* (malaria) and *Mycobacterium tuberculosis* (tuberculosis) (Röhrich et al. 2012). The hemolymph of *H. axyridis* has potent antimicrobial activity even in the absence of an immune response induced by infection or wounding (Schmidberg et al. 2013) and this is much higher than the standing antimicrobial activity in *C. septempunctata* hemolymph (Fig. 3B). Following infection, the antimicrobial activity of the *H. axyridis* hemolymph falls dramatically, but increases slightly in *C. septempunctata* (Fig. 3B). This change in antimicrobial activity does not appear to involve AMPs because the hemolymph protein content increases in both species following the challenge indicating that the internal innate immune system is activated (Fig. 3C). The high antimicrobial activity of the *H. axyridis* hemolymph prior to infection correlates with its high harmonine content, whereas AMPs are present at minimal levels (Schmidberg et al. 2013). The harmonine content declines during infection, whereas the AMPs are induced. This polarized behavior may once again indicate a trade-off between chemical defense and the internal innate immune defense system.

Whereas in other host–symbiont systems, the mainly bacterial symbionts appear to be controlled by host AMPs, the microsporidia in *H. axyridis* hemolymph appear to be kept under control by harmonine as a rather broad, constitutive mechanism. The presumably more costly production of AMPs, which also provide more specific immunity (Ebert and Schmid-Hempel 2003, Haine et al. 2008, Schmid-Hempel 2011), is only invoked to deal with bacteria that breach earlier defense levels [circumvented artificially in this case by the act of bacterial injection, in line with the findings reported by Haine and colleagues (2008)]. Nevertheless, additional experimental data, as well as field studies of invasive and native *H. axyridis* populations, are needed to confirm that parasites can promote the invasive success of their hosts, and to understand the implications of such a strategy at the community level.

Perspectives

Our examples show that understanding the trade-offs between different levels of immune defense is necessary to complete the picture of insect immune defense. Furthermore, the findings described here strengthen our approach on immune defense from an evolutionary and ecological perspective, and confirm that insect immunity is broader than innate immune defense alone. Knowledge concerning the ecology and habitat of insect species, as well as their parasites, pathogens, symbionts and microbiota, should be the base line for research on immune defense (Fig. 4). Searching for costs and benefits, we must consider not only the usual suspects, such as reproduction (Siva-Jothy et al. 1998, Hosken 2001,

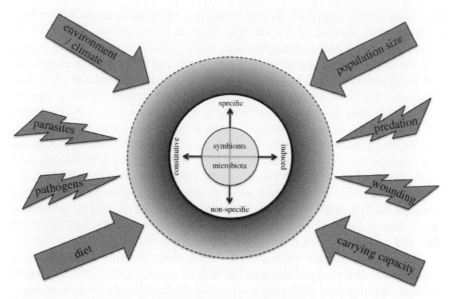

Figure 4. Factors contributing to the insect extended defense system. The central cross represents the continuous axis of the internal immune defense according to Schmid-Hempel and Ebert (2003; compare Schmid-Hempel 2011 for chart, and Boughton et al. 2011 for levels of defense), which also controls and maintains the extended internal immune defense system, i.e., symbionts, microbiota and associated parasites (inner circle). The thicker solid circle represents the physicochemical barrier, i.e., the cuticle itself and glandular secretions that cover it. The latter may diffuse into the environment and thereby extend the external immune defense (black-gray shading). Here, we also find behavioral defenses, such as grooming or avoidance. Together, behavioral defenses and external secretions, contribute to the extended external immune defense system. All these may be shaped by environmental factors, parasitism and predation.

McNamara et al. 2013), but also unanticipated benefits such as increased larval survival in *T. castaneum* in the combined presence of parental external immune defense secretions and parasites (Joop et al. 2014), or the microsporidia in *H. axyrides* serving as potential defense against predation by other ladybirds (Vilcinskas et al. 2013a). Both of these phenomena, i.e., a self-toxic external immune defense and the tolerance of large numbers of parasites, may have been rated as pure costs without the ecological context (compare Mondal 1990 in *T. castaneum*).

We must also be aware that trade-offs and antagonistic pleiotropic phenomena (Kraaijeveld and Godfray 1997) are highly dependent on environmental impact, including parasites and pathogens (Barribeau and Schmid-Hempel 2013, Joop et al. 2014), symbionts (Dion et al. 2011), microbiota (Koch and Schmid-Hempel 2011), population size (Gokhale et al. 2013) or climate (Thomas and Blanford 2003, Versace et al. 2014). We propose that the most inclusive and promising, although also the most elaborate, approach is experimental evolution, especially when incorporating several of the defense aspects listed above. This may be achieved using a correlated approach, where selection acts on one aspect and changes in other aspects are measured at the same time (Kraaijeveld and Godfray 1997, Rolff and Kraaijeveld 2003, Kraaijeveld et al. 2012). Alternatively, and potentially with a different outcome, several aspects may be applied in parallel, providing results that are more complex but presumably closer to nature (reviewed in Kerstes and Martin 2013). Ideally, the underlying genetics and gene expression data should be analyzed for all potential antagonists (e.g., host and all parasites) over all treatments using a careful experimental design (Kofler and Schlötterer 2013), making use of the latest sequencing technologies (Barribeau and Gerardo 2012) and modeling approaches (e.g., Mostowy and Engelstädter 2010). This is because immune defense, resistance and susceptibility are always a combined trait representing all players (Lambrechts et al. 2006).

The results of our experiments using flour beetles and ladybirds suggest that different parasites or pathogens are associated with different lines of defense and different trade-offs. At least in these two species, and also proposed for herbivorous insects in general (Cory and Hoover 2006), AMPs alone are not a sufficient defense against microsporidia and are not sufficient to maintain microsporidia as symbionts, thus requiring support from the extended immune defense system, in this case the chemical weaponry either self-produced or taken up from host plants. We agree with Chambers and Schneider (2012) that every process affecting immunity should be considered as part of the immune response and recommend looking at the big picture rather than snapshots to achieve a full understanding of the trade-offs between different levels of immunity (Fig. 4). This will help to discover, understand and characterize the broad phenomena in immune defense. The evolutionary background of such broad phenomena again is fundamental for studying host–parasite interactions in agriculture and human health.

Acknowledgements

The authors thank Thomas Degenkolb, Andreas Mitschke, Charlotte Rafaluk, Henrike Schmidtberg and Antje Thomas for sharing unpublished pictures and data, Oliver Otti for fruitfull discussions on the topic, and Christian Kollewe for support with SDS-PAGE interpretation. The authors acknowledge funding by the German Research Foundation (DFG) within the Priority Program 1399 on host-parasite coevolution with grants to G. Joop (JO 962/1-1) and A. Vilcinskas (VI 219/3-1; VI 219/3-2), by the Hessian Ministry of Science and Art's excellence initiative founding for the LOEWE Centre for Insect

Biotechnology & Bioresources, and by the Volkswagen Foundation with a grant to G. Joop (I/84778).

Keywords: trade-off, genotype, phenotype, extended immune defense, antimicrobial peptides, chemical defense, fitness costs, experimental evolution

References

Adamo, S.A. 1998. The specificity of behavioral fever in the cricket *Acheta domesticus*. J. Parasitol. 84: 529–533.

Adamo, S.A., M. Jensen and M. Younger. 2001. Changes in lifetime immunocompetence in male and female *Gryllus texensis* (formerly *G. integer*): trade-offs between immunity and reproduction. Anim. Behav. 62: 417–425.

Ahmed, A.M., S.L. Baggott, R. Maingon and H. Hurd. 2002. The costs of mounting an immune response are reflected in the reproductive fitness of the mosquito *Anopheles gambiae*. Oikos 97: 371–377.

Altincicek, B. and A. Vilcinskas. 2007. Identification of immune-related genes from an apterygote insect, the firebrat *Thermobia domestica*. Insect Biochem. Mol. Biol. 37: 726–731.

Altincicek, B., E. Knorr and A. Vilcinskas. 2008. Beetle immunity: Identification of immune-inducible genes from the model insect *Tribolium castaneum*. Dev. Comp. Immunol. 32: 585–595.

Armitage, S.A.O., R.Y. Freiburg, J. Kurtz and I.G. Bravo. 2012. The evolution of Dscam genes across the arthropods. BMC Evol. Biol. 12: 53.

Azambuja, P., E.S. Garcia and N.A. Ratcliffe. 2005. Gut microbiota and parasite transmission by insect vectors. Trends Parasitol. 21: 568–572.

Barribeau, S.M. and N.M. Gerardo. 2012. An evolutionarily and ecologically focused strategy for genome sequencing efforts. Heredity 108: 577–580.

Barribeau, S.M. and P. Schmid-Hempel. 2013. Qualitatively different immune response of the bumblebee host, *Bombus terrestris*, to infection by different genotypes of the trypanosome gut parasite, *Crithidia bombi*. Infect. Gen. Evol. 20: 249–256.

Bashir-Tanoli, S. and M.C. Tinsley. 2014. Immune response costs are associated with changes in resource acquisition and not resource reallocation. Funct. Ecol. 28: 1011–1019.

Bateman, A.J. 1948. Intra-sexual selection in *Drosophila*. Heredity 2: 349–368.

Bérénos, C., P. Schmid-Hempel and K.M. Wegner. 2009. Evolution of host resistance and trade-offs between virulence and transmission potential in an obligately killing parasite. J. Evol. Biol. 22: 2049–2056.

Boughton, R.K., G. Joop and S.A.O. Armitage. 2011. Outdoor immunology: methodological considerations for ecologists. Funct. Ecol. 25: 81–100.

Castella, G., M. Chapuisat, Y. Moret and P. Christe. 2008. The presence of conifer resin decreases the use of the immune system in wood ants. Ecol. Entomol. 33: 408–412.

Cerenius, L. and K. Söderhäll. 2004. The prophenoloxidase-activating system in invertebrates. Immunol. Rev. 198: 116–126.

Chambers, M.C. and D.S. Schneider. 2012. Pioneering immunology: insect style. Curr. Opin. Immunol. 24: 10–14.

Chen, Y.P. and Z.Y. Huang. 2010. *Nosema ceranae*, a newly identified pathogen of *Apis mellifera* in the USA and Asia. Apidologie 41: 364–374.

Christophides, G.K., E. Zdobnov, C. Barillas-Mury, E. Birney, S. Blandin, C. Blass, P.T. Brey, F.H. Collins, A. Danielli, G. Dimopoulos, C. Hetru, N.T. Hoa, J.A. Hoffmann, S.M. Kanok, I. Letunic, E.A. Levashina, T.G. Loukeris, G. Lycett, S. Meister, K. Michel, L.F. Moita, H.-M. Müller, M.A. Osta, S.M. Paskewitz, J.-M. Reichhart, A. Rzhetsky, L. Troxler,

K.D. Vernick, D. Vlachou, J. Volz, C. von Mering, J. Xu, L. Zheng, P. Bork and F.C. Kafatos. 2002. Immunity-related genes and gene families in *Anopheles gambiae*. Science 298: 159–165.

Colla, S.R., M.C. Otterstatter, R.J. Gegear and J.D. Thomson. 2006. Plight of the bumble bee: Pathogen spillover from commercial to wild populations. Biol. Cons. 129: 461–467.

Cory, J.S. and K. Hoover. 2006. Plant-mediated effects in insect–pathogen interactions. Trends Ecol. Evol. 21: 278–286.

Cotter, S.C., L.E.B. Kruuk and K. Wilson. 2004. Costs of resistance: genetic correlations and potential trade-offs in an insect immune system. J. Evol. Biol. 17: 421–429.

Cotter, S.C. and R.M. Kilner. 2010a. Sexual division of antibacterial resource defence in breeding burying beetles, *Nicrophorus vespilloides*. J. Anim. Ecol. 79: 35–43.

Cotter, S.C. and R.M. Kilner. 2010b. Personal immunity versus social immunity. Behav. Ecol. 21: 663–668.

Cotter, S.C., J.E. Littlefair, P.J. Grantham and R.M. Kilner. 2013. A direct physiological trade-off between personal and social immunity. J. Anim. Ecol. 82: 846–853.

Cremer, S., S.A.O. Armitage and P. Schmid-Hempel. 2007. Social immunity. Curr. Biol. 17: R693–R702.

Degenkolb, T., R.-A. Düring and A. Vilcinskas. 2011. Secondary metabolites released by the burying beetle *Nicrophorus vespilloides*: Chemical analyses and possible ecological functions. J. Chem. Ecol. 37: 724–735.

Dion, E., F. Zélé, J.-C. Simon and S. Outreman. 2011. Rapid evolution of parasitoids when faced with the symbiont-mediated resistance in their hosts. J. Evol. Biol. 24: 741–750.

Dong, Y., F. Manfredini and G. Dimopoulos. 2009. Implication of the mosquito midgut microbiota in the defense against Malaria parasites. PLoS Pathog. 5: e1000423.

Easton, D.M., A. Nijnik, M.L. Mayer and R.E.W. Hancock. 2009. Potential of immunomodulatory host defense peptides as novel anti-infectives. Trends Biotechnol. 27: 582–590.

Engel, P. and N.A. Moran. 2013. The gut microbiota of insects—diversity in structure and function. FEMS Microbiol. Rev. 37: 699–735.

Evans, J.D., K. Aronstein, Y.P. Chen, C. Hetru, J.-L. Imler, H. Jiang, M. Kanost, G.J. Thompson, Z. Zou and D. Hultmark. 2006. Immune pathways and defence mechanisms in honey bees *Apis mellifera*. Insect Mol. Biol. 15: 645–656.

Feldhaar, H. 2011. Bacterial symbionts as mediators of ecologically important traits of insect hosts. Ecol. Entomol. 36: 533–543.

Fernandez-Marin, H., J.K. Zimmerman, D.R. Nash, J.J. Boomsma and W.T. Wcislo. 2009. Reduced biological control and enhanced chemical pest management in the evolution of fungus farming in ants. Proc. R. Soc. Lond. B-Biol. Sci. 276: 2263–2269.

Freitak, D., D.G. Heckel and H. Vogel. 2009. Dietary-dependent trans-generational immune priming in an insect herbivore. Proc. R. Soc. Lond. B-Biol. Sci. 276: 2617–2624.

Freitak, D., H. Schmidtberg, F. Dickel, G. Lochnit, H. Vogel and A. Vilcinskas. 2014. The maternal transfer of bacteria can mediate trans-generational immune priming in insects. Virulence 5(4): 1–8.

Gardiner, M.M., L.A. Leslie, P.M.J. Brown, J.E. Losey, H.E. Roy and R. Rice Smyth. 2012. Lessons from lady beetles: accuracy of monitoring data from US and UK citizen-science programs. Front. Ecol. Environm. 10: 471–476.

Gasch, T., M. Schott, C. Wehrenfennig, R.-A. Düring and A. Vilcinskas. 2013. Multifunctional weaponry: The chemical defenses of earwigs. J. Insect Physiol. 59: 1186–1193.

Gerardo, N.M., B. Altincicek, C. Anselme, H. Atamian, S.M. Barribeau, M. de Vos, E.J. Duncan, J.D. Evans, T. Gabaldon, M. Ghanim, A. Heddi, I. Kaloshian, A. Latorre, A. Moya, A. Nakabachi, B.J. Parker, V. Pérez-Brocal, M. Pignatelli, Y. Rahbé, J.S. Ramsey, C.J. Spragg, J. Tamames, D. Tamarit, C. Tamborindeguy, C. Voncent-Monegat and A. Vilcinskas. 2010. Immunity and other defenses in pea aphids, *Acyrthosiphon pisum*. Genome Biol. 11: R21.

Gokhale, C.S., A. Papkou, A. Traulsen and H. Schulenburg. 2013. Lotka–Volterra dynamics kills the red queen: population size fluctuations and associated stochasticity dramatically change host-parasite coevolution. BMC Evol. Biol. 13: 254.

Grindstaff, J.L., D. Hasselquist, J.-Å. Nilsson, M. Sandell, H.G. Smith and M. Stjernman. 2006. Transgenerational priming of immunity: maternal exposure to a bacterial antigen enhances offspring humoral immunity. Proc. R. Soc. Lond. B-Biol. Sci. 273: 2551–2557.

Gross, J., K. Schumacher, H. Schmidtberg and A. Vilcinskas. 2008. Protected by fumigants: Beetle perfumes in antimicrobial defense. J. Chem. Ecol. 34: 179–188.

Haine, E.R., Y. Moret, M.T. Siva-Jothy and J. Rolff. 2008. Antimicrobial defense and persistent infection in insects. Science 322: 1257–1259.

Hamilton, C., B.T. Lejeune and R.B. Rosengaus. 2011. Trophallaxis and prophylaxis: social immunity in the carpenter ant *Camponotus pennsylvanicus*. Biol. Lett. 7: 89–92.

Hangartner, S., S.H. Sbilordo, Ł. Michalczyk, M.J.G. Gage and O.Y. Martin. 2013. Are there genetic trade-offs between immune and reproductive investments in *Tribolium castaneum*? Infect. Gen. Evol. 19: 45–50.

Horak, P., L. Saks, I. Ots, T. Kullissaar, H. Kollist and M. Zilmer. 2003. Physiological effects of immune challenge in captive greenfinches (*Carduelis chloris*). Can. J. Zool. 81: 371–379.

Hosken, D.J. 2001. Sex and death: microevolutinary trade-offs between reproductive and immune investment in dung flies. Curr. Biol. 11: R379–380.

Huffman, M.A. 2001. Self-medicative behavior in the African great apes: an evolutionary perspective into the origins of human traditional medicine. BioSci. 51: 651–661.

Jakubowska, A.K., H. Vogel and S. Herrero. 2013. Increase in gut microbiota after immune suppression in baculovirus-infected larvae. PLoS Pathog. 9: e1003379.

Jokela, J. 2010. Transgenerational immune priming as cryptic parental care. J. Anim. Ecol. 79: 305–307.

Joop, G., A. Mitschke, J. Rolff and M.T. Siva-Jothy. 2006. Immune function and parasite resistance in male and polymorphic female *Coenagrion puella*. BMC Evol. Biol. 6: 19.

Joop, G., O. Roth, P. Scmid-Hempel and J. Kurtz. 2014. Experimental evolution of external immune defence in the red flour beetle. J. Evol. Biol. 27: 1562–1571.

Kajita, Y.J.J. Obrycki, J.J. Sloggett and K.F. Haynes. 2010. Intraspecific alkaloid variation in ladybird eggs and its effects on con- and hetero-specific intraguild predators. Oecologia 163: 313–322.

Kerstes, N.A.G. and O.Y. Martin. 2014. Insect host–parasite coevolution in the light of experimental evolution. Insect Sci. 21: 401–414.

Kim, D.H. and F.M. Ausubel. 2005. Evolutionary perspectives on innate immunity from the study of *Caenorhabditis elegans*. Curr. Opin. Immunol. 17: 4–10.

Kirsch, R., H. Vogel, A. Muck, A. Vilcinskas, J.M. Pasteels and W. Boland. 2011. To be or not to be convergent in salicin-based defence in chrysomeline leaf beetle larvae: Evidence from *Phratora vitellinae* salicyl alcohol oxidase. Proc. R. Soc. Lond. Ser. B-Biol. Sci. 278: 3225–3232.

Kofler, R. and C. Schlötterer. 2014. A guide for the design of evolve and resequencing studies. Mol. Biol. Evol. 31: 474–483.

Koch, H. and P. Schmid-Hempel. 2011. Socially transmitted gut microbiota protect bumble bees against an intestinal parasite. Proc. Natl. Acad. Sci. USA 108: 19288–19292.

König, C. and P. Schmid-Hempel. 1995. Foraging activity and immunocompetence in workers of the bumble bee, *Bombus terrestris* L. Proc. R. Soc. Lond. Ser. B-Biol. Sci. 260: 225–227.

Kraaijeveld, A.R. and H.C.J. Godfray. 1997. Trade-off between parasitoid resistance and larval competitive ability in *Drosophila melanogaster*. Nature 389: 278–280.

Kraaijeveld, A.R., S.J. Layen, P.H. Futerman and H.C.J. Godfray. 2012. Lack of phenotypic and evolutionary cross-resistance against parasitoids and pathogens in *Drosophila melanogaster*. PLoS One 7: e53002.

Lambrechts, L., J.-M. Chavatte, G. Snounou and J.C. Koella. 2006. Environmental influence on the gentic basis of mosquito resistance to malaria parasites. Proc. R. Soc. Lond. Ser. B Biol. Sci. 273: 1501–1506.

Lazzaro, B.P. and T.J. Little. 2009. Immunity in a variable world. Phil. Trans. R. Soc. Lond. Ser. B Biol. Sci. 364: 15–26.

Le Conte, Y., C. Alaux, J.-F. Martin, J.R. Harbo, J.W. Harris, C. Dantec, D. Séverac, S. Cros-Arteil and M. Navajas. 2011. Social immunity in honeybees (*Apis mellifera*): Transcriptome analysis of *Varroa*-hygienic behaviour. Insect Mol. Biol. 20: 399–408.

Lee, K.A. and K.C. Klasing. 2004. A role for immunology in invasion biology. Trends Ecol. Evol. 19: 523–529.

Lee, K.P., J.S. Cory, K. Wilson, D. Raubenheimer and S.J. Simpson. 2006. Flexible diet choice offsets protein costs of pathogen resistance in a caterpillar. Proc. R. Soc. Lond. Ser. B-Biol. Sci. 273: 823–829.

Lemaitre, B. and J. Hoffmann. 2007. The host defense of *Drosophila melanogaster*. Annu. Rev. Immunol. 25: 697–743.

Li, J., S. Lehmann, B. Weißbecker, I. Ojeda Naharros, S. Schütz, G. Joop and E.A. Wimmer. 2013. Odoriferous defensive stink gland transcriptome to identify novel genes necessary for quinone synthesis in the red flour beetle, *Tribolium castaneum*. PLoS Genet 9: e1003596.

Lochmiller, R.L. and C. Deerenberg. 2000. Trade-offs in evolutionary immunology: just what is the cost of immunity? Oikos 88: 87–98.

Loconti, J.D. and L.M. Roth. 1953. Composition of the odorous secretion of *Tribolium castaneum*. Ann. Entomol. Soc. 46: 281–289.

Login, F.H., S. Balmand, A. Vallier, C. Vincent-Monégat, A. Vigneron, M. Weiss-Gayet, D. Rochat and A. Heddi. 2011. Antimicrobial peptides keep insect endosymbionts under control. Science 334: 362–365.

Martin-Vivaldi, M., A. Pena, J.M. Peralta-Sanchez, L. Sanchez, S. Ananou, M. Ruiz-Rodriguez and J.J. Soler. 2010. Antimicrobial chemicals in hoopoe preen secretions are produced by symbiotic bacteria. Proc. R. Soc. Lond. Ser. B-Biol. Sci. 277: 123–130.

Masri, L., R.D. Schulte, N. Timmermeyer, S. Thanisch, L.L. Crummenerl, G. Jansen, N.K. Michiels and H. Schulenburg. 2013. Sex differences in host defence interfere with parasite-mediated selection for outcrossing during host–parasite coevolution. Ecol. Lett. 16: 461–468.

McNamara, K.B., N. Wedell and L.W. Simmons. 2013. Experimental evolution reveals trade-offs between mating and immunity. Biol. Lett. 9: 20130262.

Mondal, K.A.M.S.H. 1990. Combined action of methylquinone, aggregation pheromone and pirimiphos-methyl on *Tribolium castaneum* larval mortality. Pakistan J. Zool. 22: 249–255.

Moret, Y. 2006. 'Trans-generational immune priming': specific enhancement of the antimicrobial immune response in the mealworm beetle, *Tenebrio molitor*. Proc. R. Soc. Lond. B-Biol. Sci. 273: 1399–1405.

Moret, Y. and P. Schmid-Hempel. 2000. Survival for immunity: the price of immune system activation for bumblebee workers. Science 290: 1166–1168.

Mostowy, R. and J. Engelstädter. 2011. The impact of environmental change on host–parasite coevolutionary dynamics. Proc. R. Soc. Lond. B-Biol. Sci. 278: 2283–2292.

Mukherjee, K., R. Fischer and A. Vilcinskas. 2012. Histone acetylation mediates epigenetic regulation of transcriptional reprogramming in insects during metamorphosis, wounding and infection. Front. Zool. 9: 25.

Nunn, C. and S. Altizer. 2006. Infectious Diseases in Primates: Behavior, Ecology, and Evolution. Oxford University Press, Oxford.

Park, S.-C., Y. Park and K.-S. Hahm. 2011. The role of antimicrobial peptides in preventing multidrug-resistant bacterial infections and biofilm formation. Int. J. Mol. Sci. 12: 5971–5992.

Pham, L.N., M.S. Dionne, M. Shirasu-Hiza and D.S. Schneider. 2007. A specific primed immune response in *Drosophila* is dependent on phagocytes. PLoS Pathog 3: e26.

Piel, J. 2002. A polyketide synthase-peptide synthetase gene cluster from an uncultured bacterial symbiont of *Paederus* beetles. Proc. Natl. Acad. Sci. USA 99: 14002–14007.

Prokkola, J., D. Roff, T. Kärkkäinen, I. Krams and M.J. Rantala. 2013. Genetic and phenotypic relationships between immune defense, melanism and life-history traits at different temperatures and sexes in *Tenebrio molitor*. Heredity 111: 89–96.

Pukowski, E. 1933. Ecological investigation of *Necrophorus* F. Zeitschr. Morphol. Oekol. Tiere 27: 518–586.

Ridley, E.V., A.C.-N. Wong, S. Westmiller and A.E. Douglas. 2012. Impact of the resident microbiota on the nutritional phenotype of *Drosophila melanogaster*. PLoS One 7: e36765.

Rohner, C., C.J. Krebs, D.B. Hunter and D.C. Currie. 2000. Roost site selection of great horned owls in relation to black fly activity: An anti-parasite behavior? Condor 102: 950–955.

Röhrich, C.R., C.J. Ngwa, J. Wiesner, H. Schmidtberg, T. Degenkolb, C. Kollewe, R. Fischer, G. Pradel and A. Vilcinskas. 2012. Harmonine, a defence compound from the harlequin ladybird, inhibits mycobacterial growth and demonstrates multi-stage antimalarial activity. Biol. Lett. 8: 308–311.

Rolff, J. 2002. Bateman's principle and immunity. Proc. R. Soc. Lond. Ser. B Biol. Sci. 269: 867–872.

Rolff, J. and A.R. Kraaijeveld. 2003. Selection for parasitoid resistance alters mating success in *Drosophila*. Proc. R. Soc. Lond. Ser. B Biol. Sci. 270: S154–155.

Rolff, J. and M.T. Siva-Jothy. 2002. Copulation corrupts immunity: a mechanism for a cost of mating in insects. Proc. Natl. Acad. Sci. USA 99: 9916–9918.

Rosengaus, R.B., T. Malak and C. MacKintosh. 2013. Immune-priming in ant larvae: social immunity does not undermine individual immunity. Biol. Lett. 9: 20130563.

Roth, O., B.M. Sadd, P. Schmid-Hempel and J. Kurtz. 2009. Strain-specific priming of resistance in the red flour beetle, *Tribolium castaneum*. Proc. R. Soc. Lond. Ser. B Biol. Sci. 276: 145–151.

Roth, O., G. Joop, H. Eggert, J. Hilbert, J. Daniel, P. Schmid-Hempel and J. Kurtz. 2010. Paternally derived immune priming for offspring in the red flour beetle, *Tribolium castaneum*. J. Anim. Ecol. 79: 403–413.

Roy, H.E. and E. Wajnberg. 2008. From Biological Control to Invasion: the Ladybird *Harmonia axyridis* as a Model Species. Springer, Netherlands.

Sackton, T.B., B.P. Lazzaro, T.A. Schlenke, J.D. Evans, D. Hultmark and A.G. Clark. 2007. Dynamic evolution of the innate immune system in *Drosophila*. Nature Gen. 39: 1461–1468.

Sadd, B.M. and P. Schmid-Hempel. 2006. Insect immunity shows specificity in protection upon secondary pathogen exposure. Curr. Biol. 16: 1206–1210.

Sadd, B.M., Y. Kleinvogel, R. Schmid-Hempel and P. Schmid-Hempel. 2005. Transgenerational immune priming in a social insect. Biol. Lett. 1: 386–388.

Schmidtberg, H., C. Röhrich, H. Vogel and A. Vilcinskas. 2013. A switch from constitutive chemical defence to inducible innate immune responses in the invasive ladybird *Harmonia axyridis*. Biol. Lett. 9: 20130006.

Schmid-Hempel, P. 2011. Evolutionary Parasitology. The Integrated Study of Infections, Immunology, Ecology, and Genetics. OUP, Oxford, UK.

Schmid-Hempel, P. and D. Ebert. 2003. On the evolutionary ecology of specific immune defence. Trends Ecol. Evol. 18: 27–32.

Schulte, R.D., C. Makus, B. Hasert, N.K. Michiels and H. Schulenburg. 2010. Multiple reciprocal adaptations and rapid genetic change upon experimental coevolution of an animal host and its microbial parasite. Proc. Natl. Acad. Sci. USA 107: 7359–7364.

Schulte, R.D., C. Makus and H. Schulenburg. 2013. Host–parasite coevolution favours parasite genetic diversity and horizontal gene transfer. J. Evol. Biol. 26: 1836–1840.

Seebacher, F. and C.E. Franklin. 2005. Physiological mechanisms of thermoregulation in reptiles: a review. J. Comp. Physiol. B-Biochem. Syst. Environm. Physiol. 175: 533–541.

Sheldon, B.C. and S. Verhulst. 1996. Ecological immunology: costly parasite defences and trade-offs in evolutionary ecology. Trends Ecol. Evol. 11: 317–321.

Simone, M., J.D. Evans and M. Spivak. 2009. Resin collection and social immunity in honeybees. Int. J. Org. Evol. 63: 3016–3022.

Siva-Jothy, M.T., Y. Tsubaki and R.E. Hooper. 1998. Decreased immune response as a proximate cost of copulation and oviposition in a damselfly. Physiol. Entomol. 23: 274–277.

Sokoloff, A. 1974. The Biology of *Tribolium* with Special Emphasis on Genetic Aspects. Vol. 2. OUP, Oxford, UK.

Steiger, S., S.N. Gershman, A.M. Pettinger, A.-K. Eggert and S.K. Sakaluk. 2011. Sex differences in immunity and rapid upregulation of immune defence during parental care in the burying beetle, *Nicrophorus orbicollis*. Funct. Ecol. 25: 1368–1378.

Suen, G., C. Teiling, L. Li, C. Holt, E. Abouheif, E. Bornberg-Bauer, P. Bouffard, E.J. Caldera, E. Cash, A. Cavanaugh, O. Denas, E. Elhaik, M.-J. Favé, J. Gadau, J.D. Gibson, D. Graur, K.J. Grubbs, D.E. Hagen, T.T. Harkins, M. Helmkampf, H. Hu, B.R. Johnson, J. Kim, S.E. Marsh, J.A. Moeller, M.C. Munoz-Torres, M.C. Murphy, M.C. Naughton, S. Nigam, R. Overson, R. Rajakumar, J.T. Reese, J.J. Scott, C.R. Smith, S. Tao, N.D. Tsutsui, L. Viljakainen, L. Wissler, M.D. Yandell, F. Zimmer, J. Taylor, S.C. Slater, S.W. Clifton, W.C. Warren, C.G. Elsik, C.D. Smith, G.M. Weinstock, N.M. Gerardo and C.R. Currie. 2011. The genome sequence of the leaf-cutter ant *Atta cephalotes* reveals insights into its obligate symbiotic lifestyle. PLoS Genet. 7: e1002007.

Thomas, M.B. and S. Blanford. 2003. Thermal biology in insect-parasite interactions. Trends Ecol. Evol. 18: 344–350.

Tidbury, H.J., A.B. Pedersen and M. Boots. 2011. Within and transgenerational immune priming in an insect to a DNA virus. Proc. R. Soc. Lond. Ser. B Biol. Sci. 278: 871–876.

Unruh, L.M., R. Xu and K.J. Kramer. 1998. Benzoquinone levels as a function of age and gender of the red flour beetle, *Tribolium castaneum*. Insect Biochem. Mol. Biol. 28: 969–977.

VanValen, L. 1973. A new evolutionary law. Evolut. Theory 1: 1–30.

Versace, E., V. Nolte, R.V. Pandey, R. Tobler and C. Schlötterer. 2014. Experimental evolution reveals habitat-specific fitness dynamics among *Wolbachia* clades in *Drosophila melanogaster*. Mol. Ecol. 23: 802–814.

Vilcinskas, A. 2013. Evolutionary plasticity of insect immunity. J. Insect Physiol. 59: 123–129.

Vilcinskas, A., K. Stoecker, H. Schmidtberg, C.R. Röhrich and H. Vogel. 2013a. Invasive harlequin ladybird carries biological weapons against native competitors. Science 340: 862–863.

Vilcinskas, A., K. Mukherjee and H. Vogel. 2013b. Expansion of the antimicrobial peptide repertoire in the invasive ladybird *Harmonia axyridis*. Proc. R. Soc. Lond. Ser. B Biol. Sci. 280: 20122113.

Villaverde, M.L., M.P. Juárez and S. Mijailovsky. 2007. Detection of *Tribolium castaneum* (Herbst) volatile defensive secretions by solid phase microextraction-capillary gas chromatography (SPME-CGC). J. Stored Prod. Res. 43: 540–545.

Viney, M.E., E.M. Riley and K.L. Buchanan. 2005. Optimal immune responses: immunocompetence revisited. Trends Ecol. Evol. 20: 665–669.

Wegner, K.M., C. Berenos and P. Schmid-Hempel. 2008. Nonadditive genetic components in resistance of the red flour beetle *Tribolium castaneum* against parasite infection. Evolution 62: 2381–2392.

Wiesner, J. and A. Vilcinskas. 2010. Antimicrobial peptides: The ancient arm of the human immune system. Virulence 1: 440–464.

Yezerski, A., C. Ciccone, J. Rozitski and B. Volingavage. 2007. The effects of a naturally produced benzoquinone on microbes common to flour. J. Chem. Ecol. 33: 1217–1225.

Zasloff, M. 2002. Antimicrobial peptides of multicellular organisms. Nature 415: 389–395.

Zou, Z., J.D. Evans, Z. Lu, P. Zhao, M. Williams, N. Sumathipala, C. Hetru, D. Hultmark and H. Jiang. 2007. Comparative genomic analysis of the *Tribolium* immune system. Genome Biol. 8: R177–R177.

10

The Coleopteran Gut and Targets for Pest Control

Brenda Oppert,[1,*] *Asieh Rasoolizadeh*[2,a] and *Dominique Michaud*[2,b]

Introduction

The rapid evolution of high-throughput sequencing and proteomics has enabled new approaches to the study of complex organisms, from tissue to whole body. Entomology-related fields are rapidly adopting these "next generation" tools to solve pest problems. These tools are prompting entomologists to reexamine and redefine the control of some of the most problematic insect pests, those belonging to the Order Coleoptera.

Molecular studies on Coleoptera have been limited by the availability of genomic data. Until recently, the only sequenced coleopteran genome was that of the genetic model, the red flour beetle, *Tribolium castaneum* (*Tribolium* Genome Sequencing Consortium 2008). The i5K project seeks to sequence the genome of 5,000 insect species over a five-year period, with species nominated by the open research community (http://www. arthropodgenomes.org/wiki/i5K). The i5K project has 69 coleopteran species nominated to date. These projects will fill in data gaps and rapidly expand the application of functional genomics to insect-related research.

[1] USDA Agricultural Research Service Center for Grain and Animal Health Research, 1515 College Ave., Manhattan, KS 66502 USA.
  Email: brenda.oppert@ars.usda.gov
[2] Département de phytologie, Université Laval, Pavillon Envirotron, 2480 boul. Hochelaga, Québec City, QC G1V 0A6, Canada.
[a] Email: asieh.rasoolizadeh.1@ulaval.ca
[b] Email: dominique.michaud@fsaa.ulaval.ca
* Corresponding author

Recently, the second coleopteran genome sequence has been released, that of the mountain pine beetle, *Dendroctonus ponderosae*, a serious pest of American pine trees (Keeling et al. 2013). A focus of this study was to identify enzymes that allow the beetle to survive and thrive in a wide range of temperatures and often nutrient-limiting environments. Interestingly, a comparison of coding sequences between *D. ponderosae* and *T. castaneum* revealed some conserved shared synteny despite the prediction of divergence more than 200,000,000 years ago.

A variety of approaches are being used to study coleopteran insects, including whole genome, targeted genome, and transcriptome sequencing. An example is a comparison of the sequences of 13 mitochondrial genomes from four coleopteran suborders demonstrating conservation in the size and organization of the coding region from the ancestral genome (Sheffield et al. 2008).

RNA-Seq refers to the quantitative sequencing of genetic transcripts. The molecular basis of the ability of insects to adapt to toxic plant allelochemicals was found in a large RNA-Seq study (Zhen et al. 2012), for which RNA-Seq data are publicly available (http://genomics-pubs. princeton.edu/insect_genomics/data.shtml). Amino acid substitutions in ATPase proteins clustered with host plant specialization in insect species from three orders, including those representing the coleopteran families Cerambycidae, Chrysomelidae and Curculionidae. Moreover, four independent duplications of ATPases from 21 species converged in tissue-specific expression patterns.

Tissue-specific transcriptome sequencing can identify specific functions, such as the transcriptome sequence of antennae of bark beetles, *Ips typographus* and *D. ponderosae* (Andersson et al. 2013). In evaluating chemosensory genes in the bark beetles and other insect genomes, some gene expansions were found, but conservation of some odorant receptors was also noted.

Probably the most serious economic threat from a coleopteran is that of the corn rootworm, *Diabrotica* spp., which is thus the subject of many commercial and academic research projects. A small transcriptome study found transcripts encoding putative proteins with largely catalytic or binding functions in the larval gut of *D. virgifera virgifera* (Siegfried et al. 2005). However, we find no large-scale molecular studies to date about this insect, likely due to the difficulty in obtaining a sequenced genome. The problems with sequencing a large and complex genome, such as *Diabrotica* spp., have been enumerated (Miller et al. 2010). While genomic data are likely available given the large sources of funding allocated to rootworm studies, there may be limitations in sharing these data due to intellectual property issues.

As agricultural research increasingly turns to functional genomics, RNA interference (RNAi) has been used both as a diagnostic tool and a new method for pest control (see articles by Orchard and Lange and Hoffmann et al. this book). A recent transcriptome study in the Colorado potato beetle, *Leptinotarsa decemlineata*, identified all known genes of the RNA interference (RNAi) pathways previously found in other organisms (Swevers et al. 2013). These types of studies will enable the validation of genes and are being incorporated as novel plant protectants (Bachman et al. 2013).

Here we present two "case studies" of the coleopteran gut, both using rapidly-evolving molecular approaches. One study is a transcriptome study of the larval gut of the genetic model *T. castaneum* and another tenebrionid, *Tenebrio molitor* (yellow mealworm), with a review of previous data. The other is a proteomic study of the larval gut of the Colorado potato beetle, *L. decemlineata*, and new approaches to the use of protease inhibitors to control coleopteran pests.

Case Study No. 1: *Tribolium castaneum* and *Tenebrio molitor*

Previous Studies

We have used the tenebrionids *T. molitor* and *T. castaneum* as biochemical and genetic models for the study of coleopteran digestion. The use of tissue-specific gene expression analysis was first applied to the gut of *T. castaneum* (Morris et al. 2009). Using custom genome microarrays, the more highly expressed genes in the larval gut were identified, and relative expression was compared with that of the head and carcass (i.e., all tissue without gut and head). Proteomics analysis identified 98 proteins in the larval gut, with 80 proteins overlapping with those identified as highly-expressed in the microarray analysis. These data have been used to study the relative expression of digestive and non-digestive proteases in larvae exposed to protease inhibitors (Oppert et al. 2010). The data have also been mined to discover a cellulase from the larval gut that may have applications in biofuel development (Willis et al. 2011, Shirley et al. 2013).

Understanding the response of insects to the insecticidal Cry toxins from the bacterium *Bacillus thuringiensis* has driven much of the research in gut-specific sequencing, although little is available for coleopteran insects. While *T. castaneum* has a sequenced genome, it is not sensitive to Bt toxins, whereas *T. molitor* is sensitive to Cry3Aa, with an LD_{50} similar to Cry1A toxins in lepidopterans (Oppert et al. 2011). Therefore, transcription profiling and microarrays were used to identify differential and temporal transcript expression in Cry3Aa-intoxicated *T. molitor* larvae (Oppert et al. 2012a). Overall the data suggested that intoxicated larvae show metabolic conservation, with increases in functions associated with respiration, signaling, cell and membrane integrity, protein synthesis, and glycosyl hydrolases. The temporal study indicated that gene expression was most different between 6 and 12 hours post intoxication, providing a time point for more intensive sampling.

One of the effects of Cry3Aa intoxication in *T. molitor* was a dramatic change in the expression of transcripts of serine (Ser) proteases in the larval gut (Oppert et al. 2012b). *T. molitor* larvae rely on a complex and highly compartmentalized digestive system, with mostly cysteine (Cys) proteases in the anterior and Ser proteases in the posterior midgut (Vinokurov et al. 2006, Prabhakar et al. 2007). Of 48 Ser protease transcripts obtained from the gut of Cry3Aa-intoxicated *T. molitor* larvae, most were severely repressed, from 2- to 15-fold. However, expression of a few transcripts encoding putative chymotrypsin or other Ser proteases (undefined due to a lack of complete sequence data) were only expressed in intoxicated larvae, and at 38–56-fold greater relative expression than other peptidase transcripts. Although Cys proteases are found in the anterior part of the midgut and also are critical to digestion, transcript expression for Cys proteases only varied less than 2.5-fold in Cry3Aa-intoxicated larvae, indicating that these highly-evolved proteases are resilient to dietary toxicants.

Deciphering the Coleopteran Gut

To further probe the coleopteran gut, high throughput sequencing was used to compare the larval gut transcriptome of *T. castaneum* and *T. molitor* (Table 1). Using this approach, we have further refined the model of *T. castaneum*, and we can begin to evaluate commonalities and differences in the digestive physiology among the tenebrionid insects.

The larval gut transcriptome of T. castaneum

With a reference genome for *T. castaneum*, reads from the larval gut were mapped to 8,986 genes in the Tcas3 genome (SeqmanNGen, DNAStar, Madison, WI). Coverage depth (the median depth of sequence coverage) for each transcript was exported for comparison (SeqManPro, DNAStar). Only 660 (7% of mapped genes) had coverage depth >1000; those genes were more likely to be more highly expressed in the larval gut. As a comparison, gene expression in the larval gut was previously estimated via microarray analysis as approximately 13% of the genome (Morris et al. 2009), an over-estimation compared to the RNA-Seq data. For the microarray analysis, approximately one-fourth of the predicted genes

Table 1. Sequencing strategies and associated data from gut tissue of *T. castaneum* and *T. molitor* larvae. For all samples: artificial diet was 85% wheat germ, 10% flour, 5% brewer's yeast; total RNA was isolated using the Absolutely RNA Kit with on-column DNase treatment (Agilent Technologies, La Jolla, CA USA); sequencing was 2 x 36 bp paired end on Illumina HiSeq 2000; average insert was about 200 bp.

Species	Larval Age	Larval mass (n)	Reads	Bioinformatics
Tenebrio molitor	1 month	20 mg (4)	>60,000,000	SeqmanNGen templated assembly (DNAStar, Tcas3)
Tribolilum castaneum	2 weeks	0.003 mg (30)	344,476,216	ArrayStar (custom assembly), Blast2GO annotation

in the *T. castaneum* reference genome (mostly housekeeping and highly conserved) were not included in the array due to the constraint of 12,000 probes (the Tcas3 reference genome has >16,000 genes); many of these deleted genes were highly expressed in the *T. castaneum* larval gut by RNA-Seq analysis (Tables 2 and 4).

In examining transcripts that were found only in the transcriptome data from *T. castaneum* (without orthologs in the *T. molitor* transcriptome data), most notable was the prevalence of protease genes (Table 2, top 50 genes in expression are listed). Two cathepsin L genes, LOC659441 and LOC659502, are the most highly expressed in the larval transcriptome. *T. castaneum* larvae rely on predominantly Cys proteases in the anterior midgut for protein digestion, accounting for approximately 80% of total digestion (Vinokurov et al. 2009). *T. molitor* larvae have more of a balance of Cys protease activity in the anterior and Ser protease activity in the posterior midgut (Vinokurov et al. 2006). While the data are preliminary, these Cys and Ser proteases in the *T. castaneum* gut may reflect the differences in digestive compartmentalization in these tenebrionid insects. We previously annotated 25 Cys cathepsin genes in the *Tribolium* genome (*Tribolium* Genome Sequencing Consortium 2008), and 12 were expressed in the larval gut according to microarray analysis (Morris et al. 2009). There are approximately 150 Ser protease genes in the *T. castaneum* genome, with approximately half expressed in the larval gut (Morris et al. 2009). The prevalence of protease gene expansion groups, and particularly Cys cathepsins in some coleopterans, has been proposed as an evolutionary adaptation to cereal protease inhibitors. However, we also found that some adaptations may have occurred to allow more efficient digestion of specialized and intractable proteins found in cereals (Goptar et al. 2012). Nonetheless, a complete understanding of the complement of digestive proteases in these insects will lead to more effective, targeted control.

Table 2. *T. castaneum* larval gut genes (top 50 in expression); corresponding genes were not found in the *T. molitor* larval gut transcriptome. Table is sorted by decreasing coverage depth (SeqManLaserGen, DNAStar) of *T. castaneum* larval gut transcripts mapped to the *T. castaneum* Tcas3 genome version. Gut rank is from microarray gene expression analysis (Morris et al. 2009). Protease/peptidase genes are shaded grey.

GeneID	Description	Coverage Depth	Gut Rank
LOC659441	cathepsin L-like proteinase	915637	82
LOC659502	cathepsin L precursor	282127	14
LOC100142511	uncharacterized LOC100142511	257121	#N/A
LOC100141537	uncharacterized LOC100141537	228196	#N/A
LOC664022	alpha amylase-like	160080	#N/A
LOC664009	alpha amylase-like	155296	44
LOC655909	similar to conserved hypothetical protein	149543	#N/A
Cht11	chitinase 11	109035	#N/A
LOC662108	uncharacterized LOC662108	106052	45
LOC664577	beta-glucosidase-like	69219	55
LOC100142085	peritrophic matrix protein 9	68361	#N/A
Pmp5-b	peritrophic matrix protein 5-B	62625	60
P166	serine protease P166	58933	47
Ctlp-5a	chymotrypsin-like proteinase 5A	55014	39
COX2	cytochrome c oxidase subunit II	49636	#N/A
Arp1	actin related protein 1	37593	39
LOC664385	alpha amylase-like	34899	2
LOC660129	AGAP007300-PA-like (alkaline phosphatase)	33866	42
H118	serine protease H118	33222	33
P121	serine protease P121	30992	17
P80	serine protease P80	23858	47
LOC660368	cathepsin L precursor	21708	44
LOC664120	GA13362-PA-like (trehalase/maltase)	21456	43
LOC656983	similar to CG4367-PA (chitin-binding domain)	21215	#N/A
H110	serine protease H110	20932	44
LOC660864	Calmodulin-like	20925	34
H112	serine protease H112	18891	43
ATP8	ATP synthase F0 subunit 8	18505	#N/A
LOC662840	uncharacterized LOC662840	18219	44
P149	serine protease P149	17650	36
LOC660583	similar to CG13315-PA	17343	#N/A
Ctlp-6a	chymotrypsin-like proteinase 6A precursor	14436	39
LOC100141903	similar to CG11538	13699	#N/A
P160	serine protease P160	13696	43
LOC661621	similar to CG10175-PC (caboxylesterase)	12394	#N/A
chic	chickadee (profilin)	12151	18
LOC656991	Actin-87E-like	12065	10
HEX1A	arylphorin	11790	51
26-29-p	cathepsin L	11790	0
Cht9	chitinase 9	11491	12
LOC661206	uncharacterized LOC661206	11363	#N/A
LOC656897	similar to GA18137-PA (chitin-binding domain)	10760	#N/A
ND5	NADH dehydrogenase subunit 5	10450	#N/A
LOC663117	cathepsin B precursor	10168	46
HEX2	hexamerin 2	9992	24
H105	serine protease H105	9664	6
P37	serine protease P37	9292	15
LOC663471	CRALBP-like (cellular retinaldehyde binding protein)	9059	9
LOC657303	AGAP001819-PA-like (chitin-binding domain)	9029	46
LOC659989	putative carboxypeptidase A-like-like	8634	4

There were a number of genes in the top 50 found only in *T. castaneum* larval gut that lack annotation (Table 2, uncharacterized). These genes are highly expressed in the gut and will be important to characterize functionally as we begin to build a complete model of the coleopteran gut. There were also carbohydrase, chitin-related, and structural genes commonly associated with gut function and integrity.

The larval gut transcriptome of T. molitor

Without a reference genome for *T. molitor*, we took a different bioinformatic approach. Reads from previous sequencing projects were combined with the paired-end data to generate a new *de novo* assembly using SeqmanNGen (DNAStar), generating 14,338 contigs. Contigs were exported to Blast2GO (Valencia, Spain) for annotation. The new assembly was used as a template to map the paired-end reads from the *T. molitor* larval gut and determine RPKM for each contig (QSeq/ArrayStar, DNAStar), resulting in 1,219 mapped contigs; 361 had no hits in the Blastx analysis.

Unfortunately, many of the top 50 expressed genes from the *T. molitor* larval gut were part of the "no hit" group (Table 3). However, the most highly expressed transcript in the *T. molitor* larval gut (comprised of many contigs and lacking an ortholog in *T. castaneum*) encoded a hypothetical protein; interestingly, this gene appears to be an ortholog to an immune-related gene characterized from the firebrat, *Thermobia domestica* (Altincicek and Vilcinskas 2007). As with the *T. castaneum* dataset, genes encoding other hypothetical proteins were found, as well as storage, carbohydrase, mucin, and immune-related genes.

Common elements of the T. castaneum and T. molitor larval gut transcriptome

A comparison of the mapped reads from *T. castaneum* and contigs from *T. molitor* resulted in 448 putative orthologs; the top 50 in gene expression are provided in Table 4. Many orthologs were conserved mitochondrial genes, as expected. In this group, a number of digestive enzymes, including carbohydrases and five Ser proteases, were common to the gut of both insect larvae; no Cys cathepsins were found in this highly-expressed group. The relative expression levels were usually different between the two insects, but this may be due to the different bioinformatic analyses and fewer reads for *T. molitor*. There were genes without annotation (hypothetical proteins) and yet highly expressed in the gut of both insects.

Table 3. *T. molitor* larval gut genes (top 50 in expression, TmContig); corresponding genes were not found in the *T. castaneum* larval gut transcriptome. Table is sorted by decreasing RPKM (reads per kilobase per million reads, ArrayStar, DNAStar) for transcripts from the *T. molitor* larval gut. Protease/peptidase genes are shaded grey.

Accession	TmContig	Seq. Description	RPKM
AM495109	Contigs56, 76, 77, 79, 84, 94, 96, 134, 350, 561	hypothetical protein	42321
no hits	Contig819	none	4470
ABC87267	Contigs37, 2405	muscle lim protein	3797
no hits	Contig2605	none	3514
no hits	Contig340	none	2572
ABC88769	Contigs44, 170, 234, 226, 475,	cathepsin l precursor	1856
no hits	Contig11833	none	1768
HEX4	Contigs315, 1932	arylphorin subunit alpha	1426
no hits	Contig328	none	1302
no hits	Contig494	none	1284
no hits	Contig17908	none	1233
no hits	Contig128	none	1220
EFA08766	Contigs110, 691, 768, 950, 1532, 2327, 2418	lysosomal alpha-mannosidase	1002
no hits	Contig11905	none	952
no hits	Contig1069	none	768
no hits	Contig11897	none	736
no hits	Contig11856	none	691
EJY57844	Contig13030	hypothetical protein	603
no hits	Contig510	none	590
no hits	Contig11973	none	556
	Contigs20, 735	heat shock protein 70	487
no hits	Contig4097	none	483
	Contig17989	unknown	477
no hits	Contig7845	none	464
no hits	Contig1975	none	414
no hits	Contig11965	none	389
P127	Contigs57, 4250	serine protease P127	374
XP_976343	Contigs65, 175, 325, 374, 964, 1180, 1861, 11854	PREDICTED: hypothetical protein	352
ML4	Contigs101, 390, 356, 375	mpa2 allergen	332
no hits	Contig327	none	321
EFA13164	Contig700	hypothetical protein TcasGA2_TC002105	320
no hits	Contig391	none	319
no hits	Contig1418	none	313
no hits	Contig1783	none	312
no hits	Contig17901	none	283
Pmp14	Contigs15, 3444	mucin-like protein	264
no hits	Contig11880	none	204
EFA10280	Contigs1016, 10557, 13859	high density lipoprotien binding protein vigilin	200
Pmp3	Contigs121, 4086	mucin-like peritrophin	191
JC1348	Contig88	jc1348hypothetical 18k protein	191
YP_003331371	Contig276	atp synthase f0 subunit 6	190
no hits	Contig1443	none	189
ACR44208	Contig14	cytochrome oxidase subunit i	186
no hits	Contig1328	none	179
ABC88754	Contig236	serine proteinase	178
no hits	Contig1656	none	167
	Contig172, 227, 615, 1059	56 kda early-staged encapsulation-inducing protein	166
no hits	Contig11825	none	156
no hits	Contig13278	none	154
	Contig1326	hypothetical protein	154

Table 4. *T. molitor* larval gut genes (top 50 in expression, TmContig) that correspond to genes in the *T. castaneum* larval gut transcriptome (TcGeneID). Table is sorted by decreasing RPKM (reads per kilobase per million reads, ArrayStar, DNAStar) for transcripts from the *T. molitor* larval gut, with the corresponding coverage depth from transcripts from the *T. castaneum* larval gut mapped to the Tcas3 genome (SeqManLaserGen, DNAStar). Gut rank is from microarray gene expression analysis (Morris et al. 2009). Protease/peptidase genes are shaded grey.

TcGeneID	TmContig	Description	RPKM	Coverage Depth	Gut Rank
COX1	Contigs1, 2, 4, 10, 16, 17, 34, 11782	cytochrome c oxidase subunit I	13125	65947	#N/A
LOC660916	Contig323	autotransporter beta-domain	6517	29181	19
CYTB	Contigs5, 9, 41, 11792	cytochrome b	4803	49087	#N/A
COX3	Contigs11, 12, 42, 46, 78	cytochrome c oxidase subunit III	3531	34057	#N/A
LOC663954	Contigs187, 543, 430, 1454, 168, 191, 8	alpha amylase isoform 1	3114	182164	53
ATP6	Contigs40, 164	ATP synthase F0 subunit 6	2406	47732	
LOC661072	Contigs261, 749, 290	serine proteinase	1249	287	#N/A
LOC658268	Contigs69, 87, 694, 388, 2889, 337, 702	fatty acid binding protein 4	951	61920	48
LOC660229	Contigs220, 326, 3535, 1746, 1135, 11818, 2113, 11909, 5079, 1478, 3527, 968	cockroach allergen-like protein	923	47931	36
LOC658401	Contigs147, 205, 1120, 2045, 2976, 3299, 17273, 17596	cytochrome c oxidase subunit ii	796	219	11
ND5	Contig43	NADH dehydrogenase subunit 5	624	10450	6
Cda6	Contigs55, 790, 900, 58	chitin deacetylase 1	518	5180	14
LOC662037	Contigs231, 483, 653	PREDICTED: hypothetical protein	514	6600	23
LOC661814	Contig105	28 kDa desiccation stress protein-like	451	326	6
LOC661778	Contigs142, 360, 1598, 288, 380, 1118	serine proteinase	441	933	3
LOC655492	Contigs4045, 361, 310, 341	ferritin 2 isoform 1	421	2190	15
P166	Contigs209, 566, 1250	serine protease P166	404	58933	47
LOC660577	Contigs195, 722, 260, 775, 11838, 762, 1626, 116	PREDICTED: hypothetical protein isoform 2	369	64134	50
P76	Contig248	serine protease P76	353	38833	45
LOC660377	Contigs425, 413, 1079	PREDICTED: hypothetical protein	340	5608	13
LOC658001	Contigs849, 12813, 452, 1329	peroxisomal acyl-coenzyme a oxidase 3-like	276	245	1
LOC664315	Contigs3776, 703	40s ribosomal protein s2-like	273	6856	27
LOC657776	Contigs47, 28, 135	polyadenylate-binding protein 1-like isoform 1	273	7293	8
ND4	Contig13	nadh dehydrogenase subunit 1	267	4522	#N/A

Table 4. contd....

Table 4. contd.

TcGeneID	TmContig	Description	RPKM	Coverage Depth	Gut Rank
H118	Contig501	serine protease H118	242	33222	33
ND4	Contigs19, 23, 153	NADH dehydrogenase subunit 4	236	4522	#N/A
Pmp5-b	Contigs383, 951, 963, 1085	peritrophic matrix protein 5-B	231	62625	60
LOC656849	Contig1573	sarcosine dehydrogenase	216	45	2
LOC660479	Contigs830, 719	arginine kinase	203	20084	38
Pmp2-b	Contigs235, 176, 484, 2210, 3331	peritrophic matrix protein 2-B	194	50984	47
LOC655958	Contig3545	glucosyl glucuronosyl transferases	166	266	2
Cda7	Contig1277,Contig1775	chitin deacetylase 1	163	1928	#N/A
LOC660970	Contigs448, 575	splicing factor subunit 2 isoform 1	149	249397	39
Cda9	Contigs3544, 100, 358, 152	chitin deacetylase-like 9	137	16720	37
LOC656927	Contigs30, 54	PREDICTED: hypothetical protein	136	10377	1
LOC656327	Contigs410, 207, 629	vacuolar h	134	13391	11
LOC656653	Contigs393, 1435	60s acidic ribosomal protein p0	116	8493	49
LOC664592	Contigs486, 500, 4614, 1651	similar to CG6845 CG6845-PA	113	977	2
Cda8	Contigs165, 11862, 11990	chitin deacetylase 1	108	8254	15
LOC662403	Contigs562, 2007, 1740	renin receptor-like isoform 1	106	2362	5
LOC660146	Contigs600, 1172, 4000	PREDICTED: hypothetical protein	100	142	1
LOC656534	Contigs11816, 578, 423, 115	lysosomal alpha-mannosidase	100	1672	#N/A
LOC655909	Contigs758, 1477	isoform a (chitin-binding)	89	82775	53
LOC662040	Contigs1491, 654, 2339	mitochondrial adp atp carrier protein	88	10213	35
LOC657472	Contig245	ubiquitin ribosomal protein s27ae fusion protein	85	4312	23
LOC663500	Contigs1851, 2922, 3221	survival motor neuron protein	84	260	3
LOC655736	Contig311,Contig1834	isoform b (gelsolin)	80	4159	#N/A
LOC659942	Contig836	vacuolar protein sorting-associated protein 13d	75	43	1
LOC655420	Contigs1706, 1654, 1960	ferritin subunit	71	10488	2
LOC659219	Contig282	actin	70	6775	39

Other genes were typical for functions in the insect gut, including those related to allergen and stress, chitin- and iron-binding, and structure.

Where do we go from here?

These models will be improved with additional sequence data from the larval gut of other coleopterans, and with additional reference genomes, including *T. molitor*. In addition, a number of hypothetical genes are conserved among tenebrionids and other insects; some also are conserved across phyla. Enumeration of the functions of these genes is critical to defining the gut physiology of the Coleoptera.

Our focus has been on insect digestion, particularly proteases in the larval gut of these insects. We now have a good understanding of the type and relative expression patterns of proteases in the *T. castaneum* larval gut; future manuscripts will provide details. Additionally, we are supplementing an extensive biochemical analysis of the *T. molitor* larval gut with genetic data. We continue to use these data to probe how tenebrionids respond to perturbation (i.e., toxins and protease inhibitors). These data are being compiled to identify biological inhibitors and other control products, such as dsRNA, that can be used judiciously to specifically and effectively target pests such as *T. castaneum* without current associated problems, such as resistance and environmental harm (see Borovsky this book).

Case Study No. 2: *Leptinotarsa decemlineata*

General Context

L. decemlineata has been nominated in the i5K genome sequencing project (i5K Consortium, 2013), and a first draft of its genome has been released (http://arthropodgenomes.org/). A significant task now is to annotate this genome, useful to address complex biological questions and issues of practical relevance. An interesting study was recently published towards this end, reporting the sequencing, *de novo* assembly, and systematic annotation of a large dataset of midgut transcript reads associated with diapause and resistance to chemical and biological pesticides (Kumar et al. 2014). Additional sequencing studies involving homology-based gene cloning, low-throughput EST sequencing, or high-throughput 454 pyrosequencing were published with the same insect, addressing different aspects of its biology and metabolic status in response to pesticides (Yocum et al. 2009a,b, Pauchet et al. 2010, Petek et al. 2012). Current knowledge of the *L. decemlineata* genome is still fragmentary compared to what is known about the genome of a model species such as *T. castaneum* (*Tribolium*

Genome Sequencing Consortium 2008), but the rapid developments in genome and transcriptome bioinformatics, the availability of a draft genome and the economic importance of potato beetles worldwide should readily establish the usefulness of genomic and transcriptomic endeavors dedicated to this important pest.

Proteomics, one step downstream along the protein biosynthetic pathway, also shows potential for a better understanding of *L. decemlineata* responses to pesticides and environmental challenges. Comparative proteomic studies have assessed the proteome of phytophagous Coleoptera in response to various factors including chemical pesticides, Bt toxins, antidigestive protease inhibitors and adverse climate conditions (Table 5). Functional proteomic strategies have also been devised recently to decipher the complex set of digestive Cys proteases in *L. decemlineata* in relation to the existence of multiple gene coding sequences for these enzymes and with their differential susceptibility to rapidly evolving Cys protease inhibitors in their plant hosts (Sainsbury et al. 2012a, Smid et al. 2013). Proteomic tools will be useful in coming years to elucidate basic physiological processes and complex protein networks in Coleoptera. They should be useful, as well, for the identification of relevant protein targets and the characterization of protein-ligand interactions of practical interest. Our goal here is to illustrate how genomic data and functional proteomics may be used to solve questions of interest in crop protection, using as a 'case study' the identification of inhibitor candidates for an effective, broad-spectrum inhibition of *L. decemlineata* digestive Cys proteases.

Harnessing the Potential of Protease Inhibitors in Plant Protection–A Timely, but Challenging Task

Dozens of studies have discussed the potential of plant protease inhibitors in crop protection since the publication of two seminal papers by Hilder et al. (1987) and Johnson et al. (1989) some 25 years ago, reporting the engineering of protease inhibitor-expressing tobacco plants resistant to the lepidopteran models *Heliothis virescens* and *Manduca sexta*. Many studies, over the past two decades, have described the potential of trypsin and chymotrypsin inhibitors to develop transgenic crops resistant to Lepidoptera (Haq et al. 2004), and the usefulness of these inhibitors in field conditions has been documented (Huang et al. 2005, Qiu 2008). Most protease inhibitors in plants are competitive inhibitors acting as pseudo-protein substrates to penetrate the active site cleft of target proteases (Birk 2003). The inhibited enzymes can no longer cleave peptide bonds

Table 5. The proteome of phytophagous Coleoptera—A selection of recent studies.

Model species	Biological question	Reference
Callosobruchus maculatus	Midgut proteome rebalancing upon protease inhibitor ingestion	Nogueira et al. 2012
Cucujus clavipes puniceus	Physiological and biochemical determinants of overwintering capacity	Carrasco et al. 2011, 2012
Dendroctonus frontalis	Characterization of pronotum functions in relation with fungal symbiosis	Pechanova et al. 2008
Dendroctonus ponderosae	Physiological and biochemical determinants of overwintering capacity	Bonnett et al. 2012
Leptinotarsa decemlineata	Midgut proteases susceptible to antidigestive protease inhibitors	Sainsbury et al. 2012a, Smid et al. 2013
Phaedon cochleariae	Plant cell wall-degrading enzymes in phytophagous Coleoptera	Kirsch et al. 2012
Tribolium casteneum	Larval susceptibility to *Bacillus thuringiensis* strains	Contreras et al. 2013
	Enzyme determinants of sperm protection in seminal fluid	Xu et al. 2013
	Response/symptoms to a benzoylphenyl urea insecticide	Merzendorfer et al. 2012
	Characterization of midgut proteins	Morris et al. 2009

and drive protein digestion in the insect midgut, resulting in detrimental growth delays presumably associated with amino acid depletion and shortage (Broadway 2000).

Promising protective effects against insect herbivores have been reported for a number of plants expressing recombinant protease inhibitors, but these proteins have proved inefficient in many other instances. Herbivorous insects have developed various strategies to elude the negative effects of dietary protease inhibitors, typically involving digestive systems with proteases from distinct functional families, overexpression of susceptible proteases to outnumber the ingested inhibitors, secretion of protease isoforms weakly sensitive to these inhibitors, and/or the up-regulation of proteases from alternative functional classes (Broadway 2000, Zhu-Salzman et al. 2003, Srinivasan et al. 2006). It is now well known that recombinant protease inhibitors expressed in plants not only alter the protein-hydrolyzing activity of constitutively expressed target proteases in the midgut of naive insects, but also induce an adjustment of the midgut transcriptome, an alteration of proteome patterns and a rebalancing of the digestive protease complement to sustain protein digestion (Brunelle et al. 2004, Liu et al. 2004, Rivard et al. 2004, Chi et al. 2009, Nogueira et al. 2012).

A straightforward way to harness the potential of protease inhibitors in plant protection consists of using combinations of inhibitors active against different sets of proteases in the target pest, in such a way as to broaden the range of susceptible proteases and minimize compensation by 'insensitive', non-inhibited proteases following inhibitor intake (Oppert et al. 2005). Transgene stacking approaches for the simultaneous expression of two or more inhibitors allowed to implement effective insect resistance in different plants, hardly attainable using the same inhibitors expressed alone (Abdeen et al. 2005, Dunse et al. 2010a, Senthilkumar et al. 2010). In a similar way, fusion proteins integrating coding sequences of aspartate (Asp), Ser and/or Cys protease inhibitors showed improved potential against target herbivores compared to the inhibitors alone (Urwin et al. 1998, Inanaga et al. 2001, Outchkourov et al. 2004, Brunelle et al. 2005, Benchabane et al. 2008, Sainsbury et al. 2012b). An example of this was provided for *L. decemlineata*, where a translational fusion protein integrating inhibitory functions of the Cys-type inhibitor corn cystatin II and the broad-spectrum inhibitor of Asp and Ser proteases, tomato cathepsin D inhibitor, significantly delayed growth of 3rd instars, in sharp contrast with single inhibitors showing no measurable effects (Brunelle et al. 2005). While protease inhibitors have been mostly considered until now as a possible complement to Bt toxins in agricultural fields (Huang et al. 2005), recent advances towards the development of potent, 'pesticide-like' inhibitors could eventually make these proteins an attractive alternative to currently used toxins in major plant–pest systems, including—but not restricted to—those cases where no effective toxins are available for the pest to control (Chougule and Bonning 2012).

Keeping the Balance Between Inhibitory Efficiency and Specificity

Broadening the inhibitory range of protease inhibitors to different protease classes, however, raises issues of specificity and compatibility in an environmental context that implicates extracellular protein digestion at different trophic levels (Schlüter et al. 2010). The expression of Ser-type inhibitors in food crops may also raise concerns on food safety and quality, at least from a regulatory standpoint, given their antidigestive effects and allergenic potential in humans (Gilani et al. 2005). By comparison, no Cys protease targets are found in the human gut, and the innocuity of plant cystatins in transgenic crops has been documented (Atkinson et al. 2004). In this perspective, an ecologically acceptable way of using protease inhibitors in pest control might be to identify protein inhibitor candidates, such as Cys-type inhibitors, that allow for an adequate protective effect by the inhibition of a minimal, but vital set of proteases in the target midgut. For *L. decemlineata*, combining Cys protease inhibitors with complementary inhibitory spectra in such a way as to broaden the range of Cys proteases inhibited in the midgut without affecting the activity of remaining, non-Cys proteases, may be an effective strategy.

L. decemlineata larvae and adults secrete proteases from several functional classes to digest proteins (Michaud et al. 1995, Novillo et al. 1997), including an array of ~30-kDa Cys proteases, the so-called 'intestains' (Gruden et al. 2003), which account together for about 90% of protein-hydrolyzing activities in the midgut (Michaud et al. 1993). These proteases are sensitive overall to the specific, low-molecular-weight inhibitor of Cys proteases trans-epoxysuccinyl-L-leucylamido-(4-guanidino) butane (E-64), but differentially sensitive to a variety of plant and animal cystatins considered for plant genetic improvement (Michaud et al. 1993, 1995, Gruden et al. 1998, 2003, 2004, Goulet et al. 2008). Whereas E-64 ingestion is toxic to potato beetle larvae (Bolter and Latoszek-Green 1997), recombinant cystatins expressed in potato leaves show no adverse effects on growth and reproduction of both larvae and adults (Cloutier et al. 1999, 2000), despite the establishment of strong cystatin–Cys protease complexes *in vivo* (Bouchard et al. 2003).

Different groups have been involved in the *in silico* modeling of pest protease–plant inhibitor complexes, the identification of protease inhibitors with complementary effects against herbivorous pest proteases, or the molecular engineering of inhibitor variants with improved potency against these enzymes (e.g., Urwin et al. 1995, Koiwa et al. 2001, Ceci et al. 2003, Melo et al. 2003, Outchkourov et al. 2004, Kiggundu et al.

2006, Goulet et al. 2008, Dunse et al. 2010b). Research efforts should also be directed, at present, to the definition of rational frameworks for the selection of broad-spectrum or complementary inhibitor variants that take into account the complex dynamics of protease–inhibitor interactions and the striking diversity of midgut proteases encoded in herbivore insect genomes (*Tribolium* Genome Sequencing Consortium 2008, Sainsbury et al. 2012a). Until now, most protein engineering and modeling work aimed at developing protease inhibitors useful in pest control has been directed towards the inhibition of specific (model) proteases or protease activities, with little attention paid to non- or weakly inhibited proteases related to the model protease, that could allow protein digestion in the target insect despite the strong inhibition of (a) specific protease(s) *in vivo*.

A Promising Combination of Cys-type Inhibitors for L. decemlineata *Intestains*

We have engineered single variants of tomato cystatin SlCYS8 by single mutations at functionally relevant amino acid sites (Kiggundu et al. 2006), which exhibit strongly improved activity against Z-Phe-Arg-methylcoumarin (MCA) and Z-Arg-Arg-MCA hydrolyzing intestains (Goulet et al. 2008). Some variants, such as 'P2F' bearing a phenylalanine in place of a proline at position 2 or 'T6R' bearing a threonine in place of an arginine at position 6, showed strongly improved inhibitory effects against both populations of intestains *in vitro* (Goulet et al. 2008). It is not yet clear, still, whether these improved activities were the consequence of a broadened inhibitory range including E-64-sensitive intestains insensitive to SlCYS8 or, instead, of a stronger effect on intestain forms already inhibited by the wild-type inhibitor.

Here we show, using a recently described functional proteomic procedure for the capture and identification of cystatin-sensitive Cys protease targets (Sainsbury et al. 2012a), that wild-type SlCYS8 and improved variants such as P2F indeed inhibit the same complement of beetle intestains, despite an apparently broader inhibitory range inferred for the mutant protein based on quantitative data with synthetic peptide substrates (Goulet et al. 2008). The procedure consists of using biotinylated versions of the cystatins expressed in *Escherichia coli* to capture target proteases in midgut protein extracts. Cystatin–intestain complexes in the protein mixtures were purified by affinity chromatography on an avidin-agarose column, and the captured intestains identified and quantified by tandem mass spectrometry (MS/MS) following SDS-PAGE, Coomassie blue staining and trypsin digestion of an enriched 30-kDa protein band corresponding to the intestains. A simplified version of this procedure was first used here, in which the 30-kDa protein band for MS/MS

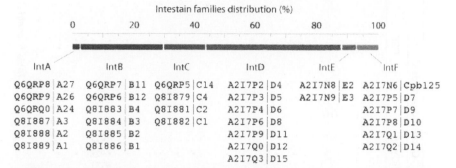

Figure 1. Intestain forms in the midgut of *L. decemlineata* larvae fed potato plants, cv. Kennebec. Distribution percentages are based on the total number of unique peptide spectral counts assigned to each intestain family. Intestains names and assignment to specific families are from Gruden et al. (2004), modified by Sainsbury et al. (2012a). Inclusion in the list is based on the detection of unique MS-MS peptides shared by different intestain forms within each family, and is thus subject to false positives.

analysis was collected without prior Cys protease capture, to get a whole picture of the E-64-sensitive intestain complement in the insect midgut (Fig. 1). Thirty-six sequences are found in the NCBI/GenBank database for *L. decemlineata* intestains (http://ncbi.nlm.nih.gov/), categorized into six functional families—the IntA, IntB, IntC, IntD, IntE and IntF families—based on primary sequence homologies and specific conserved motifs in functionally relevant areas of the encoded polypeptide (Sainsbury et al. 2012a). About 120 unique peptides were detected following MS/MS, that could each be assigned to a single intestain family (see Table 6 for a list of the most abundant peptides). Peptide sequences found in 31 distinct intestains were detected overall, both indicating the absence of some genome-encoded intestains in the midgut and suggesting the constitutive occurrence of intestain members from the six functional families, albeit at different relative levels based on spectral count numbers (Fig. 1).

The inhibitory ranges of SlCYS8 and P2F towards intestains were then compared, based on supplementary MS/MS data provided online by Sainsbury et al. (2012a). Both cystatin variants allowed for the identification of multiple unique peptides specific to IntB, IntC, IntD, IntE and IntF family members (see Table 6 for examples). An identical set of peptides was obtained with SlCYS8 and P2F, suggesting the capture of the same intestains by, and hence a comparable inhibitory range for, the two inhibitors. Spectrum numbers counted for unique peptides of the IntB, IntC and IntF families were comparable for the two inhibitors, in line with similar consolidated counts reported for each of these families taken as a whole (Sainsbury et al. 2012a). Also in line with consolidated numbers, spectral counts three to 20 times greater than with SlCYS8

Table 6. Intestain family assignment of the most abundant unique peptides detected following MS/MS.[1]

Family	Unique peptide	Intestains bearing the peptide[2]	Captured peptides[3]	
			SICYS8	P2F
IntA	NSWGTTWGEDGYFR	A1, A2, A26, A27		
	LLADEDELKK	A1, A2, A24, A26, A27		
	KLLADEDELKK	A1, A2, A24, A26, A27		
	VKNSWGTTWGEDGYFR	A1, A2, A26, A27		
IntB	GIEADSSYPYK	B1, B3, B11, B12	•	•
	GAVLDVK	B11, B12	•	•
	SATGALEGQNAIVNNVK	B1, B2, B3, B4, B11, B12	•	•
	GIDTPCQYDAK	B1, B2, B11, B12	•	•
	DWGEQGYFR	B2, B3, B11, B12	•	•
IntC	KNDEIDLQK	C1, C2, C14	•	•
	LISLSEQQLVDCVK	C1, C2, C14	•	•
	NADNQCGIATR	C1, C4, C14	•	•
IntD	GCSHDLDHGVLVVGYGK	D4, D5, D6, D8, D11	•	•••
	EGGDMSAAFEYVR	D5, D6, D15	•	•••
	GAVLEVK	D4, D5, D6, D15	•	•••
	KQTECQYDASK	D4, D5, D6, D15	•	•••
	IWGENGYFR	D4, D5, D6, D8, D11, D12	•	•••

Table 6. contd....

	Peptide	Intestains[2]	[3]
IntE	AVGTIGPISVAVSSEHLR	E2, E3	•••
	GIEAGSSYPYQGR	E2, E3	•••
	DLDHAVLAVGYGSENGR	E2, E3	•••
IntF	GAVLDIK	D13, D14	••
	DYGIEAEESYPYK	D13, D14	••
	SIETVCQYDASK	D13, D14	••
	GCTDELDHGVLAVGYGEVSQSSGNTK	D7	•
	GCSDELDHAVLAVGYGEVSQSSAK	D9	•

[1] Unique peptides specific to intestains A were detected only for the modified analytic procedure with no protease capture step (see text). Intestain electrophoresis, trypsin digestion and MS/MS analysis were carried out as described in Sainsbury et al. (2012a). Data for SlCYS8 and its P2F mutant were inferred from Supplementary Table 3 of the same study.

[2] Intestain nomenclature based on Gruden et al. (2004) and subsequent gene sequence submissions to the GenBank/NCBI database (http://ncbi.nlm.nih.gov/).

[3] ••• indicates a mean number of peptide spectral counts 3 to 20 times greater for P2F than for wild-type SlCYS8 (•). Blank spaces indicate no capture of the unique peptide for either inhibitor.

were obtained with P2F for the IntD- and IntE-specific peptides (Table 6). These observations support previous K_d value data showing an improved inhibitory potency for P2F towards Z-Phe-Arg-MCA and Z-Arg-Arg-MCA hydrolyzing intestain activities, compared to the original inhibitor (Goulet et al. 2008). These data also link, however, the improved potency of the mutant cystatin with a stronger effect against already inhibited (IntD and IntE) intestains, rather than with a broadened effect including the non-inhibited, E-64-sensitive IntA family members.

Proteomic data were recently published by Smid et al. (2013), which suggest the potential of macrocypins, a new type of Cys protease inhibitors from the edible mushroom *Macrolepiota procera* (Sabotic et al. 2009), as a complement to plant cystatins for *L. decemlineata* intestain inhibition. Macrocypins and homologues from related mushrooms rely on three versatile loops to inhibit proteases (Renko et al. 2010), via an inhibitory mechanism distinct from that of plant cystatins, which involve two inhibitory loops and a flexible N-terminal trunk for enzyme–inhibitor complex stabilization (Benchabane et al. 2010). A recombinant form of Macrocypin 1 (Mcp1) was expressed in *E. coli* and chemically linked to a Sepharose chromatography column. Midgut protein extracts from potato beetle 4th instars were then passed through the column, and the captured intestains eluted at low pH, resolved by SDS-PAGE, digested with trypsin and submitted to ESI-MS/MS peptide mass fingerprinting. Intestain-specific peptides were finally identified and assigned to specific intestain families based on the classification of Sainsbury et al. (2012a). The procedure allowed overall for the capture and identification of unique peptides specific to the IntA, IntB, IntD, IntE and IntF intestain families (Smid et al. 2013). Figure 2 compares the capture profiles of SlCYS8 and Mcp1 in terms of intestains in the insect genome that include the peptide sequences identified. Based on this, we propose that the two inhibitors have the potential to inhibit, together, the whole set of intestains eventually found in the midgut, and hence to reproduce, in combination, the broad Cys protease inhibitory effect of E-64 (Fig. 3) and the strong detrimental impact of this chemical on *L. decemlineata* growth and development (Bolter and Latoszek-Green 1997). We reported some years ago the strong negative impact of two plant cystatins on larval growth of the banana weevil *Cosmopolites sordidus*, which digests banana plant proteins using Cys proteases strongly inhibited by both plant cystatins and the E-64 chemical inhibitor (Kiggundu et al. 2010). Work is underway to determine whether comparable detrimental effects may be observed with

	Uniprot | Clone	TOTAL	SICYS8	MCP1
IntA	Q6QRP8 | A27	•		•
	Q6QRP9 | A26	•		•
	Q6QRQ0 | A24	•		•
	Q8I887 | A3	•		•
	Q8I888 | A2	•		•
	Q8I889 | A1	•		•
IntB	Q6QRP7 | B11	•	•	•
	Q6QRP6 | B12	•	•	•
	Q8I883 | B4	•	•	•
	Q8I884 | B3	•	•	•
	Q8I885 | B2	•	•	•
	Q8I886 | B1	•	•	•
IntC	Q6QRP5 | C14	•	•	
	Q8I879 | C4	•	•	
	Q8I880 | C3			
	Q8I881 | C2	•	•	
	Q8I882 | C1	•	•	
IntD	*A2I7N7 | D9*			
	A2I7P2 | D4	•	•	•
	A2I7P3 | D5	•	•	•
	A2I7P4 | D6	•	•	•
	A2I7P6 | D8	•	•	•
	A2I7P9 | D11	•	•	•
	A2I7Q0 | D12	•	•	•
	A2I7Q3 | D15	•	•	•
IntE	*A2I7N5 | Cpb77*			
	A2I7N8 | E2	•	•	•
	A2I7N9 | E3	•	•	•
	A2I7P0 | E4			
	A2I7P1 | E5			
IntF	A2I7N6 | Cpb125	•	•	
	A2I7P5 | D7	•	•	•
	A2I7P7 | D9	•	•	•
	A2I7P8 | D10	•	•	•
	A2I7Q1 | D13	•	•	
	A2I7Q2 | D14	•	•	

Figure 2. MS/MS detection of *L. decemlineata* intestains in midgut protein extracts of 3rd or 4th instars fed potato plants, cv. Kennebec. MS/MS analysis was conducted on enriched intestain samples after capture with wild-type SlCYS8, its P2F variant, or mushroom macrocypin Mcp1. 'Total' refers to intestains detected in crude extracts not treated with the inhibitors (Fig. 1). Inclusion in the list is based on the detection of unique MS/MS peptides shared by different intestain forms within each family, and is thus subject to false positives.

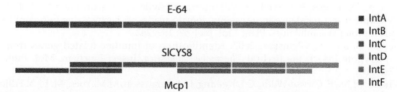

Figure 3. Schematic representation for the inhibitory range of SlCYS8 and macrocypin Mcp1 against *L. decemlineata* intestains. Inferences are based on functional proteomic data from Sainsbury et al. (2012a) and Smid et al. (2013). The two inhibitors reproduce, together, the broad inhibitory range reported earlier for the low-molecular-weight chemical inhibitor E-64 (Goulet et al. 2008).

Color image of this figure appears in the color plate section at the end of the book.

L. decemlineata when combining plant cystatins and mushroom macrocypins, using dual translational fusions integrating different macrocypins and SlCYS8 variants.

Summary

It is an exciting time to work in the area of insect control, particularly Coleoptera. Targeting insect proteases to control problematic species has been challenging in the past, and we hope that we have convinced readers that knowledge of the complement of proteases in the target pest is essential to a successful strategy. Understanding the complex responses of highly evolved insect pests to toxins and inhibitors is also critical to the design process. Incorporating sequencing and proteomic tools will be an integral part of the research to identify novel insect control proteins for the future.

Acknowledgements

Mention of trade names or commercial products in this publication is solely for the purpose of providing specific information and does not imply recommendation or endorsement by the U.S. Department of Agriculture. USDA is an equal opportunity provider and employer. Data from the transcriptome study are available on request from BO. Work on *L. decemlineata* and protease inhibitors in DM's lab is funded by the Natural Science and Engineering Research Council of Canada and by the Fonds de Recherche Québec–Nature et Technologies.

Keywords: Coleoptera, functional genomics, insect control, insect gut, insect proteases, protease inhibitors, proteomics

References

Abdeen, A., A. Virgos, E. Olivella, J. Villanueva, X. Aviles, R. Gabarra and S. Prat. 2005. Multiple insect resistance in transgenic tomato plants over-expressing two families of plant proteinase inhibitors. Plant Mol. Biol. 57: 189–202.

Altincicek, B. and A. Vilcinskas. 2007. Identification of immune-related genes from an apterygote insect, the firebrat *Thermobia domestica*. Insect Biochem. Mol. Biol. 37: 726–731.

Andersson, M.N., E. Grosse-Wilde, C.I. Keeling, J.M. Bengtsson, M.M. Yuen, M. Li, Y. Hillbur, J. Bohlmann, B.S. Hansson and F. Schlyter. 2013. Antennal transcriptome analysis of the chemosensory gene families in the tree killing bark beetles, *Ips typographus* and *Dendroctonus ponderosae* (Coleoptera: Curculionidae: Scolytinae). BMC Genomics 14: 198.

Atkinson, H.J., K.A. Johnston and M. Robbins. 2004. *Prima facie* evidence that a phytocystatin for transgenic plant resistance to nematodes is not a toxic risk in the human diet. J. Nutr. 134: 431–434.

Benchabane, M., U. Schlüter, J. Vorster, M.-C. Goulet and D. Michaud. 2010. Plant cystatins. Biochimie 92: 1657–1666.

Benchabane, M., M.-C. Goulet, C. Dallaire, P.-L. Côté and D. Michaud. 2008. Hybrid protease inhibitors for pest and pathogen control—a functional cost for the fusion partners? Plant Physiol. Biochem. 46: 701–708.

Birk, Y. 2003. Plant Protease Inhibitors. Springer, New York.

Bachman, P.M., R. Bolognesi, W.J. Moar, G.M. Mueller, M.S. Paradise, P. Ramaseshadri, J. Tan, J.P. Uffman, J. Warren, B.E. Wiggins and S.L. Levine. 2013. Characterization of the spectrum of insecticidal activity of a double-stranded RNA with targeted activity against Western Corn Rootworm (*Diabrotica virgifera virgifera* LeConte). Transgenic Res. 22: 1207–1222.

Bolter, C.J. and M. Latoszek-Green. 1997. Effect of chronic ingestion of the cysteine proteinase inhibitor, E-64, on Colorado potato beetle gut proteinases. Entomol. Exp. Appl. 83: 295–303.

Bonnett, T.R., J.A. Robert, C. Pitt, J.D. Fraser, C.I. Keeling, J. Bohlmann and D.P.W. Huber. 2012. Global and comparative proteomic profiling of overwintering and developing mountain pine beetle, *Dendroctonus ponderosae* (Coleoptera: Curculionidae) larvae. Insect Biochem. Mol. Biol. 42: 890–901.

Bouchard, É., C. Cloutier and D. Michaud. 2003. Oryzacystatin I expressed in transgenic potato induces digestive compensation in an insect natural predator via its herbivorous prey feeding on the plant. Mol. Ecol. 12: 2439–2446.

Broadway, R.M. 2000. The adaptation of insects to protease inhibitors. pp. 80–88. *In*: D. Michaud (ed.). Recombinant Protease Inhibitors in Plants. Landes Bioscience, Georgetown, Texas.

Brunelle, F., C. Girard, C. Cloutier and D. Michaud. 2005. A hybrid, broad-spectrum inhibitor of Colorado potato beetle aspartate and cysteine proteinases. Arch. Insect Biochem. Physiol. 60: 20–31.

Brunelle, F., C. Cloutier and D. Michaud. 2004. Colorado potato beetles compensate for tomato cathepsin D inhibitor expressed in transgenic potato. Arch. Insect Biochem. Physiol. 55: 103–113.

Carrasco, M.A., S.A. Buechler, R.J. Arnold, T. Sformo, T.M. Barnes and J.G. Duman. 2012. Investigating the deep supercooling ability of an Alaskan beetle, *Cucujus clavipes puniceus*, via high throughput proteomics. J. Proteom. 75: 1220–1234.

Carrasco, M.A., S.A. Buechler, R.J. Arnold, T. Sformo, T.M. Barnes and J.G. Duman. 2011. Elucidating the biochemical overwintering adaptations of larval *Cucujus clavipes puniceus*, a nonmodel organism, via high throughput proteomics. J. Proteome Res. 10: 4634–4646.

Ceci, L.R., M. Volpicella, Y. Rahbé, R. Gallerani, J. Beekwilder and M.A. Jongsma. 2003. Selection by phage display of a variant mustard trypsin inhibitor toxic against aphids. Plant J. 33: 557–566.

Chi, Y.H., R.A. Salzman, S. Balfe, J.-E. Ahn, W. Sun, J. Moon, D.-J. Yun, S.Y. Lee, T.J.V. Higgins, B. Pittendrigh, L.L. Murdock and K. Zhu-Salzman. 2009. Cowpea bruchid midgut transcirptome response to a soybean cystatin—costs and benefits of counter-defense. Insect Mol. Biol. 18: 97–110.

Chougule, N.P. and B.C. Bonning. 2012. Toxins for transgenic resistance to hemipteran pests. Toxins 4: 405–429.

Cloutier, C., C. Jean, M. Fournier, S. Yelle and D. Michaud. 2000. Adult Colorado potato beetles, *Leptinotarsa decemlineata* compensate for nutritional stress on oryzacystatin I transgenic potato plants by hypertrophic behavior and over-production of insensitive proteases. Arch. Insect Biochem. Physiol. 44: 69–81.

Cloutier, C., M. Fournier, C. Jean, S. Yelle and D. Michaud. 1999. Growth compensation and faster development of Colorado potato beetle (Coleoptera: Chrysomelidae) feeding on potato foliage expressing oryzacystatin I. Arch. Insect Biochem. Physiol. 40: 69–79.

Contreras, E., C. Rausell and M.D. Real. 2013. Proteome response of *Tribolium castaneum* larvae to *Bacillus thuringiensis* toxin producing strains. PLoS One 8: e55330.

Dunse, K.M., J.A. Stevens, F.T. Lay, Y.M. Gaspar, R.L. Heath and M.A. Anderson. 2010a. Coexpression of potato type I and II proteinase inhibitors gives cotton plants protection against insect damage in the field. Proc. Natl. Acad. Sci. USA 107: 15011–15015.

Dunse, K.M., Q. Kaas, R.F. Guarino, P.A. Barton, D.J. Craik and M.A. Anderson. 2010b. Molecular basis for the resistance of an insect chymotrypsin to a potato type II proteinase inhibitor. Proc. Natl. Acad. Sci. USA 107: 15016–15021.

Gilani, G.S., K.A. Cockell and E. Sepher. 2005. Effects of antinutritional factors on protein digestibility and amino acid availability in foods. J. AOAC Int. 88: 967–987.

Goptar, I.A., T. Semashko, S.A. Danilenko, E.N. Lysogorskaya, E.S. Oksenoit, D.P. Zhuzhikov, M.A. Belozersky, Y.E. Dunaevsky, B. Oppert, I.Y. Filippova and E.N. Elpidina. 2012. Cysteine digestive peptidases function as post-glutamine cleaving enzymes in tenebrionid stored product pests. Comp. Biochem. Physiol. 161B: 148–154.

Goulet, M.-C., C. Dallaire, L.-P. Vaillancourt, M. Khalf, A.M. Badri, A. Preradov, M.-O. Duceppe, C. Goulet, C. Cloutier and D. Michaud. 2008. Tailoring the specificity of a plant cystatin toward herbivorous insect digestive cysteine proteases by single mutations at positively selected amino acid sites. Plant Physiol. 146: 1010–1019.

Gruden, K., A.G.J. Kuipers, G. Guncar, N. Slapar, B. Strukelj and M.A. Jongsma. 2004. Molecular basis of Colorado potato beetle adaptation to potato plant defence at the level of digestive cysteine proteinases. Insect Biochem. Mol. Biol. 34: 365–375.

Gruden, K., T. Popovic, N. Cimerman, I. Krizaj and B. Strukelj. 2003. Diverse enzymatic specificities of digestive proteases, 'intestains', enable Colorado potato beetle larvae to counteract the potato defence mechanism. Biol. Chem. 384: 305–310.

Gruden, K., B.S. Strukelj, T. Popovic, B. Lenarcic, T. Bevec, J. Brzin, I. Kregar, J. Herzog-Velikonja, W.J. Stiekema, D. Bosch and M.A. Jongsma. 1998. The cysteine protease activity of Colorado potato beetle (*Leptinotarsa decemineata* Say) guts, which is insensitive to potato protease inhibitors, is inhibited by thyroglobulin type-I domain inhibitors. Insect Biochem. Mol. Biol. 28: 549–560.

Haq, S.K., S.M. Atif and R.H. Khan. 2004. Protein proteinase inhibitor genes in combat against insects, pests, and pathogens: natural and engineered phytoprotection. Arch. Biochem. Biophys. 431: 145–159.

Hilder, V.A., A.M.R. Gatehouse, S.E. Sheerman, R.F. Barker and D. Boulter. 1987. A novel mechanism of insect resistance engineered into tobacco. Nature 330: 160–163.

Huang, J., R. Hu, S. Rozelle and C. Pray. 2005. Insect-resistant GM rice in farmers' fields: assessing productivity and health effects in China. Science 388: 688–690.

i5K Consortium. 2013. The i5K Initiative: advancing arthropod genomics for knowledge, human health, agriculture, and the environment. J. Hered. 104: 595–600.

Inanaga, H., D. Kobayasi, Y. Kouzuma, C. Aoki-Yasunaga, K. Iiyama and K. Kimura. 2001. Protein engineering of novel proteinase inhibitors and their effects on the growth of *Spodoptera exigua* larvae. Biosci. Biotechnol. Biochem. 65: 2259–2264.

Johnson, R., J. Narvaez, G. An and C.A. Ryan. 1989. Expression of proteinase inhibitors I and II in transgenic tobacco plants: effects on natural defense against *Manduca sexta* larvae. Proc. Natl. Acad. Sci. USA 86: 9871–9875.

Keeling, C.I., M.M.S. Yuen, N.Y. Liao, T.R. Docking, S.K. Chan, G.A. Taylor, D.L. Palmquist, S.D. Jackman, A. Nguyen, M. Li, H. Henderson, J.K. Janes, Y. Zhao, P. Pandoh, R. Moore, F.A.H. Sperling, D.P.W. Huber, I. Birol, S.J.M. Jones and J. Bohlmann. 2013. Draft genome of the mountain pine beetle, *Dendroctonus ponderosae* Hopkins, a major forest pest. Genome Biol. 14: R27.

Kiggundu, A., J. Muchwezi, C. Van der Vyver, A. Viljoen, J. Vorster, U. Schlüter, K. Kunert and D. Michaud. 2010. Deleterious effects of plant cystatins against the banana weevil *Cosmopolites sordidus*. Arch. Insect Biochem. Physiol. 73: 87–105.

Kiggundu, A., M.-C. Goulet, C. Goulet, J.-F. Dubuc, D. Rivard, M. Benchabane, G. Pépin, C. van der Vyver, K. Kunert and D. Michaud. 2006. Modulating the proteinase inhibitory profile of a plant cystatin by single mutations at positively selected amino acid sites. Plant J. 48: 403–413.

Kirsch, R., N. Wielsch, H. Vogel, A. Svatos, D.G. Heckel and Y. Pauchet. 2012. Combining proteomics and transcriptome sequencing to identify active plant-cell-wall-degrading enzymes in a leaf beetle. BMC Genom. 13: 587.

Koiwa, H., M.P. D'Urzo, I. Assfalg-Machleidt, K. Zhu-Salzman, R.E. Shade, H. An, L.L. Murdock, W. Machleidt, R.A. Bressan and P.M. Hasegawa. 2001. Phage display selection of hairpin loop soyacystatin variants that mediate high affinity inhibition of a cysteine proteinase. Plant J. 27: 383–391.

Kumar, A., L. Congiu, L. Lindström, S. Piiroinen, M. Vodotto and A. Grapputo. 2014. Sequencing, *de novo* assembly and annotation of the Colorado potato beetle, *Leptinotarsa decemlineata*, transcriptome. PLoS One 9: e86012.

Liu, Y., R.A. Salzman, T. Pankiw and K. Zhu-Salzman. 2004. Transcriptional regulation in southern corn rootworm larvae challenged by soyacystatin N. Insect Biochem. Mol. Biol. 34: 1069–1077.

Melo, F.R., M.O. Mello, O.L. Franco, D.J. Rigden, L.V. Mello, A.M. Genu, M.C. Silva-Filho, S. Gleddie and M.F. Grossi-de-Sa. 2003. Use of phage display to select novel cystatins specific for *Acanthoscelides obtectus* cysteine proteinases. Biochim. Biophys. Acta 1651: 146–152.

Merzendorfer, H., H.S. Kim, S.S. Chaudhari, M. Kumari, C.A. Specht, S. Butcher, S.J. Brown, J.R. Manak, R.W. Beeman, K.J. Kramer and S. Muthukrishnan. 2012. Genomic and proteomic studies on the effects of the insect growth regulator diflubenzuron in the model beetle species *Tribolium castaneum*. Insect Biochem. Mol. Biol. 42: 264–276.

Michaud, D., N. Bernier-Vadnais, S. Overney and S. Yelle. 1995. Constitutive expression of digestive cysteine proteinase forms during development of the Colorado potato beetle, *Leptinotarsa decemlineata* Say (Coleoptera: Chrysomelidae). Insect Biochem. Mol. Biol. 25: 1041–1048.

Michaud, D., B. Nguyen-Quoc and S. Yelle. 1993. Selective inhibition of Colorado potato beetle cathepsin H by oryzacystatins I and II. FEBS Lett. 331: 173–176.

Miller, N.J., S. Richards and T.W. Sappington. 2010. The prospects for sequencing the western corn rootworm genome. J. Appl. Entomol. 134: 420–428.

Morris, K., M.D. Lorenzen, Y. Hiromasa, J.M. Tomich, C. Oppert, E.N. Elpidina, V. Vinokurov, J.L. Jurat-Fuentes, J. Fabrick and B. Oppert. 2009. *Tribolium castaneum* larval gut transcriptome and proteome: A resource for the study of the coleopteran gut. J. Proteome Res. 8: 3889–3898.

Nogueira, F.C.S., C.P. Silva, D. Alexandre, R.I. Sumuels, E.L. Soares, F.J.L. Aragao, G. Palmisano, G.B. Domont, P. Roepstorff and F.A.P. Campos. 2012. Global proteome changes in larvae of *Callosobruchus maculatus* (Coleoptera: Chrysomelidae: Bruchinae) following ingestion of a cysteine proteinase inhibitor. Proteomics 12: 2704–2715.

Novillo, C., P. Castanera and F. Ortego. 1997. Characterization and distribution of chymotrypsin-like and other digestive proteases in Colorado potato beetle. Arch. Insect Biochem. Physiol. 36: 181–201.

Oppert, B., T.D. Morgan, K. Hartzer and K.J. Kramer. 2005. Compensatory proteolytic responses to dietary proteinase inhibitors in the red flour beetle, *Tribolium castaneum* (Coleoptera: Tenebrionidae). Comp. Biochem. Physiol. 140C: 53–58.

Oppert, B., E.N. Elpidina, M. Toutges and S. Mazumdar-Leighton. 2010. Microarray analysis reveals strategies of *Tribolium castaneum* larvae to compensate for cysteine and serine protease inhibitors. Comp. Biochem. Physiol. 5D: 280–287.

Oppert, B., T. Morgan and K.J. Kramer. 2011. Efficacy of *Bacillus thuringiensis* Cry3Aa protoxin and protease inhibitors toward coleopteran storage pests. Pest Manag. Sci. 67: 568–573.

Oppert, B., S.E. Dowd, P. Bouffard, L. Li, A. Conesa, M.D. Lorenzen, M. Toutges, J. Marshall, D.L. Huestis, J. Fabrick, C. Oppert and J.L. Jurat-Fuentes. 2012a. Transcriptome profiling of the intoxication response of *Tenebrio molitor* larvae to *Bacillus thuringiensis* Cry3Aa protoxin. PLoS One: e34624.

Oppert, B., A.G. Martynov and E.N. Elpidina. 2012b. *Bacillus thuringiensis* Cry3Aa intoxication of *Tenebrio molitor* induces widespread changes in the expression of serine peptidase transcripts. Comp. Biochem. Physiol. 7D: 233–242.

Outchkourov, N.S., W.J. de Kogel, G.L. Wiegers, M. Abrahamson and M.A. Jongsma. 2004. Engineered multidomain cysteine protease inhibitors yield resistance against western flower thrips (*Frankliniella occidentalis*) in greenhouse trials. Plant Biotechnol. J. 2: 449–458.

Pauchet, Y., P. Wilkinson, R. Chauhan and R.H. ffrench-Constant. 2010. Diversity of beetle genes encoding novel plant cell wall degrading enzymes. PLoS One 5: e15635.

Pechanova, O., W.D. Stone, W. Monroe, T.E. Nebeker, K.D. Klepzig and C. Yuceer. 2008. Global and comparative protein profiles of the pronotum of the southern pine beetle, *Dendroctonus frontalis*. Insect Mol. Biol. 17: 261–277.

Petek, M., N. Turnsek, M.B. Gasparic, M.P. Novak, K. Gruden, N. Slapar, T. Popovic, B. Strukelj and M.A. Jongsma. 2012. A complex of genes involved in adaptation of *Leptinotarsa decemlineata* larvae to induced potato defense. Arch. Insect Biochem. Physiol. 79: 153–181.

Prabhakar, S., M.S. Chen, E.N. Elpidina, K.S. Vinokurov, C.M. Smith, J. Marshall and B. Oppert. 2007. Sequence analysis and molecular characterization of larval midgut cDNA transcripts encoding peptidases from the yellow mealworm, *Tenebrio molitor* L. Insect Mol. Biol. 16: 455–468.

Qiu, J. 2008. Is China ready for GM rice? Nature 455: 850–852.

Renko, M., J. Sabotic, M. Mihelic, J. Brzin, J. Kos and D. Turk. 2010. Versatile loops in mycocypins inhibit three protease families. J. Biol. Chem. 285: 308–316.

Rivard, D., C. Cloutier and D. Michaud. 2004. Colorado potato beetles show differential digestive compensatory responses to host plants expressing distinct sets of defense proteins. Arch. Insect Biochem. Physiol. 55: 114–123.

Sabotic, J., T. Popovic, V. Puizdar and J. Brzin. 2009. Macrocypins, a family of cysteine protease inhibitors from the basidiomycete *Macrolepiota procera*. FEBS J. 276: 4334–4345.

Sainsbury, F., A.-J. Rhéaume, M.-C. Goulet, J. Vorster and D. Michaud. 2012a. Discrimination of differentially inhibited cysteine proteases by activity-based profiling using cystatin variants with tailored specificities. J. Proteome Res. 11: 5983–5993.

Sainsbury, F., M. Benchabane, M.-C. Goulet and D. Michaud. 2012b. Multimodal protein constructs for herbivore insect control. Toxins 4: 455–475.

Schlüter, U., M. Benchabane, A. Munger, A. Kiggundu, J. Vorster, M.-C. Goulet, C. Cloutier and D. Michaud. 2010. Recombinant protease inhibitors for herbivore pest control: a multitrophic perspective. J. Exp. Bot. 61: 4169–4183.

Senthilkumar, R., C.-P. Cheng and K.-W. Yeh. 2010. Genetically pyramiding protease-inhibitor genes for dual broad-spectrum resistance against insect and phytopathogens in transgenic tobacco. Plant Biotechnol. J. 8: 65–75.

Sheffield, N.C., H. Song, S.L. Cameron and M.F. Whiting. 2008. A comparative analysis of mitochondrial genomes in Coleoptera (Arthropoda: Insecta) and genome descriptions of six new beetles. Mol. Biol. Evol. 25: 2499–2509.

Shirley, D., C. Oppert, T. Reynolds, B. Miracle, B. Oppert, W.E. Klingeman and J.L. Jurat-Fuentes. 2013. Expression and functional characterization of an endoglucanase from *Tribolium castaneum* (TcEG1) in *Saccharomyces cerevisiae*. Insect Sci. (in press). DOI:10.1111/1744-7917.12069.

Siegfried, B.D., N. Waterfield and R.H. ffrench-Constant. 2005. Expressed sequence tags from *Diabrotica virgifera virgifera* midgut identify a coleopteran cadherin and a diversity of cathepsins. 14: 137–143.

Smid, I., K. Gruden, M.B. Gasparic, K. Koruza, M. Petek, J. Pohleven, J. Brzin, J. Kos, J. Zel and J. Sabotic. 2013. Inhibition of the growth of Colorado potato beetle larvae by macrocypins, protease inhibitors from the Parasol Mushroom. J. Agric. Food Chem. 61: 12499–12509.

Srinivasan, A., A.P. Giri and V.S. Gupta. 2006. Structural and functional diversities in lepidopteran serine proteases. Cell. Mol. Biol. Lett. 11: 132–154.

Swevers, L., H. Huvenne, G. Menschaert, D. Kontogiannatos, A. Kourti, Y. Pauchet, R. ffrench-Constant and G. Smagghe. 2013. Colorado potato beetle (Coleoptera) gut transcriptome analysis: expression of RNA interference-related genes. Insect Mol. Biol. 22: 668–684.

Tribolium Genome Sequencing Consortium. 2008. The genome of the model beetle and pest *Tribolium castaneum*. Nature 452: 949–955.

Urwin, P.E., M.J. McPherson and H.J. Atkinson. 1998. Enhanced transgenic plant resistance to nematodes by dual proteinase inhibitor constructs. Planta 204: 472–479.

Urwin, P.E., H.J. Atkinson, D.A. Waller and M.J. McPherson. 1995. Engineered oryzacystatin-I expressed in transgenic hairy roots confers resistance to *Globodera pallida*. Plant J. 8: 121–131.

Vinokurov, K.S., E.N. Elpidina, B. Oppert, S. Prabhakar, D.P. Zhuzhikov, Y.E. Dunaevsky and M.A. Belozersky. 2006. Diversity of digestive proteinases in *Tenebrio molitor* (Coleoptera: Tenebrionidae) larvae. Comp. Biochem. Physiol. 145B: 126–137.

Vinokurov, K.S., E.N. Elpidina, D.P. Zhuzhikov, B. Oppert, D. Kodrik and F. Sehnal. 2009. Digestive proteolysis organization in two closely related tenebrionid beetles: red flour beetle (*Tribolium castaneum*) and confused flour beetle (*Tribolium confusum*). Arch. Insect Biochem. Physiol. 70: 254–279.

Willis, J.D., B. Oppert, C. Oppert, W.E. Klingeman and J.L. Jurat-Fuentes. 2011. Identification, cloning, and expression of a GHF9 cellulase from *Tribolium castaneum* (Coleoptera: Tenebrionidae). J. Insect Physiol. 57: 300–306.

Xu, J., J. Baulding and S.R. Palli. 2013. Proteomics of *Tribolium castaneum* seminal fluid proteins: identification of an angiotensin-converting enzyme as a key player in regulation of reproduction. J. Proteom. 78: 83–93.

Yocum, G.D., J.P. Rinehart, A. Chirumamilla-Chapara and M.L. Larson. 2009a. Characterization of gene expression patterns during the initiation and maintenance phases of diapause in the Colorado potato beetle, *Leptinotarsa decemlineata*. J. Insect Physiol. 55: 32–39.

Yocum, G.D., J.P. Rinehart and M.L. Larson. 2009b. Down regulation of gene expression between the diapause initiation and maintenance phases of the Colorado potato beetle, *Leptinotarsa decemlineata* (Coleoptera: Chrysomelidae). Eur. J. Entomol. 106: 471–476.

Zhen, Y., M.L. Aardema, E.M. Medina, M. Schumer and P. Andolfatto. 2012. Parallel molecular evolution in an herbivore community. Science 337: 1634–1637.

Zhu-Salzman, K., H. Koiwa, R.A. Salzman, R.E. Shade and J.-E. Ahn. 2003. Cowpea bruchid *Callosobruchus maculatus* uses a three-component strategy to overcome a plant defensive cysteine protease inhibitor. Insect Mol. Biol. 12: 135–145.

11

Trypsin Modulating Oostatic Factor (TMOF) and Insect Biotechnology

Dov Borovsky

Introduction

Peptide hormones that are insect-specific are synthesized by agricultural pest insects, as well as, by mosquitoes that are important vectors of detrimental human infectious diseases such as malaria, dengue, encephalitis and yellow fever. An excellent personal overview on the diseases that mosquitoes transmit and the impact on our lives, society and historical background has been published by Spielman and D'Antonio (2001) which is not the subject of this review. Nevertheless, I would like briefly to mention that malaria kills more than one million people each year (mostly African children) and 300–500 million people are infected annually (WHO 1998). While tropical diseases spread by mosquitoes cause misery and death, agricultural pest insects cause major economic problems in growing and cultivating plants. Traditional controls using chemical insecticide to control mosquitoes and agricultural pests often cause environmental and human health problems, as well as, the development of resistance. Thus, new approaches are urgently needed to

Consultant to the USDA horticulture research lab, Ft. Pierce FL, President and CEO of Borovsky Consulting, Founder and partner of Vector Busters International.
Email: dovborovsky@gmail.com

overcome these challenges to health and agriculture. Development of new insecticides is expensive and lengthy. It takes 7–10 years and over US$ 50 million to develop and register a new insecticide (Zaim and Guillet 2002). Although biotechnological approach offers an alternative to traditional control of mosquitoes and agricultural pest insects, the Cry and Cyt toxin genes that are expressed by the *Bacillus thuringiensis* (*Bt*) and its subsp. *israelensis* (*Bti*) are currently an effective alternative for chemical pesticides. However, indiscriminate use of *Bti* will eventually cause resistance if single toxin is used rather than a toxin cocktail. Even though *Bti* is highly specific against mosquito and black fly larvae, there are reports that it also affects non-target species (see review by Boisvert and Boisvert 2000). *Bt* toxins are effective against lepidopteran, dipteran and coleopteran insects and most are still used as microbial formulations (Metz 2003). While the use of *Bt* sprays and genetically modified (GMO) crops, especially cotton, seems to control agricultural pest insects, we cannot ignore the fact that insects are capable of rapidly developing resistance against *Bt* toxins (e.g., diamondback moth *Plutella xylostella* and the tobacco budworm *Heliothis virescens*) and new approaches should continuously be developed to control insects.

One approach is to utilize insect-specific peptide hormones to selectively control different insects; these hormones control diverse functions in insects such as: digestion, reproduction, water balance, feeding behavior, metamorphosis and sex attraction (Gäde and Goldworthy 2003, Orchard and Lange this book, Hoffmann et al. this book). The advantage is that these peptide hormones are insect-specific and they control vital functions in the life cycle of insects. Disruption of these processes causes irreversible damage and, eventually, death. Because these hormones are peptides, and are found naturally in insects, they are not xenobiotic and would not cause harm to the environment. The peptidic nature of these compounds prevents spraying them on insect's cuticle unless they are attached to lipolytic moieties that will allow their transport through the cuticle into the insect's hemolymph. They also exhibit short residual activity, photolability of certain amino acids (e.g., tyrosine), pH sensitivity and rapid degradation in the environment unless they are protected. Since many of these hormones are blocked either at the amino terminus as pyroglutamic acid derivatives or are amidated at the carboxylic terminus, the peptides cannot be expressed in baculovirus, bacterial, yeast, or plant cells, because molecular engineering of these cells to enable them to amidate or to block the amino terminus with pyroglutamic acid derivative has not been developed yet. Several of these peptide hormones, however, in the absence of the blocked carboxylic and amino termini have reduced activities that are still effective (e.g., diuretic hormone and ovarian ecdysiotropic hormone). Many of these peptide hormones have a high degree of sequence conservation among agricultural pest insects and economically may be good candidates for future plant protection. Most have not been cloned and expressed in plants or tested on agricultural pest insects except for Trypsin Modulating Oostatic Factor (TMOF) and Pheromone Biosynthesis Activating Neuropeptide (PBAN) from *Helicoverpa zea*, which was cloned and expressed in baculovirus and reduced neonate and 3rd instar survival time of *Trichoplusia ni* larvae by 26% and 19%, respectively (Ma et al. 1998). TMOF, an unblocked decapeptide, isolated from the ovaries of female *Aedes aegypti* (Borovsky 1985), down regulates tryspin biosynthesis in the midgut of adult female mosquitoes and *H. virescens* (Borovsky et al. 1989, Nauen et al. 2001) and also affects black cutworm, *Agrotis segetum*, cotton boll weevil, *Anthonomus grandis*, citrus weevil, *Diaprepes abbreviatus* and the gypsy moth, *Lymantria dispar* (Borovsky 2007) and may offer an alternative approach in controlling mosquitoes and agricultural pest insects.

Background

Factors that inhibit egg development (oostatic hormones and antigonadotropins) have been demonstrated in the cockroach, *Blattella germanica* (Iwanov and Mescherskaya 1935), decapod crustaceans (Carlise and Knowles 1959), and the housefly, *Musca domestica* (Adams et al. 1968, Kelly et al. 1984). In mosquitoes, the ovary secretes a humoral factor during oogenesis that inhibits yolk deposition in less developed follicles (Else and Judson 1972, Meola and Lea 1972). In *Rhodnius prolixus*, oostatic hormone produced by the abdominal neurosecretory organs is a small peptide of M_r 1,411 that inhibits the action of juvenile hormone (JH) on vitellogenic follicle cells and prevents the ovary from accumulating vitellogenin from the hemolymph (Liu and Davey 1974, Davey 1978, Davey and Kunster 1981). In the house fly *M. domestica*, oostatic hormone inhibits the release or synthesis of egg developmental neurosecretory hormone (EDNH) (Adams 1981), but in mosquito it was proposed that the hormone acts directly on the ovary (Meola and Lea 1972). Kelly et al. (1984) injected a crude extract of oostatic hormone from *M. domestica* into the autogenous mosquito *Aedes atropalpus*, and demonstrated inhibition of both egg development and ecdysteroid biosynthesis.

These reports indicated that an ovarian factor synthesized by the ovary controls oogenesis in diverse insects. Borovsky (1985) demonstrated that the mosquito ovary is a rich source of "oostatic hormone". Injection of the hormone into decapitated and ovariectomized females stopped trypsin biosynthesis and blood digestion in the female guts. Although blood fed decapitated female mosquitoes do not synthesize ecdysteroids or develop eggs, their guts synthesize proteases. "Oostatic hormone", therefore, appears to affect the activity of those midgut cells that synthesize trypsin, but not ovarian or endocrine tissues (Borovsky 1988). The hormone is not species-specific, as injection of the *Ae. aegypti* hormone inhibits egg development and trypsin biosynthesis in *Culex quinquefasciatus*, *Cx. nigripalpus* and *Anopheles albimanus* (Borovsky 1988). Because the target tissue of the hormone is the mosquito midgut, and not the ovary or the brain, the hormone was named "Trypsin Modulating Oostatic Factor" (TMOF). The hormone has been purified, sequenced and characterized by mass spectroscopy as unblocked decapeptides (YDPAPPPPPP and DYPAPPPPPP) (Borovsky et al. 1990). Various synthetic peptide analogs that were synthesized and purified exhibited TMOF biological activity when tested with adults and larval mosquitoes (Borovsky et al. 1990, 1991, 1993, Borovsky and Meola 2004). NMR analyses (Curto et al. 1993) confirmed computer-modeling suggestions that the polyproline portion of TMOF in solution is a left-handed alpha helix (Borovsky et al. 1990, 1993).

Aedae-TMOF inhibits the synthesis of serine proteases in the cat flea *(Ctenocephalides felis)*, stable fly *(Stomoxys calcitrans)*, house fly *(Musca domestica)*, and the midge *(Culicoides variipennis)* (Borovsky et al. 1990, 1993). The hormone, however, does not inhibit trypsin biosynthesis in flesh flies; the *Neobu*-TMOF (NPTNLH) of these insects, in turn, does not inhibit trypsin biosynthesis in mosquitoes (Borovsky et al. 1990, 1991, 1993, Bylemans et al. 1994, DeLoof et al. 1995).

TMOF affects also important agricultural pest insects. Topical treatment of larval citrus weevils *(D. abbreviatus)* inhibits weight gain and trypsin synthesis, as does feeding it to larval *D. abbreviatus* (Yan et al. 1999). Larval *H. virescens* secrete trypsin- and chymotrypsin-like enzymes in their guts, and feeding or injecting *Aedae*-TMOF into larval *H. virescens* inhibits biosynthesis of these enzymes (Nauen et al. 2001). The mechanism of trypsin synthesis in mosquitoes, therefore, may be similar to that in certain beetles and moths (Borovsky 2007).

Biological Activity

Aedae-TMOF, at concentrations of 3×10^{-9} M and 6.8×10^{-6} M, injected into the hemolymph of ligated blood-fed female *Ae. aegypti* mosquitoes, inhibits 50 and 90% of midgut trypsin-like enzyme biosynthesis, respectively (Borovsky et al. 1993). TMOF is present at 33 and 37 ng in the hemolymph of non-treated mosquitoes 30 h and 38 h after the blood meal (Fig. 1), an amount that is at least 30-fold greater that is needed, *in vivo*, to inhibit 90% of trypsin biosynthesis (Borovsky et al. 1992, 1993). Kinetics of TMOF activity follows a similar pattern in female *C. quinquefasciatus* (Borovsky unpublished observations). TMOF does not act as a classical

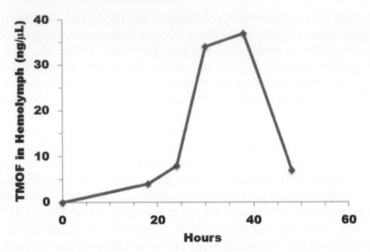

Figure 1. Concentration of TMOF in the hemolymph of *Ae. aegypti* after the blood meal. Each point represents an average of 2 determinations of hemolymph withdrawn from seventy females at different intervals after the blood meal and assayed by ELISA (Borovsky et al. 1992).

trypsin inhibitor (e.g., TLCK, TPCK and Soybean trypsin inhibitor) that binds to the active site of serine proteases and prevents protein hydrolysis. Instead, TMOF binds to a specific gut epithelial cell receptor before stopping trypsin biosynthesis (Borovsky et al. 1990, 1994a).

Biosynthesis and Secretion

Aedae-TMOF is synthesized by the mosquito ovary 18 h after the blood meal, and is rapidly secreted into the hemolymph. The synthesis is greatest at 33 h, and rapidly declines, thereafter, to a minimum at 48 h after the blood meal (Borovsky et al. 1994a, Fig. 1). About 33% of the TMOF that is synthesized by the ovary at 33 h is secreted into the hemolymph (37 ng or 3.5×10^{-5} M) (Borovsky et al. 1994a, Fig. 1). Cytoimmunochemical studies indicate that the site of synthesis of *Aedae*-TMOF in mosquitoes and flesh flies is the follicular epithelium of the ovary. Because initially *Aedae*-TMOF was not detectable in brains of mosquitoes, the hormone was assumed not to be a neuropeptide (Borovsky et al. 1994a, Bylemans et al. 1996). Recent collaborative studies, with Professor Dick Nässel (University of Stockholm) and Dr. Shirlee Meola (USDA, College Station TX), however, identified *Aedae*-TMOF reactive cells in the brain of larval *Drosophila melanogaster* (Borovsky and Nässel unpublished observations) and in the brain, abdomimal ganglia and upper gut of adult female *Ae. aegypti*. In larvae, TMOF reactive cells are present in the prothoracic and abdominal ganglia, suggesting that TMOF is also a neurosecretory hormone that mediates trypsin biosynthesis in adult female mosquitoes by the ovaries and in larvae by the neurosecretory cells (Borovsky and Meola 2004).

Larvacidal Potency

Peptide Analogs

Borovsky et al. (1989) and Borovsky and Mahmood (1995) suggested that TMOF can be used as a larvacide because the hormone can traverse through the gut into the hemolymph and mosquito larvae also use trypsin and chymotrypsin to digest food (Yang and Davies 1971). The presence of trypsin and chymotrypsin-like enzymes in larvae and pupae was confirmed when guts of larvae and pupae at different developmental stages were analyzed for trypsin and chymotrypsin using [³H]DFP (Borovsky and Schlein 1988, Table 1). When female mosquitoes were given an enema of *Aedae*-TMOF mixed with blood, both egg development and trypsin biosynthesis were inhibited. Similar results were observed when *Aedae*-TMOF was mixed with the blood meal and fed to female mosquitoes through a membrane. Radioactively labeled hormone that was fed to adult female *Ae. aegypti* diffused from the gut into the hemolymph increasing linearly for the first 24 h after the feeding. These results suggest that TMOF and its analogs can traverse through the gut epithelial cells into the hemolymph, bind TMOF gut receptor(s), and modulate trypsin biosynthesis (Borovsky et al. 1994b, Borovsky and Mahmood 1995).

Feeding *Ae. aegypti* and *Cx. quinquefasciatus* larvae *Aedae*-TMOF that was adsorbed onto yeast cells inhibited trypsin biosynthesis 88% and 91.7%, respectively and the larvae stopped growing (Borovsky and Meola 2004). These results indicate that shutting off trypsin biosynthesis with *Aedae*-TMOF can be used to control larval growth and development, as a new biorational insecticide, that is desperately needed (Zaim and Guillet 2002). Synthesis of twenty-nine different peptidic analogs of *Aedae*-TMOF

Table 1. Synthesis of trypsin and chymotrypsin-like enzymes in *Ae. aegypti* larvae and pupae.

Developmental Stage	Trypsin/gut (ng ± SEM)	Chymotrypsin/gut (ng ± SEM)
1st instar larva	2.4 ± 0.3	5.5 ± 0.3
2nd instar larva	17 ± 2.5	20 ± 4
3rd instar larva	59 ± 14	16 ± 5
4th instar larva	56 ± 5	25 ± 11
Pupa	52 ± 17	17 ± 9

Three groups of larvae or pupae guts (5 per group) were analyzed for trypsin and chymotrypsin biosynthesis at different developmental stages using [³H]DFP (Borovsky and Schlein 1988, Nauwelaers 2002).

and feeding them to mosquito larvae indicate that the biological activity of the hormone resides at the N-terminus of the peptide. The tetrapeptide YDPA exhibited full biological activity thus, it is possible to synthesize short active analogs of the hormone to control mosquito larvae. TMOF-K-methyl (ethyleneglycol)$_7$-O-propionyl (TMOF-K-PEG$_7$P) increased the biological activity of TMOF by 6–10 fold as compared to TMOF or TMOF-K when neonates of *Ae. albopictus*, *H. virescens* and *H. zea* were tested indicating that the hormone could be used to protect crops in the field by spraying (Jeffers et al. 2012). Nauen et al. (2001) have reported that *H. virescens* (Lepidoptera) larvae control their trypsin biosynthesis with a hormone that is similar to *Aedae*-TMOF. Injections of *Aede*-TMOF or feeding it to 4th instar *H. virescens* larvae caused inhibition of trypsin biosynthesis and larval growth. *Aedae*-TMOF also showed enhanced activity against the cotton boll weevil *Anthonomus grandis* by retarding its growth rate and enhanced its mortality and the black cutworm *Agrotis segetum* by retarding its growth rate (Borovsky 2007). The hormone also affected ecdysteroid production in the prothoracic glands of the gypsy moth *Lymantria dispar* (Gelman and Borovsky 2000). No effect was observed on *Spodoptera litura*, *P. xylostella* and *Cydia pomonella* indicating that these insects may have a different TMOF that regulates their digestion. Feeding *Aedae*-TMOF to adult mosquitoes to control adult populations in the field is not practical. On the other hand, larvae that are aquatic and mainly feed on decaying particles in the field are an easier target group to control. Applying synthetic *Aedae*-TMOF or PEG conjugated *Aedae*-TMOF (Jeffers et al. 2012) to control mosquito larvae in their natural habitat would be expensive and inefficient because TMOF is water-soluble and larvae are mainly filter and particle feeders (Clements 1992). A possible way to overcome this problem is to express TMOF in organisms that are readily eaten by mosquito larvae like *Chlorella*, yeast and bacteria or to synthesize organic analogs and modify the hormone so it can easily penetrate through the cuticle of target insects.

Organic Analogs

A 3D modeling approach was developed to study TMOF and used to synthesize several novel aromatic, aliphatic organic acid and ester analogs of the hormone (Vanderherchen et al. 2005). The effect of these analogs on *C. pipiens* complex was studied by feeding. Four of these analogs were as effective as *Aedae*-TMOF and three of the analogs were 1.2 to 2.5-fold more active than *Aedae*-TMOF. Injection of several of the organic analogs to *Manduca sexta* caused a toxic effect and *Aedae*-TMOF and its peptidic analogue FDPAP did not affect trypsin biosynthesis, larval growth or mortality indicating that *Aedae*-TMOF is not biologically active on larval

M. sexta. Three *Aedae*-TMOF analogs that were highly toxic by injection into *M. sexta* were also toxic to fourth instars of the tobacco budworm, *H. virescens*, and the cotton bollworm, *H. zea*, as well as to the adult male German cockroaches, *B. germanica* indicating that some of the organic analogs are generally highly toxic (Vanderherchen et al. 2005). Borovsky and Nauen (2007) tested ten putative mimics of *Aedae*-TMOF (Fig. 2) that were assayed by NMR and purified by chromatography (purity > 95%) and stock solutions of each compound were prepared for testing in acetone for lepidopteran and in DMSO for larval mosquitoes. Compounds IBI-152, 156, 163 and 169 were tested against both diamondback moth and tobacco budworm larvae. The diamondback moth larvae did not eat the leaf disks starting at day one and all starved to death indicating that these *Aedae*-TMOF mimics have an insecticidal potential against *P. xylostella*. No activity, however, was observed against third-instar *Heliothis* larvae, they fed the whole treated leaf disks without a toxic effect. Experiments that were carried out with compounds IBI-171, 215, 216, 217, 219 and 220 (Fig. 2) by feeding diamondback moth larvae on treated leaf disks showed that the most active compound is IBI-215 and the least active the biphenyl-derivative IBI-217 (Table 2). Mortality values show a strong reverse-correlation with the percentage of each leaf disk

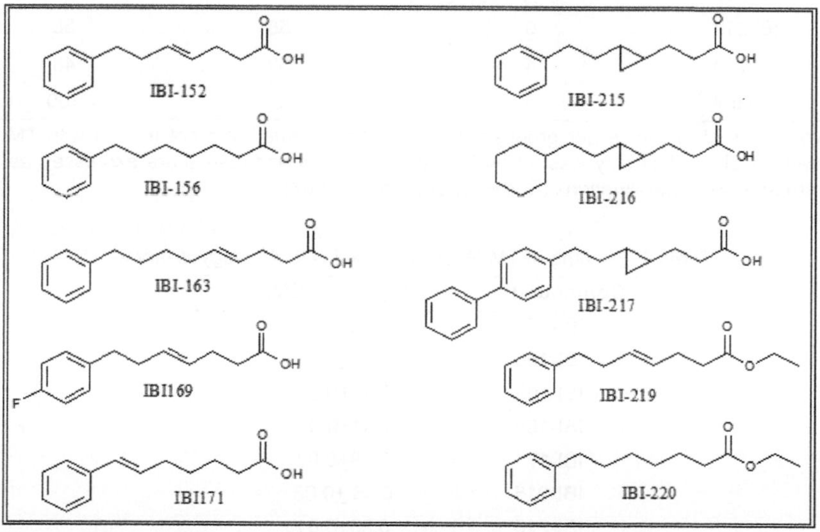

Figure 2. TMOF organo-synthetic mimics: IBI-152=*E-7-phenylhept-4-enoic acid*; IBI-156 = *7-phenylheptanoic acid*; IBI-163=*9-phenyl-non-4-enoic acid*; IBI-169=*7-(4-fluoro-phenyl)-hept-4-enoic acid*; IBI-171=*7-phenyl-hept-6-enoic acid*; IBI-215= *3-(2-phenethyl-cyclopropyl)-propionic acid*; IBI-216= *3-[2-(2-cyclohexyl-ethyl)-cyclopropyl]-propionic acid*; IBI-217 = *3-[2-(2-biphenyl-4-yl-ethyl)-cyclopropyl]-propionic acid*; IBI-219=*E-7-phenyl-hept-4-enoic acid ethyl ester*; IBI-220=*7-phenyl-heptanoic acid ethyl ester*.

that was consumed by the larvae. The more effective mimics caused higher larval mortality and protected the leaf disks from extensive larval damage (Table 2). The cyclopropane-containing compounds IBI-215 and 216 were most effective causing 96% and 71% mortalities, respectively. Compounds IBI-215 and 216 are novel, however, most of the other compounds are known, or have been described in international patent disclosures, e.g., IBI-219 (EP (1988)-305189 and US (1988)-5134128). Several of the *Aedae*-TMOF organic mimics were tested on *Ae. aegypti* larvae. Feeding the mimics with Brewer's yeast to mosquito larvae caused larval mortality within 5–6 days of the feeding (Table 3), in a similar fashion that was observed when TMOF peptide analogs were fed to first instar *Ae. aegypti* larvae (Borovsky and Meola 2004). IBI-156, 163, 169, 171, 215 and 217 exhibited similar LC_{50} values even though

Table 2. Effect of IBI-compounds on *P. xylostella* larvae.

Compound	Mortality (%) 1d	Mortality (%) 3d	Feeding (%) 3d
IBI-171	4	54	59
IBI-215	8	96	19
IBI-216	17	71	36
IBI-217	4	25	71
IBI-219	0	58	50
IBI-220	4	58	48
Control	0	0	100

Two groups of larvae (8 per group) fed on leaf disks treated and not treated with TMOF organic analogs. Mortality was followed at daily intervals and results are expressed as an average of two determinations (Borovsky and Nauen 2007).

Table 3. Feeding of TMOF organic mimics to *Ae. aegypti* larvae.

Compound	LC_{50} (mM±SEM)
IBI-152	0.78±0.08
IBI-156	0.19±0.02
IBI-163	0.17±0.01
IBI-169	0.11±0.01
IBI-171	0.18±0.02
IBI-215	0.18 ±0.03
IBI-216	0.30±0.01
IBI-217	0.13±0.04

Three groups of first instar *Ae. aegypti* larvae (12 per group), were fed different concentration of *Aedae*-TMOF organic analogs for 5–6 days. Larval mortality was followed daily and is expressed as LC_{50} of 3 determinations ± SEM using probit analyses (Borovsky and Nauen 2007).

IBI-217 is a biphenyl-derivative and IBI-215 has a cyclopropyl group instead of a double bond. When the double bond was placed in the middle of the molecule (IBI-152) and the benzene ring replaced by cyclohexane (IBI-216) the activities were greatly reduced (Table 3), due to different steric properties.

The trypsin activity of the different TMOF mimics were followed in larval *P. xylosella* with BApNA and is closely correlated with the observed mortality that these compounds induced by starving larval *Plutella* to death. These results strongly suggest that larval mortality caused by the action of the synthetic compounds is correlated with the reduction in trypsin biosynthesis in the midgut of lepidopteran larvae, demonstrating that 'in principal' synthetic mimics of TMOF could protect crops against agricultural pest insects.

Effect on Bacterial Motility

Many peptide hormones have a predictable mode of action that is broadly accepted, however, some have unexpected effects including *Aedae*-TMOF. Southwick and Purich (1995) reported that *Aedae*-TMOF resembles the primary structure of oligoproline-rich regions within actin A (ActA) (235-DFPPPPTDE, 269-FEFPPPPTDE, 304-FEFPPPPTED, 350-DFPIPTEE, where the numbers indicate the amino acid residue location on the protein). ActA is a bacterial surface protein required for *Listeria* motility in host cells. Injecting *Aedae*-TMOF into *Listeria* infected PtK2 cells blocked the *Listeria* induced actin rocket tail assembly and intracellular locomotion of this pathogen. Different concentrations of *Aedae*-TMOF (90 nM and 0.9 μM) caused either short or permanent arrest, respectively. Epifluorescence microscopy showed that cells that were injected with *Aedae*-TMOF lost all the actin stress fibers and accumulated F-actin in regions of membrane retraction. Addition of profilin negated the effect of *Aedae*-TMOF on rocket tail formation and bacterial locomotion. Since profilin is assumed to directly be involved in intracellular pathogen locomotion and reorganization of actin cytoskeleton of the host cell peripheral membrane, and *Aedae*-TMOF B is cleaved from the vitelline membrane in mosquitoes (Lin et al. 1993) it appears that *Aedae*-TMOF is perfectly suited peptide to displace profilin and could act like growth inhibiting peptides that are cleaved from collagen-like precursors (De Loof et al. 1995).

Molecular-Characterization and Expression of TMOF

The initial purification and mass spectra analyses studies of TMOF identified two forms YDPAPPPPP and DYPAPPPPP with similar activities, but with different abundances. *Aedae*-TMOF B (DYPAPPPPPP, Borovsky et al. 1990), and its truncated analogue DYP, are expressed by *Ae. aegypti* vitelline membrane genes (GenBank accession numbers S54556 and S54555, respectively, Lin et al. 1993). *Aedae*-TMOF B and DYP are flanked by signal peptides and protease cleavage sites that allow them to be cleaved from the vitelline membrane and secreted into the hemolymph as was shown for *Aedae*-TMOF (Fig. 1) and for growth inhibiting peptides like Neobu-colloostatin that is cleaved from a collagen-like precursor molecule (De Loof et al. 1995). The gene coding for *Aedae*-TMOF A (YDPAPPPPPP) has not as yet been identified, however, the gene for *Drosophila* TMOF has been recently reported (Liu et al. 2006) and the gene coding for the nematodes peptide (YDPLPPPPPP; GenBank accession number CEY37D8A.21) was reported by Borovsky and Nauen (2007). The effect of TMOF on the trypsin gene was first studied in *Neobellieria*. After injecting *Neobu*-TMOF into these flies, the biosynthesis of trypsin mRNA was followed using Northern blot analysis (Borovsky et al. 1996). Feeding these flies a liver meal causes degradation of the endogenous trypsin early mRNA, and synthesis of a new mRNA that corresponds to late trypsin biosynthesis associated with post meal digestion. Flesh flies injected with *Neobellieria* TMOF (10^{-9} M), maintained their early trypsin message, however, the late trypsin transcript was not translated. TMOF, therefore, controls the translation of the late trypsin mRNA, as would be expected for a hormone that is released after trypsin mRNA has already been transcribed (Borovsky et al. 1996). Similar results were reported when TMOF was expressed on the coat protein of Tobacco Mosaic Virus and the recombinant virus was fed to larval *H. virescens*. Although the larvae transcribed a normal level of trypsin message, it was not translated (Borovsky et al. 2006). Injecting TMOF into female *Ae. aegypti* or *Cx. quinquefasciatus* and following the late trypsin message by RT-PCR and Northern blot analysis confirmed that in mosquitoes the trypsin message is transcribed but not translated (Borovsky unpublished observations) as was shown for *Neobellieria* and *Heliothis* (Borovsky et al. 1996, Borovsky et al. 2006).

Genetic Engineering of TMOF

Expression in E. coli

Aedae-TMOF was initially cloned in *E. coli* based on the amino acids sequence of *Aedae*-TMOF. The bacteria was fermented and analyzed for *Aedae*-TMOF using ELISA, HPLC and biological activity. The fermented *E. coli* produced the N-formylmethionine form of the hormone as expected for short peptides that are synthesized in bacterial cells. The

N-formylmethionine was then cleaved using CNBr and the N-terminal cleaved *Aedae*-TMOF activity and structural identity was confirmed (Carlson et al. 1994). Although recombinant *E. coli* can produce TMOF, it cannot be used in the field because of bacterial endotoxins that may cause harm to the environment. These initial results prompted examination of alternative expression cells that are not harmful to the environment and can be used as a good food source to control mosquito larvae.

Expression in Chlorella sp.

Selectable Markers

To achieve stable transformation in *Chlorella* sp. several selectable markers using the strong promoter of cauliflower mosaic virus CaMV 35S promoter (Jarvis and Brown 1991) were developed; the nitrate reductase (NR) gene (Dawson et al. 1997), the hygromaycin B resistant gene (Chow et al. 1999) and the Neo^r/Kan^r gene confirming resistance to geneticin (G418) (El Sheek 1999).

Transformation Techniques

Number of techniques for delivering DNA into *Chlorella* sp. cells have been tried. Protoplasting, which involves removal of the cell wall allowing the transfection of foreign DNA, incubated in the presence of carrier DNA, PEG and/or calcium, achieved only transient expression of reporter genes (Jarvis and Brown 1991, Hawkins et al. 1999). Stable transformations were observed only after using particle bombardment technique (Dawson et al. 1997, El Sheekh 1999).

Cloning and Expressing Aedae-TMOF and GFP-Aedae-TMOF

For efficient transformation of *Chlorella* (NC64A) the cell wall was removed with cellulose and pectinase and the protoplast was treated with PEG and CaCl$_2$ to allow uptake of DNA into the cell. pKYLX71, a 12 Kbp plasmid, originally designed to transform plant cells, was used (Fig. 3). The plasmid contains a strong constitutive cauliflower mosaic virus (CaMV) 35S promoter followed by a multiple cloning site, a rbcS-E9 (ribulose-1, 5 biphosphate carboxylase small subunit gene) and a polyadenylation signal.

The plasmid contains a bacterial origin of replication (RK2) and a tetracycline resistant gene for maintenance and selection in *E. coli*. A nos (nopaline synthetase gene) promoter and a polyadenylation signal

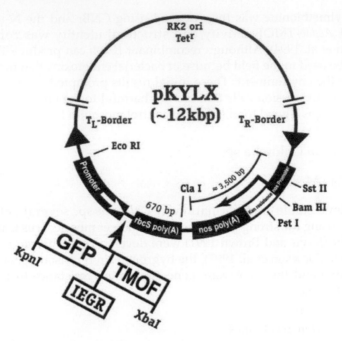

Figure 3. Structure of pKYLX71:35S, including the cloning site for the GFP-TMOF gene with a trypsin cleavage site (IEGR) on the fusion protein, direction of replication and Kanamycin resistant gene.

flanking a Kanamycin resistant gene allows selection in eukaryotes (Schardl et al. 1987). The plasmid was opened with restriction enzymes and ligated with *tmfA* (*Aedae*-TMOF gene) or *tmfA-gfp* (green fluorescent protein gene). After transformation, cells were selected with Kanamycin and tested for biological activity by feeding them to mosquito larvae. TMOF transformed algal cells that were fed to *Ae. aegypti* larvae caused starvation and death (Table 4). The recombinant *Chlorella* (NC64A) was also effective against *Cx. quinquefasciatus*, killing all the larvae within 72 h, as well as *An. albimanus* (results not shown). *Aedae*-TMOF and GFP-*Aedae*-TMOF produced by the transformed *Chlorella* (NC64A) were followed by ELISA (Borovsky et al. 1992) showing that clone A and B synthesized 44 µg and 24 µg TMOF per 3×10^8 cells, respectively and eight clones that synthesized GFP-*Aedae*-TMOF produced an average of 23 ± 1 µg TMOF in the same number of cells killing (83% to 100%) of the fed larvae in 4 days (Borovsky et al. 1998, Nauwelaers 2002, Borovsky 2007).

These results suggest that GFP-*Aedae*-TMOF is cleaved in the larval gut by proteases releasing *Aedae*-TMOF that traverses the gut, enters the hemolymph, binds a TMOF receptor and shuts off the translation of the trypsin message and the biosynthesis of trypsin in the gut causing anorexia

Table 4. Survival of *Ae. aegypti* larvae fed TMOF-*Chlorella.*

Larvae were fed *Chlorella* with	No. of larvae surviving after feeding for				
	0h	24 h	48 h	72 h	144 h
pKYLX + TMOF	96	92	78	0	0
pKYLX - TMOF	96	92	92	92	91
Control without *pKYLX*	96	92	91	91	91

First instar larvae (96 per group) were fed recombinant *Chlorella* (NC64A), with or without *pKylx* TMOF gene. Survival was determined at 24 h intervals.

and death (Borovsky et al. 1994b, Borovsky et al. 1996). The possibility that GFP is toxic to mosquito larvae was also investigated by feeding them recombinant yeast cells producing GFP with no toxic effect. GFP has also been extensively used as a marker gene for transgenic mosquitoes and other insects, with no toxic effect (Berghammer et al. 1999, Catteruccia et al. 2000, Allen et al. 2001). Although *Chlorella* (NC64A) was successfully transformed, *tmfA* and *tmfA-gfp* did not integrate into the cell genome and the cells had to be maintained on antibiotics. Adding yeast 2 μ origin of replication to pKYLX71 could maintain an autonomous replication in the cell nucleus under selective pressure (Hasnain et al. 1985). If pKYLX71, in the future, is engineered with a native *Chlorella* gene like nitrate reductase (Dawson et al. 1997), it is possible to integrate by homologous recombination *tmfA* into *Chlorella* (NC64A) as was successfully done with *Saccharomyces cerevisiae* and *Pichia pastoris.*

Expression by Tobacco Mosaic Virus and Alfalfa

Because *Aedae*-TMOF affects the synthesis of trypsin like enzymes in the larval gut of mosquito and *H. virescens*, it was fused with the coat protein (CP) of Tobacco mosaic virus (TMV) which produces large quantities of RNA and protein in infected plant cells without causing damage. Virions of TMVU1 were used to present *Aedae*-TMOF on the surface using a read-through sequence so that the TMOF peptide was presented from the C-terminal of 5% of the viral CP subunits (Borovsky et al. 2006, Fig. 4). This approach allows efficient assembly of the viral CP subunits in spite of the steric hindrance that the left-handed helical portion of *Aedae*-TMOF exerts (Fig. 5). When it was expressed on each CP subunit, virion assembly was prevented (Borovsy, unpublished observations). The same strategy was successfully used in producing angiotensin-I-converting enzyme on the CP of TMV (Hamamoto et al. 1993).

Aedae-TMOF is released in the larval gut from the virions by trypsin allowing a rapid transport of *Aedae*-TMOF through the gut epithelial cells into the hemolymph (Borovsky and Mahmood 1995). In the hemolymph,

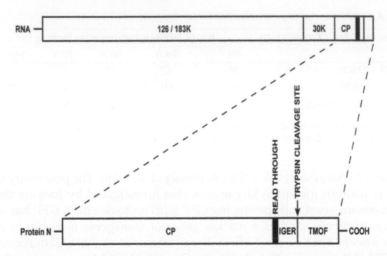

Figure 4. Genome organization of TMV-TMOF containing a read through sequence. TMV-U1 ORF representing the replicase of 126 and 183 kDa (126K/183K), the cell to cell movement protein 30 kDa (30K) and the viral coat protein (CP) was fused to a read through sequence (5'- TAGCAATTA-3'), a trypsin cleavage site (IEGR) and *Aedae*-TMOF (YDPAPPPPPP) sequence behind a T7 promoter at the 5'-end of the construct. The TMV-*Aedae*-TMOF RNA was transcribed using T7 RNA polymerase and the RNA was mechanically inoculated onto *Nicotiana tabacum* (L.) cv. xanthu. Gene distances are not proportional to gene sizes.

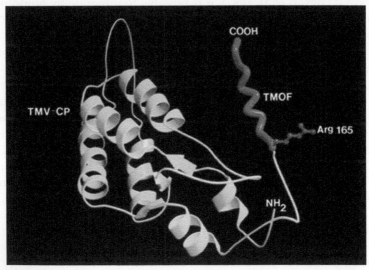

Figure 5. Ribbon drawing by means of MOLSCRIPT, of TMV coat protein (CP) fused to *Aedae*-TMOF. The TMV-CP is shown in yellow and the N-terminus is shown as NH₂. TMOF exhibiting a left-handed helix is shown in blue with a trypsin cleavage site at Arg165 in red. The C-terminus of the fusion protein is at the carboxyl-end of *Aedae*-TMOF (COOH) (Borovsky et al. 2006).

Color image of this figure appears in the color plate section at the end of the book.

the hormone binds its gut-receptor located at the hemolymph side of the gut (Borovsky et al. 1994b), and causes cessation of trypsin biosynthesis in the gut, starvation and death (Borovsky and Meola 2004).

This mode of action is different from *B. thuringiensis* toxins that bind to receptors located in the larval midgut epithelial brush border membrane, insert into the membrane, causing formation of pores or ion channels and osmotic pressure imbalance that kill insects (Knight et al. 2004). The ability of the CP-*Aedae*-TMOF chimera to starve and kill mosquito larvae was demonstrated when CP-*Aedae*-TMOF virions were adsorbed onto yeast particles and fed to *Ae. aegypti* larvae for 5–6 days (Fig. 6). The estimated lethal dose of TMOF that fused to the CP and caused 87.5% mortality was about 26.4 µM, this concentration is 7.5-fold lower than when TMOF alone is adsorbed onto yeast particles and fed to larvae (Borovsky and Meola 2004). It is possible that the CP-*Aedae*-TMOF virions bind better to the yeast particles or that the CP-*Aedae*-TMOF virions aggregate and then bind to the yeast cells making it easier for the larvae to internalize because mosquito larvae are filter feeders and swallow particles (Clements 1992).

Similar results were observed when *H. virescens* larvae were fed tobacco leaf discs that were infected with TMV-*Aedae*-TMOF. Larval weight gain was 2.3-fold lower than controls (*p*<0.001) and trypsin and chymotrypsins activities in the midgut were 2.2- and 2.6-fold lower, respectively at day 4. On the other hand, control groups that were fed uninfected tobacco leaf discs were not affected. The inhibition of trypsin biosynthesis is due to the down regulation effect of *Aedae*-TMOF and not to the CP that was fused to TMOF. Earlier observations showed that when TMOF was mixed with artificial food and directly fed to *H. virescens*

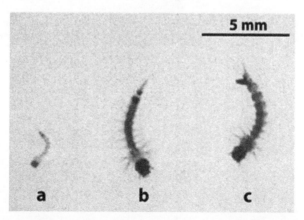

Figure 6. Feeding *Ae. aegypti* larvae TMV coat protein-*Aedae*-TMOF virions containing: **a.** coat protein-*Aedae*-TMOF, **b.** coat protein and **c.** Brewer's yeast. First instar larvae were fed for 6 days on a diet containing **a, b,** or **c.** Horizontal bar denotes larval length of 5 mm (Borovsky et al. 2006).

larvae, trypsin biosynthesis and weight gain were greatly reduced (Nauen et al. 2001). Northern blot analysis by Borovsky et al. (2006) strengthens these observations and suggests that the trypsin gene is down regulated by *Aedae*-TMOF in *H. virescens* through a translational control mechanism. Such a mechanism would be expected for a hormone that is released after the trypsin message has already being transcribed. Thus, it is now possible to suggest that *Aedae*-TMOF down regulates trypsin biosynthesis by a translational control mechanism in *H. virescens* and possibly in larval *Ae. aegypti* as was shown for *N. bullata* (Borovsky et al. 1996).

Borovsky et al. (2005) cloned and expressed TMOF in alfalfa plants using pKylx71 and *A. tumefaciens* transformation. Fifty plants were assayed by qPCR and the results were confirmed by Northern blot analyses showing a strong *Aedae*-TMOF transcript at about 0.32 Kb in plants C12 and D24 and faint to moderate *Aedae*-TMOF transcripts in C9 and D38 (Fig. 7). Non-transformed alfalfa leaf discs were coated with recombinant yeast cells (*P. pastoris*) synthesizing *Aedae*-TMOF, and compared with untreated wild type alfalfa and several *tmfA* transformed plants. Larvae that fed on leaf discs that were treated with *Pichia-Aedae*-TMOF did not cause damage to the discs, and trypsin biosynthesis was inhibited by 84%. From five recombinant plants that were tested, only two plants (C9 and B23) showed moderate leaf damage of 56% and 36%, respectively, and inhibition of trypsin biosynthesis of 59% and 41.5%, respectively (Table 5) (Borovsky et al. 2005). Because the level of *Aedae*-TMOF transcripts in C9 (Fig. 7) is low, and also in C13 and B23 (results not shown), feeding these leaf discs to *H. virescens* is not expected to completely stop trypsin biosynthesis in *H. virscens*.

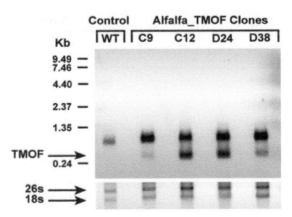

Figure 7. Northern blot analysis of alfalfa plants transformed with pKylx71-*Aea*TMOF using *A. tumefaciens* transformation. TMOF transcript is shown by arrow at about 0.32 Kb. The fidelity of the transfer was followed with 26S and 18S RNA probe. WT=wild type (Borovsky, Shatters and Powell unpublished observations).

Table 5. The effect of feeding 1st instar *H. virescens* larvae on alfalfa-*tmfA*.

H. virescens larvae were fed	N	Trypsin (nmol/min/gut ± SEM)	Inhibition (%)
TMOF (on leaf)	3	0.11 ± 0.001^a	84
Recombinant plants			
C-2	8	0.72 ± 0.11	0
C-9	5	0.29 ± 0.05^b	59
B-5	6	1.13 ± 0.07	0
B-23	4	0.34 ± 0.09^c	41.5
Control (wild type)	6	$0.79 \pm 0.08^{a,b,c}$	0

First instar *H. virescens* larvae 24 h after emergence were fed for 3 days on alfalfa leaf discs that were transformed with pKylx-*tmfA* plasmid using agrobacterium. After the feeding, larvae were assayed for trypsin activity in their guts using BApNA (Borovsky and Schlein 1988). Control leaf discs with yeast cells without *Aedae*-TMOF did not inhibit trypsin biosynthesis (data not shown). Same letters results indicate significant difference from control by student's *t*-test. $^a p<0.0002$, $^b p<0.0003$, $^c p<0.003$ (Borovsky et al. 2005).

The ability to express foreign genes in plants will allow future plant protection to mainly use biological and environmental friendly natural proteins and peptides that can be designed to exert specific control on agricultural pest insects. TMV infected plants are ideal for high production of proteins (e.g., TMOF) that can be harvested, isolated and used against vector insects such as mosquitoes that are of medical importance because they transmit diseases in many parts of the world. Plant RNA virus-based vector such as tobamoviruses have a wide host range, they move easily between plant cells, they rapidly and systematically spread in the infected plant, the RNA replicates at high level and the infection is maintained for the lifetime of the plant (Donson et al. 1991). In addition, the virus can be selected or manipulated so it does not have an adverse effect on the plant. The ease of manipulating the TMV genome as a DNA copy before transcribing it, *in vitro*, into infectious RNA makes it an ideal system for expressing foreign proteins fused to the CP. TMV-*Aedae*-TMOF produced 4.8 micromoles TMOF per Kg of leaf extract (1.3% of total soluble proteins). On the other hand, expressing *Aedae*-TMOF in tobacco plants as a fusion with tomato prosystemin produced 325-fold less TMOF (0.004% of total soluble protein) and low inhibition (4%) of *H. virescens* larval growth (Tortiglione et al. 2003). *Aedae*-TMOF was cloned and express in about 100 alfalfa plants, 50 of those plants showed biological activity against *H. virescens* and were analyzed by real time qPCR and by Northern blot analyses (Borovsky et al. 2005). Five of the plants that produced the highest *Aedae*-TMOF message were assayed by ELISA and mass spectrometry showing that alfalfa plants produce 117-fold less *Aedae*-TMOF (41 nmol per Kg of leaf extract) than the TMV-*Aedae*-TMOF recombinant (Borovsky, Powell and Shatters unpublished observations). Because TMV has a wide host range, this technique of expressing TMOF on the coat protein of TMV can be used as a general method to protect plants against agricultural pest- insects that use serine proteases as their main digestive enzymes. TMV-*Aedae*-TMOF can be easily harvested from infected tobacco leaves by rapid and simple extraction steps, in future industrial production of the CP chimera to control mosquito larvae and agricultural pest insects that cause extensive crop damage.

Expression in Saccharomyces cerevisiae

The budding yeast *Saccharomyces cerevisiae* is used to express foreign proteins for research, industrial and medicinal use. Yeast cells grow rapidly achieving high cell densities on simple media. Yeast is a eukaryote, capable of post translational modifications of disulfide bond formation, glycosylation and protein folding as seen in higher eukaryotes. Yet, it can be manipulated almost as readily as *E. coli*. Since yeast neither contains pathogens, pyrogens, viral inclusions, nor produces toxins, it is an ideal cell for genetic engineering to produce therapeutic compounds like insulin and hepatitis B vaccines.

Most yeast expression vectors are based on multi copy 2μ plasmids and can be maintained in *E. coli* and yeast. They all have a yeast promoter and a termination sequence for efficient transcription of foreign genes. A selection marker is also present and routinely used to maintain foreign genes expression preventing the cell from excising cloned foreign genes. Nutrition deficient markers like *LEU2, URA3, TRY1* and *HIS3* are used in mutant strains that lack leucine, uracil,tryptophan and histidine (Romanos et al. 1992). Antibiotic-resistant markers are used to increase the range of host strains especially the G418 resistant marker that encodes for the *E. coli Tn903* transposon (Wach et al. 1994) as well as hygromycin B and chlorophenicol-resistant markers.

To make the cell wall more permeable for DNA vector transformation, the following techniques are used: generation of spheroplasts, lithium salt treatment and electroporation. Episomal DNA vectors use the 2μ circle of *Saccharomyces*, are very stable and are present in 10 to 40 copies per cell. Because they are not equally distributed among the cells high variance in plasmid copy per cell is found in the population (Schneider and Guarente 1991). Vectors can be integrated into the yeast genome by homologous recombination through a single cross over between homologous sequences of the vector and genome. The vector is opened at a unique restriction site on the homologous DNA sequence to allow efficient transformation and targets integration. Using this technique multi-copy integrations occur at very low frequency and most of the recombinant cells will carry one copy of the integrated gene, hence, the use of *S. cerevisiae* for large scale production of foreign proteins often results in low yields. However, *S. cerevisiae* is easy to work with and manipulate and was the first yeast to be used for the production of recombinant proteins. Because yeast is a favorite food of mosquito larvae in the laboratory, *S. cerevisiae* was initially used to express *Aedae*-TMOF and GFP-*Aedae*-TMOF (Nauwelaers and Borovsky 2002). Two transformation approaches were used: homologous recombination and cell transformation using high copy episomal plasmids. Although homologous transformation is preferred because integration of foreign DNA into the yeast genome is stable, but only one copy of the gene can be integrated. For homologous recombination Nauwelaers and Borovsky (2002) used plasmid pYES2 (5.9 Kb, Invitrogen, CA USA), and digested it with restriction enzymes to remove the 2μ origin of replication. The new plasmid pYDB2 (4.1 Kb) was engineered and unidirectional cloned with: *tmfA-gfp* possessing a trypsin cleavage site (IEGR), *tmfA*, *gfp* and *gfp* with a trypsin cleavage site (IEGR) (Fig. 8). The plasmids were opened with *Apa*I and competent yeast cells were transformed by homologous recombination grown for 48 h and stimulated with galactose

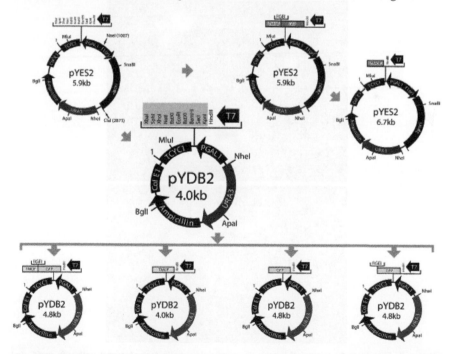

Figure 8. Construction of pYDB2 from pYES2 includes the positions of the restriction enzymes *Nae*I and *Cla*I on pYES2, the positions of GFP-TMOF, TMOF, GFP and GFP-IEGR in the multiple cloning sites. IEGR denotes trypsin cleavage site and the restriction enzyme *Apa*I for homologous recombination and the direction of replication from the GAL1 promoter are also included.

(2%) and tested for GFP-*Aedae*-TMOF synthesis under a fluorescence microscope. For multicopy plasmid expression pYES2 containing 2µ origin of replication was cloned with *tmfA-gfp* and *tmfA* and yeast cells transformed, grown and stimulated as above. Cells that were transformed by homologous recombination reached a peak of synthesis at 8 h and synthesized 570 ng of *Aedae*-TMOF determined by ELISA (Nauwelaers 2002, Nauwelaers and Borovksy 2002). Similar amount of *Aedae*-TMOF was also synthesized when cells were transformed with free plasmid and stimulated for shorter time. Recombinant cells that were transformed with *gfp* are easily observed under a fluorescence microscope and larvae that ate the cells also fluoresce under UV (Fig. 9).

Mosquito larvae that were fed recombinant yeast cells synthesizing GFP-*Aedae*-TMOF or *Aedae*-TMOF alone stopped growing and starved to death. Cells expressing GFP-*Aedae*-TMOF were more effective than cells expressing *Aedae*-TMOF causing 83% and 38% mortality, respectively (Table 6). Because larvae that are fed GFP-*Aedae*-TMOF are easily detected

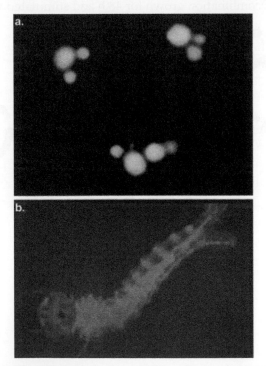

Figure 9. Feeding *Ae. aegypti* larvae *S. cerevisiae* cells transformed with *gfp-tmfA*. (a) Transformed *S. cerevisiae* cells expressing GFP-*Aedae*-TMOF observed under a fluorescent microscope. (b) *Ae. aegypti* larva fed for 2 days recombinant *S. cerevisiae* cells expressing GFP-*Aedae*-TMOF observed under a fluorescent microscope.

Color image of this figure appears in the color plate section at the end of the book.

Table 6. Feeding *Ae. aegypti* larvae with recombinant *S. cerevisiae*.

S. cerevisiae sc1245 Transformed by:	N	GAL Stimulation (h)	Dead larvae ± SEM	Mortality (% ± SEM)
Homologous recombination:				
pYDB2-GFP-*Aea*TMOF	3	4	6.6 ± 0.66[a,d]	83 ± 8
pYDB2-*Aea*TMOF	3	6	3.0 ± 0[a,d]	38 ± 0
pYDB2-GFP (control)	3	8	0.66 ± 0.33[a]	0 ± 0
pYDB2-GFP-IEGR (control)	3	10	0 ± 0	0 ± 0
Free plasmid:				
pYES2-GFP-*Aea*TMOF	3	2	5.3 ± 0.33[b,c]	67 ± 4
pYES2-*Aea*TMOF	3	2	3.66 ± 0.66[b,c]	46 ± 8
Control:				
Non-transformed	3	4	0.33 ± 0.33[c,d]	4 ± 4

Three groups of *Ae. aegypti* larvae (8 per group) were fed 10^7 yeast cells for 12 days. Larval mortality was daily determined and results are expressed as means of 3 determinations ± SEM. *S. cerevisiae* sc1245 is a haploid strain with genotype *MATα*, *leu*2-3, 112 *ura*3-52. [a,c,d] Significant difference by ANOVA and student's *t*-test $p<0.05$ (Nauwelaers and Borovsky 2002).

by fluorescence microscopy it is possible to monitor larval populations in the field and find out if enough yeast GFP-*Aedae*-TMOF is available for larval control.

Expression in Pichia pastoris *and* Anabaena

Pichia pastoris is ascomycetous yeast with cells that can readily mate with each other making it homothallic. Unlike homothallic strains of *S. cerevisiae*, which are diploid, *P. pastoris* remains haploid unless forced to mate. The key feature of the *P. pastoris* life cycle that permits genetic manipulation is its physiological regulation of mating and high stability in vegetative haploid state, a great advantage in the isolation and phenotypic characterization of mutants (Cregg and Higgins 1995, Cregg et al. 1998). *Pichia* gained popularity as an effective organism for foreign gene expression because of the presence of strong inducible alcohol oxidase promoters (P_{AOX1} and P_{AOX2}) that can be used for heterologous gene expression. These promoters depend strongly on methanol and are expressed in media containing other carbon sourced such as glycerol, glucose and sorbitol (Thorpe et al. 1999). A variety of proteins from different sources such as humans, invertebrates, bacteria, fungi, virus and plants have been expressed in this system (Farber et al. 1995, Gelissen 2000). Foreign genes can be directed into a specific genomic site either by gene replacement or by additive integration through homologous recombination. Linearizing the vectors either at the

AOX1 or the *HIS4* locus helps homologous recombination by a single crossover between sequences shared by the vector and host genome. Multiple gene insertion at a single locus (*AOX1* or *his4*) is an event that occurs spontaneously at a low but detectable frequency (between 1 to 10% of the transformants).

Aedae-TMOF, GFP-*Aedae*-TMOF and (DPAR)$_4$ analogue, that showed high potency when fed to mosquito larvae (Borovsky and Meola 2004), were cloned and expressed in *P. pastoris* engineering pPICZ B (Invitrogen, CA USA) (Fig. 10). The recombinant cells that were cloned with TMOF and analyzed by Southern blot analyses show that the recombinant cells carry 10 copies of *tmfA* (Nauwelaers 2002). The recombinant cells were fermented for 96 h using large scale fermenter (150 liters).

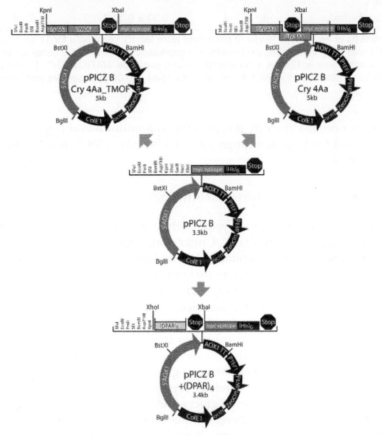

Figure 10. Construction of *P. pastoris* plasmid pPICZ B, carrying TMOF, GFP-TMOF and (DPAR)$_4$. The positions of GFP-TMOF, TMOF (DPAR)$_4$ in the multiple cloning site are indicated. The restriction enzyme *Bst*XI for homologous recombination and the direction of replication from the *AOX1* promoter are also included.

Different concentrations of *Aedae*-TMOF, determined by ELISA, in the fermented cells were fed to *Ae. aegypti* larvae and the lethal concentrations at 50 and 90% (LC_{50} and LC_{90}) were found to be 66–67 ng and 130–180 ng, respectively (Table 7) indicating that low concentrations of TMOF are sufficient to starve and kill mosquito larvae.

Bacillus thuringiensis subsp. *israelensis* (*Bti*) is the first subspecies of *B. thurengiensis* (*Bt*) that was found and used as an effective biological control agent against larvae of many mosquito and black fly species (Goldberg and Margalit 1977). Its Cry toxins are *per os* poisons, they bind mosquito midgut cell membranes causing rapid death. They are species specific, and hence are considered safe larvicides and do not cause damage to the environment. Field persistence of current preparations of *Bti* is low, but could be improved by expressing the *Bti* toxin genes in heterologous organisms that are readily eaten by mosquito larvae. The recombinant cells should be killed and formulated before they are released into the environment and the larvacide inside the cells would be protected from degradation by sunlight. Synergy has been described for bacterial endochitinase and δ-endotoxins against *Ae. aegypti* larvae (Sirichotpakorn 2001) prompting Borovsky et al. (2010) to investigate whether TMOF and the δ-endotoxins of *Bti* can also synergize. They observed first that starved mosquito larvae are 35-fold more sensitive to *Bti* toxins. Feeding *E. coli* 10^5 cells expressing *cry4Aa*, *cry11Aa*, *cyt1Aa* and *p20* alone did not cause larval mortality in the presence of non-recombinant *Pichia* cells. However, if *P. pastoris* cells expressing low amount of TMOF (75 nM) that does not affect mosquito larvae was fed with the recombinant *E. coli* (10^5 cells) larval mortality reached 95% indicating that TMOF enhances the effect of *Bti*. Following these observations, Borovsky et al. (2010) cloned and expressed in *P. pastoris* glutathione-S-transferase gene (*gst*) fused to *cry11Aa-tmfA* (Fig. 11). The addition of SUMO (small ubiquitin-related modifier) or *gst* genes allows foreign proteins synthesized in recombinant organism to be more soluble in the cell cytoplasm (Malakhov et al. 2004). Apparently, GST also protected Cry11Aa-*Aedae*-TMOF from heat inactivation at 50°C for 3 h before the recombinant cells were fed to *Ae. aegypti* larvae. Cell extract of *P. pastoris* gst-*cry11Aa-tmfA* cells when fed, *in vitro*, killed

Table 7. Effect of feeding recombinant *P. pastoris* expressing TMOF on *Ae. aegypti* larvae.

Fermentation	N	LC_{50} (nM ± SEM)	LC_{90} (nM ± SEM)
A	3	66 ± 18	180 ± 38
B	3	67 ± 8	130 ± 11

Three groups of *Ae. aegypti* larvae (20 per group) in 160 mL water were fed different concentrations of recombinant *Pichia* cells expressing *tmfA* for 12 days. Larval survival was followed daily and LC_{50} and LC_{90} calculated using probit analyses.

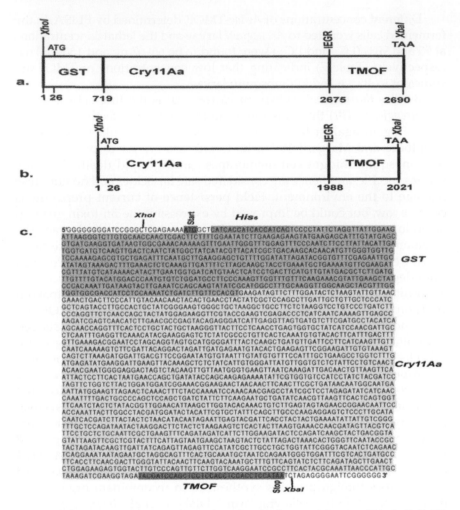

Figure 11. Genomic organization of *cry11Aa-tmf* synthetic gene. (a) *gst-cry11Aa-tmfA* (2,690 nt) genetic map with restriction enzymes cleavage sites for cloning, the ATG start signal, the IEGR trypsin cleavage site and the TAA stop signal. (b) *cry11Aa-tmfA* (2,021 nt) genetic map with restriction enzymes cleavage sites for cloning, IEGR trypsin cleavage site and start (ATG) and stop (TAA) signals. (c) *gst-cry11Aa-tmf* optimized synthetic gene sequence: His_6-*gst*, *cry11Aa* and *tmfA*, with the stop and start signals.

40% of the tested *Ae. aegypti* larvae in 24 h, and mass spectra analyses of purified cell extract identified Cry11Aa. Borovsky et al. (2011) used Cry4Aa fragment of 678 amino acids (60 kDa), cloned and expressed it in *P. pastoris*. The yeast cells were transformed at two loci (*AOX1* and *HIS4*) using plasmids pPICZ B and pPIC3.5k (Invitrogen, CA USA) engineering the following *P. pastoris* genotypes: *AOX1:tmfA; HIS4:gst-cry4Aa-tmfA* and *AOX1:gst-tmfA; HIS4:gst-cry4Aa-tmfA*. The transformed cells carried a single *gst*ligate to *cry4Aa* or two *gst* ligated to *tmfA* and to *cry4Aa-tmfA*. Feeding of heat inactivated (56°C for 3 h) or non-heat inactivated cells caused 72% and 92% larval mortality, respectively (Borovsky et al. 2011). Because TMOF is more effective against 1st and 2nd instar larvae (Borovsky and Meola 2004) the use of Cry toxins and *Aedae*-TMOF in early larval stages will enhance the effect of low *Bti* toxin doses and reduce the amount of *Bti* required in sewage treatment ponds in which it is not effective alone.

Although expression of several Cry toxins was reported in *Anabaena*, however, when *Anabaena* was cloned with *tmfA* it produced low amount of *Aedae-TMOF* (Borovsky et al. 2010). Perhaps the decapeptide precipitated in the inclusion bodies or it needs to be fused with a larger protein like GFP or GST in order to be efficiently synthesized in bacteria.

Expression in the Entomopathogenic Fungus Beauveria bassiana

Entomopathogenic fungi, such as *Metarhizium anisopliae* and *Beauveria bassiana*, both EPA approved biological control agents, offer an environmentally friendly alternative to chemical insecticides. One limitation to the use of entomopathogenic fungi is the relatively long time (6–12 days) it takes for the fungus to kill target insects. Expression of a scorpion toxin in *M. anisopliae* increased fungal toxicity about 9-fold towards *Ae. aegypti* (Wang and St Leger 2007). Expression of non-species specific toxins to control mosquitoes may promote the development of toxin-resistance and non-specific kill of beneficial insects. To decrease likelihood of developing resistance and increase the virulence of entomopathogenic fungi, two insect peptides, *M. sexta* diuretic hormone (MSDH) and *Aedae*-TMOF, were genetically engineered in *B. bassiana* in such a way that the fungus expressed and secreted the hormones only when it infected its host making it more susceptible to the invading fungus. MSDH-Gly (42 amino acids) and *Aedae*-TMOF, were expressed in *B. bassiana* via transformation of expression vectors containing a constitutive *B. bassiana*-derived *gpd*-promoter, and the nucleotide sequence corresponding to the MSDH or *Aedae*-TMOF fused to a 28-amino acid signal sequence derived from the *B. bassiana* chitinase (*chit1*) gene to produce strains Bb::spMSDH and Bb::spAedae-TMOF (Fan et al. 2012). While expression of MSDH by *B. bassiana* increased the virulence (LD_{50}) 8-fold towards a lepidopteran host, the greater waxmoth, *Galleria mellonella*. Bb::spMSDH was not significantly better than WT (wild type) against *Ae. aegypti* adults or larvae. In contrast, Bb::spAedae-TMOF, was much more potent than its wild-type parent against blood fed female adults; 6-fold fewer conidia of Bb::spAedae-TMOF were needed to obtain the same level of control as WT. Expressing

Aedae-TMOF also resulted in a 25% reduction in the survival time of the target mosquitoes. Modulation of trypsin activity was confirmed after fungal infection with Bb::spAedae-TMOF; infected females showed a 50% reduction in trypsin activity. In addition, a dramatic drop in fecundity was noted, with a 40% reduction in the number of eggs laid by TMOF infected females as compared to WT infected females. Microscopic examination of the ovaries of Bb::spAedae-TMOF infected females showed that many of the oocytes were smaller and underdeveloped as compared with WT and control, which were essentially indistinguishable. Bb::spAedae-TMOF also displayed increased virulence towards *Ae. aegypti* larvae. The LT_{50} value was dramatically reduced for Bb::spAedae-TMOF as compared to the WT, dropping from 6.5 to 4.2 days ($p < 0.01$). In addition, Bb::spAedae-TMOF treated larvae were significantly smaller than WT treated ones and their normal development was impaired (Fan et al. 2012). Similar results were also shown for *An. gambiae* by Kamareddine et al. (2013).

Future Work

Although it is hard to predict the future of genetically modified organisms that express peptide hormones, I would like to briefly mention an ongoing project of genetically engineered maize pollen in collaboration with the laboratories of the late Professor Andrew Spielman and Dr. Richard Pollack at Harvard University, Professor Charles Powell at the University of Florida, Dr. Robert Shatters Jr. at the USDA horticulture Research laboratory in Florida, Professor Arieh Zaritsky at the Ben Gurion University of the Negev and my laboratory at the University of Florida. Preliminary studies transforming maize pollen with *Bti-tmfA* to control mosquito larvae were recently reported (Zaritsky et al. 2010, Shatters et al. 2013). These earlier results are promising, however, final selection of homogeneous transgenic plants expressing *cry4Aa-tmfA*, *cry11Aa-tmfA* and *cyt1Aa-tmfA* in the pollen has not been achieved yet.

Acknowledgements

I thank all my collaborators during the 30 year journey of studying a hormone that at the beginning no one believed that it was a real hormone, because it was not found in the mosquito brain. When I named the peptide I was hesitant to name it Trypsin Modulating Oostatic Hormone, and instead I named it Trypsin Modulating Oostatic Factor. I would like to thank my good friend and collaborator David Carlson at the USDA in Gainesville Florida and my dear friend of over 20 years Professor Dr. Arnold De Loof at the KU of Leuven that immediately recognized the importance of the hormone even though others were uncertain. Thanks are also extended to my long-time colleagues and collaborators Robert Shatters, Jr. at the USDA Horticulture research lab in Florida, Professor Charles Powell at the University of Florida, Professor Arieh Zaritsky who introduced me to Bti at The Ben Gurion University in Israel, and to all my students, postdoctoral trainees and visiting scientists that participated and are part of the TMOF research story. I would like also to thank a relatively new colleague that joined the TMOF research group, Dr. Nemat Keyhani at the University of Florida. A special thank is also extended to Professor Hunt at the university of Virginia for sequencing the peptide by mass spectrometry, for the first time, showing that it is a non-blocked peptide. I am also grateful for the support that I received from Insect Biotechnology Inc., Bayer, KU of Leuven, Lady Davis award, NIH, BSF, Florida Mosquito Control Organization, USDA, NATO, University of Florida and the Florida Citrus Industry. Finally, I would also like to thank the secretary of Agriculture Dan Glickman that recognized in 1993 that TMOF is an important hormone and awarded me the silver plow medal for the TMOF technology.

Keywords: Mosquito, *Ae. aegypti*, *Bacillus thurengiensis israelensis*, *Pichia pastoris*, *Saccharomyces cerevisiae*, genetic engineering, Trypsin Modulating Oostatic Factor, yeast transformation, plant transformation, Cry and Cyt toxins, synergy, *Anabaena*, larval control, Alfalfa transformation, Tobacco Mosaic Virus, *Beauveria bassiana*, *H. virescens*

References

Adams, T.S. 1981. The role of ovarian hormone in maintaining cyclical egg production in insects. pp. 109–125. *In*: W.H. Clark, Jr. and T.S. Adams (eds.). Advances in Invertebrate Reproduction. Elsevier Science, Amsterdam.

Adams, T.S., A.M. Hintz and J.G. Pomonis. 1968. Oostatic hormone production in houseflies, *Musca domestica*, with developing ovaries. J. Insect Physiol. 14: 983–993.

Allen, M.L., D.A. O'Brochta, P.W. Atkinson and C.S. Levesque. 2001. Stable, germ-line transformation of *Culex quinquefeciatus* (Diptera: Culicidae). J. Med. Ent. 38: 701–710.

Berghammer, A.J., M. Klingler and E.A. Wimmer. 1999. A universal marker for transgenic insects. Nature 402: 370–371.

Boisvert, M. and J. Boisvert. 2000. Effects of *Bacillus thuringiensis* var. *israelensis* on target and nontargert organisms: a review of laboratory and field experiments. Biocontr. Sci. Technol. 10: 517–561.

Borovsky, D. 1985. Isolation and characterization of highly purified mosquito oostatic hormone. Arch. Insect Biochem. Physiol. 2: 333–349.

Borovsky, D. 1988. Oostatic hormone inhibits biosynthesis of midgut proteolytic enzymes and egg development in mosquitoes. Arch. Insect Biochem. Physiol. 7: 187–210.

Borovsky, D. 2007. Trypsin Modulating Oostatic Factor for developing resistant crops. pp. 135–149. *In*: I. Ishaaya, R. Nauen and R. Horowitz (eds.). Insecticides Design Using Advanced Technologies. Springer-Verlag, Berlin-Heidelberg.

Borovsky, D. and F. Mahmood. 1995. Feeding the mosquito *Aedes aegypti* with TMOF and its analogs; effect on trypsin biosynthesis and egg development. Reg. Pept. 57: 273–281.

Borovsky, D. and S.M. Meola. 2004. Biochemical and cytoimmunological evidence for the control of *Aedes aegypti* larval trypsin with *Aea*-TMOF. Arch. Insect Biochem. Physiol. 55: 124–139.

Borovsky, D. and R. Nauen. 2007. Biological and biochemical effects of organo-synthetic analogues of Trypsin Modulating Oostatic Factor (TMOF) on *Aedes aegypti*, *Heliothis virescens* and *Plutella xylostella*. Pestycyde/Pesticides 3-4: 17–26.

Borovsky, D. and Y. Schlein. 1988. Quantitative determination of trypsin-like and chymotrypsin-like enzymes in insects. Arch. Insect Biochem. Physiol. 8: 249–260.

Borovsky, D., F. Mahmood and D.A. Carlson. 1989. Mosquito oostatic hormone, and its potential as a unique adulticide and larvicide. J. Fla. Anti-Mosq. Assoc. 60: 66–70.

Borovsky, D., D.A. Carlson, P.R. Griffin, J. Shabanowitz and D.F. Hunt. 1990. Mosquito oostatic factor a novel decapeptide modulating trypsin-like enzyme biosynthesis in the midgut. FASEB J. 4: 3015–3020.

Borovsky, D., D.A. Carlson and D.F. Hunt. 1991. Mosquito oostatic hormone a trypsin modulating oostatic factor. pp. 133–142. *In*: J.J. Menn, T.J. Kelly and E.P. Masler (eds.). Insect Neuropeptides Chemistry, Biology and Action. ACS Symposium Series 453, Washington, DC.

Borovsky, D., C.A. Powell and D.A. Carlson. 1992. Development of specific RIA and ELISA to study trypsin modulating oostatic factor in mosquitoes. Arch. Insect Biochem. Physiol. 21: 13–21.

Borovsky, D., D.A. Carlson, P.R. Griffin, J. Shabanowit and D.F. Hunt. 1993. Sequence analysis, synthesis and characterization of *Aedes aegypti* trypsin modulating oostatic factor (TMOF) and its analogs. Insect Biochem. Mol. Biol. 23: 703–712.

Borovsky, D., Q. Song, M. Ma and D.A. Carlson. 1994a. Biosynthesis, secretion and cytoimmunochemistry of trypsin modulating oostatic factor of *Aedes aegypti*. Arch. Insect Biochem. Physiol. 27: 27–38.

Borovsky, D., C.A. Powell, J.K. Nayar, J.E. Blalock and T.K. Hayes. 1994b. Characterization and localization of mosquito-gut receptors for trypsin modulating oostatic factor (TMOF) using a complementary peptide and immunocytochemistry. FASEB J. 8: 350–355.

Borovsky, D., I. Janssen, J. Vanden Broeck, R. Huybrechts, P. Verhaert, H.L. DeBondt, D. Bylemans and A. DeLoof. 1996. Molecular sequencing and modeling of *Neobellieria bullata* trypsin: Evidence for translational control with Neb TMOF. Eur. J. Biochem. 237: 279–287.

Borovsky, D., C.A. Powell, W.O. Dawson, S. Shivprasad, D. Lewandowski, H.L. DeBondt, C. DeRanter and A. DeLoof. 1998. Trypsin modulating oostatic factor (TMOF): A new biorational insecticide against mosquitoes. pp. 131–140. *In*: D. Konopinska, G. Goldsworthy, R.J. Nachman, J. Nawrot, I. Orchard and G. Rosinski (eds.). Insects, Chemical Physiological and Environmental Aspects. University of Wroclaw, Wroclaw.

Borovsky, D., D.A. Iannotti, R.G. Shatters Jr. and C.A. Powell. 2005. Effect of recombinant Aea-TMOF on *Heliothis virescens*. Pestycyde/Pesticides 3: 79–85.

Borovsky, D., S. Rabindran, W.O. Dawson, C.R. Powell, D. Iannotti, T. Morris, J. Shabanowitz, D.F. Hunt, H.L. De Bondt and A. De Loof. 2006. Expression of *Aedes* TMOF on the virion of TMV: potential larvicide. Proc. Nat. Acad. Sci. USA 103: 18963–18968.

Borovsky, D., V. Khasdan, S. Nauwelaers, C. Theunis, L. Bertier, E. Ben-Dov and A. Zaritsky. 2010. Synergy between *Bacillus thuringiensis* subsp. *israelensis* crystal proteins and *Aedes aegypti* Trypsin Modulating Oostatic Factor. Open Toxicol. J. 3: 116–125.

Borovsky, D., S. Nauwelaers, A. Van Mileghem, Y. Meyvis, A. Laeremans, C. Theunis, L. Bertier and E. Boons. 2011. Control of mosquito larvae with TMOF and 60 kDa Cry4Aa expressed in *Pichia* pastoris. Pestycyde/Pesticides 1-4: 5–15.

Bylemans, D., D. Borovsky, D.F. Hunt, L. Grauwels and A. DeLoof. 1994. Sequencing and characterization of oostatic factor (TMOF) from the ovaries of the grey fleshfly *Neobellieria* (*Sarcophaga*) *bullata*. Reg. Pept. 50: 61–72.

Bylemans, D., P. Verhaert, I. Janssen, J. Vanden Broeck, D. Borovsky, M. Ma and A. DeLoof. 1996. Immunolocalization of the oostatic and prothoracicostatic peptide, Neb-TMOF, in the fleshfly *Neobellieria bullata*. Gen. Comp. Endocrinol. 103: 273–280.

Carlisle, D.B. and F. Knowles. 1959. Endocrine control in crustaceans. Cambridge Monogr. Exp. Biol. 10: 10–120.

Carlson, D.A., C.M. Krueger, G. Hehman, D. Borovsky, A.M. Rossi, A.F. Cockburn, J. Shabanowitz and D.F. Hunt. 1994. Cloning and partial sequencing of the trypsin modulating oostatic factor gene from mosquito genomic libraries, and expression of the TMOF gene in *E. coli*. pp. 626–630. *In*: K.G. Davey, R.E. Peter and S.S. Tobe (eds.). Perspectives in Comparative Endocrinology. National Research Council Canada, Ottawa.

Catteruccia, F., T. Nolan, T.G. Loukeris, C. Blass, C. Savakis, F.C. Kafatos and A. Crisanti. 2000. Stable germline transformation of the malaria mosquito *Anopheles stephensi*. Nature 405: 959–962.

Chow, K.–C. and W.L. Tung. 1999. Electrotransformation of *Chlorella vulgaris*. Plant Cell Rep. 18: 778–780.

Clements, A.N. 1992. The Biology of Mosquitoes, Vol. 1: Development Nutrition and Reproduction. Chapman and Hall, New York.

Cregg, J.M. and D.R. Higgins 1995. Production of foreign proteins in the yeast *Pichia pastoris*. Can. J. Bot. 73 (Suppl. 1): S891–S897.

Cregg, J.M., S. Shen, M. Johnson and H.R. Waterham. 1998. Classical genetic manipulation: *Pichia* protocols. Methods Mol. Biol. 103: 17–26.

Curto, E.V., M.A. Jarpe, J.B. Blalock, D. Borovsky and N.R. Krishna.1993. Solution structure of trypsin modulating oostatic factor is a left-handed helix. Biochem. Biophy. Res. Comm. 193: 688–693.

Davey, K.G. 1978. Hormonal stimulation and inhibition in the ovary of an insect, *Rhodnius prolixus*. pp. 13–15. *In*: P.J. Gaillard and H.H. Boer (eds.). Comparative Endocrinology. Elsevier Science, Amsterdam.

Davey, K.G. and J.E. Kunster. 1981. The source of an antigonadotropinin the female of *Rhodnius prolixus* Stal. Can. J. Zool. 59: 761–764.

Dawson, H.N., R. Burlingame and A.C. Cannons. 1997. Stable transformation of *Chlorella*: rescue of nitrate reductase deficient mutants with the nitrate reductase gene. Curr. Microbiol. 35: 356–362.

De Loof, A., D. Bylemans, L. Schoofs, I. Janssen, K. Spittaels, J. Vanden Broeck, R. Huybrechts, D. Borovsky, Y. Hua, J. Koolman and S.A. Sower. 1995. Folliculostatins, gonadotropins and a model for control of growth in the grey fleshfly, *Neobellieria (Sarcophaga) bullata*. Insect Biochem. Molec. Biol. 25: 661–667.

Donson, J., C.M. Kearney, M.E. Hilf and W.O. Dawson. 1991. Systemic expression of a bacterial gene by a tobacco mosaic virus-based vector. Proc. Natl. Acad. Sci. USA 88: 7204–7208.

Else, J.G. and C.L. Judson. 1972. Enforced egg-retention and its effect on vitellogenesis in the mosquito *Aedes aegypti*. J. Med. Entomol. 9: 527–530.

El-Sheekh, M.M. 1999. Stable transformation of the intact cells of *Chlorella kessleri* with high velocity microprojectiles. Biologia Plantarum 42(2): 209–216.

Fan, Y., D. Borovsky, C. Hawkings, A. Ortiz-Urquiza and N.O. Keyhani. 2012. Exploiting host molecules to augment mycoinsecticide virulence. Nature Biotechnol. 30: 35–37.

Faber, K.N., W. Harder, G. Ab and M. Veenhuis. 1995. Review: Methylotropic yeasts as factories for the production of foreign proteins. Yeast 11: 1331–1344.

Gäde, G. and G.J. Goldworthy. 2003. Insect peptide hormones: a selective review of their physiology and potential application for pest control. Pest Manag. Sci. 59: 1063–1075.

Gelissen, G. 2000. Heterologous protein production in methylotrophic yeasts. App. Microbiol. Biotechnol. 54: 741–750.

Gelman, D.B. and D. Borovsky. 2000. *Aedes aegypti* TMOF modulates ecdysteroid production by prothoracic glands of the gypsy moth *Lymantria dispar*. Arch. Insect. Biochem. Physiol. 45: 60–68.

Goldberg, L.J. and J. Margalit. 1977. A bacterial spore demonstrating rapid larvicidal activity against *Anopheles sergentii, Uranataenia unguiculata, Culex univittatus, Aedes aegypti* and *Culex pipiens*. Mosq. News 37: 355–358.

Hamamoto, H., Y. Sugiyama, N. Nakagawa, E. Hashida, Y. Matsunaga, S. Takemoto, Y. Watanabe and Y. Okada. 1993. A new tobacco mosaic virus vector and its use for the systemic production of angiotensin-I-converting enzyme inhibitor in transgenic tobacco and tomato. Biotechnology 11: 930–932.

Hasnain, S.E., E.K. Manavathu and W.C. Leung. 1985. DNA-mediated transformation of *Chlamydomonas reinhardi* cells: used of aminoglycoside 3'-phosphotransferase as a selectable marker. Mol. Cell. Biol. 5: 3647–3650.

Hawkins, R.L. and M. Nakamura. 1999. Expression of human growth hormone by the eukaryotic alga, *Chlorella*. Curr. Microbiol. 38: 335–341.

Iwanov, P.P. and K.A. Mescherskaya. 1935. Die physiologischen Besonderheiten der geschlechtlich unreifen Insektenovarien und die zyklischen Veränderungen ihrer Eigenschaften. Zool. Jb. Physiol. 55: 281–348.

Jarvis, E.E. and L.M. Brown. 1991. Transient expression of firefly luciferase in protoplasts of the green algae *Chlorella ellipsoidea*. Curr. Genet. 19: 317–321.

Jeffers, L.A., H. Shen, S. Khalil, B.W. Bissinger, A. Brandt, T.B. Gunnoe and R.M. Roe. 2012. Enhanced activity of an insecticidal protein, trypsin modulating oostatic factor (TMOF), through conjugation with aliphatic polyethylene glycol. Pest Manag. Sci. 68: 49–59.

Kamareddine, L., Y. Fan, M.A. Osta and N.O. Keyhani. 2013. Expression of trypsin modulating oostatic factor (TMOF) in an entomopathogenic fungus increases its virulence towards *Anopheles gambiae*. Parasites & Vectors 6: 22–26.

Kelly, T.J., M.J. Birnbaum, C.W. Woods and A.B. Borkovec. 1984. Effects of housefly oostatic hormone on egg development neurosecretory hormone action in *Aedes atropalpus*. J. Exp. Zool. 229: 491–496.

Knight, P.J.K., J. Carroll and D.J. Ellar. 2004. Analysis of glycan structures on the 120 kDa aminopeptidase N of *Manduca sexta* and their interactions with *Bacillus thuringiensis* Cry1Ac toxin. Insect Biochem. Mol. Biol. 34: 101–112.

Lin, Y., M.T. Hamblin, M.J. Edwards, C. Barillas-Mury, M.R. Kanost, D.C. Knipple, M.F. Wolfner and H.H. Hagedorn. 1993. Structure, expression, and hormonal control of genes from the mosquito, *Aedes aegypti*, which encode proteins similar to the vitelline membrane proteins of *Drosophila melanogaster*. Dev. Biol. 155: 558–568.

Liu, F., G. Baggerman, W. D'Hertog, P. Verleyen, L. Schoof and G. Wets. 2006. *In silico* identification of new secretory peptide genes in *Drosophila melanogaster*. Mol. Cell. Proteomics 5: 510–522.

Liu, T.P. and K.G. Davey. 1974. Partial characterization of a proposed antigonadotropin from the ovaries of the insect *Rhodnius prolixus* Stal. Gen. Comp. Endocrinol. 24: 405–408.

Ma, P.W.K., T.R. Davis, H.A. Wood, D.C. Knipple and W.L. Roelofs. 1998. Baculovirus expression of an insect gene that encodes multiple neuropeptides. Insect Biochem. Mol. Biol. 28: 239–249.

Malakhov, M.P., M.R. Mattern, O.A. Malakhova, M. Drinker, S. Weeks and T.R. Butt. 2004. SUMO fusions and SUMO-specific protease for efficient expression and purification of proteins. J. Struct. Funct. Genomics 5: 75–86.

Meola, R. and A.O. Lea. 1972. Humoral inhibition of egg development in mosquitoes. J. Med. Entomol. 9: 99–103.

Metz, M. 2003. *Bacillus thuringiensis*. A Cornerstone of Modern Agriculture. Food Products Press, Binghamton, New York.

Nauen, R., D. Sorge, A. Sterner and D. Borovsky. 2001. TMOF like factor controls the biosynthesis of serine proteases in the larval gut of *Heliothis virescens*. Arch. Insect Biochem. Physiol. 47: 169–180.

Nauwelaers, S. 2002. Bioengineering of *Chlorella species*, *Saccharomyces cerevisiae* and *Pichia pastoris* with *Aedes aegypti* Trypsin Modulating Oostatic Factor (TMOF). Ph.D. Thesis. Catholic University of Leuven, Belgium.

Nauwelaers, S. and D. Borovsky. 2002. Cloning and expression of trypsin modulating oostatic factor (Aea-TMOF) and green fluorescent protein (GFP) in *Saccharomyces cerevisiae*. pp. 138–140. *In*: D. Konopinska, G. Coast, G. Goldsworthy, R.J. Nachman and G. Rosinski (eds.). Arthropods 2001. Wydawnictawa Uniwersytetu, Wroclawskiego.

Romanos, M.A., C.A. Scorer and J.J. Clare. 1992. Foreign gene expression in yeast: A Review. Yeast 8: 423–488.

Schardl, C., A.D. Byrd, G.B. Benzion, M.A. Altschuler, D.F. Hildebrand and A.G. Hunt. 1987. Design and construction of a versatile system for the expression of foreign genes in plants. Gene 61: 1–11.

Schneider, J.C. and L. Guarente. 1991. Vectors for expression of clonal genes in yeast: regulation, overproduction, and underproduction. Meth. Enzymol. 194: 373–388.

Shatters, Jr. R.G., A. Zaritsky, E. Ben-Dov, C.A. Powell and D. Borovsky. 2013. Maize pollen to control mosquitoes. GMO integrated Plant Production IOBC-WPRS Bulletin 97: 121–124.

Sirichotpakorn, N., P. Rongnoparut, K. Choosang and W. Panbangred. 2001. Coexpression of chitinase and the *cry11Aa1* toxin genes in *Bacillus thuringiensis* serovar. *israelensis*. J. Invertebr. Pathol. 78: 160–9.

Southwick, F.S. and D.L. Purich. 1995. Inhibition of *Listeria* locomotion by mosquito oostatic factor, a natural oligoproline peptide uncoupler of profiling action. Infect. Immun. 6: 182–190.

Spielman, A. and M. D'Antonio. 2001. Mosquito: a Natural History of our Most Persistent and Deadly Foe. Hyperion, New York.

Thorpe, E.D., M.C. d'Anjou and A.J. Daigulis. 1999. Sorbitol as a non-repressing carbon source for fed-batch fermentation of recombinant *Pichia pastoris*. Biotech. Lett. 21: 669–672.

Tortiglione, C., P. Fanti, F. Pennacchio, C. Malva, M. Breuer, A. De Loof, L.M. Monti, E. Tremblay and R. Rao. 2002. The expression in tobacco plants of *Aedes aegypti* trypsin modulating oostatic factor (Aea-TMOF) alters growth and development of the tobacco budworm, *Heliothis virescens*. Molec. Breeding 9: 159–169.

Vanderherchen, M.B., M. Isherwood, D.M. Thompson, R.J. Linderman and R.M. Roe. 2005. Toxicity of novel aromatic and aliphatic organic acid and ester analogs of trypsin modulating oostatic factor to larvae of the northern house mosquito, *Culex pipiens* complex, and the tobacco hornworm, *Manduca sexta*. Pestic. Biochem. Physiol. 81: 71–84.

Wach, A., A. Brachat, R. Pohlmann and P. Philippsen. 1994. New heterologous modules for classical or PCR-based gene disruptions in *Saccharomyces cerevisiae*. Yeast 10: 1793–1808.

Wang, C.S. and R.J. St Leger. 2007. A scorpion neurotoxin increases the potency of a fungal insecticide. Nature Biotechnol. 25: 1455–1456.

[WHO] World Health Organization. 1998. Roll Back Malaria. Fact sheet 203.

Xin-Hua, Y., H.L. DeBond, C.A. Powell, R.C. Bullock and D. Borovsky. 1999. Sequencing and characterization of the citrus weevil *Diaprepes abbreviatus*, trypsin cDNA; effect of *Aedes* trypsin modulating oostatic factor on trypsin biosynthesis. Eur. J. Biochem. 262: 627–636.

Yang, Y.J. and D.M. Davies. 1971. Digestive enzymes in the excreta of *Aedes aegypti* larvae. J. Insect Physiol. 17: 2119–2123.

Zaim, M. and P. Guillet. 2002. Alternative insecticides: an urgent need. Trends Parasitol. 18: 161–163.

Zaritsky, A., E. Ben-Dov, D. Borovsky, S. Boussiba, M. Einav, G. Gindin, A.R. Horowitz, M. Kolot, O. Melnikov, Z. Mendel and E. Yagil. 2010. Transgenic organisms expressing genes from *Bacillus thuringiensis* to combat insect pests. Bioengin. Bugs 1(5): 341–344.

12

RNAi Based Functional Analysis of Biosynthetic Enzymes and Transport Proteins Involved in the Chemical Defense of Juvenile Leaf Beetles

Antje Burse[a],* and *Wilhelm Boland*[b]

Introduction

Plant-herbivore interactions dominate the planet's terrestrial ecology. Phytophagous insects account for more than double of the non-herbivorous taxa. This disparity became especially pronounced with the increasing diversity of angiosperms in the post-Cretaceous period. In response to herbivores, plants developed several morphological and biochemical adaptations which allowed them to wage a kind of chemical warfare; one strategy of this war was based on toxic secondary metabolite production, storage and eventually release (Macias et al. 2007). As some insects became adapted to these metabolites, interactions between the

Max Planck Institute for Chemical Ecology, Department of Bioorganic Chemistry, Beutenberg Campus, Hans-Knoell-Str. 8, D-07745 Jena, Germany.
[a] Email: aburse@ice.mpg.de
[b] Email: boland@ice.mpg.de
* Corresponding author

two organismic groups occasionally led to highly specific relationships (Agrawal et al. 2012, Hare 2012). When taking the habitats of herbivores into account, it becomes clear that insects face a geographical assortment of chemical environments, from non-toxic to highly toxic plants (Ibanez et al. 2012, Fürstenberg-Hägg et al. 2013). Therefore, the resistance traits may vary with the probability of meeting a toxin. Moreover, other pressures, such as the presence or absence of competitors, predators, parasitoids, or pathogens can also influence the selection of particular resistance traits (Ode 2006). Thus, the complexity of the local community composition is a key factor in maintaining the diversity of adaptive mechanisms to plant toxins.

Particularly beetles (Coleoptera) are among the most versatile evolutionary innovators when it comes to use plants as a food source. The reciprocal adaptation processes have contributed to the enormous biodiversity that is found today in both organismic groups (Ehrlich and Raven 1964, Beccera 1997, Farrel 1998, Farrell and Sequeira 2004). The herbivorous beetles, named as taxon "Phytophaga", are constituted by the Chrysomelidae (leaf beetles) together with the Cerambycidae (longhorn beetles), and the Curculionoidea (weevils). They form the largest beetle radiation representing roughly 40 percent of all 350,000 known species (Fernandez and Hilker 2007, Gómez-Zurita et al. 2007). The success of the Phytophaga can be attributed also to the ability to disarm chemical plant defense combined with often astonishing defensive strategies against their enemies. A host chemical defense can be disarmed by contact and ingestion avoidance, intestinal uptake prevention, excretion, toxin degradation, target-site mutation, or sequestration (Sorensen and Dearing 2006, Li et al. 2007, Boeckler et al. 2011, Winde and Wittstock 2011, Dobler et al. 2012). Sequestration is one of the most widespread adaptive mechanisms to plant toxins prevalent not only in beetles but in many orders of the Insecta (Duffey 1980, Nishida 2002, Opitz and Mueller 2009). It includes the uptake and concentration of phytochemicals into hemolymph, cuticle, specialized tissues or glands frequently for the benefit of herbivores in trophic networks.

Leaf beetles are known for their ability to sequester structurally different phytochemicals, such as β-amyrin (Laurent et al. 2003b), cucurbitacins (Gillespie et al. 2003), pyrrolizidine alkaloids (Hartmann 2004), phenolglucosides (Pasteels et al. 1983), naphthaleneglucosides (Pasteels et al. 1990), glucosidically bound aliphatic alcohols (Schulz et al. 1997), or irdoid glucosides (Willinger and Dobler 2001) especially for the purpose of chemical defense against their enemies. Not only the adults but also the larvae are known to sequester phytochemicals (Pasteels 1993, Blum 1994, Schulz 1998, Laurent et al. 2003a, Laurent et al. 2005). The larvae of the leaf beetle subtribe Chrysomelina, for example, developed for the production of defensive secretions from plant toxins specialized epidermis derived exocrine glands (Hinton 1951, Renner 1970, Pasteels and Rowell-Rahier 1989). The glands, herein after referred to as defensive glands, consist of small pouches repleted with fluid, ordinarily kept withdrawn, but promptly everted when a larva is disturbed (Fig. 1).

In these glands not only exogenous but also endogenous metabolites are stored and enzymatically transformed into deterrents. During the Chrysomelina evolution, the *de novo* production of deterrents pre-dates the phenomenon of sequestration. Therefore, the Chrysomelina represents an unrivalled insect taxon to study both, (*i*) the evolution of host dependent deterrent synthesis from deterrent *de novo* production in related species, and (*ii*) the correlation of deterrent synthesis strategies with host plant affiliation.

In this book chapter we focus on the ecological relevant leaf beetle taxon Chyrsomelina and its larval chemical defense. At first, we provide an overview of the different strategies of deterrent production evolved by the beetles. Next, we describe the use of RNA interference (RNAi) to discover components relevant for the deterrents in the larvae *in vivo*. Expressional, translational, chemical, functional and developmental effects are discussed in silenced insects compared to control groups. By means of the iridoid synthesis, we show the value of the method to complement *in vitro* data in order to resolve biosynthetic pathways and manipulate metabolic pattern in defensive secretions. We used the gene silencing technique also to create phenotypes lacking translocation activities important for the sequestration process. Finally, we bring our results in context of the development of metabolite diversity in chrysomelids as well as the relevance of sequestration as a process for enhancing the adaptive radiation observed in plants and beetles.

Figure 1. Larval chemical defense of the poplar leaf beetle, *Chrysomela populi*. (A) non-stimulated larva of the third instar compared to (B) forceps stimulated larva demonstrating the release of droplets of secretions (black triangle) from the everted reservoirs of the defensive glands which are arranged segmentally in nine pairs along the back. (C) drawing of a dissected defensive gland according to Hinton (1951).

The Synthetic Strategies to Produce Chemical Defenses of the Juvenile Chrysomelina

Insects frequently developed glands to store and to release deleterious exudates in order to circumvent auto-intoxicative effects. The defensive glands of the juvenile Chrysomelina species are segmentally arranged in nine pairs on their backs. According to morphological studies, each defensive gland is composed of a number of enlarged secretory cells, which are in turn connected to a chitin-lined reservoir. The secretory cells are always accompanied by two canal cells that form a cuticular canal, which connects the secretory cell with the reservoir (Noirot and Quennedey 1974). In Chrysomelina larvae, all compounds reaching the glandular reservoir via the hemolymph are glucosides (*de novo* produced or sequestered) that are converted enzymatically into the biologically active form within the reservoir (Pasteels and Rowell-Rahier 1989, Pasteels et al. 1990). Thus, the glands also secrete enzymes for the final metabolic conversion of precursors into defensive compounds in the reservoir.

Besides an almost identical gland anatomy, all Chrysomelina larvae have the first enzymatic reactions occurring in the reservoir in common. They possess an unspecific glucosidase activity removing the sugar moiety from glucosides which are either sequestered from the food plant or synthesized *de novo* (Fig. 2) (Pasteels et al. 1990, Soetens et al. 1993).

Subsequently, a specific oxidase reaction catalyzes the formation of aldehydes or dialdehydes (Veith et al. 1997, Brueckmann et al. 2002). A single change in oxidase specificity could permit different glucosides to be used as precursors for defensive compounds after hydrolysis. Based on that, Pasteels et al. (1990) postulated *de novo* synthesis of defensive

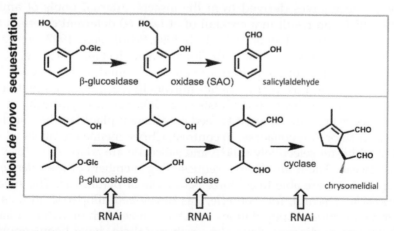

Figure 2. Steps of the deterrent biosynthesis in the larval gland reservoir of *C. populi* (sequestration) and *P. cochleariae* (*de novo*).

compounds as the primitive state harboring a set of enzymes that allows the evolution of utilizing plant derived glucosides for defense.

According to phylogenetic analyses by Termonia et al. (2001), *de novo* production of defensive secretions fortified with iridoids (cylopentanoidmonoterpenoids) is considered the ancestral strategy (Meinwald et al. 1977, Blum et al. 1978, Pasteels et al. 1990) (Fig. 3A).

The larvae of the Chrysomelina species feeding on Salicaceae sequester phenolglucosides such as salicin or salicortin (Pasteels et al. 1983, Pasteels et al. 1986, Rowell-Rahier and Pasteels 1986, Kuhn et al. 2004) (Fig. 3B). The glucosides serve as precursors for the repellent salicylaldehyde (Michalski et al. 2008, Kirsch et al. 2011a, Kirsch et al. 2011b). This hydrophobic aldehyde forms an organic layer, accounting for 15 percent of the total discharge volume, while the aqueous phase constitutes 85 percent (Brueckmann et al. 2002). The latter contains the precursor salicin which is cleaved into salicyl alcohol and glucose. The alcohol is a substrate for a flavine-dependent salicyl alcohol oxidase (SAO) which belongs to the glucose-methanol-choline (GMC) family of oxidoreductases (Cavener 1992). The SAO uses molecular oxygen as an electron acceptor for alcohol oxidation, yielding the aldehyde and hydrogen peroxide (Brueckmann et al. 2002, Michalski et al. 2008, Kirsch et al. 2011a). Salicylaldehyde is considered to be a potent repellent against generalist predators (Wallace and Blum 1969, Pasteels et al. 1983, Pasteels et al. 1986) and a powerful antimicrobial agent (Gross et al. 2002).

In contrast to the incorporation of only few plant-derived compounds, larvae of the monophyletic *interrupta*–group of the taxon Chrysomelina are able to take up a wide variety of glucosidically bound leaf alcohols. After deglucosylation, these compounds are further esterified with butyrate derivatives derived from the insects' internal pools of amino acids, which can result in a cocktail of at least 60 deterrent esters in the defensive secretions (Hilker and Schulz 1994, Schulz et al. 1997, Termonia and Pasteels 1999, Kuhn et al. 2007, Tolzin-Banasch et al. 2011) (Fig. 3C).

A combination of the pattern of deterrents found in the juvenile leaf beetles with their host plant families mirrors the reciprocal adaptation of Chrysomelina beetles to their hosts (Fig. 3D). Species synthesizing the deterrents *de novo* are adapted to feed on different plant families, such as Brassicaceae or Polygonaceae. In contrast, Chrysomelina members whose larvae sequester selectively salicin and salicortin are adapted exclusively to Salicaceae. The non-selective sequestering members of the *interrupta*-group, in turn, are able to colonize Salicaceae or Betulaceae. The species *Chrysomela lapponica*, for example, has developed allopatric and, as an exception, sympatric populations which colonize birch or willow plants. Populations feeding on *Salix* spp., rich in salicin, have been reported to produce salicylaldehyde with minor amounts of esters in the larval

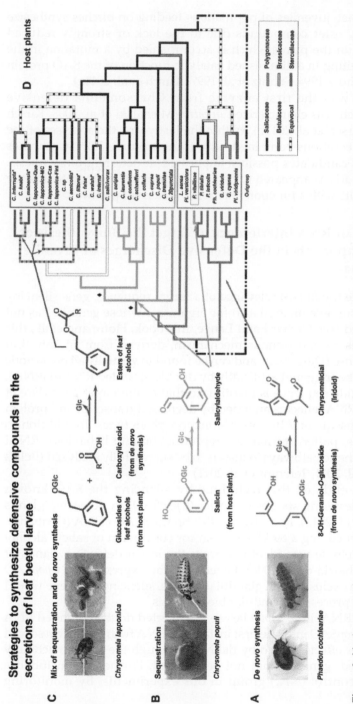

Figure 3. Maximum parsimony reconstruction of the evolution of the taxon Chrysomelina considering the synthesis of deterrents in the defensive glands of the larvae (A–C) and the affiliations of host plants (D) (adapted from Termonia et al. 2001).

Color image of this figure appears in the color plate section at the end of the book.

stage. In contrast, juveniles of populations feeding on birches synthesize predominantly ester compounds due to the lack or strongly reduced level of salicin in the plants which is accompanied by a mutation in the SAO gene resulting in a diminished catalytic function of the SAO protein (Hilker and Schulz 1994, Schulz et al. 1997, Kirsch et al. 2011a).

The SAO was the first enzyme from Chrysomelina defensive secretions which was characterized *in vitro* (Michalski et al. 2008, Kirsch et al. 2011a, Kirsch et al. 2011b). Within the last years, the analyses of the proteome in secretions combined with transcriptome library searches revealed other candidates possibly implicated in deterrent biosynthesis. In our studies, RNAi appeared to be an indispensable method to link the screening results with a function *in vivo* in a deterrent pathway.

Introduction of RNA Interference to Target the Biosynthesis of Discrete Components in the Defensive Discharges of Juvenile Chrysomelina

To demonstrate the *in vivo* relevance of a target sequence, gene silencing by RNAi is a suitable method also for organisms whose genome has not been sequenced (see Orchard and Lange, this book, Hoffmann et al., this book). RNAi is an endogenous mechanism, derived from an anti-viral immune response (Ding 2010), and can be found in virtually all eukaryotic species. It can be triggered artificially by double-stranded RNA (dsRNA), whose nucleotide sequence is identical to that of the target gene (Fire et al. 1998). The RNAi effect is attended by decreased transcript and protein levels, and consequently by loss-of-function phenotypes. In addition to embryogenesis, pattern formation, reproduction and behaviour, RNAi allows biosynthetic pathways in insects to be successfully analyzed (Belles 2010, Mito et al. 2011, Terenius et al. 2011).

We first validated this technique by silencing the SAO known from the sequestering species *C. populi* (*CpopSAO*) in defensive glands (Bodemann et al. 2012). Recently an 1872 bp *CpopSAO* cDNA (Genbank/ HQ245154.1) encoding a 69 kDa protein for conversion of salicyl alcohol into salicylaldehyde was identified from the larval defensive glands of *C. populi* (Michalski et al. 2008). Given that the expression of *CpopSAO* was detectable exclusively in glandular tissues, silencing this gene would affect only the process of glandular biosynthesis.

To induce RNAi in *C. populi* larvae, we injected double-stranded RNA (dsRNA) of *CpopSAO* into late first instar. A dsRNA fragment of *gfp* served as a control for effects caused by dsRNA; although the RNAi machinery will be induced, genes should not be silenced. Further, we included a non-injected control (NIC) group in our experiments. By monitoring the developmental traits and the secretion production in C. *populi* and comparing the results with those from control groups, we found that silencing of *CpopSAO* did not influence either growth rate or pupae weight. But the larvae treated with *CpopSAO*-dsRNA produced slightly more secretions than the larvae of the control groups, which might be due to the different osmotic characteristics between salicyl alcohol (which is now almost exclusively present) and its aldehyde (Brueckmann et al. 2002). Because we did not detect significant differences between NIC and *gfp* controls in any experiments delineated below, we continue discussing only the data of the *gfp* controls (Bodemann et al. 2012).

In accordance with the literature, SAO corresponds to the dominant band at 70 kDa in the secretions of C. *populi* (Brueckmann et al. 2002, Michalski et al. 2008). The composition of the secretome after dsRNA treatment was monitored in a time series in silver-stained 1D-SDS gels. Due to the silencing effect, the quantity of *CpopSAO* was apparently reduced just two days after dsRNA injection, and the protein was barely visible after day 5 (Fig. 4B).

Next, the effects on the biosynthesis of salicylaldehyde in the defensive secretions were determined by GC/MS analysis. We detected salicyl alcohol in the defensive secretions just two days after the injection of *CpopSAO* dsRNA (Fig. 4C). The precursor was not detectable in *gfp* control secretions. In addition, no unexpected chemical compound arose due to the dsRNA treatments. By setting the peak area of salicylaldehyde in ratio equalling the sum of the main peak areas, a diagram of the RNAi-dependent reduction of the aldehyde can be plotted (Fig. 4D). We have tested dsRNA amounts ranging from 0.1 to 3.0 μg. After RNAi induction, significantly less aldehyde was observed for the 3.0 μg *CpopSAO* group ($p=0.015$) on the fourth day and for all tested *CpopSAO* groups on the fifth day (0.1 μg, $p=0.016$; 1 μg, $p=0.002$; 3 μg, $p=<<0.001$). In summary, we show that we are able to manipulate the metabolic pattern in defensive secretions which provides a base to build on further studies of biosynthetic pathways.

The Iridoid de novo Synthesis

After successfully introducing the "lack of function approach" to the defensive secretions of C. *populi* by silencing an enzyme for which we had a clear expectation of a resulting phenotype, we used the method to screen for the function of unknown proteins in deterrent biosynthesis. The exploration of the iridoid metabolism in Chrysomelina larvae is an excellent example to show the value of RNAi technique for the complementation of *in vitro* data which enabled us to confirm the previously proposed iridoid

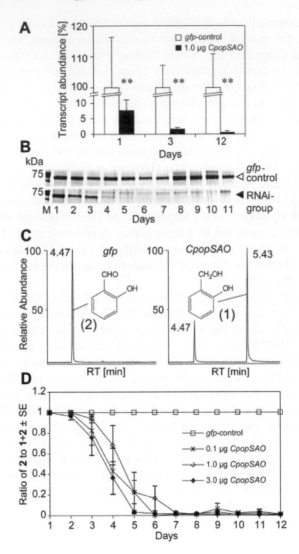

Figure 4. RNAi effects in juvenile *C. populi*. (A) white bars: transcript abundance of *CpopSAO* after injecting 1.0 µg *gfp*-dsRNA, n=3 ± SD; black bars: transcript abundance of *CpopSAO* after injecting 1.0 µg *CpopSAO*-dsRNA, n=3. 100% = ΔCq of *gfp* control. Asterisks indicate level of significance: ** p< 0.01. (B) *Cpop*SAO protein abundance in defense secretions was monitored over time. 0.85 mg secretions/lane was separated on silver-stained SDS-gels. The 70 kDa band corresponds to *Cpop*SAO. Secretions originating from control treatment with 1 µg dsRNA of *gfp* (white arrowhead) and RNAi treatment with 1 µg dsRNA of *CpopSAO* (black arrowhead) are shown. (C) GC-chromatogram of secretions on day 5 after treatment; left: injecting 1.0 µg *gfp*-dsRNA resulted in the production of salicylaldehyde (**2**), right: injecting 1.0 µg *CpopSAO*-dsRNA resulted in the accumulation of salicyl alcohol (**1**) and **2**. (D) GC-chromatogram peak-area-based plot of secretions after dsRNA injection of *gfp* and different amounts of *CpopSAO* (n=5). Adapted from Bodemann et al. (2012).

pathway. The larvae have the ability for *de novo* synthesis but at the same time they can sequester 8-hydroxygeraniol-*O*-β-D-glucoside as precursor for the production of deterrent monoterpenes if present in the diet (Feld et al. 2001, Kuhn et al. 2004, Soe et al. 2004, Kuhn et al. 2007, Kunert et al. 2008). This was proven by experiments including ^{13}C-labelling of the terpenoids in the food plant, larval feeding on leaves impregnated with analogs and labelled precursors of 8-hydroxygeraniol-*O*-β-D-glucoside, and injection of the precursors into the hemocoel followed by mass spectroscopic analysis of their distribution in the hemolymph, defensive secretion, and feces. Not only glucosides but also agluca can pass membranes and reach the hemocoel of the insects. To minimize damage by uncontrolled release of aglycons the compounds are re-glucosylated in the fat body as demonstrated by the injection of labelled d_5-Ger-8-OH (Kunert et al. 2008). Accordingly, the larvae are able to detoxify putatively harmful agluca by selective glucosylation which is also important for im- and export. Hence, the deeper knowledge of the iridoid biosynthesis developed by ancestral species is a prerequisite to understand how the Chrysomelina beetles developed sequestration later in their evolution.

The iridoid biosynthesis starts in the fat body tissue from the mevalonate pathway and proceeds *via* the intermediates of the terpene metabolism, isopentenyl diphosphate and dimethylallyl diphosphate, to geranyl diphosphate (GDP). GDP is converted into 8-hydroxygeraniol most likely by a ω-hydroxylation followed by glucosylation to obtain 8-hydroxygeraniol-*O*-β-D-glucoside (Fig. 5) (Daloze and Pasteels 1994, Veith et al. 1997, Burse et al. 2009). These early steps of the iridoid production take place in the fat body tissue from where the glucosides are transferred via a hemolymph passage into the defensive glands for further conversion (Burse et al. 2007). To date two enzymes have been identified in this part of the pathway: the 3-hydroxy-3-methylglutaryl-CoA reductase (HMGR) and the isoprenyl diphosphate synthase (IDS) whose properties are described below in more detail.

HMGR is the key regulatory enzyme of the mevalonate pathway (Goldstein and Brown 1990, Friesen and Rodwell 2004). The enzyme utilizes two molecules of NADPH to mediate the four-electron reduction of HMG-CoA to the carboxylic acid mevalonate. Its activity can be attenuated in enzyme assays by 8-hydroxygeraniol whereas no effect has been observed by addition of the glucoside or geraniol (Burse et al. 2008). Iridoid producing larvae are potentially able to sequester glucosidically bound 8-hydroxygeraniol which after cleavage of the sugar moiety results in 8-hydroxygeraniol. Therefore, HMGR may represent a regulator in maintenance of homeostasis between *de novo* produced and sequestered intermediates of iridoid metabolism.

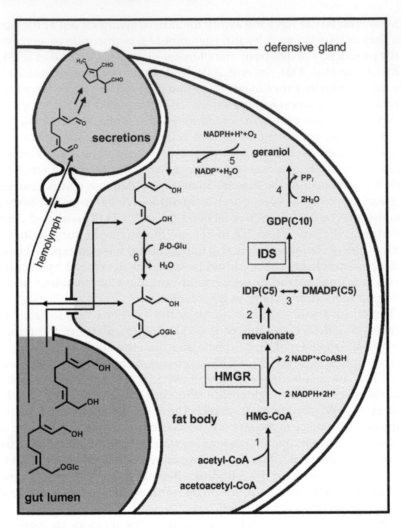

Figure 5. Biosynthesis of deterrent iridoids in the larvae of *P. cochleariae*. 1, 3-hydroxy-3-methylglutaryl CoA synthetase; HMGR, 3-hydroxy-3-methylglutaryl-CoA reductase; 2,mevalonate kinase (phosphomevalonate kinase, diphosphomevalonate decarboxylase);3,isopentenyl-diphosphate Δ-isomerase; IDS, isoprenyldiphosphate synthase; 4, phosphatase; 5, cytochrome P-450 mixed-function oxygenase; 6, β-glucosidase. Adapted from Pasteels et al. (1990) and Burse et al. (2007).

The second characterized enzyme, the IDSs catalyze the alkylation of the homoallylic isopentenyl diphosphate (C_5-IDP) by the allylic dimethylallyl diphosphate (C_5-DMADP) resulting in geranyl diphosphate (GDP), the ubiquitous C_{10}-building block of many monoterpenes (Wang and Ohnuma 2000, Liang et al. 2002, Vandermoten et al. 2009). This step can be controlled by the presence of divalent cations as it has been shown by means of the IDS1 identified from *P. cochleariae* (Frick et al. 2013). The enzyme *Pc*IDS1 produces 96 percent C_{10}-geranyl diphosphate (GDP) and only 4 percent C_{15}-farnesyl diphosphate (FDP) in the presence of Co^{2+} or Mn^{2+} as co-factor, whereas it yields only 18 percent C_{10} GDP, but 82 percent C_{15} FDP in the presence of Mg^{2+}. As a GDP synthase, *Pc*IDS1 could be associated with the formation of monoterpenes, such as iridoids; as FDP synthase with the formation of sesquiterpenes, such as juvenile hormones.

RNAi targeting *PcIDS1* revealed the participation of this enzyme in the *de novo* synthesis of defensive monoterpenoids in the beetle larvae. In larvae challenged by *PcIDS1*-dsRNA, the relative weight of defensive secretions decreased (by 52 percent; $p<0.001$). Detailed analyses of the relative amount of chrysomelidial per larva revealed a significant decline of this iridoid (by 78 percent; $p<0.001$) in *PcIDS1*-treated larvae compared to the controls after five days; this decline proceeded until the complete loss of secretions.

To determine if the chrysomelidial reduction is correlated with a decrease of the precursor, 8-hydroxygeraniol-*O*-β-D-glucoside, we analyzed the hemolymph and fat body tissue of larvae treated with *PcIDS1*dsRNA and controls. Seven days after *PcIDS1*-dsRNA injection, the level of precursor in the hemolymph was significantly reduced (by 89 percent; $p<0.001$) compared to corresponding controls. This effect continued further with a reduction of 97 percent on day 11. A similar effect was observed in the fat body tissue where the amount of 8-hydroxygeraniol-*O*-β-D-glucoside was diminished (by 64.5 ± 14.08 percent) after seven days. In addition to the reduction of chrysomelidial and 8-hydroxygeraniol-*O*-β-D-glucoside, we also observed a significant loss of the overall IDS activity (by 93 percent; $p<0.001$) in the fat body tissue of *Pc*IDS1-silenced larvae seven days after injection compared to the controls.

After translocation of 8-hydroxygeraniol-*O*-β-D-glucoside into the defensive glands, its final transformation into iridoids occurs in the secretions. This transformation involves the hydrolysis of the glucoside and oxidation of the two primary hydroxy groups to produce dialdehyde 8-oxogeranial which is followed by a cyclisation (Fig. 2) (Veith et al. 1994, Oldham et al. 1996, Kunert et al. 2013).

LC-MS analysis of the proteome in the secretions identified putative glucosidases from the transcript catalogue of *P. cochleariae*. Subsequent silencing of these selected candidates revealed a ß-glucosidase belonging to the glycoside hydrolase family 1 whose knocking down resulted in the accumulation of 8-hydroxygeraniol-*O*-β-D-glucoside indicating implication of this protein in the hydrolysis of the glucoside (Häger 2013). With few exceptions ß-glucosidases are known to possess generally a broad substrate spectrum for β-D-glucosides. The determination of the kinetic parameters of the heterologous expressed enzyme from *P. cochleariae* showed hydroxylation of 8-hydroxygeraniol-*O*-β-D-glucoside (Km = 5.6 mM) and also of salicin (Km = 29.2 mM), the sequestered precursor for salicylaldehyde produced in *C. populi* larvae. Obviously, the intrinsic wide substrate selectivity of the enzyme does not require changes in the catalytic center to allow conversion of different plant derived compounds which may contribute to the development of sequestration in the Chrysomelina evolution.

The following reaction, the oxidation is known to be catalysed from SAOs. However, from the related iridoid producing species an oxidizing enzyme has not been identified as yet. Recently, we found a candidate gene whose silencing resulted in an accumulation of 8-hydroxygeraniol indicating that this precursor is a substrate of the *Pc*8HGO enzyme, which, in turn, catalyses the oxidation to the chrysomelidial biosynthesis intermediate 8-oxogeranial (Kugel 2012). Indeed, enzyme assays with purified recombinant protein revealed conversion of 8-hydroxygeranial to 8-oxogeranial. In addition, the substrate specificity of *Pc*8HGO was tested by incubating the oxidase with salicyl alcohol, the substrate of chrysomeline SAOs. No enzyme-based conversion to salicylaldehyde could be detected, indicating this particular enzyme does not react with salicyl alcohol. Subsequent phylogenetic analysis of chrysomeline GMCs revealed that the oxidase of the iridoid pathway originated from a GMC clade other than that of the SAOs. Thus, the evolution of host independent chemical defense followed by a shift to host dependent chemical defense in Chrysomelina beetles coincided with independent recruitments of genes from different GMC subfamilies. These findings illustrate the importance of the GMC multigene family for the underlying evolutionary dynamics of host-plant affiliation.

Whereas deglucosylation and oxidation reactions are found also in sequestering species, the final cyclization of acyclic dialdehydes to generate iridoids proceeds exclusively in *de novo* producing species. Interestingly, with the same precursor, isotopic tracing studies have shown that there are two mechanistically different cyclisation modes in different leaf beetle groups (Kunert et al. 2013). When deuterium atom labeled [^{2}H$_5$] Ger-8-OH was used for the feeding experiments, the precursor lost a single deuterium atom from C(4) in *Phaedon cochleariae*, *Hydrothassa marginella* and *Phratora vulgatissima*. In contrast, in *Gastrophysa cyanea*, *G. polygoni*, *G. atrocyanea* and *G. viridula*, [^{2}H$_5$]Ger-8-OH was observed to exchange all three deuterium atoms from the methyl group at C (3). Moreover, the absolute configuration and optical purity of chrysomelidial secreted by different families was determined by GS-MS. Curiously, except for those in *H. marginella* and *P. vulgatissima* which are (5*S*,8*S*)-chrysomelidial, secretions in *P. cochleariae* and all investigated members of the genus *Gastrophysa* contain (5R,8R)-chrysomelidial.

Since proteins responsible for the cyclization have not been identified before, we used LC-MS of the proteome in the secretions of larval *P. cochleariae* to select candidates for this reaction. Subsequent RNAi targeting the expression of the selected proteins revealed a protein, named *Pc*To-like, associated with the cyclization monoterpene precursors into iridoids (Bodemann et al. 2012). GC-MS analyses of the composition of low molecular weight compounds in the secretions 5 days after dsRNA injection of *PcTo-like* showed clearly the accumulation of 8-oxogeranial in addition to a decrease of chrysomelidial. *Pc*-To-like could be classified as a member of the juvenile hormone binding protein superfamily which comprises ligand binding proteins for juvenile hormones or similar hydrophobic terpenoids. The detailed biochemical characterization of *Pc*To-like *in vitro* is in progress. Our studies of the iridoid metabolism illustrate the value of RNAi to disentangle biosynthetic pathways.

Introduction of RNA Interference to Target the Membrane Transport Proteins Involved in the Sequestration of Phytochemicals of Juvenile Chrysomelina

Transport proteins take up a central position when investigating evolutionary and ecological consequences of sequestration in insects. They control, on one hand, the uptake and excretion of plant derived metabolites from the gut lumen into and out of the hemocoel. On the other hand, they take up a key position in the manipulation of deterrents' diversity in the secretions. The composition of the defensive secretions, however, has an effect on different trophic levels. Secretions can be either beneficial for the insect to affiliate enemy and competitor-free space or costly if these secretions increase the insects' attractiveness to parasites or predators. Therefore, transport proteins have an impact on the diversification not only of plants and phytophagous insects but of all the interaction partners in an ecological network. Based on this importance we have started to study these integral membrane proteins and to apply RNAi to study their function in the sequestration process in juvenile Chrysomelina.

Recently, we described the silencing of an ATP binding cassette (ABC) transporter, herein after referred to as *Cp*MRP, in the defensive glands of the poplar leaf beetle, *C. populi* (Strauss et al. 2013). The transporter is localized inside the secretory cell in intracellular membranes forming vesicular storage compartments. To link the protein with a function in the sequestration process of salicin we have carried out RNAi silencing in juvenile poplar leaf beetles. By comparing developmental traits among

C. populi larvae after injecting *Cpmrp*-dsRNA or *gfp*-dsRNA as a control, we found that the silencing of *Cpmrp* had no influence on larval growth (Fig. 6A).

About 10 days post-injection, however, the *CpMRP* knock-down larvae completely lost their ability to respond to stimulation with droplets of defensive secretion (Fig. 6B). The secretions began to diminish at day 8. On the basis of transcript abundance *CpMRP* mRNA was reduced to a basal level of 15–20 percent within 2–3 days and persisted until the larvae pupated (Fig. 6C). Figure 6D summarizes the immunohistochemical analysis of *CpMRP* expression in the secretory cells. At day 3 post-injection, both the *CpMRP* knocked-down and control secretory cells displayed a similar pattern of *Cp*MRP distribution (Fig. 6D a,b). At later time points, however, the *Cp*MRP expression in the RNAi group was strongly reduced in comparison to the *gfp* control (Fig. 6D c-f). *Cp*MRP decayed exponentially to a relatively low basal level with a half-life of about 1 day, suggesting a degradation of *Cp*MRP that is linear proportional to its concentration. Our results demonstrate that knocking down of *Cp*MRP created a depletion of membrane protein together with a defenseless phenotype lacking defensive secretions, which indicated its key function for the sequestration of salicin, the plant derived precursor for the deterrent salicylaldehyde.

As ATP is needed to fuel the ABC transporter, but also the cellular metabolism in the defensive glands, sugars have to be delivered by sugar transporters to drive ATP production in this tissue. Therefore, we assumed sugar transporters to be essential for the production of defensive secretions (Augustin 2010, Wright et al. 2011, Mueckler and Thorens 2013). In the study by Stock et al. (2013), we delineated the identification of six candidate sequences encoding members of the facilitated sugar transporter family SLC2 with high expression in the defensive glands. Silencing by RNAi resulted in two cases in a reduced production of deterrents suggesting implication of these transporters in the production of defensive secretions. In contrast to the silencing effect of the ABC transporter in the defensive glands, which caused a total loss of defensive secretions (Strauss et al. 2013), RNAi targeting SLC2 transporters could not shut down the production of defensive exudates completely. This may be due to the turn-over rate of the integral-membrane proteins which may diminish the silencing effect or other transporters may be expressed to take over the function to achieve homeostasis in the tissue. To test the latter hypothesis, we have gathered RNA-seq data from larvae silenced in SLC2 members exhibiting normally a high expression in the defensive glands (Stock et al. 2013). We have analyzed the differential SLC2 transporter expression in samples obtained from SLC2 members silenced larvae compared to samples from *gfp*-treated larvae. The data

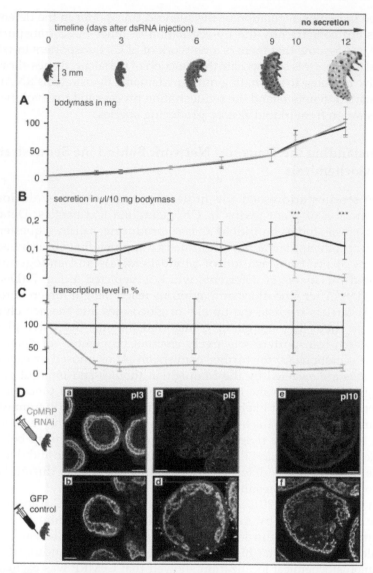

Figure 6. Timeline of different *Cp*MRP knockdown effects in *C. populi* larvae. *Cp*MRP knockdown effects were colored in gray and the effects of *gfp*-injected control larvae in black. (A) larval body mass was not influenced by *CpMRP* knockdown (n > 10, mean ± SD). (B) *CpMRP* knockdown larvae lack defensive secretions 10 days after dsRNA injection (n > 10, mean ± SD). (C) transcriptional level of *CpMRP* inside the glands (each time point contains n = 3 (biological replicates), mean ± SD). (D) intracellular localization of *Cp*MRP and decrease of the protein level after dsRNA injection. Scale bars, 20 μm; plx = x days post dsRNA-injection. Asterisks represent significant differences in *CpMRP*-silenced larvae compared to *gfp*-injected control larvae (*p≤0.05, ***p≤0.001). Adapted from Strauss et al. (2013).

showed the down-regulation of the silenced transporter in the defensive glands but concurrently the up-regulation of other SLC2 transporters. Hence, we assume that there is a network of SLC2 transporters in which several transporters compensate the function of the silenced ones allowing adaptive response to alternating nutrient demand. By using this RNAi, we have started to understand the sequestration process and the evolution of sequestration from iridoid *de novo* producing species.

Understanding the Transport Network Behind the Sequestration of Phytochemicals

Recent studies addressed the molecular basis of the evolution of sequestration of plant toxins in Chrysomelina leaf beetles. Detailed physiological studies on iridoid *de novo* producing, salicin sequestering, and ester producing larvae have indicated a complex influx-efflux transport network of the translocation of phytochemicals and endogenously produced precursors of deterrents which is depicted in Fig. 7 (Discher et al. 2009). After glucosides and aglucons reach the gut lumen with the food (1) carriers mediate the uptake of glucosides into the hemolymph (2). From there they are either transferred into the defensive glands, from which transporters selectively channel progenitors of deterrents into the secretions (3) for further conversion (4) or unused or excessive glucosides are excreted by the Malpighian tubules and hindgut organs (5). Agluca may reach the hemocoel by diffusion (6) and precursors of the deterrents are most likely glucosylated in the fat body (7) from where they are channeled *via* the hemolymph either into the defensive secretion (3) or to the excretory tissue (5). Most likely, the relationship between uptake into the gland and the excretion process is dynamic and depends on the capacity of the defensive system; this can change during larval development and as a result of disturbance by predators.

In Strauss et al. (2013), we describe the role of ABC transporters in the sequestration by means of *Cp*MRP identified from *C. populi*. Screening of expression levels of *Cp*MRP in different larval tissues revealed an exceptional high transcript level in the glandular tissue. Immunohistochemical localization showed that *Cp*MRP was exclusively localized inside the secretory cells in storage vesicles. There it dictates (pacemaker function) the transport rate of a still unknown selective transporter for salicin in the plasma membrane by a constant accumulation of salicin in these intracellular vesicles. The vesicles are targeted by exocytosis to the reservoir where the enzymatic conversion of salicin to salicylaldehyde takes place.

Transport studies in *Xenopus laevis* oocytes revealed *Cp*MRP as a transporter for salicin, the naturally sequestered host-plant precursor of

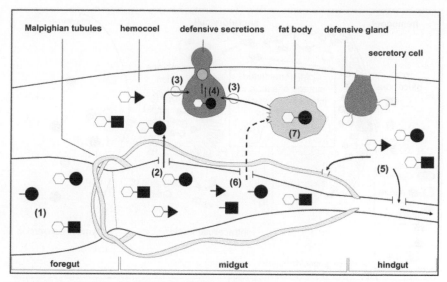

Figure 7. A network of transport processes is implicated in the sequestration of secondary metabolites from the host plant.

C. populi. In order to test if *Cp*MRP has substrate prevalence for salicin we carried out comparative transport assays. Among plant precursor and non-precursor glucosides an adequate selection of glucosides was chosen. We applied an equimolar mixture of salicin, 8-hydroxygeraniol-*O*-β-D-glucoside, the early precursor of the iridoid pathway found in *P. cochleariae* and phenylethyl-*S*-glucoside which represents a substrate mimicking an *O*-glucoside sequestered by *C. lapponica*. This mixture was already tested in feeding and hemolymph injection experiments on *C. populi* and revealed the specific transport of salicin into the reservoir (Discher et al. 2009). However, *Cp*MRP did not discriminate between the substrates indicating broader substrate selectivity of the ABC transporter.

Transporter sequences highly similar to *Cp*MRP have been identified in the larval glands of other Chrysomelina species, such as *P. cochleariae* (86 percent amino acid identity to *Cp*MRP) and *C. lapponica* (93 percent amino acid identity to *Cp*MRP). Both sequences share the same transcription pattern like *Cp*MRP being highly expressed in the larval glands. The involvement of broad-spectrum ABC transporters in the sequestration of plant-derived metabolites seems to be a mechanism common in the taxon Chrysomelina. In addition, RNA sequencing data of the tissue of the red flour beetle, *Tribolium castanaeum*, which possesses two pairs of odiferous or stink glands for chemical defense showed high expression of transcripts encoding putative ABC transporters of the ABCC family in the glands compared to control tissue (Li et al. 2013). RNAi

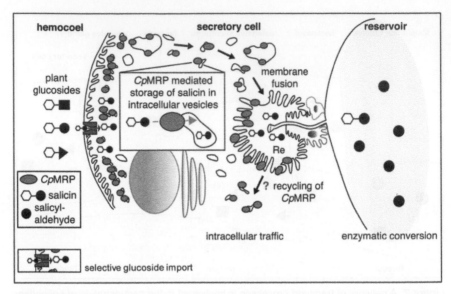

Figure 8. Suggested transportation route of phytochemicals from the larval hemolymph into the defensive glandular reservoir through a secretory cell. Deterrent precursors are filtered (e.g., salicin, in the case of *C. populi*) at the hemolymph-exposed plasma membrane of the secretory cell, which represents a bottleneck for controlling the metabolic diversity in the secretions and then are trapped by the ABC transporter in storage vesicles as soon as they enter the secretory cells. The apical part facing the lumen of the gland is a brush border membrane where storage vesicles are secreted (exocytosis) into the extracellular room which is connected with the reservoir by the canal. Adapted from Strauss et al. (2013).

of one of these ABC transporters resulted in a phenotype impaired in the production of defensive compounds. A hypothesis can be advanced that ABC transporters are more widespread for translocation processes in defensive glands of insects. While such a hypothesis is necessarily speculative, it may contribute to developing models for how insects set up chemical defense and how they adapt to changing biotic and abiotic conditions in ecological networks.

Further, we correlated the function of predicted sugar transporters of the SLC2 family with the production of defensive secretion in the defensive glands. Their function seems to be important for the sugar supply of the glandular cells. However, the substrate spectrum of the SLC2 transporters needs to be determined *in vitro*. Our preliminary results of differential expression of putative ABC transporters in different tissues of juvenile *C. populi* revealed that these proteins occur in all tested tissues. In particular, in Malpighian tubules we have identified high expression levels (≥ 50 reads per base) of 14 ABC transporters which cluster into the ABCC, B, and G subfamilies containing multi-drug resistance proteins (Sarkadi et al. 1992, Gottesman et al. 2002, Holland 2011, Slot et al. 2011). Therefore we consider ABC transporters to be involved in the non-selective excretion of phytochemicals. Detailed analyses are still in progress.

From our results we conclude that the sequestration of Chrysomelina larvae is the result of the presence of several barriers with various degrees of selectivity: (*i*) those controlling the non-selective uptake of phenolglucosides from gut lumen into the hemolymph and excretion by Malpighian tubules (most likely involving ABC transporters), (*ii*) those controlling the selective transfer from the hemolymph into the secretory cells, and (*iii*) those where the broad-spectrum ABC transporter shuttles pre-filtered metabolites from inside the secretory cell into the defensive secretions.

The Evolution of Sequestration from *de novo* Synthesis

The uniform architecture and morphology and the enzymes of the defensive system in Chrysomelina larvae lead us to expect functional similarities on the molecular level of transport proteins as well. In the ancestral species *P. cochleariae*, we have identified a sequence with 86 percent amino acid identity to *Cp*MRP which is highly expressed in the larval glands (Strauss et al. 2013). Transport coupled with enzymatic activity of glucosidases and oxidases in the defensive secretions shows that already in the iridoid *de novo* producers the basic principles are present for evolutionary transition to sequestration. Hence, although the use of plant compounds as anti-predator defenses appears to be a spectacular evolutionary innovation, it requires only a few modifications from ancestral processes.

Although salicin sequestration makes the insect tightly dependent on the chemistry of the host plant, it did not prevent a Chrysomelina lineage from escaping this subordination by a development of a chemical defense potentially compatible with a large number of new host-plant associations. Some species in the monophyletic *interrupta*-group (Fig. 3C) evolved the ability to sequester in addition to salicin a variety of glucosylated plant-alcohols which are esterified with endogenous metabolites yielding a complex mixture of esters and aldehyde in the secretions. It has been proposed that the evolution of ester biosynthesis allowed species such as *C. lapponica* to shift from salicin-rich Salicacae to the salicin-depleted plant family of Betulaceae into an enemy-free space along a "phytochemical bridge" seminally postulated by Ehrlich and Raven (Ehrlich and Raven 1964). We have identified sequences highly identical (96 percent at the amino acid level) to *Cp*MRP in the larval glands of willow- and birch-feeding populations of *C. lapponica* (Strauss et al. 2013). The amino acid sequences between the two populations were almost identical (99 percent), which suggests that willow-feeders are already pre-adapted to overcome the chemical constraints of another host plant. Consequently, the involvement of broad-spectrum ABC transporters into the sequestration of plant derived metabolites may contribute to a dynamic host plant use by insects.

Final Thoughts

Taken our results together, the arrangement of non-selective and selective transport processes as well as enzymatic reactions in the synthesis of defensive compounds creates a flexible base to build on a host plant dependent chemical defense not leading into an evolutionary dead-end. From our data it is clear that only few selective barriers have to be adjusted to variations in the chemical compositions of hosts that Chrysomelina beetles may encounter, for example, during host plant shifts which can have large-scale consequences for the networking organisms in an ecosystem.

Acknowledgments

We wish to thank Prof. Jacques M. Pasteels for helpful discussions on aspects of this work and Kerstin Ploss for critical reading of the manuscript. This work has been supported by the German Research Funding Agency (BU 1862/2-1) and the Max Planck Society.

Keywords: RNA interference, transcriptome, secretome, leaf beetles, Chrysomelina, evolution, plant-insect interaction, chemical defense, defensive glands, terpenoids, iridoid biosynthesis, secondary metabolites, sequestration, transport proteins, ABC transporter

References

Agrawal, A.A., A.P. Hastings, M.T.J. Johnson, J.L. Maron and J.P. Salminen. 2012. Insect herbivores drive real-time ecological and evolutionary change in plant populations. Science 338: 113–116.

Augustin, R. 2010. The protein family of glucose transport facilitators: It's not only about glucose after all. Iubmb Life 62: 315–333.

Becerra, J.X. 1997. Insects on plants: Macroevolutionary chemical trends in host use. Science 276: 253–256.

Belles, X. 2010. Beyond *Drosophila*: RNAi *in vivo* and functional genomics in insects. Annu. Rev. Entomol. 55: 111–128.

Blum, M.S. 1994. Antipredatory devices in larvae of the Chrysomelidae: a unidentified synthesis for defensive electicism. pp. 277–288. *In*: P.H. Jolivet, M.L. Cox and E. Petitpierre (eds.). Novel Aspects of the Biology of the Chrysomelidae. Kluwer Academics Publishers, Dordrecht.

Blum, M.S., J.B. Wallace, R.M. Duffield, J.M. Brand, H.M. Fales and E.A. Sokoloski. 1978. Chrysomelidial in the defensive secretion of the leaf beetle *Gastrophysa cyanea*. J. Chem. Ecol. 4: 47–53.

Bodemann, R.R., P. Rahfeld, M. Stock, M. Kunert, N. Wielsch, M. Groth, S. Frick, W. Boland and A. Burse. 2012. Precise RNAi-mediated silencing of metabolically active proteins in the defence secretions of juvenile leaf beetles. Proc. R. Soc. Lond., Ser. B: Biol. Sci. 279: 4126–4134.

Boeckler, G.A., J. Gershenzon and S.B. Unsicker. 2011. Phenolic glycosides of the Salicaceae and their role as anti-herbivore defenses. Phytochemistry 72: 1497–1509.

Brueckmann, M., A. Termonia, J.M. Pasteels and T. Hartmann. 2002. Characterization of an extracellular salicyl alcohol oxidase from larval defensive secretions of *Chrysomela populi* and *Phratora vitellinae* (Chrysomelina). Insect Biochem. Mol. Biol. 32: 1517–1523.

Burse, A., S. Frick, S. Discher, K. Tolzin-Banasch, R. Kirsch, A. Strauss, M. Kunert and W. Boland. 2009. Always being well prepared for defense: The production of deterrents by juvenile Chrysomelina beetles (Chrysomelidae). Phytochemistry 70: 1899–1909.

Burse, A., S. Frick, A. Schmidt, R. Buechler, M. Kunert, J. Gershenzon, W.G. Brandt and W. Boland. 2008. Implication of HMGR in homeostasis of sequestered and *de novo* produced precursors of the iridoid biosynthesis in leaf beetle larvae. Insect Biochem. Mol. Biol. 38: 76–88.

Burse, A., A. Schmidt, S. Frick, J. Kuhn, J. Gershenzon and W. Boland. 2007. Iridoid biosynthesis in Chrysomelina larvae: Fat body produces early terpenoid precursors. Insect Biochem. Mol. Biol. 37: 255–265.

Cavener, D.R. 1992. GMC oxidoreductases-a newly defined family of homologous proteins with diverse catalytic activities. J. Mol. Biol. 223: 811–814.

Daloze, D. and J. Pasteels. 1994. Isolation of 8-hydroxygeraniol-8-O-β-D-glucoside, a probable intermediate in biosynthesis of iridoid monoterpenes, from defensive secretions of *Plagiodera versicolora* and *Gastrophysa viridula* (Coleoptera: Chrysomelidae). J. Chem. Ecol. 20: 2089–2097.

Ding, S.W. 2010. RNA-based antiviral immunity. Nat. Rev. Immunol. 10: 632–644.

Discher, S., A. Burse, K. Tolzin-Banasch, S.H. Heinemann, J.M. Pasteels and W. Boland. 2009. A versatile transport network for sequestering and excreting plant glycosides in leaf beetles provides an evolutionary flexible defense strategy. Chem. Bio. Chem. 10: 2223–2229.

Dobler, S., S. Dalla, V. Wagschal and A.A. Agrawal. 2012. Community-wide convergent evolution in insect adaptation to toxic cardenolides by substitutions in the Na,K-ATPase. Proc. Natl. Acad. Sci. USA 109: 13040–13045.

Duffey, S.S. 1980. Sequestration of plant natural-products by insects. Annu. Rev. Entomol. 25: 447–477.

Ehrlich, P.R. and P.H. Raven. 1964. Butterflies and plants-a study in coevolution. Evolution 18: 586–608.

Farrell, B.D. 1998. Inordinate fondness explained-why are there so many beetles. Science 281: 555–559.

Farrell, B.D. and A.S. Sequeira. 2004. Evolutionary rates in the adaptive radiation of beetles on plants. Evolution 58: 1984–2001.

Feld, B.K., J.M. Pasteels and W. Boland. 2001. *Phaedon cochleariae* and *Gastrophysa viridula* (Coleoptera:Chrysomelidae) produce defensive iridoid monoterpenes *de novo* and are able to sequester glycosidically bound terpenoid precursors. Chemoecology 11: 191–198.

Fernandez, P. and M. Hilker. 2007. Host plant location by Chrysomelidae. Basic Appl. Ecol. 8: 97–116.

Fire, A., S.Q. Xu, M.K. Montgomery, S.A. Kostas, S.E. Driver and C.C. Mello. 1998. Potent and specific genetic interference by double-stranded RNA in *Caenorhabditis elegans*. Nature 391: 806–811.

Frick, S., R. Nagel, A. Schmidt, R.R. Bodemann, P. Rahfeld, G. Pauls, W. Brandt, J. Gershenzon, W. Boland and A. Burse. 2013. Metal ions control product specificity of isoprenyl diphosphate synthases in the insect terpenoid pathway. Proc. Natl. Acad. Sci. USA 110: 4194–4199.

Friesen, J. and V. Rodwell. 2004. The 3-hydroxy-3-methylglutaryl coenzyme-A (HMG-CoA) reductases. Genome Biol. 5: 248–248.247.

Fürstenberg-Hägg, J., M. Zagrobelny and S. Bak. 2013. Plant defense against insect herbivores. Int. J. Mol. Sci. 14: 10242–10297.

Gillespie, J.J., K.M. Kjer, C.N. Duckett and D.W. Tallamy. 2003. Convergent evolution of cucurbitacin feeding in spatially isolated rootworm taxa (Coleoptera: Chrysomelidae; Galerucinae, Luperini). Mol. Phylogenet. Evol. 29: 161–175.

Goldstein, J.L. and M.S. Brown. 1990. Regulation of the mevalonate pathway. Nature 343: 425–430.

Gómez-Zurita, J., T. Hunt, F. Kopliku and A.P. Vogler. 2007. Recalibrated tree of leaf beetles (Chrysomelidae) indicates independent diversification of angiosperms and their insect herbivores. PloS One 2: e360.

Gottesman, M.M., T. Fojo and S.E. Bates. 2002. Multidrug resistance in cancer: Role of ATP-dependent transporters. Nature Rev. Cancer 2: 48–58.

Gross, J., L. Podsiadlowski and M. Hilker. 2002. Antimicrobial activity of exocrine glandular secretion of *Chrysomela* larvae. J. Chem. Ecol. 28: 317–331.

Häger, W. 2013. Charakterisierung von β-Glucosidasen aus dem Wehrsekret der juvenilen Blattkäfer *Phaedon cochleariae* und *Chrysomela populi*. Bachelor Thesis, Friedrich Schiller University Jena, Jena.

Hare, J.D. 2012. How insect herbivores drive the evolution of plants. Science 338: 50–51.

Hartmann, T. 2004. Plant-derived secondary metabolites as defensive chemicals in herbivorous insects: a case study in chemical ecology. Planta 219: 1–4.

Hilker, M. and S. Schulz. 1994. Composition of larval secretion of *Chrysomela lapponica* (Coleoptera, Chrysomelidae) and its dependence on host plant. J. Chem. Ecol. 20: 1075–1093.

Hinton, H.E. 1951. On a little-known protective device of some chrysomelid pupae (Coleoptera). Proc. R. Soc. Lond. Ser. A 26: 67–73.

Holland, I.B. 2011. ABC transporters, mechanisms and biology: an overview. pp. 1–17. In: F.J. Sharom (ed.). Essays in Biochemistry: ABC Transporters. Portland Press Ltd., London, UK.

Ibanez, S., C. Gallet and L. Despres. 2012. Plant insecticidal toxins in ecological networks. Toxins 4: 228–243.

Kirsch, R., H. Vogel, A. Muck, K. Reichwald, J.M. Pasteels and W. Boland. 2011a. Host plant shifts affect a major defense enzyme in *Chrysomela lapponica*. Proc. Natl. Acad. Sci. USA 108: 4897–4901.

Kirsch, R., H. Vogel, A. Muck, A. Vilcinskas, J.M. Pasteels and W. Boland. 2011b. To be or not to be convergent in salicin-based defence in chrysomeline leaf beetle larvae: evidence from *Phratora vitellinae* salicyl alcohol oxidase. Proc. R. Soc. Lond. Ser: B. Biol. Sci. 278: 3225–3232.

Kugel, S. 2012. GMC-Oxidoreduktasen aus Blattkäfern. Bachelor Thesis, Friedrich Schiller University Jena, Jena, Germany.

Kuhn, J., E.M. Pettersson, B.K. Feld, A. Burse, A. Termonia, J.M. Pasteels and W. Boland. 2004. Selective transport systems mediate sequestration of plant glucosides in leaf beetles: A molecular basis for adaptation and evolution. Proc. Natl. Acad. Sci. USA 101: 13808–13813.

Kuhn, J., E.M. Pettersson, B.K. Feld, L. Nie, K. Tolzin-Banasch, S.M. M'Rabet, J. Pasteels and W. Boland. 2007. Sequestration of plant-derived phenolglucosides by larvae of the leaf beetle *Chrysomela lapponica*: Thioglucosides as mechanistic probes. J. Chem. Ecol. 33: 5–24.

Kunert, M., P. Rahfeld, K.H. Shaker, B. Schneider, A. David, K. Dettner, J.M. Pasteels and W. Boland. 2013. Beetles do it differently: Two stereodivergent cyclisation modes in iridoid-producing leaf-beetle larvae. Chem. Bio. Chem. 14: 353–360.

Kunert, M., A. Soe, S. Bartram, S. Discher, K. Tolzin-Banasch, L. Nie, A. David, J. Pasteels and W. Boland. 2008. *De novo* biosynthesis *versus* sequestration: A network of transport systems supports in iridoid producing leaf beetle larvae both modes of defense. Insect Biochem. Mol. Biol. 38: 895–904.

Laurent, P., J.C. Braekman, D. Daloze and J. Pasteels. 2003a. Biosynthesis of defensive compounds from beetles and ants. Eur. J. Org. Chem. 2003: 2733–2743.

Laurent, P., J.C. Braekman and S. Daloze. 2005. Insect chemical defense. pp. 167–229. *In*: S. Schulz (ed.). Chemistry of Pheromones and other Semiochemicals II. Springer-Verlag Berlin, Berlin.

Laurent, P., C. Dooms, J.C. Braekman, D. Daloze, J.L. Habib-Jiwan, R. Rozenberg, A. Termonia and J.M. Pasteels. 2003b. Recycling plant wax constituents for chemical defense: hemi-biosynthesis of triterpene saponins from beta-amyrin in a leaf beetle. Naturwissenschaften 90: 524–527.

Li, J.W., S. Lehmann, B. Weissbecker, I.O. Naharros, S. Schutz, G. Joop and E.A. Wimmer. 2013. Odoriferous defensive stink gland transcriptome to identify novel genes necessary for quinone synthesis in the red flour beetle, *Tribolium castaneum*. PLoS Genet. 9: e1003596.

Li, X.C., M.A. Schuler and M.R. Berenbaum. 2007. Molecular mechanisms of metabolic resistance to synthetic and natural xenobiotics. Annu. Rev. Entomol. 52: 231–253.

Liang, P.H., T.P. Ko and A.H.J. Wang. 2002. Structure, mechanism and function of prenyltransferases. Eur. J. Biochem. 269: 3339–3354.

Macias, F.A., J.L.G. Galindo and J.C.G. Galindo. 2007. Evolution and current status of ecological phytochemistry. Phytochemistry 68: 2917–2936.

Meinwald, J., T.H. Jones, T. Eisner and K. Hicks. 1977. Defense-mechanisms of arthropods: New methylcyclopentanoid terpenes from larval defensive secretion of a chrysomelid beetle (*Plagiodera versicolora*). Proc. Natl. Acad. Sci. USA 74: 2189–2193.

Michalski, C., H. Mohagheghi, M. Nimtz, J.M. Pasteels and D. Ober. 2008. Salicyl alcohol oxidase of the chemical defense secretion of two chrysomelid leaf beetles-Molecular and functional characterization of two new members of the glucose-methanol-choline oxidoreductase gene family. J. Biol. Chem. 283: 19219–11928.

Mito, T., T. Nakamura, T. Bando, H. Ohuchi and S. Noji. 2011. The advent of RNA interference in Entomology. Entomol. Sci. 14: 1–8.

Mueckler, M. and B. Thorens. 2013. The SLC2 (GLUT) family of membrane transporters. Mol. Aspects Med. 34: 121–138.

Nishida, R. 2002. Sequestration of defensive substances from plants by Lepidoptera. Annu. Rev. Entomol. 47: 57–92.

Noirot, C. and A. Quennedey. 1974. Fine-structure of insect epidermal glands. Annu. Rev. Entomol. 19: 61–80.

Ode, P.J. 2006. Plant chemistry and natural enemy fitness: Effects on herbivore and natural enemy interactions. Annu. Rev. Entomol. 51: 163–185.

Oldham, N.J., M. Veith, W. Boland and K. Dettner. 1996. Iridoid monoterpene biosynthesis in insects: Evidence for a *de novo* pathway occurring in the defensive glands of *Phaedon armoraciae* (Chrysomelidae) leaf beetle larvae. Naturwissenschaften 83: 470–473.

Opitz, S.E.W. and C. Mueller. 2009. Plant chemistry and insect sequestration. Chemoecology 19: 117–154.

Pasteels, J.M. 1993. The value of defensive compounds as taxonomic characters in the classification of leaf beetles. Biochem. Syst. Ecol. 21: 135–142.

Pasteels, J.M., D. Daloze and M. Rowell-Rahier. 1986. Chemical defense in chrysomelid eggs and neonate larvae. Physiol. Entomol. 11: 29–38.

Pasteels, J.M., S. Duffey and M. Rowell-Rahier. 1990. Toxins in chrysomelid beetles possible evolutionary sequence from *de novo* synthesis to derivation from food-plant chemicals. J. Chem. Ecol. 16: 211–222.

Pasteels, J.M. and M. Rowell-Rahier. 1989. Defensive glands and secretions as taxonomical tools in the Chrysomelidae. Entomography 6: 423–432.

Pasteels, J.M., M. Rowell-Rahier, J.C. Braekman and A. Dupont. 1983. Salicin from host plant as precursor of salicyl aldehyde in defensive secretion of chrysomeline larvae. Physiol. Entomol. 8: 307–314.

Renner, K. 1970. Über die ausstülpbaren Hautblasen der Larven von *Gastroidea viridula* De Geer und ihre ökologische Bedeutung (Coleoptera: Chrysomelidae). Beitr. Entomol. 20: 527–533.

Rowell-Rahier, M. and J.M. Pasteels. 1986. Economics of chemical defense in Chrysomelinae. J. Chem. Ecol. 12: 1189–1203.

Sarkadi, B., E.M. Price, R.C. Boucher, U.A. Germann and G.A. Scarborough. 1992. Expression of the human multidrug resistance cDNA in insect cells generates a high-activity drug-stimulated membrane ATPase. J. Biol. Chem. 267: 4854–4858.

Schulz, S. 1998. Insect-plant interactions-Metabolism of plant compounds to pheromones and allomones by Lepidoptera and leaf beetles. Eur. J. Org. Chem. 1: 13–20.

Schulz, S., J. Gross and M. Hilker. 1997. Origin of the defensive secretion of the leaf beetle *Chrysomela lapponica*. Tetrahedron 53: 9203–9212.

Slot, A.J., S.V. Molinski and S.P.C. Cole. 2011. Mammalian multidrug-resistance proteins (MRPs). pp. 179–207. *In*: F.J. Sharom (ed.). Essays in Biochemistry: ABC Transporters. Portland Press Ltd., London, UK.

Soe, A.R.B., S. Bartram, N. Gatto and W. Boland. 2004. Are iridoids in leaf beetle larvae synthesized *de novo* or derived from plant precursors? A methodological approach. Isotopes Environ. Health Stud. 40: 175–180.

Soetens, P., J.M. Pasteels and D. Daloze. 1993. A simple method for *in vivo* testing of glandular enzymatic activity on potential precursors of larval defensive compounds in *Phratora* species (Coleoptera: Chrysomelinae). Experientia 49: 1024–1026.

Sorensen, J.S. and M.D. Dearing. 2006. Efflux transporters as a novel herbivore countermechanism to plant chemical defenses. J. Chem. Ecol. 32: 1181–1196.

Stock, M., R.R. Gretscher, M. Groth, S. Eiserloh, W. Boland and A. Burse. 2013. Putative sugar transporters of the mustard leaf beetle *Phaedon cochleariae*: Their phylogeny and role for nutrient supply in larval defensive glands. PloS One 8: e84461.

Strauss, A.S., S. Peters, W. Boland and A. Burse. 2013. ABC transporter functions as a pacemaker for sequestration of plant glucosides in leaf beetles. eLife 2: e01096.

Terenius, O., A. Papanicolaou, J.S. Garbutt, I. Eleftherianos, H. Huvenne, S. Kanginakudru, M. Albrechtsen, C.J. An, J.L. Aymeric, A. Barthel, P. Bebas, K. Bitra, A. Bravo, F. Chevalieri, D.P. Collinge, C.M. Crava, R.A. de Maagd, B. Duvic, M. Erlandson, I. Faye, G. Felfoldi, H. Fujiwara, R. Futahashi, A.S. Gandhe, H.S. Gatehouse, L.N. Gatehouse, J.M. Giebultowicz, I. Gomez, C.J.P. Grimmelikhuijzen, A.T. Groot, F. Hauser, D.G. Heckel, D.D. Hegedus, S. Hrycaj, L.H. Huang, J.J. Hull, K. Iatrou, M. Iga, M.R. Kanost, J. Kotwica, C.Y. Li, J.H. Li, J.S. Liu, M. Lundmark, S. Matsumoto, M. Meyering-Vos, P.J. Millichap, A. Monteiro, N. Mrinal, T. Niimi, D. Nowara, A. Ohnishi, V. Oostra, K. Ozaki, M. Papakonstantinou, A. Popadic, M.V. Rajam, S. Saenko, R.M. Simpson, M. Soberon, M.R. Strand, S. Tomita, U. Toprak, P. Wang, C.W. Wee, S. Whyard, W.Q. Zhang, J. Nagaraju, R.H. Ffrench-Constant, S. Herrero, K. Gordon, L. Swelters and G. Smagghe. 2011. RNA interference in Lepidoptera: An overview of successful and unsuccessful studies and implications for experimental design. J. Insect Physiol. 57: 231–245.

Termonia, A., T.H. Hsiao, J.M. Pasteels and M.C. Milinkovitch. 2001. Feeding specialization and host-derived chemical defense in Chrysomeline leaf beetles did not lead to an evolutionary dead end. Proc. Natl. Acad. Sci. USA 98: 3909–3914.

Termonia, A. and J.M. Pasteels. 1999. Larval chemical defence and evolution of host shifts in *Chrysomela* leaf beetles. Chemoecology 9: 13–23.

Tolzin-Banasch, K., E. Dagvadorj, U. Sammer, M. Kunert, R. Kirsch, K. Ploss, J.M. Pasteels and W. Boland. 2011. Glucose and glucose esters in the larval secretion of *Chrysomela lapponica*; selectivity of the glucoside import system from host plant leaves. J. Chem. Ecol. 37: 195–204.

Vandermoten, S., E. Haubruge and M. Cusson. 2009. New insights into short-chain prenyltransferases: structural features, evolutionary history and potential for selective inhibition. Cell. Mol. Life Sci. 66: 3685–3695.

Veith, M., M. Lorenz, W. Boland, H. Simon and K. Dettner. 1994. Biosynthesis of iridoid monoterpenes in insects: defensive secretions from larvae of leaf beetles (Coleoptera: Chrysomelidae). Tetrahedron 50: 6859–6874.

Veith, M., N.J. Oldham, K. Dettner, J.M. Pasteels and W. Boland. 1997. Biosynthesis of defensive allomones in leaf beetle larvae: Stereochemistry of salicylalcohol oxidation in *Phratora vitellinae* and comparison of enzyme substrate and stereospecificity with alcohol oxidases from several iridoid producing leaf beetles. J. Chem. Ecol. 23: 429–443.

Wallace, J.B. and M.S. Blum. 1969. Refined defensive mechanisms in *Chrysomela scripta*. Ann. Entomol. Soc. Am. 62: 503–506.

Wang, K.C. and S. Ohnuma. 2000. Isoprenyl diphosphate synthases. BBA-Mol. Cell Biol. L. 1529: 33–48.

Willinger, G. and S. Dobler. 2001. Selective sequestration of iridoid glycosides from their host plants in *Longitarsus* flea beetles. Biochem. Syst. Ecol. 29: 335–346.

Winde, I. and U. Wittstock. 2011. Insect herbivore counteradaptations to the plant glucosinolate-myrosinase system. Phytochemistry 72: 1566–1575.

Wright, E.M., D.D.F. Loo and B.A. Hirayama. 2011. Biology of human sodium glucose transporters. Physiol. Rev. 91: 733–794.

13

Silks from Insects—From Natural Diversity to Applications

M. Neuenfeldt[a] and *T. Scheibel*[b]

Introduction

During evolution, silk has evolved multiple times (Sutherland et al. 2010). Since all known silks are heterogeneous in both molecular structure and biological function, it is difficult to postulate a distinct definition of silk: Silk belongs to the family of structural extracorporeal proteins with highly repetitive sequences (Craig 1997), which are processed starting with a concentrated silk solution. This definition excludes, e.g., hair which is fabricated out of low concentrations of keratin. Further, silk glands are required to store the concentrated silk dope. In particular, silks are fibers, which distinguish them from other secreted structural proteins such as glue produced by marine organisms (Waite 2002). On a molecular level, a specific characteristic of silk is the semicrystalline structure, based on a distinct amino acid sequence (Craig 1997). The crystallites contribute to the unique mechanical properties of silk (Porter and Vollrath 2009). Throughout all known silk types, the fraction of crystalline structures varies from low to high amounts (Walker et al. 2013). Finally, a spinning process is associated with silk, which describes a highly controlled phase transition of the liquid silk dope towards a solid fiber induced by shear forces (Porter and Vollrath 2009, Greving et al. 2012). In most cases, these

Lehrstuhl Biomaterialien, University of Bayreuth, 95440 Bayreuth, Germany.
[a] Email: martin.neuenfeldt@bm.uni-bayreuth.de
[b] Email: thomas.scheibel@bm.uni-bayreuth.de

shear forces are applied within the organism's body, and the proteins are assembled in form of a fiber. However, the spinning process can also take place after the excretion of the silk dope (Weisman et al. 2009, Ashton et al. 2012); therefore, the assembly trigger can be generalized as rheological stress.

Silks are only produced by arthropods; the silkworm (larva of *Bombyx mori*) and spiders (Araneae) are probably the most prominent producers of silk (Zhou et al. 2001, Vepari and Kaplan 2007, Heim et al. 2010). In marine systems, secreted structural proteins often function as glue and are widely produced by mollusks and annelids (Silverman and Roberto 2007, Stewart et al. 2011). In contrast, only few examples of aquatic silks are known, e.g., the marine silk of amphipods (Kronenberger et al. 2012), or the insect silk produced under water by the larvae of midges and caddisflies (Case et al. 1997, Stewart and Wang 2010). The major types of silks are produced in terrestrial environments, and during evolution an impressive variety of different silks has emerged especially within the class of insects (Sutherland et al. 2010).

Since ancient times mankind has been fascinated by silk, originating in its esthetic appearance and its unique mechanical properties. Silk can possess a combination of different mechanical characteristics leading to a very high toughness; therein, it is superior to all known synthetic fibers (Heidebrecht and Scheibel 2013, Table 1). Within the last decades, the emerging field of biotechnology offered the opportunity to recombinantly produce silk proteins and materials thereof (Scheibel 2004, Omenetto

Table 1. Mechanical properties of selected insect silks compared to spider silk and other materials. n.d.: not determined (Adapted from Hepburn et al. 1979, Bauer and Scheibel 2012, Heidebrecht and Scheibel 2013).

Material	Young's modulus [GPa]	Strength [MPa]	Extensibility [%]	Toughness [MJ/m³]
Bombyx mori cocoon	7	600	18	70
Chrysopa carnea egg stalk (30% relative humidity)	5.8 ± 1.3	68 ± 19	2 ± 1	1.2 ± 0.72
Chrysopa carnea egg stalk (70% relative humidity)	3.2 ± 1.0	155 ± 75	210 ± 100	87 ± 49
Apis mellifera cocoon	n.d.	0.4	204	n.d.
Araneus diadematus dragline	6	700	30	150
Elastin	0.001	2	15	2
Nylon 6.6	5	950	18	80
Kevlar 49	130	3600	2.7	50
Steel	200	1500	0.8	6
Carbon fiber	300	4000	1.3	25

and Kaplan 2010). At the end of this chapter, an overview of possible applications of silk products is described.

Molecular Insights into Insect Silk

All known insect silk proteins have glycine, serine and alanine residues as the most abundant amino acids in common. These three amino acids exhibit an intermediate hydrophobicity, ensuring a sufficient solubility of the silk proteins in the silk gland, yet leading to a high insolubility in their ordered functional state as a fiber (Sutherland et al. 2010).

In proteins, the amino acid sequence (primary structure) can fold into ordered structural elements with defined secondary structures, and further into tertiary and quaternary structures. In silk crystallites are five types of ordered structure: Coiled coil, extended beta sheet, cross beta sheet, collagen triple helix and polyglycine II (cf. Fig. 1). Beta strands are often arranged as extended sheets, but in some cases, the strands are connected by beta turns and, thus, generate consecutive sheets which are arranged perpendicular to the fiber axis (cross beta structure), being better known from, e.g., neurodegenerative diseases, where these structures cause the formation of amyloid fibrils (Jahn et al. 2010). Probably due to their potential to form amyloids, cross beta sheets are far less common in nature as a functional structure (Weisman et al. 2009) So far, only in capture threads of glowworms (*Arachnocampa luminosa*), in cocoons of weevils (*Hypera* spp.), in silken rafts of water beetle eggs (*Hydrophilus piceus*) as well as in egg stalks of lacewings (*Mallada signata*), cross beta structures have been detected (Parker and Rudall 1957, Kenchington 1983, Weisman et al. 2009). Although collagen itself is not considered to be a silk, there are a few examples of insect silks with collagen-like helical assembly, e.g., in sawflies (Sehnal and Sutherland 2008).

During evolution, several amino acid motifs have evolved, which fulfill the premises of each structural element. The formation of coiled coils only occurs when the individual alpha-helices contain repeats of seven amino acids (called heptad motif). This motif shows the general pattern HPPHPPP, where H reflects in general hydrophobic residues and P in general polar residues (Sutherland et al. 2012). Furthermore, beta strands can be formed by recurring amino acid motifs such as GAGAGS[1] or poly-alanine regions, both found in antiparallel beta-sheets of two different silkworm fibroins (Yukuhiro et al. 1997, Wilson et al. 2000). The selection of amino acids present in beta sheets is adjusted to proper interactions between the side chains, which ensure the stability of the whole structural element (Smith and Regan 1995). Additionally, beta

---

[1] all amino acids stated in the text are according to the IUPAC single letter code.

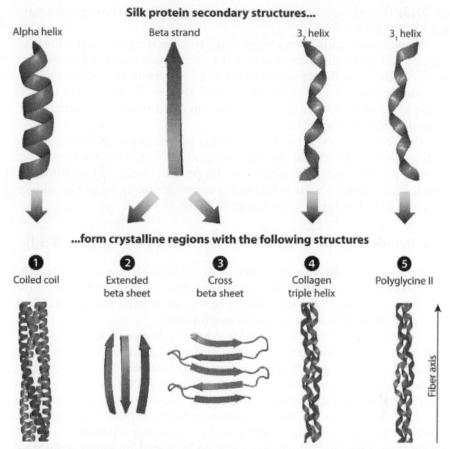

Figure 1. The five types of ordered silk structures. 1. Coiled coil: Two to five alpha helices (always right-handed) wind around each other to form a superhelix. 2. Extended beta sheet: Multiple planar beta strands form sheets which are stacked face to face. 3. Cross-beta sheet: Beta strands of uniform length alternating with turns at which the protein chain direction reverses. This is the only structural element which is aligned perpendicular to the fiber axis. 4. Collagen triple helix: Three left-handed helices with three amino acids per turn (3_2 helix) are intertwined to form a superhelix. 5. Polyglycine II: Three right-handed helices with three amino acids per turn (3_1 helix) are intertwined to form likewise a superhelix. (Adapted from Sutherland et al. 2010).

Color image of this figure appears in the color plate section at the end of the book.

strands can be connected by beta turns which are always represented by four amino acids (Hutchinson and Thornton 1994). Finally, proteins with a repetitive GX_1X_2- (with proline as most probable residue at position X_1) or GGX_2-motif will fold into triple-helical structures yielding a collagen-like or polyglycine II structure, respectively (Wilson et al. 2000, Sutherland et al. 2013). Polyglycine II helices are able to assemble into hexagonal arrays (Crick and Rich 1955).

Some amino acids exert a bigger influence on structural elements than others: Due to the steric constraint of the pyrrolidine ring, proline residues disfavor the formation of alpha helices and beta sheets and induce random coil structures and turn elements instead (Liu et al. 2008). Due to their bulky sidechains, aromatic amino acids can also impede the formation of densely packed ordered structures.

Glycine residues exhibit ambivalent properties in inducing either ordered or unordered structures (Dicko et al. 2008). On the one hand, they allow a conformational flexibility of the protein chain, but on the other hand, depending on their neighboring amino acids, they can also induce amyloid-like structures (Rauscher et al. 2006).

The Relation between Structure and Biological Function of Silk

As mentioned above, the existence of crystalline elements is one of the key characteristics of silk. These crystalline regions are responsible for the resulting strength of the silk fiber, whereas unordered regions (random coil) contribute to flexible, amorphous properties.

In both ordered and unordered fractions hydrogen bonds are of major importance for the mechanical properties: In crystalline regions, hydrogen bonds exist between amide groups (Porter and Vollrath 2009). The resulting densely packed crystals exhibit a pronounced resistance against applied forces (observed as strength on a macroscopic level). Even when breaking these hydrogen bonds, the repetitive domains expose hidden length allowing the fiber to dissipate additional strain energy (Ashton et al. 2013). In contrast, random coil elements can be hydrated by water molecules. This lowers the glass transition temperature of the region and imparts silk with elastomeric properties (Porter and Vollrath 2008). Incidentally, this hydration depicts an important difference between natural silk and conventional synthetic fibers (e.g., nylon). The combination of strength and elasticity results in exceptional high values of toughness, which illustrates the unique mechanical properties of silk (Table 1).

Posttranslational modifications can induce further interesting properties of silk. For example, phosphorylation of side chains represents a common modification of silk (Chen et al. 2010, Ashton et al. 2012). In Trichoptera, the extensively phosphorylated serine residues of H-fibroin increase the adhesive character of the silk and, thus, adapt the material to the aqueous environment. Especially in such polyionic protein systems, inorganic counter ions such as Ca^{2+} may play an important role (Stewart and Wang 2010). In collagen-like silk, hydroxyproline is not present, although it is abundant in real collagen to stabilize the triple helix. Instead, hydroxylysine was found as a modified amino acid in collagen-like silk, while it remains unclear if this amino acid functions, e.g., as glycosylation site (Sehnal and Sutherland 2008).

Furthermore, intra- and intermolecular disulfide bridges can stabilize silk structures. As an extreme example, aquatic larvae of the midge produce silk with unusual high amounts of cysteine residues (Case et al. 1997). The containing CXCXC-motif is highly conserved and probably contributes to intermolecular crosslinking of the silk. In *Bombyx mori*, an intermolecular disulfide bridge between the light- and heavy-chain of fibroin is essential for the efficient secretion of the silk (Takei et al. 1987). Additionally, several tanning reactions based on phenolic crosslinking have been observed in silk (Brunet and Coles 1974, Sutherland et al. 2012), e.g., dityrosine crosslinking in silk fibroin of *Antheraea pernyi* (Raven et al. 1971). Interestingly, unlike in marine glue systems, there are no reports about dihydroxyphenylalanine-(DOPA)-residues in insect silk so far. Finally, insect silk can contain acetylated amino acid residues as well as lysine-mediated crosslinks (Chen et al. 2010). Posttranslational modifications can also occur via glycosylation (Hoffman et al. 1996, Tanaka et al. 1999a). Due to the intrinsic hydration properties, these residues probably contribute to the elastomeric properties of silk. Even organic compounds such as flavonoids can be embedded in silk structures, but recent studies indicate that this feature is primarily a side effect of a metabolic response to dietary toxins (Xu et al. 2013).

So far, several structural elements and their impact on the material properties have been identified. This leads to the essential question of the correlation between structure and function. In other words: How necessary is the presence of certain structural elements in silk to fulfill its purpose?

The cement-like silk of the grey silverfish (*Ctenolepisma longicaudata*) is a good example for little requirements on material properties (Walker et al. 2013). Male silverfish produce the silk as a tactile cue during mating, which navigates the female during the uptake of spermatophores. The silk contains high fractions of random coil structures and only little alpha helical content. It becomes brittle after drying and exhibits a low chemical stability, but since the mating only lasts for approximately thirty minutes, it fulfills its purpose.

As a completely different example, female lacewings deposit their eggs by attaching them individually to a stalk of silk (Weisman et al. 2009). The stalk protects the egg from predators like ants. These egg stalks require a bending stiffness to withstand environmental influences over time periods of several days. Therefore, the high content of densely packed cross-beta structures found in the stalks seems to be a logical consequence due to tight interactions between the beta strands which are aligned perpendicular to the fiber axis.

However, in general the correlation between structure and function is considered to be low, illustrated by the fact that all five types of ordered silk structure are generally used to make cocoons (Sutherland et al. 2010). It also has to be mentioned that there is a low correlation between the function of silk and the glands in which they are produced: For example, in labial glands all five known silk protein structures can be produced depending on the respective species (Sutherland et al. 2010). Importantly, several glands exist such as Malpighian tubule or dermal glands that can produce silk proteins (cf. Table 2).

Evolutionary Conservation of Silk

Silk is believed to have evolved multiple times: 23 different "silk lineages" have been proposed, as summed up in Table 2 and Fig. 2 (Sutherland et al. 2010). As already discussed, a crystalline structure in silk is strictly based on the primary sequence of the proteins (such as periodicity of the comprising repetitive elements). Once one type of ordered structure has been emerged, it seems very unlikely that another ordered crystalline structure can arise from it during evolution, since only a few mutations on the genetic level are able to destroy the symmetric conditions of the already existing silk type. Only collagen-like silk could theoretically arise from polyglycine structures, since both silk types have a very similar amino acid periodicity (Sutherland et al. 2010).

The degree of tolerated variety in the primary sequence may depend on the type of crystalline structure: While extended beta sheets seem to compensate occasional sequence irregularities to a certain amount (Zhou et al. 2001), cross beta structures are apparently much more determined by the regularity of their consecutive repeats. This may also be due to the fact that mutations in this silk type can cause amyloidogenic properties of the protein.

The homologous tetrameric coiled coil silk of bees and ants (Apoidea and Vespoidae) is an example of highest sequence divergence (Sutherland et al. 2007a): Here, only the character of amino acid residues, in particular heptad positions, is conserved instead of actual sequence repeats. As long as the coiled coil structure maintains, an extensive variety in the primary sequence seems to be allowed. Interestingly, ancestral species of the two mentioned superfamilies formerly produced beta-sheet silk, though it is claimed that the transition towards coiled coil silk occurred by gene duplication of a silk-unrelated gene and not by accumulated mutations within the silk genes encoding for beta-sheet structures (Sutherland et al. 2006).

Table 2. Overview of silks in the class Insecta.[2,3] (Modified after Sutherland et al. 2010).

Common name of insect group [higher classification] (references)	Purpose of silk	Life stage/gland
A) Silk proteins predominantly in beta-sheet conformation; mRNA of dominant silk gene approximately 2 kb; partial silk sequences contain reiterating GS- or GA-repeats; silk fiber contains 4% alanine, 31% serine, 44% glycine.		
1. Jumping bristletails, silverfish [Archeognatha: families Machilidae and Meinertellidae; Zygentoma] (Sturm 1992, Grimaldi and Engel 2005)	Indirect sperm transfer, silken threads guiding the female to spermatophore, silken stalks or mats holding droplets of sperm above the ground	Adult males/Type III secretory units
2. Mayflies [Ephemeroptera: family Polymitarcyidae] (Sattler 1967)	Lining for U-shaped tunnels in submerged wood	Larvae/apparently in Malpighian tubules
3. Dragonflies [Odonata: family Gomphidae] (Gambles 1956)	Presumably anchoring eggs; bundles of fibers attached to eggs that uncoil upon exposure to water	Adult female/unknown
4. Webspinners [Embiidina] (Edgerly 1987, Okada et al. 2008)	Tunnels and egg coatings	All stages/Type III secretory units in prothoracic tarsomeres
B) Silk proteins predominantly in beta-sheet conformation; silk is used for shelters; silk fiber contains 12% alanine, 32% serine, 13% glycine.		
5. Crickets [Orthoptera: Stenopelmatoidea in the families Gryllacrididae and Anostostomatidae] (Rentz and Weissman 1973, Morton and Rentz 1983, Bland and Rentz 1991)	Binding leaves together for construction of cocoon-like nests, linings for sand burrows	All stages/labial glands
6. Book lice [Psocoptera: suborders Psocomorpha and Troctomorpha] (New 1987)	Egg coverings and nests	Adult females/labial glands

[2] Silks produced by animals in the class Entognatha (orders Protura, Diplura, ard Collembola) are not included because of the scarcity of information on silk production in these animals.
[3] The number pertaining to each insect group also refers to its respective illustration in Fig. 2.

Table 2. contd....

Table 2. contd.

Common name of insect group [higher classification] (references)	Purpose of silk	Life stage/gland
7. Thrips [Thysanoptera: Terebrantia in the families Heterothripidae and Aeolothripidae; Tubulifera, family Phlaeothripidae in the subfamily Phlaeothripinae] (Mound and Morris 2001, Izzo et al. 2002)	Cocoons or tent-like shelters for protection against predators, extreme temperatures, and low humidity	Larvae and possibly adults/secretions from the anal region, likely from Malpighian tubules
8. *Kahaono montana* Evans [Hemiptera in this single species] (Chang et al. 2006)	Protective shelters	Unknown
C) Silk proteins in a cross-beta conformation with X-ray diffraction patterns that are markedly different from the proteins in the neuropteran cross-beta-sheet silks but similar to supercontracted keratins.		
9. Water beetles [Coleoptera: family Hydrophilidae] (Rudall 1962)	Silken rafts to support eggs	Adult female/colleterial glands
D) *Hypera* species cocoon silk proteins adopt a cross-beta structure 10 amino acids in width.		
10. Plant-eating beetles [Coleoptera: Cucujiformia] (Kenchington 1983, Oberprieler et al. 2007)	Terrestrial or aquatic cocoons, silk-lined tunnels in the sand	Larvae/Malpighian tubules
E) Species in Chrysopidae and Nymphidae have silk proteins in a cross-beta structure 8 amino acids in width; silk fiber contains 20% alanine, 41% serine, 24% glycine.		
11. Lacewings [Neuroptera, found within four of the six superfamilies] (Lucas et al. 1957, Parker and Rudall 1957, Duelli 1986, Weisman et al. 2009)	Egg stalks or egg coverings	Adult females/colleterial glands

Table 6. contd....

Table 6. contd.

Description	Silk use	Source
F) Cocoon of the green lacewing *Mallada signata* is composed of small (49 kDa) alpha-helical proteins and lipids; silk fiber contains 40% alanine, 7% serine, 11% glycine. **12. Lacewings and ant lions** [Neuroptera] (Rudall and Kenchington 1971, Weisman et al. 2008)	Cocoons	Larvae/Malpighian tubules
G) Silks with an extended beta-sheet structure; the phylogenetic distribution suggests that this structure is plesiomorphic to the Hymenoptera with other structures seen in this order (see below) representing derived conditions; a partial silk gene sequence (785 nucleotides) corresponding to a large (>500 kDa) cocoon protein from the parasitic wasp. *Cotesia glomerata* (Ichneumonoidea) encodes a protein containing a 28-amino-acid repeat sequence rich in consecutive DS-dipeptide repeats; silk fiber contains 36–56% alanine, 8–40% serine, 2–31% glycine. **13. Sawflies and parasitic wasps** [Hymenoptera] (Rudall and Kenchington 1971, Quicke et al. 2004, Sutherland et al. 2007a)	Cocoons, nests, and webs	Larvae/labial gland
H) Silk fibers contain four small (30–50 kDa) homologous alpha-helical proteins that adopt a tetrameric coiled-coil structure; silk fiber contains 23–34% alanine, 10–17% serine, 5–6% glycine. **14. Parasitic wasps** [Hymenoptera: Chalcidoidea in several Eupelmidae and one Signiphoridae species] (Woolley and Vet 1981)	Egg covering and host covering	Adult female/abdomen secretion
15. Bees, ants, and wasps [Hymenoptera: Apoidea and Vespoidea] (Sutherland et al. 2006, Sezutsu et al. 2007, Sutherland et al. 2007a)	Nests and cocoons	Larvae/labial gland
I) Polyglycine II structured proteins contain high levels of glycine (66%) and are present in the cocoons of some Allantinae, Heterarthrinae, and Blennocampinae species; related collagen structures are found in cocoons from the tribe Nematini (Nematinae) as well as species from the Blennocampinae and Tenthredininae; the amino acid hydroxyproline which is usually abundant in collagens is not found in silk, but hydroxylysine (3.3–4%) is; collagen silk fiber contains 28% alanine, 14% serine, 11% glycine. **16. Sawflies** [Hymenoptera: Tenthredinoidea in the family Tenthredinidae] (Rudall and Kenchington 1971)	Cocoons	Larvae/labial gland

Table 2. contd....

Table 2. contd.

Common name of insect group [higher classification] (references)	Purpose of silk	Life stage/gland
J) Silk fiber contains 14–20% alanine, 16–19% serine, 10–27% glycine (total 49–57%).		
17. Wasps [Hymenoptera: Apoidea in the subfamily Pemphredoninae, some species in *Microstigmus*, *Spilomena* and *Arpactophilus*] [Matthews and Starr 1984, Serrao 2005)	Rope fabricated from plant fibers and silk suspending the nest	Adult females/Type III secretory units in posterior metasoma
18. Wasps [Hymenoptera: Vespoidea in the family Vespidae; the species *Dolichovespula maculate*, *Polistes metricus*, *P. annularis*, and *Ropalidia opifex*] [Mcgovern et al. 1988, Espelie and Himmelbach 1990, Maschwitz et al. 1990)	Nests constructed from fine silk fibers binding plant fibers	Adult females/labial glands
K) Proteins in the cocoon of *Xenopsylla cheopis* are in an alpha-helical conformation; the X-ray diffraction patterns are similar to known coiled-coil structures.		
19. Fleas [Siphonaptera] [Rudall and Kenchington 1971, Silverman et al. 1981, Lawrence and Foil 2002)	Cocoons	Larvae/labial glands
L) A partial protein sequence (766 amino acids) identified from mRNA extracted from male silk glands and corresponding to a large (220 kDa) silk protein present in the silk-producing basitarsi is repetitive with high levels of aspartic acid (21.9%) and glycine (13.9%).		
20. Dance flies [Diptera: family Empididae in the subfamily Empidinae] (Young and Merritt 2003, Sutherland et al. 2007b)	Silk-wrapped nuptial gifts	Adult males/Type III secretory units in prothoracic basal tarsomeres

Table 6. contd....

Table 6. contd.

M) The threads of *Arachnocampa luminosa* contain proteins with cross-beta crystalline structure.		
21. Glowworms [Diptera: family Keroplatidae] (Rudall 1962)	Nests/prey capture threads	Larvae/labial glands
N) The silk of Chironomidae is produced from proteins (40 kDa up to very high molecular weight) containing high levels of cysteine; silk fiber contains 6% alanine, 12% serine, 10% glycine.		
22. Midges [Diptera: Chironomoidea in the families Chironomidae and Simulidae] [Macgregor and Mackie 1967, Oliver 1971, Case et al. 1994, Kiel and Roder 2002)	Aquatic silken tubes, adhesive lifelines, cocoons	Larvae/labial gland
O) Silk consists of two filaments glued together by serine-rich (>15% serine) proteins known as sericins; the filaments consist of three proteins: heavy fibroin (>200 kDa), light fibroin (25–30 kDa), and the glycoprotein P25 (~25 kDa); around 90% of the sequence of the heavy fibroin protein of *Bombyx mori* contains consecutive GX-repeats (X = alanine, serine, or tyrosine), which form extended beta-sheet structures, interspersed with amorphous motifs; other lepidopteran silks contain homologs of these proteins; trichopteran silk contains homologs of only the heavy and light fibroin proteins; silk fiber contains 5–57% alanine, 0–22% serine, 19–48% glycine.		
23. Butterflies, moths, caddisflies [Lepidoptera, Trichoptera] [Mita et al. 1994, Tanaka et al. 1999a, Tanaka et al. 1999b, Zhou et al. 2000, Tanaka and Mizuno 2001, Yonemura and Sehnal 2006, Yonemura et al. 2006, He et al. 2012)	Cocoons (aquatic and terrestrial), tunnels, retreats, communal webs, prey capture nets	Larvae/labial gland

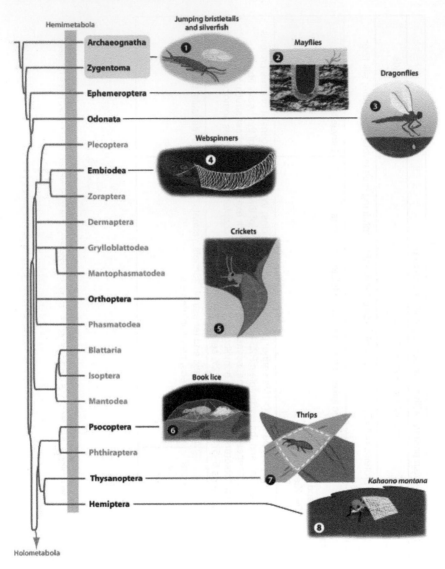

Figure 2. contd....

Figure 2. contd.

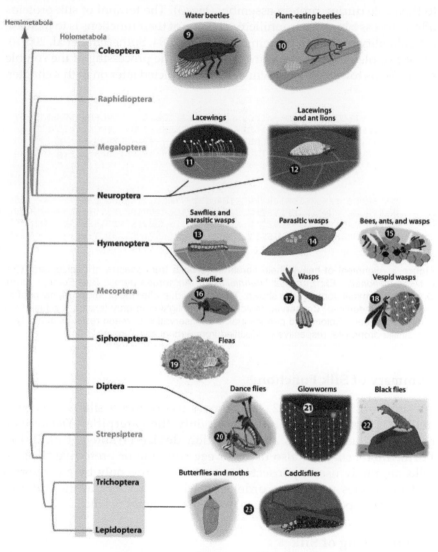

Sutherland TD, et al. 2010.
Annu. Rev. Entomol. 55:171–88

Figure 2. Phylogenetic relationship of all silk-producing insects (from Sutherland et al. 2010, with permission).

Color image of this figure appears in the color plate section at the end of the book.

The aspect of conservation in terminal sequences is strongly connected to their role during protein assembly (Fig. 3). The termini of silk proteins often possess a high hydrophilicity, and one of their functions is to increase the solubility of the silk proteins (Bini et al. 2004, Sutherland et al. 2007a). They also play a very important role during the processing of the soluble silk proteins towards a fiber which will be depicted later on in this chapter.

```
              I          I          I          I          I          I
Gm   MRVTTFVILCCALQYVTAD--AIDDSLLNFNNENFIEIGESTTAEVDVE-NGTLVERETT 57
Ek   MRVTTFVILCCALQYVAAA--DILET--GILGGNVREVSESTTDNFTTDGNGNVTEVKTT 56
Ap   MRVIAFVILCCALQYATAK--NLRHH----------------DEYVDNHGQLVERFTT 40
Bm   MRVKTFVILCCALQYVAYTNANINDFDEDYFGSDVTVQSSNTTDEIIRDASGAVIEEQIT 60
     ***  :**********.:        :  .              :    :  * : *    *

              I          I          I          I          I          I
Gm   RKKYERDGDITPNISGEDKIVRTFVIETDASGHETVYEEDVVIKRKPGQQ---------- 107
Ek   HKEFRRQGEVPNNISGEDKLVRTFVIETDASGNEVIYEEDVVIHKVPGTS---------- 106
Ap   RKHFERNAATRPHLSGNERLVETIVLEEDPYGHEDIYEEDVVIKRVPGASSSAAAASSAS 100
Bm   TKKMQRKNKNHGILGKNEKMIKTFVITTDSDGNESIVEEDVLMKTLSDGT---------- 110
     *.  .*.        :.  :::::.*:*:  *.  *:* :  ****:::      ..
```

Figure 3. Alignment of heavy-chain fibroin sequences from species of Lepidoptera (Gm: *Galleria mellonella*; Ek: *Ephestia kuehniella*; Ap: *Antheraea pernyi*; Bm: *Bombyx mori*). Only the N-terminal domains are shown. The first residue after the predicted signal peptide cleavage is underlined. Conserved residues are highlighted in gray (except residues from signal sequences). Colons and periods indicate conservation between residues of highly or less similar properties, respectively. (Modified from Sehnal and Zurovec 2004).

Summary of Silk Function

As summarized in Table 2, the primary use of insect silk is as cocoon material. Besides silk of silverfish, only the caterpillar *Yponomeuta cagnagellus* uses silk as an orientation device (Walker et al. 2013). Furthermore, silks are also used for egg covering or protective shelters. Silks are rarely used by insects for capturing prey, only by glowworms and by species within the order of Trichoptera. Only in the order of Neuroptera, species can produce two different types of silk.

Bioprocessing of Silk

The high translation rates of silk mRNA can affect the genetic level in silk glands: In many insects, e.g., dipteran species, gland cells undergo several rounds of endoreplication (DNA replication without subsequent mitosis) which results in high copies of connected sister chromatids encoding silk proteins (Botella and Edstrom 1991, Wieslander 1994). These polytene chromosomes contain sites of intensive transcription (named Balbiani rings in salivary gland cells of chironomid larvae). Since the silk of

Chironomidae is either used for protective tunnels or pupation tubes, the composition of the silk varies depending on the larval instar (Sehnal and Sutherland 2008).

Especially when silk functions as cocoon material, the regulation of silk production is strongly linked to metamorphosis (Jindra et al. 2013). Both molting and metamorphosis is controlled by a complex interaction of juvenile hormone and ecdysteroids. Juvenile hormone is synthesized in the corpora allata, the removal of this endocrine gland (allatectomy) in early penultimate-instar silkworms results in precocious metamorphosis (Sehnal and Akai 1990) (see also chapters by Orchard and Orchard, this book, Hoffmann et al., this book). This and other transplantation experiments led to the conclusion that juvenile hormone suppresses both growth and activity of silk glands (Sehnal and Akai 1990). Additionally, juvenile hormone-deficient *Bombyx mori* strains are known, which exhibit precocious metamorphosis in the third or fourth larval instar instead of the regular fifth (Daimon et al. 2012).

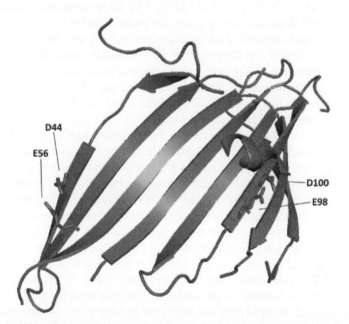

Figure 4. Position of conserved acidic residues in the structure of the heavy-chain fibroin N-terminal domain (*Bombyx mori*, pdb-file: 3UA0). One pair of acidic residues is located on the same beta strand (E98 and D100), one pair is located on two adjacent strands (D44 and E56). At neutral pH, the charge repulsion caused by these residues impedes the dimerization of N-terminal domains. (Adapted from He et al. 2012).

Color image of this figure appears in the color plate section at the end of the book.

In the silk glands of the silkworm, the natural spinning process was analyzed most extensively compared to other species (Iizuka 1985, Iizuka 1988, Matsumoto et al. 2007). After fibroin synthesis in the posterior region of the glands (approx. 0.6 mm in diameter and 200 mm in length), the protein is secreted into the lumen (Iizuka 1988). The initial protein concentration amounts 12 to 15 percent by weight, and the fibroin assembles into micelles as indicated by experiments with regenerated silk fibroin (Jin and Kaplan 2003). The aqueous solution is then extruded by peristaltic motions to the duct of the middle gland region (2 mm thick and 60 mm long), where it is concentrated for storage at 20 to 30 percent by weight (Matsumoto et al. 2007). At this concentration, the micelles presumably condense into globular structures (Jin and Kaplan 2003). In the middle region of the silk gland, sericin is secreted covering the wall of the lumen and absorbing water from the fibroin (Iizuka 1988, Jin and Kaplan 2003). In the lumen of the anterior glands (35 mm in length and narrowing toward the end to 50 μm in diameter), the pH drops from near neutral to 4.9 at the point of spinning and the salt concentration increases (Matsumoto et al. 2007, He et al. 2012). The final spinning process into water-insoluble fibers occurs by shear forces applied by the spinneret. The transition towards a solid state emerges out of a lyotropic liquid crystal phase which enables a relatively low flow viscosity and allows uniform molecular alignment to a preexisting molecular orientation (Kerkam et al. 1991, Jin and Kaplan 2003). The sericin envelopes the fibroin fibrils and may also promote the elasticity of the spun fiber and function as a donor for divalent cations (Matsumoto et al. 2007).

Recent studies revealed the role of the terminal domains of silk fibroin during fiber assembly (He et al. 2012). In the N-terminal domain of the fibroin heavy chain, random coil is the predominant structural element at neutral pH. Additionally, acidic residues of the domain impede the formation of homodimers by charge repulsion (Fig. 4). During the decrease of pH in the anterior gland, these residues get protonated, and along with emerging beta-sheets a dimer entanglement is enabled. The respective acidic residues which prevent premature beta sheet formation at neutral pH are conserved throughout the orders of Lepidoptera and Trichoptera (Sehnal and Zurovec 2004, Yonemura et al. 2009, Fig. 3). As an intriguing fact, the same mechanism of pH-induced dimerization by acidic residues is also involved in spidroin assembly of spider silk, though both silk proteins have been evolved independently, and the spider silk domains stay alpha-helical throughout the entire process (Eisoldt et al. 2012). This is a fascinating example of convergent evolution on a molecular level (He et al. 2012).

The spinning speed of silkworms ranges from 0.4 to 1.5 cm/sec, the critical shear rate has been estimated to be from 100 to 1000 sec^{-1} (Iizuka

1988). Artificial reeling of silkworm silk under steady conditions results in shifted mechanical properties towards increasing either strength or extensibility, indicating the crucial role of spinning conditions on the mechanical properties of the resulting silk (Shao and Vollrath 2002).

In the order of Trichoptera, the morphological alignment of silk proteins is similar to the one described for silkworm silk (Engster 1976, Ashton et al. 2012). The peripheral protein layer consists of a glycoprotein, but the layer is much thinner than in silkworm silk, and structural similarities to silk sericin remain unclear so far (Engster 1976). Additionally, the protein P25 which ensures the solubility of the silkworm fibroin during the processing is absent in silk of caddisflies (Inoue et al. 2000, Yonemura et al. 2009).

In the order of Hymenoptera, the concentrated silk dope is organized in two possible ways within the gland lumen: The protein molecules can either be organized into spindle-shaped tactoids with a length up to 40 μm or they can form fibrous bars (Flower and Kenchington 1967). Tactoids were observed in silk glands of honey bees and ants, whereas in bumble bees and colonial wasps the proteins are mainly arranged into fibrous bars (Gomes et al. 2004, Sutherland et al. 2012). As in the silk gland of *Bombyx mori*, the macromolecular arrangement in the order of Hymenoptera lowers the flow viscosity of the silk dope when it passes through the spinneret.

Biomedical and Technical Applications of Insect Silk

The knowledge on natural silk processing is the premise for a biotechnological approach of silk production (Jin and Kaplan 2003). In this context, in analogy to the biomimetic principle of extracting the "key idea" of a biological concept, various types of natural silk have been mimicked by a modular design (Huang et al. 2003, Scheibel 2004). Briefly, consensus sequences were defined based on repetitive motifs of natural silk proteins. Genes encoding for the consensus motifs were multimerized to a desired length, and the artificial silk protein can be produced in host organisms like *Escherichia coli* (Heidebrecht and Scheibel 2013).

After purification of the silk proteins, several methods of artificial fiber formation have been reported: Fibers can be produced by wet-spinning in a coagulation bath (Matsumoto et al. 1996) or by electrospinning, where the solvent evaporates in an electric field (Jin et al. 2002). The last mentioned method can result in non-woven scaffolds depending on the method of fiber collection. Additionally, artificial silk proteins can generally be processed into films, foams, hydrogels, particles or capsules (Mori and Tsukada 2000, Spiess et al. 2010).

Unlike spider silk, the manufacturing process of silkworm silk is well established since ancient times. For distinct applications, the usage of reconstituted silk fibroin out of natural cocoons is often a cost-efficient method instead of the biotechnological production described above. However, since sericin is the major cause for an impaired biocompatibility, additional processing methods are required to make silkworm silk suitable especially for biomedical applications (Altman et al. 2003). In general, sericin can be removed by boiling of silkworm silk fibers in aqueous solutions (Jin and Kaplan 2003). Subsequent coating with waxes or silicone reduces frying and results in frequently used suture materials (Altman et al. 2003).

Especially silk scaffolds exhibit beneficial properties for tissue engineering: Mechanical integrity makes silk scaffolds suitable for engineering of, e.g., bone or ligament tissue (Altman et al. 2003). Due to its slow proteolytic degradation, silk fibers can retain their initial tensile strength for the duration of tissue development (Altman et al. 2003). Thus, in several studies the potential of silk scaffolds for regenerating tendons, ligaments and skin tissue has been evaluated (Mori and Tsukada 2000, Kundu et al. 2013). Furthermore, a promising application is the use as a scaffolding material for injuries of the peripheral nervous system: Conduits comprising silkworm silk are able to bridge nerve gaps up to 13 mm (Huang et al. 2012).

Silk particles and films are widely tested as drug delivery systems (Hardy et al. 2013, Kundu et al. 2013). For example, particles of defined size and surface charge can be noncovalently loaded with a drug of opposite charge. After endocytosis by a cell, the silk particles often accumulate in the lysosomes (Duncan and Richardson 2012). The drug release can be controlled by a stimulus-responsive silk polymer: For example, proteolysis of silk particles ensures a drug release specifically in the slightly acidic environment of lysosomes (Seib et al. 2013). If the uptake of the drug particles is controlled by additional signaling pathways, e.g., by decorating them with cell surface receptor-specific ligands such as the peptide RGD, the modified drug delivery system allows to increase its specificity.

In general, chemical modifications can enhance, e.g., cell proliferation or cell adhesion (Altman et al. 2003). Post-treatment of artificially processed silk with organic solvents can alter the proportions of the scaffold's secondary structures; e.g., methanol induces beta-sheet formation, resulting in water insensitive silk materials (Wilson et al. 2000). Introducing the peptide motif RGD can enhance cell adhesion on the respective silk scaffold (Hardy et al. 2013).

Beyond biomedical applications, there are several approaches towards technical utilizations of the unique silk properties. Due to proportions of densely packed crystalline structures, many silks already exhibit a chemical and thermal stability satisfying most common textile demands. Functionalization of silk can introduce additional properties such as water repellency, thermal stability or antibacterial properties (Li et al. 2012, Fei et al. 2013). Furthermore, the exceptional bending stiffness inherent in lacewing egg stalks could be utilized for filter materials where resistance against lateral forces is required. Compared to synthetic filter materials, much less amounts of silk are needed for an equivalent filter performance (Lang et al. 2013).

In conclusion, insect silk represents a material which is produced by a vast number of species. The mechanical properties of silks are highly adapted to both functional requirements and environmental conditions in which they are produced. Besides unique material performance, silk exhibits non-toxic, biocompatible and biodegradable properties and can therefore be used for versatile biomedical and technical applications.

Acknowledgement

Funding was derived from StMUV (U8793-2012/6-2).

Keywords: insect silk, protein fiber, protein assembly, biomaterial, biotechnology

References

Altman, G.H., F. Diaz, C. Jakuba, T. Calabro, R.L. Horan, J. Chen, H. Lu, J. Richmond and D.L. Kaplan. 2003. Silk-based biomaterials. Biomaterials 24: 401–416.

Ashton, N.N., D.S. Taggart and R.J. Stewart. 2012. Silk tape nanostructure and silk gland anatomy of trichoptera. Biopolymers 97: 432–445.

Ashton, N.N., D.R. Roe, R.B. Weiss, T.E. Cheatham 3rd and R.J. Stewart. 2013. Self-tensioning aquatic caddisfly silk: ca (2+)-dependent structure, strength, and load cycle hysteresis. Biomacromolecules 14: 3668–3681.

Bini, E., D.P. Knight and D.L. Kaplan. 2004. Mapping domain structures in silks from insects and spiders related to protein assembly. J. Mol. Biol. 335: 27–40.

Bland, R.G. and D.C.F. Rentz. 1991. External morphology and abundance of mouthpart sensilla in Australian Gryllacrididae, Stenopelmatidae, and Tettigoniidae. J. Morphol. 207: 315–325.

Botella, L.M. and J.E. Edstrom. 1991. The Balbiani ring 6 induction in *Chironomus*. Eur. Cell Biol. Org. 71: 11–16.

Brunet, P.C. and B.C. Coles. 1974. Tanned silks. Proc. R. Soc. Lond. Ser. B 187: 133–170.

Case, S.T., J. Powers, R. Hamilton and M.J. Burton. 1994. Silk and silk proteins from 2 aquatic insects. Silk Polym. 544: 80–90.

Case, S.T., C. Cox, W.C. Bell, R.T. Hoffman, J. Martin and R. Hamilton. 1997. Extraordinary conservation of cysteines among homologous *Chironomus* silk proteins sp185 and sp220. J. Mol. Evol. 44: 452–462.

Chang, J.C., G.M. Gurr, M.J. Fletcher and R.G. Gilbert. 2006. Structure-property and structure-function relations of leafhopper (*Kahaono montana*) silk. Austral. J. Chem. 59: 579–585.

Chen, W.Q., H. Priewalder, J.P. John and G. Lubec. 2010. Silk cocoon of *Bombyx mori*: proteins and posttranslational modifications—heavy phosphorylation and evidence for lysine-mediated cross links. Proteomics 10: 369–379.

Craig, C.L. 1997. Evolution of arthropod silks. Annu. Rev. Entomol. 42: 231–267.

Crick, F.H. and A. Rich. 1955. Structure of polyglycine II. Nature 176: 780–781.

Daimon, T., T. Kozaki, R. Niwa, I. Kobayashi, K. Furuta, T. Namiki, K. Uchino, Y. Banno, S. Katsuma, T. Tamura, K. Mita, H. Sezutsu, M. Nakayama, K. Itoyama, T. Shimada and T. Shinoda. 2012. Precocious metamorphosis in the juvenile hormone-deficient mutant of the silkworm, *Bombyx mori*. PLoS Gen. 8: e1002486.

Dicko, C., D. Porter, J. Bond, J.M. Kenney and F. Vollrath. 2008. Structural disorder in silk proteins reveals the emergence of elastomericity. Biomacromolecules 9: 216–221.

Duelli, P. 1986. A missing link in the evolution of the egg pedicel in lacewings. Experientia 42: 624–624.

Duncan, R. and S.C.W. Richardson. 2012. Endocytosis and intracellular trafficking as gateways for nanomedicine delivery: opportunities and challenges. Mol. Pharmac. 9: 2380–2402.

Edgerly, J.S. 1987. Maternal-behavior of a webspinner (Order Embiidina). Ecol. Entomol. 12: 1–11.

Eisoldt, L., C. Thamm and T. Scheibel. 2012. Review the role of terminal domains during storage and assembly of spider silk proteins. Biopolymers 97: 355–361.

Engster, M.S. 1976. Studies on silk secretion in the Trichoptera (F. Limmephilidae). II. Structure and amino acid composition of the silk. Cell Tiss. Res. 169: 77–92.

Espelie, K.E. and D.S. Himmelsbach. 1990. Characterization of pedicel, paper, and larval silk from nest of *Polistes annularis* (L.). J. Chem. Ecol. 16: 3467–3477.

Fei, X., M. Jia, X. Du, Y. Yang, R. Zhang, Z. Shao, X. Zhao and X. Chen. 2013. Green synthesis of silk fibroin-silver nanoparticle composites with effective antibacterial and biofilm-disrupting properties. Biomacromolecules 14: 4483–4488.

Flower, N.E. and W. Kenchington. 1967. Studies on insect fibrous proteins: the larval silk of *Apis*, *Bombus* and *Vespa* (Hymenoptera: Aculeata). J. Roy. Microscop. Soc. 86: 297–310.

Gambles, R.M. 1956. Eggs of *Lestinogomphus africanus* (Fraser). Nature 177: 663–663.

Gomes, G., E.C. Silva-Zacarin, F.J. Zara, R.L. Silva de Moraes and F.H. Caetano. 2004. Macromolecular array patterns of silk gland secretion in social Hymenoptera larvae. Gen. Mol. Res. 3: 309–322.

Greving, I., M. Cai, F. Vollrath and H.C. Schniepp. 2012. Shear-induced self-assembly of native silk proteins into fibrils studied by atomic force microscopy. Biomacromolecules 13: 676–682.

Grimaldi, D.A. and M.S. Engel. 2005. Evolution of the Insects. Cambridge University Press, Cambridge, New York. 755 pp.

Hardy, J.G., Leal-Egana and T.R. Scheibel. 2013. Engineered spider silk protein-based composites for drug delivery. Macromol. Biosci. 13: 1431–1437.

He, Y.X., N.N. Zhang, W.F. Li, N. Jia, B.Y. Chen, K. Zhou, J. Zhang, Y. Chen and C.Z. Zhou. 2012. N-Terminal domain of *Bombyx mori* fibroin mediates the assembly of silk in response to pH decrease. J. Mol. Biol. 418: 197–207.

Heidebrecht, A. and T. Scheibel. 2013. Recombinant production of spider silk proteins. Adv. Appl. Microbiol. 82: 115–153.

Heim, M., L. Romer and T. Scheibel. 2010. Hierarchical structures made of proteins. The complex architecture of spider webs and their constituent silk proteins. Chem. Soc. Rev. 39: 156–164.

Hepburn, H.R., H.D. Chandler and M.R. Davidoff. 1979. Extensometric properties of insect fibroins—green lacewing cross-beta, honeybee alpha-helical and greater waxmoth parallel-beta conformations. Insect Biochem. 9: 69–77.

Hoffman, R.T., E.R. Schmidt and S.T. Case. 1996. A cell-specific glycosylated silk protein from *Chironomus thummi* salivary glands. Cloning, chromosomal localization, and characterization of cDNA. J. Biol. Chem. 271: 9809–9815.

Huang, J., R. Valluzzi, E. Bini, B. Vernaglia and D.L. Kaplan. 2003. Cloning, expression, and assembly of sericin-like protein. J. Biol. Chem. 278: 46117–46123.

Huang, W., R. Begum, T. Barber, V. Ibba, N.C. Tee, M. Hussain, M. Arastoo, Q. Yang, L.G. Robson, S. Lesage, T. Gheysens, N.J. Skaer, D.P. Knight and J.V. Priestley. 2012. Regenerative potential of silk conduits in repair of peripheral nerve injury in adult rats. Biomaterials 33: 59–71.

Hutchinson, E.G. and J.M. Thornton. 1994. A revised set of potentials for beta-turn formation in proteins. Protein Sci. 3: 2207–2216.

Iizuka, E. 1985. Silk thread—mechanism of spinning and its mechanical-properties. Appl. Polymer Symp. 41: 173–185.

Iizuka, E. 1988. Properties of the liquid crystals of some biopolymers. Adv. Biophysics 24: 1–56.

Inoue, S., K. Tanaka, F. Arisaka, S. Kimura, K. Ohtomo and S. Mizuno. 2000. Silk fibroin of *Bombyx mori* is secreted, assembling a high molecular mass elementary unit consisting of H-chain, L-chain, and P25, with a 6 : 6 : 1 molar ratio. J. Biol. Chem. 275: 40517–40528.

Izzo, T.J., S.M.J. Pinent and L.A. Mound. 2002. *Aulacothrips dictyotus* (Heterothripidae), the first ectoparasitic thrips (Thysanoptera). Florida Entomol. 85: 281–283.

Jahn, T.R., O.S. Makin, K.L. Morris, K.E. Marshall, P. Tian, P. Sikorski and L.C. Serpell. 2010. The common architecture of cross-beta amyloid. J. Mol. Biol. 395: 717–727.

Jin, H.J., S.V. Fridrikh, G.C. Rutledge and D.L. Kaplan. 2002. Electrospinning *Bombyx mori* silk with poly (ethylene oxide). Biomacromolecules 3: 1233–1239.

Jin, H.J. and D.L. Kaplan. 2003. Mechanism of silk processing in insects and spiders. Nature 424: 1057–1061.

Jindra, M., S.R. Palli and L.M. Riddiford. 2013. The juvenile hormone signaling pathway in insect development. Annu. Rev. Entomol. 58: 181–204.

Kenchington, W. 1983. The larval silk of *Hypera* spp. (Coleoptera, Curculionidae)—a new example of the cross-beta protein conformation in an insect silk. J. Insect Physiol. 29: 355–361.

Kerkam, K., C. Viney, D. Kaplan and S. Lombardi. 1991. Liquid crystallinity of natural silk secretions. Nature 349: 596–598.

Kiel, E. and T. Roder. 2002. Gelelectrophoretic studies on labial gland secretions of immature blackflies (Simuliidae, Diptera). Limnologica 32: 201–205.

Kronenberger, K., C. Dicko and F. Vollrath. 2012. A novel marine silk. Naturwissenschaften 99: 3–10.

Kundu, B., R. Rajkhowa, S.C. Kundu and X. Wang. 2013. Silk fibroin biomaterials for tissue regenerations. Adv. Drug Deliv. Rev. 65: 457–470.

Lang, G., S. Jokisch and T. Scheibel. 2013. Air filter devices including nonwoven meshes of electrospun recombinant spider silk proteins. J. Visual. Exp. 75: e50492.

Lawrence, W. and L.D. Foil. 2002. The effects of diet upon pupal development and cocoon formation by the cat flea (Siphonaptera: Pulicidae). J. Vector Ecol. 27: 39–43.

Li, G.H., H. Liu, T.D. Li and J.Y. Wang. 2012. Surface modification and functionalization of silk fibroin fibers/fabric toward high performance applications. Mat. Sci. Engin. C-Mat. Biol. Appl. 32: 627–636.

Liu, Y., A. Sponner, D. Porter and F. Vollrath. 2008. Proline and processing of spider silks. Biomacromolecules 9: 116–121.

Lucas, F., J.T.B. Shaw and S.G. Smith. 1957. Amino-acid composition of the silk of *Chrysopa* egg-stalks. Nature 179: 906–907.

Macgregor, H.C. and J.B. Mackie. 1967. Fine structure of the cytoplasm in salivary glands of *Simulium*. J. Cell Sci. 2: 137–144.

Maschwitz, U., W.H.O. Dorow and T. Botz. 1990. Chemical-composition of the nest walls, and nesting-behavior, of *Ropalidia- (Icarielia)-opifex* Vandervecht, 1962 (Hymenoptera, Vespidae), a Southeast-Asian social wasp with translucent nests. J. Nat. Hist. 24: 1311–1319.

Matsumoto, A., H. Kim, I. Tsai, X. Wang, P. Cebe and D. Kaplan. 2007. Silk. pp. pp. 383–404. *In*: M. Lewin (ed.). Handbook of Fiber Chemistry, 3rd Edit. CRC/Taylor & Francis, Boca Raton.

Matsumoto, K., H. Uejima, T. Iwasaki, Y. Sano and H. Sumino. 1996. Studies on regenerated protein fibers. 3. Production of regenerated silk fibroin fiber by the self-dialyzing wet spinning method. J. Appl. Polymer Sci. 60: 503–511.

Matthews, R.W. and C.K. Starr. 1984. *Microstigmus comes* wasps have a method of nest construction unique among social insects. Biotropica 16: 55–58.

Mcgovern, J.N., R.L. Jeanne and M.J. Effland. 1988. The nature of wasp nest paper. Tappi J. 71: 133–139.

Mita, K., S. Ichimura and T.C. James. 1994. Highly repetitive structure and its organization of the silk fibroin gene. J. Mol. Evol. 38: 583–592.

Mori, H. and M. Tsukada. 2000. New silk protein: modification of silk protein by gene engineering for production of biomaterials. J. Biotechnol. 74: 95–103.

Morton, S.R. and D.C.F. Rentz. 1983. Ecology and taxonomy of fossorial, granivorous gryllacridids (Orthoptera, Gryllacrididae) from arid Central Australia. Austral. J. Zool. 31: 557–579.

Mound, L.A. and D.C. Morris. 2001. Domicile constructing phlaeothripine Thysanoptera from *Acacia phyllodes* in Australia: *Dunatothrips moulton* and *Sartrithrips* gen. n., with a key to associated genera. Syst. Entomol. 26: 401–419.

New, T.R. 1987. Biology of the Psocoptera. Oriental Insects 21: 1–109.

Oberprieler, R.G., A.E. Marvaldi and R.S. Anderson. 2007. Weevils, weevils, weevils everywhere. Zootaxa 1668: 491–520.

Okada, S., S. Weisman, H.E. Trueman, S.T. Mudie, V.S. Haritos and T.D. Sutherland. 2008. An Australian webspinner species makes the finest known insect silk fibers. Int. J. Biol. Macromol. 43(3): 271–275.

Oliver, D.R. 1971. Life history of Chironomidae. Annu. Rev. Entomol. 16: 211–230.

Omenetto, F.G. and D.L. Kaplan. 2010. New opportunities for an ancient material. Science 329: 528–531.

Parker, K.D. and K.M. Rudall. 1957. Structure of the silk of *Chrysopa* egg-stalks. Nature 179: 905–906.

Porter, D. and F. Vollrath. 2008. The role of kinetics of water and amide bonding in protein stability. Soft Matt. 4: 328–336.

Porter, D. and F. Vollrath. 2009. Silk as a biomimetic ideal for structural polymers. Adv. Mater. 21: 487–492.

Quicke, D.L.J., M.R. Shaw, M. Takahashi and B. Yanechin. 2004. Cocoon silk chemistry of non-cyclostome Braconidae, with remarks on phylogenetic relationships within the Microgastrinae (Hymenoptera: Braconidae). J. Nat. Hist. 38: 2167–2181.

Rauscher, S., S. Baud, M. Miao, F.W. Keeley and R. Pomes. 2006. Proline and glycine control protein self-organization into elastomeric or amyloid fibrils. Structure 14: 1667–1676.

Raven, D.J., C. Earland and M. Little. 1971. Occurrence of dityrosine in Tussah silk fibroin and keratin. Biochim. Biophys. Acta 251: 96–99.

Rentz, D.C. and D.B. Weissman. 1973. Origins and affinities of Orthoptera of Channel-Islands and adjacent mainland California .1. Genus *Cnemotettix*. Proc. Acad. Nat. Sci. Philad. 125: 89–120.

Rudall, K.M. 1962. Silk and other cocoon proteins. Comp. Biochem. IV: 397–433.

Rudall, K.M. and W. Kenchington. 1971. Arthropod silks—problem of fibrous proteins in animal tissues. Annu. Rev. Entomol. 16: 73–96.

Sattler, W. 1967. Über die Lebensweise, insbesondere das Bauverhalten, neotropischer Eintagsfliegen-Larven (Ephemeroptera: Polymitarcidae). Beitr. Neutrop. Fauna 5: 89–110.

Scheibel, T. 2004. Spider silks: recombinant synthesis, assembly, spinning, and engineering of synthetic proteins. Microbiol. Cell Fact. 3: 14.

Sehnal, F. and H. Akai. 1990. Insect silk glands—their types, development and function, and effects of environmental-factors and morphogenetic hormones on them. Int. J. Insect Morphol. Embryol. 19: 79–132.

Sehnal, F. and T. Sutherland. 2008. Silks produced by insect labial glands. Prion 2: 145–153.

Sehnal, F. and M. Zurovec. 2004. Construction of silk fiber core in lepidoptera. Biomacromolecules 5: 666–674.

Seib, F.P., G.T. Jones, J. Rnjak-Kovacina, Y. Lin and D.L. Kaplan. 2013. pH-dependent anticancer drug release from silk nanoparticles. Adv. Healthcare Mater. 2: 1606–1611.

Serrao, J.E. 2005. Ultrastructure of the silk glands in three adult females of sphecid wasps of the genus *Microstigmus* (Hymenoptera: Pemphredoninae). Rev. Chil. Hist. Nat. 78: 15–21.

Sezutsu, H., H. Kajiwara, K. Kojima, K. Mita, T. Tamura, Y. Tamada and T. Kameda. 2007. Identification of four major hornet silk genes with a complex of alanine-rich and serine-rich sequences in *Vespa simillimaxanthoptera* Cameron. Biosci. Biotechnol. Biochem. 71: 2725–2734.

Shao, Z. and F. Vollrath. 2002. Surprising strength of silkworm silk. Nature 418: 741.

Silverman, H.G. and F.F. Roberto. 2007. Understanding marine mussel adhesion. Mar. Biotechnol. 9: 661–681.

Silverman, J., M.K. Rust and D.A. Reierson. 1981. Influence of temperature and humidity on survival and development of the cat flea, *Ctenocephalides felis* (Siphonaptera: Pulicidae). J. Med. Entomol. 18: 78–83.

Smith, C.K. and L. Regan. 1995. Guidelines for protein design: the energetics of beta sheet side chain interactions. Science 270: 980–982.

Spiess, K., A. Lammel and T. Scheibel. 2010. Recombinant spider silk proteins for applications in biomaterials. Macromol. Biosci. 10: 998–1007.

Stewart, R.J. and C.S. Wang. 2010. Adaptation of caddisfly larval silks to aquatic habitats by phosphorylation of h-fibroin serines. Biomacromolecules 11: 969–974.

Stewart, R.J., C.S. Wang and H. Shao. 2011. Complex coacervates as a foundation for synthetic underwater adhesives. Adv. Coll. Interf. Sci. 167: 85–93.

Sturm, H. 1992. Mating-behavior and sexual dimorphism in *Promesomachilis hispanica* Silvestri, 1923 (Machilidae, Archaeognatha, Insecta). Zool. Anz. 228: 60–73.

Sutherland, T.D., P.M. Campbell, S. Weisman, H.E. Trueman, A. Sriskantha, W.J. Wanjura and V.S. Haritos. 2006. A highly divergent gene cluster in honey bees encodes a novel silk family. Genome Res. 16: 1414–1421.

Sutherland, T.D., S. Weisman, H.E. Trueman, A. Sriskantha, J.W. Trueman and V.S. Haritos. 2007a. Conservation of essential design features in coiled coil silks. Mol. Biol. Evol. 24: 2424–2432.

Sutherland, T.D., J.H. Young, A. Sriskantha, S. Weisman, S. Okada and V.S. Haritos. 2007b. An independently evolved Dipteran silk with features common to Lepidopteran silks. Insect Biochem. Mol. Biol. 37: 1036–1043.

Sutherland, T.D., J.H. Young, S. Weisman, C.Y. Hayashi and D.J. Merritt. 2010. Insect silk: one name, many materials. Annu. Rev. Entomol. 55: 171–188.

Sutherland, T.D., S. Weisman, A.A. Walker and S.T. Mudie. 2012. The coiled coil silk of bees, ants, and hornets. Biopolymers 97: 446–454.

Sutherland, T.D., Y.Y. Peng, H.E. Trueman, S. Weisman, S. Okada, A.A. Walker, A. Sriskantha, J.F. White, M.G. Huson, J.A. Werkmeister, V. Glattauer, V. Stoichevska, S.T. Mudie, V.S. Haritos and J.A.M. Ramshaw. 2013. A new class of animal collagen masquerading as an insect silk. Sci. Reports 3: 2864.

Takei, F., Y. Kikuchi, A. Kikuchi, S. Mizuno and K. Shimura. 1987. Further evidence for importance of the subunit combination of silk fibroin in its efficient secretion from the posterior silk gland-cells. J. Cell Biol. 105: 175–180.

Tanaka, K., S. Inoue and S. Mizuno. 1999a. Hydrophobic interaction of P25, containing Asn-linked oligosaccharide chains, with the H-L complex of silk fibroin produced by *Bombyx mori*. Insect Biochem. Mol. Biol. 29: 269–276.

Tanaka, K., N. Kajiyama, K. Ishikura, S. Waga, A. Kikuchi, K. Ohtomo, T. Takagi and S. Mizuno. 1999b. Determination of the site of disulfide linkage between heavy and light chains of silk fibroin produced by *Bombyx mori*. Biochim. Biophys. Acta 1432: 92–103.

Tanaka, K. and S. Mizuno. 2001. Homologues of fibroin L-chain and P25 of *Bombyx mori* are present in *Dendrolimus spectabilis* and *Papilio xuthus* but not detectable in *Antheraea yamamai*. Insect Biochem. Mol. Biol. 31: 665–677.

Vepari, C. and D.L. Kaplan. 2007. Silk as a biomaterial. Progr. Polymer Sci. 32: 991–1007.

Waite, J.H. 2002. Adhesion a la moule. Integr. Comp. Biol. 42: 1172–1180.

Walker, A.A., J.S. Church, A.L. Woodhead and T.D. Sutherland. 2013. Silverfish silk is formed by entanglement of randomly coiled protein chains. Insect Biochem. Mol. Biol. 43: 572–579.

Weisman, S., H.E. Trueman, S.T. Mudie, J.S. Church, T.D. Sutherland and V.S. Haritos. 2008. An unlikely silk: the composite material of green lacewing cocoons. Biomacromolecules 9: 3065–3069.

Weisman, S., S. Okada, S.T. Mudie, M.G. Huson, H.E. Trueman, A. Sriskantha, V.S. Haritos and T.D. Sutherland. 2009. Fifty years later: the sequence, structure and function of lacewing cross-beta silk. J. Struct. Biol. 168: 467–475.

Wieslander, L. 1994. The Balbiani ring multigene family: coding repetitive sequences and evolution of a tissue-specific cell function. Progr. Nucleic Acid Res. Mol. Biol. 48: 275–313.

Wilson, D., R. Valluzzi and D. Kaplan. 2000. Conformational transitions in model silk peptides. Biophys. J. 78: 2690–2701.

Woolley, J. and L. Vet. 1981. Postovipositional web-spinning behaviour in a hyperparasite, *Signiphora coquiletti* Ashmead (Hymenoptera: Signiphoridae). Neth. J. Zool. 31: 627–33.

Xu, X., M. Wang, Y. Wang, Y. Sima, D. Zhang, J. Li, W. Yin and S. Xu. 2013. Green cocoons in silkworm *Bombyx mori* resulting from the quercetin 5-O-glucosyltransferase of UGT86, is an evolved response to dietary toxins. Mol. Biol. Reports 40: 3631–3639.

Yonemura, N. and F. Sehnal. 2006. The design of silk fiber composition in moths has been conserved for more than 150 million years. J. Mol. Evol. 63: 42–53.

Yonemura, N., F. Sehnal, K. Mita and T. Tamura. 2006. Protein composition of silk filaments spun under water by caddisfly larvae. Biomacromolecules 7: 3370–3378.

Yonemura, N., K. Mita, T. Tamura and F. Sehnal 2009. Conservation of silk genes in Trichoptera and Lepidoptera. J. Mol. Evol. 68: 641–653.

Young, J.H. and D.J. Merritt. 2003. The ultrastructure and function of the silk-producing basitarsus in the Hilarini (Diptera: Empididae). Arthr. Struct. Dev. 32: 157–165.

Yukuhiro, K., T. Kanda and T. Tamura. 1997. Preferential codon usage and two types of repetitive motifs in the fibroin gene of the Chinese oak silkworm, *Antheraea pernyi*. Insect Mol. Biol. 6: 89–95.

Zhou, C.Z., F. Confalonieri, N. Medina, Y. Zivanovic, C. Esnault, T. Yang, M. Jacquet, J. Janin, M. Duguet, R. Perasso and Z.G. Li. 2000. Fine organization of *Bombyx mori* fibroin heavy chain gene. Nucleic Acids Res. 28: 2413–2419.

Zhou, C.Z., F. Confalonieri, M. Jacquet, R. Perasso, Z.G. Li and J. Janin. 2001. Silk fibroin: structural implications of a remarkable amino acid sequence. Proteins-Struct. Funct. Gen. 44: 119–122.

Insect Index

Subject Index

Color Plate Section

Chapter 6

Controls

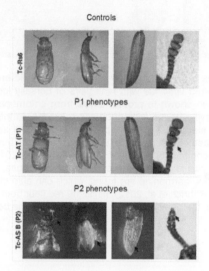

P1 phenotypes

P2 phenotypes

Figure 9. Representative adult phenotypes obtained from 4 h old *T. castaneum* pupae injected with 500 ng of dsRNA *for Trica-AT* and *Trica-AS B-type*. Controls were injected with dsRNA for *Trica-Rs6* (*T. castaneum* ribosomal protein 6). Ventral and lateral view onto the whole body of the adult beetles (left). Elytra and antenna views (right). Arrows indicate the sites of phenotype characters in knockdown animals which are different from the controls. For more data see Abdel-latief and Hoffmann (2014). With permission from Elsevier.

Chapter 9

Figure 1. Model beetles in extended external and internal defenses. (A) Larva of the chrysomelid beetle *Phratora vitellinae* with nine pairs of dorsally located exocrine glands, where secretion has been shown to protect larvae from entomopathogenic fungi (Gross et al. 2008). (B) The burying beetle *Nicrophorus vespilloides*. In this biparental beetle, 34 compounds have been identified in head space, anal and oral secretions, the majority showing antimicrobial activity, supporting the hypothesis on burying beetles actively regulating the microbial community of carrion (Degenkolb et al. 2011). Adult (C) and larva (D) of the harlequin ladybird *Harmonia axyridis*. In both life stages, high concentrations of the antimicrobial substance harmonine are found, which can also inhibit parasites such as *Plasmodium falciparum* (Röhrich et al. 2012). (E) In the red flour beetle *Tribolium castaneum* adult secretions condition their environment, potentially also to the benefit of their offspring.

Chapter 10

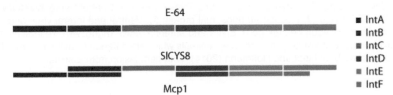

Figure 3. Schematic representation for the inhibitory range of SlCYS8 and macrocypin Mcp1 against *L. decemlineata* intestains. Inferences are based on functional proteomic data from Sainsbury et al. (2012a) and Smid et al. (2013). The two inhibitors reproduce, together, the broad inhibitory range reported earlier for the low-molecular-weight chemical inhibitor E-64 (Goulet et al. 2008).

Chapter 11

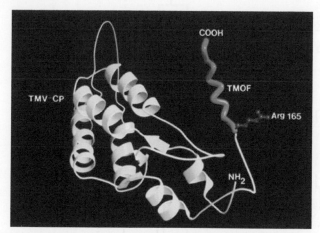

Figure 5. Ribbon drawing by means of MOLSCRIPT, of TMV coat protein (CP) fused to *Aedae*-TMOF. The TMV-CP is shown in yellow and the N-terminus is shown as NH₂. TMOF exhibiting a left-handed helix is shown in blue with a trypsin cleavage site at Arg165 in red. The C-terminus of the fusion protein is at the carboxyl-end of *Aedae*-TMOF (COOH) (Borovsky et al. 2006).

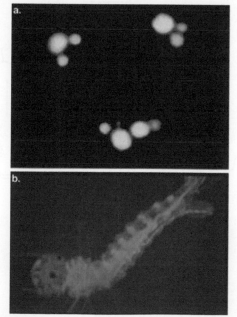

Figure 9. Feeding *Ae. aegypti* larvae *S. cerevisiae* cells transformed with *gfp-tmfA*. (a) Transformed *S. cerevisiae* cells expressing GFP-*Aedae*-TMOF observed under a fluorescent microscope. (b) *Ae. aegypti* larva fed for 2 days recombinant *S. cerevisiae* cells expressing GFP-*Aedae*-TMOF observed under a fluorescent microscope.

Chapter 12

Strategies to synthesize defensive compounds in the secretions of leaf beetle larvae

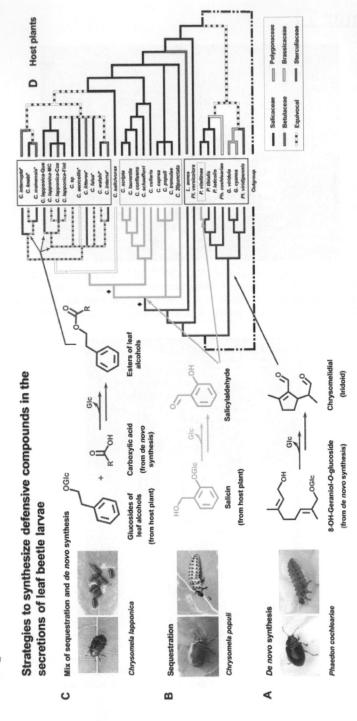

Figure 3. Maximum parsimony reconstruction of the evolution of the taxon Chrysomelina considering the synthesis of deterrents in the defensive glands of the larvae (A–C) and the affiliations of host plants (D) (adapted fromTermonia et al. 2001).

Chapter 13

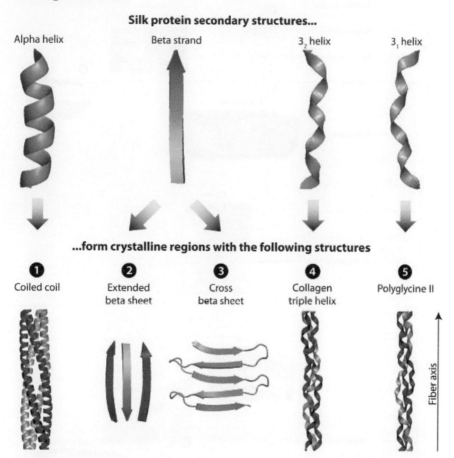

Figure 1. The five types of ordered silk structures. 1. Coiled coil: Two to five alpha helices (always right-handed) wind around each other to form a superhelix. 2. Extended beta sheet: Multiple planar beta strands form sheets which are stacked face to face. 3. Cross-beta sheet: Beta strands of uniform length alternating with turns at which the protein chain direction reverses. This is the only structural element which is aligned perpendicular to the fiber axis. 4. Collagen triple helix: Three left-handed helices with three amino acids per turn (3_2 helix) are intertwined to form a superhelix. 5. Polyglycine II: Three right-handed helices with three amino acids per turn (3_1 helix) are intertwined to form likewise a superhelix (Adapted from Sutherland et al. 2010).

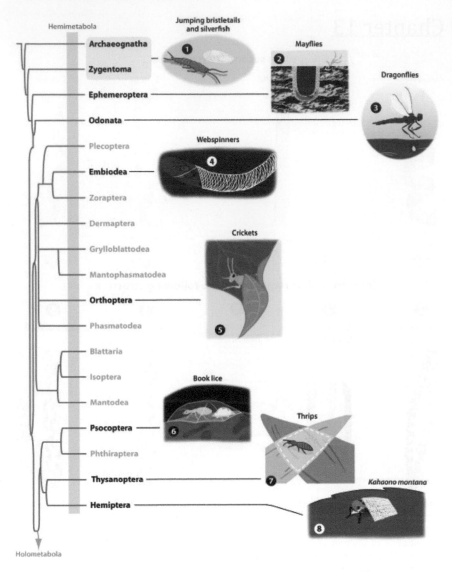

Figure 2. contd....

Figure 2. contd.

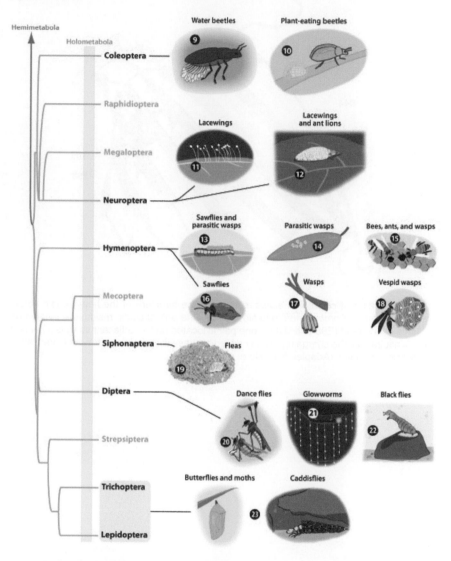

Sutherland TD, et al. 2010.
Annu. Rev. Entomol. 55:171–88

Figure 2. Phylogenetic relationship of all silk-producing insects (from Sutherland et al. 2010, with permission).

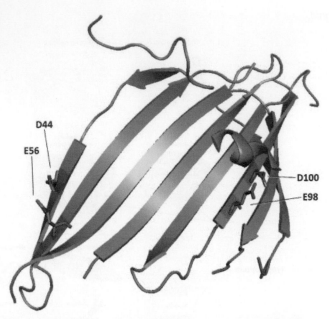

Figure 4. Position of conserved acidic residues in the structure of the heavy-chain fibroin N-terminal domain (*Bombyx mori*, pdb-file: 3UA0). One pair of acidic residues is located on the same beta strand (E98 and D100), one pair is located on two adjacent strands (D44 and E56). At neutral pH, the charge repulsion caused by these residues impedes the dimerization of N-terminal domains (Adapted from He et al. 2012).

Printed and bound by CPI Group (UK) Ltd, Croydon, CR0 4YY

24/10/2024

01778307-0015